TRANSPORT PHENOMENA IN METALLURGY

TRANSPORT PHENOMENA IN METALLURGY

G. H. GEIGER
Inland Steel Company

D. R. POIRIER
University of Arizona

ADDISON-WESLEY PUBLISHING COMPANY
Reading, Massachusetts · Menlo Park, California
London · Amsterdam · Don Mills, Ontario · Sydney

This book is in the
ADDISON-WESLEY SERIES IN METALLURGY AND MATERIALS

Consulting Editor
Morris Cohen

Second printing, December 1980

ISBN 0-201--2352-0
GHIJ-MA-89876543

FOREWORD

The student in metallurgy must develop a background in the transport of momentum, heat, and mass if he is to grapple successfully with many theoretical and practical problems of the field in the laboratory, in the pilot plant, and in industrial operations. The need for him to do so has become a pressing one because of the acceleration in the development of new processes and requirements for improving and controlling existing processes. Virtually all students in the field are sooner or later faced with problems involving transport. This text was written to prepare them to meet the need. Concepts and analytical methods are illustrated in the discussion giving examples and problems in terms of systems that are meaningful to metallurgists of all kinds. The advantages to this are that the concept or method is met on familiar ground, and at the same time there is developed a greater appreciation of the nature of these systems. Where appropriate, the text builds on an understanding of the relationships between structure and properties of matter. This is best illustrated in the chapters on diffusion in solids. Students in chemical, mechanical, and other engineering disciplines will find this text useful because it provides a coverage of systems pertinent to metals and materials in terms of familiar principles.

This book is well suited to the undergraduate student in metallurgy, and will be useful to him later in the practical world. The authors have successfully sought a middle ground in the level of the treatment of topics. The material is neither too mathematical for the reader with an average-to-good mathematical background, nor is it too descriptive so that it would fail to stimulate him. A wide range of topics is covered, so that this text can also serve as an introduction to a number of sophisticated specialized fields. A student wishing to learn more of one of these fields will find the material in this text a good base for going on to other texts devoted specifically to topics such as diffusion, fluid mechanics, heat transfer, and mass transfer.

It is assumed that the reader has a basic knowledge of chemistry, thermodynamics, calculus, and differential equations. Vector notation has been avoided. The level is designed for juniors because of the required background in thermodynamics and mathematics, and because the material presented will be of value in professional courses such as mineral processing, process metallurgy, solidification and casting, forming and joining, and heat treatment or other materials-processing subjects which the student might normally take in his junior or senior years.

Cambridge, Massachusetts   John F. Elliott
December 1972

PREFACE

Despite the fact that metallurgists and metallurgical engineers have been dealing with processing problems and metallurgical reactions for many years, there has not been any text available for metallurgical educators who would like to present their students with a unified fundamental approach to fluid dynamics, heat transfer, and mass transfer. Also, the remarkable advances made in the field of metallurgy during the past decade have mostly been limited to the scientific areas, while the engineering aspects of metallurgical engineering have largely been neglected. This book provides an up-to-date approach to transport phenomena and then goes further by discussing numerous examples, applications, and phenomena which are of particular importance to metallurgists.

Quite a few books on transport phenomena (momentum, heat, and mass transport) have appeared over the past decade; some are intended for chemical engineers and others for mechanical engineers. When we started to organize and teach a course to juniors in metallurgical engineering, we found that these were only partly satisfactory, mainly for pedagogical reasons. Students of metallurgy, when exposed to the subject matter of transport phenomena without benefit of a substantial effort to introduce them to the direct link between transport phenomena and metallurgy, could hardly gain an appreciation of the fact that transport phenomena are just as basic to the field of metallurgy as is physical metallurgy. We also felt the need for a text that would utilize and match the mathematical abilities of upper-class students of metallurgical engineering and metallurgy. Most upper-class students in metallurgical engineering have studied integral and differential calculus, but often make no use of it again, except in the most trivial applications. Since one can proceed quite a way into the field of transport phenomena without resorting to higher-level mathematics, and because we think that it is important for the student, and certainly for the practicing engineer, to visualize the physical situations, we have attempted to lead the reader through the development and solution of the necessary differential equations in a manner as clear as possible, by applying the familiar principles of conservation to numerous situations and by purposefully including many intermediate mathematical steps. We hope that students with good mathematical abilities will not feel that we are being condescending, but will realize that we are trying to make the material understandable to a group as wide as possible. The reader who is not "up" on his mathematics may still learn from the book, but must pay careful attention to the characteristics of each situation and the boundary conditions for which a particular solution is valid in order to avoid applying a wrong solution, which unfortunately happens all too often.

The book is organized in the seemingly traditional manner characteristic of texts in transport phenomena; Section I deals with the properties and mechanics of fluid motion; Section II with thermal properties and heat transfer; and Section III with mass transfer. However, we have departed from the tradition in several significant ways.

The first chapter of each section is devoted to transport properties of substances which are of interest to metallurgical engineers. In all cases, the transport properties of materials are discussed from a structural point of view; in this regard, some of the material is easily recognizable by physical metallurgists, particularly in the discussions of diffusion coefficients in solids. On the other hand, the viscosity of liquid metals and molten slags and salts, which is dealt with in the book, is hardly ever touched on in most curricula throughout this country. Viscosity is a structure-sensitive property and, as such, should be of interest to all metallurgists. To mention a few aspects of the way thermal conductivity is treated, there are the usual basic discussions of gases and liquids and also a discussion of pure single-phase metals leading to the Lorenz number. But, beyond that, some attention is given to two-phase solids, and bulk, porous materials, such as molding sand for castings, are discussed in great detail.

While the first section of the book dealing with fluid dynamics points out many concerns of the process metallurgist, such as packed beds of solids, fluidized beds, flow through pipes, meters, and other phenomena, and then gives an introduction to the engineering use of fans, blowers, high-velocity jets and vacuum systems, it should also be of interest to the physical metallurgist, mentioning such topics as elutriation characterizing of particles, and powder rolling, both of importance in powder metallurgy. In addition, one need only scan through the recent metallurgical literature to see that convection is a very important part of solidification phenomena, traditionally of prime interest to physical metallurgists.

The subject matter of heat transfer included herein covers the fundamentals of heat transfer by conduction and radiation, and heat transfer with convection. Going further, so as to stimulate metallurgists and metallurgical engineers, numerous applications to heating and cooling processes are discussed. Quenching, heating of billets, heating and cooling of strip by radiation and sprays are some examples. Chapter 10 is devoted to solidification heat transfer and Chapter 12 to heat transfer in packed beds, imparting definite metallurgical flavor to the text.

The section on mass transfer is generous in the portion devoted to diffusion in solids (metals and ionic solids as well). Mechanisms of diffusion in solids are discussed, and diffusion of metal vapors in gases at high temperatures is treated. Drawing heavily on the mathematical similarity between conduction heat transfer and diffusion in solids, numerous diffusional problems are discussed including carburizing and decarburizing of single- and two-phase steels, homogenization of cast and banded structures, and the diffusion of gases into or out of basic simple shapes. Interphase and interfluid reactions are also discussed and the concept of the "controlling step" is introduced and illustrated by examples such as a discussion of the reduction of iron oxide, metal vapor loss from melts under vacuum, and many others.

The book may be used as a supplement to several courses besides a primary course in transport processes. For example, by starting with Chapters 6, 7, and 9, Chapter 10 may be taught as a part of a course in solidification or foundry engineering. One of the authors has regularly included material from Chapters 3, 5, 12, 15, and 16 in courses in process or extractive metallurgy. Chapters 13, 14, 15, and 16 should make a good basis for a course in heterogeneous kinetics. The material in Chapters 6, 7, 8, 9, 11, and 14 has been satisfactorily used, following a course in fluid mechanics, as the basis for part of a course in heat treating.

We would like to emphasize that the book is not intended to be a comprehensive review of all the applications and current research in the field. Its aim is rather to present the basic equations and show how they can be applied to a variety of topics, thereby preparing the reader to make direct use of the information or go on to develop their applications, and also enabling him to read the current literature in the field. Each chapter is followed by Problems; many of them are numerical and we have tried very much to gather problems that not only have learning value but also a factor of interest catering to students of metallurgy. We hope that the book will be found useful and will help metallurgists improve their analyses and also provide aid to those who already are actively participating in trying to find better ways how to engineer their processes.

Chicago, Illinois   G.H.G.
Bridgeport, Connecticut   D.R.P.
April 1972

ACKNOWLEDGMENTS

The authors express their sincere thanks to Professor Richard W. Heine of the University of Wisconsin for his encouragement and support of our effort, initiated while we were in Madison. We would also like to especially thank our former mentors, Professor J. Bruce Wagner of Northwestern University and Professor Merton C. Flemings of the Massachusetts Institute of Technology, for their encouragement and guidance when we pursued our graduate studies; their influence is particularly recognizable in Chapters 10 and 14. Many of our students who had to put up with the inconvenience of studying our notes and rough manuscript deserve our thanks and belated sympathies. They had the thankless job of helping us to sift and sort out so that finally the manuscript could be completed.

We also acknowledge our indebtedness to the pioneering contributors to the field of transport phenomena; noteworthy among these are Professors R. Byron Bird, Warren E. Stewart, and Edwin N. Lightfoot of the University of Wisconsin. They have given permission to use a number of figures, tables, and examples from their classic text, *Transport Phenomena*. And it was they who, as much as anybody, have pointed out that "the subject of transport phenomena should rank along with thermodynamics, mechanics, and electromagnetism as one of the key engineering sciences." We are also grateful to Professor John F. Elliott for critically reading the manuscript and for his gracious Foreword to this book, to Professor C. Wagner for permission to use material from his notes presented for many years at M.I.T., and to Professor Morris Cohen of M.I.T. for his early endorsement of the manuscript.

Finally we owe a debt of gratitude to our wives who, after earning their Ph.T. degrees (putting husband through), had to listen to all our gripes directed to this ****** book; to our children, who were constantly bullied away from the kitchen table and "shh-shh-ed" so that we could discuss the manuscript; and to our parents who have promised to buy a copy of the book even if no one else does.

CONTENTS

Chapter 16   Interphase Mass Transfer

INDEX TO SOURCES

PART ONE

FLUID DYNAMICS

The first part of this text deals with fluids, their intrinsic properties, their behavior under various conditions, and the methods by which we can manipulate and utilize them to produce desired results. Most metallurgical processes deal with fluids at one point or another, and although the metallurgist is usually not required to be an "expert" on fluids, he should understand the fundamentals of fluid dynamics as presented in the following chapters, and be able to make intelligent use of the properties of fluids and characteristics of equipment used to manipulate and control fluids.

The behavior of fluids is also intimately related to heat and mass transport processes. For example, if a gas is hotter than a solid past which it is flowing, the solid naturally is heated. The rate at which heat is transferred to the solid's surface is dependent on the fluid's properties and its flow pattern. Similarly, if a piece of graphite is to be dissolved in a bath of molten iron, the rate of dissolution depends on the motion of the liquid iron adjacent to the graphite. Thus, to appreciate transfer of heat and/or mass, some understanding of fluid dynamics is important.

These are just two simple examples which illustrate the necessity for the student to become acquainted with the means of examining and expressing the flow of fluids and to eventually recognize the role of fluid flow in rate processes involving heat and mass transfer. Actually, if we take a fundamental approach to the study of fluid flow, then the subject matter is appropriately designated *momentum transport*. Momentum transport with *energy transport* and *mass transport* make up the subject of *transport phenomena*, which Bird, Stewart, and Lightfoot* rank as a "key engineering science," along with thermodynamics and mechanics. Transport phenomena as a key engineering science has been well received in both the academic and industrial communities of engineering.

---

* R. B. Bird, W. E. Stewart, and E. N. Lightfoot, *Transport Phenomena*, Wiley, New York, 1960.

1

PROPERTIES OF FLUIDS

1.1 TYPES OF FLUID FLOW

When fluids move through a system, either one of two different types of fluid flow may occur. We can most easily visualize the fact that there are two distinctly different types of fluid flow by referring to an experiment performed by Reynolds in 1883. Imagine a transparent pipe with water flowing through it; a threadlike stream of dye is injected parallel to the path of the water's flow. At sufficiently small velocities of water, the dye will flow in parallel, straight lines. When the velocity is increased, a point is reached at which the entire mass of water becomes colored. In other words, hypothetical individual particles of liquid, instead of flowing in an orderly manner parallel to the long axis of the pipe, flow in an erratic manner so as to cause complete mixing of the dye and water.

The first type of dye flow is called *laminar* or *streamline* flow. The significance of these terms is that the fluid's motion seems to be the sliding of laminations of infinitesimal thickness relative to adjacent layers, and that the hypothetical particles in the layers move in predictable paths or streamlines, as depicted in Fig. 1.1.

Fig. 1.1 Laminar flow.

The second (erratic) type of flow is described as *turbulent* flow. In turbulent flow, the motion of the fluid particles is irregular, and accompanied by fluctuations in velocity. This type of flow is illustrated in Fig. 1.2, in which part (a) shows the erratic path of a single particle during some time interval and part (b) demonstrates that the velocity at a fixed point in the fluid fluctuates randomly about some mean value, which is called the *temporal mean* velocity, and which is given the symbol $\overline{V}_x$.

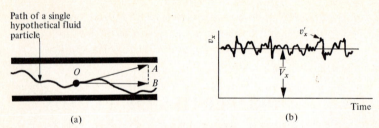

Fig. 1.2 Turbulent flow. (a) The instantaneous velocity OA varies continuously in direction and magnitude. The velocity OB is the x-directed component and is designated v'_x in (b). (b) Variation at point O of v'_x about the temporal mean velocity $\bar{V}_x$.

One of the earliest systematic investigations of turbulent flow was conducted by Reynolds, who suggested the parameter $\bar{V}D/v$ as the criterion for predicting the type of flow in round tubes, where $\bar{V}$ is the average fluid velocity, D is the pipe diameter, and v is the kinematic viscosity (to be described in the following section). In a consistent set of units, the parameter is dimensionless, and is called the *Reynolds number* (Re). The value of Re at which transition from laminar to turbulent flow occurs is approximately 2100 in the usual engineering applications of flow in pipes. In general, however, the transition Reynolds number varies with different systems, and even for a given system it may vary according to such external factors as surface roughness and initial disturbances in the fluid.

Figure 1.3 shows the distribution of velocity across the radius of a tube, for laminar and turbulent flow. The temporal mean velocity is plotted for turbulent flow. When dealing with turbulent flow, we are usually interested in the temporal

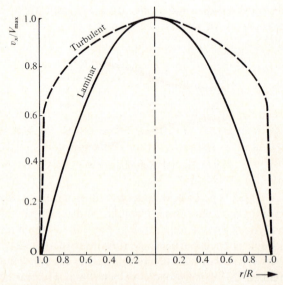

Fig. 1.3 Distribution of laminar and turbulent velocities in a tube.

mean value, so that, unless otherwise stated, the temporal mean value will be implied. Note that for both cases the velocity is zero at the fluid–wall interface. For laminar flow the velocity profile is parabolic; in turbulent flow, the curve is somewhat flattened in the middle.

1.2 NEWTONIAN FLUIDS

Consider a fluid between two parallel plates (Fig. 1.4). The upper plate is stationary and the lower one is set in motion with a velocity V at time zero. From experience we know that the fluid adjacent to the plates will have the same velocity as the plates themselves. Hence the fluid adjacent to the lower plate moves with a velocity V, while that adjacent to the upper plate has null velocity. As time proceeds, the fluid gains momentum, and after sufficient time has elapsed a steady state is reached, in which, in order to keep the lower plate in motion with the velocity V, a force F must be maintained, and an equal but opposite force is exerted on the stationary plate.

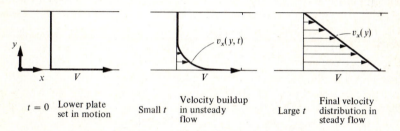

| $t = 0$ Lower plate set in motion | Small t Velocity buildup in unsteady flow | Large t Final velocity distribution in steady flow |

Fig. 1.4 Laminar flow of fluid between parallel plates.

At steady state, for plates of area A, and laminar flow, the force is expressed by

$$\frac{F}{A} = \eta \frac{V}{Y},$$
(1.1)

where Y = distance between plates and η = constant of proportionality.

The force system as described is shear, and the force per unit area (F/A) is the shear stress. At steady state, when the velocity profile is linear, V/Y can be replaced by the constant velocity gradient dv_x/dy and the shear stress τ_{yx} between any two thin layers of fluid may be expressed as

$$\tau_{yx} = -\eta \frac{dv_x}{dy}.$$
(1.2)

Equation (1.2) may alternatively be interpreted in terms of momentum transport. Picture the fluid as a series of thin layers parallel to the plates. Each layer has momentum associated with it and causes the layer directly above it to move. Thus momentum is transported in the y-direction. The subscripts of τ_{yx} refer to this direction of momentum transport (y) and the velocity component being considered (x-direction). The minus sign in Eq. (1.2) reflects the fact that momentum is transferred from the lower layers of fluid to the upper layers, that is, in the positive y-direction. In this case, dv_x/dy is negative, so that the minus sign makes τ_{yx} positive. This follows the generally accepted convention for heat transfer, in that momentum flows in the direction of decreasing velocity, just as heat flows from hot to cold.

The period between $t = 0$, when the lower plate is set into motion, and large t, when steady state is reached, is called the *transient period*. During the transient period, v_x is a function of both time and position, so that a more general relationship for τ_{yx} is used:

$$\tau_{yx} = -\eta \frac{\partial v_x}{\partial y}. \tag{1.3}$$

This empirical relationship is known as *Newton's law of viscosity*, and defines the constant of proportionality, η, as the *viscosity*.

The dimensions of viscosity are found by referring to Eq. (1.3):

$$\eta = -\frac{\tau_{yx}}{(\partial v_x/\partial y)}.$$

Units of η are:

$$\eta = \frac{\text{lb}_f\,\text{ft}^{-2}}{(\text{ft hr}^{-1})(\text{ft}^{-1})} = \text{lb}_f\,\text{hr ft}^{-2}.$$

Alternatively the English system yields the following units for η:

$$\eta = \frac{(\text{lb}_m\,\text{ft hr}^{-2})(\text{ft}^{-2})}{(\text{ft hr}^{-1})(\text{ft}^{-1})} = \text{lb}_m\,\text{hr}^{-1}\,\text{ft}^{-1}.$$

In the cgs system of units, the *poise* (P) is used, in which

$$1 \text{ poise (P)} = 1 \text{ dyn sec cm}^{-2}.$$

The *centipoise* (cP) is probably the most common unit tabulated for viscosity. It equals 0.01 poise, and is the viscosity of water at 68.4°F. Thus the value of the viscosity in centipoises is an indication of the viscosity of the fluid relative to that of water at 68.4°F.

In many problems involving viscosity, it is useful to have a value of a fluid's viscosity divided by its density ρ. Hence we define the *kinematic viscosity v* at this point as

$$v \equiv \frac{\eta}{\rho}.$$

The kinematic viscosity is a fundamental quantity, in that it is a measure of *momentum diffusivity*, analogous to thermal and mass diffusivities, which will be presented in later chapters. In the English system, kinematic viscosity is measured in $ft^2 \ hr^{-1}$, while in the cgs system, the units are $cm^2 \ sec^{-1}$, often called the *stoke*. The *centistoke* (0.01 stoke) is also commonly used.

In this text you will most often encounter the English system of units. You should, however, be able to use equations in any system, because all are in current use in the technical literature. The following are conversion factors for viscosity:

$$1 \ cP = 2.42 \ lb_m/hr \ ft,$$

$$1 \ cP = 2.09 \times 10^{-5} \ lb_f \ sec/ft^2.$$

Other conversion factors are found in Appendix IV.

Example 1.1   Two parallel plates are $\frac{1}{8}$ in. apart. The lower plate is stationary and the upper plate moves with a velocity of 5 ft/sec. A stress of 0.05 lb_f/ft^2 is needed to maintain the upper plate in motion. Find the viscosity of the fluid contained between the plates in (a) $lb_m/ft \ hr$, and (b) cP.

Solution. From Eq. (1.1), and referring to Fig. 1.4, we have

$$\eta = \frac{F/A}{V/Y},$$

$$F/A = \frac{0.05 \ lb_f}{ft^2} \left| \frac{32.2 \ lb_m \ ft/sec^2}{1 \ lb_f} \right. = 1.61 \ \frac{lb_m}{ft \ sec^2}.$$

Also

$$\frac{V}{Y} = \frac{5 \ ft/sec}{\frac{1}{96} \ ft} = 480 \ sec^{-1}.$$

Then

$$\eta = \frac{1.61 \ lb_m}{480 \ ft \ sec} \left| \frac{3600 \ sec}{1 \ hr} \right. = 12.08 \ \frac{lb_m}{ft \ hr}$$

or

$$\eta = \frac{12.08}{2.42} \ cP = 4.99 \ cP.$$

1.3 VISCOSITY OF GASES

For the purpose of explaining momentum transport in gases, we resort to the simplest treatment of the kinetic theory of gases. We utilize the concept of the mean free path, in which the molecules are idealized as billiard balls, and postulate a hypothetical "ideal" gas possessing the following features:

1.   The molecules are hard spheres resembling billiard balls, having diameter d and mass m.

2.   The molecules exert no force on one another except when they collide.

3.   The collisions are perfectly elastic and obey the classical laws of conservation of momentum and energy.

4.   The molecules are uniformly distributed in a concentration of n per unit volume throughout the gas. They are in a state of continuous motion and are separated by distances which are large compared to their diameter.

5.   All directions of molecular velocities are equally probable. The speed (magnitude of velocity) of a molecule can have any value between zero and infinity.

If we assume that the molecules possess a Maxwellian speed distribution (i.e., the thermal energy of the gas is given by the total kinetic energy of all the moving molecules), then the average speed $\bar{V}$ is given by

$$\bar{V} = \sqrt{\frac{8\kappa_B T}{\pi m}} \tag{1.4}$$

where κ_B is the Boltzmann constant and T is absolute temperature.

In addition, for such a collection of molecules, a significant parameter that governs the mechanism of momentum transfer in gases is the free path, defined as the distance traveled by a molecule between two successive collisions. At the instant of collision, the center-to-center distance of two molecules is d. Intuitively we know that the *mean free path* λ should be inversely proportional to the collision cross section πd^2, and also inversely proportional to the concentration n of the molecules. The rigorous analysis for determining λ includes these terms, along with a coefficient whose numerical value is developed by considering the random fluctuations of the colliding molecules. The final result gives

$$\lambda = \left(\frac{1}{\sqrt{2}}\right)\left(\frac{1}{\pi d^2 n}\right). \tag{1.5}$$

Now consider an imaginary plane at $y = y_1$, which is being crossed by molecules in either direction. If we examine the condition of no bulk motion (no macroscopic flow) of the gas in the y-direction, then the molecules cross the y_1 plane from both sides with equal frequency. This frequency per unit area at which molecules cross the plane at y_1 from one side is given by

$$Z = \tfrac{1}{4} n \bar{V}. \tag{1.6}$$

We may picture the molecules crossing the y_1-plane as carrying momentum characteristic of an average distance $\bar{y}$ above and below the y_1-plane at which they made their last collision. Numerically, $\bar{y}$ is not exactly equal to λ, but rather is given by

$$\bar{y} = \tfrac{2}{3} \lambda. \tag{1.7}$$

Up to this point, no macroscopic flow of the gas has been considered, so that, as stated above, the number of molecules arriving from above and below y_1 is equal, and on the average no net momentum is transferred across the plane y_1.

To determine the viscosity of the gas, consider the gas under the influence of macroscopic flow in the x-direction, with a velocity gradient dv_x/dy, as depicted in Fig. 1.5.

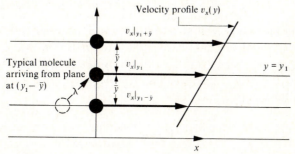

Fig. 1.5   Relation between the macroscopic velocity profile and the plane of interest at y_1.

Now if all the molecules have velocities characteristic of the plane in which they last collided, we may write the x-momentum above y_1 as

$$mv_x\bigg|_{y_1+\bar{y}} = mv_x\bigg|_{y_1} + \tfrac{2}{3}\lambda m\frac{dv_x}{dy}. \tag{1.8}$$

Similarly, for below y_1

$$mv_x\bigg|_{y_1-\bar{y}} = mv_x\bigg|_{y_1} - \tfrac{2}{3}\lambda m\frac{dv_x}{dy}. \tag{1.9}$$

We find the net rate of x-momentum crossing the plane y_1 by summing the x-momentum of molecules that cross from below and subtracting the x-momentum of those that cross from above. In this manner, we write

$$\tau_{yx} = Zm\left[v_x\bigg|_{y_1-\bar{y}} - v_x\bigg|_{y_1+\bar{y}}\right]. \tag{1.10}$$

By combining Eqs. (1.6), (1.8), and (1.9), we arrive at

$$\tau_{yx} = -\tfrac{1}{3}nm\bar{V}\lambda\frac{dv_x}{dy}. \tag{1.11}$$

In addition, by utilizing the expressions for $\bar{V}$ and λ, Eqs. (1.4) and (1.5), respectively, we write

$$\tau_{yx} = -\frac{2}{3\pi^{3/2}}\frac{(m\kappa_B T)^{1/2}}{d^2}\frac{dv_x}{dy}. \tag{1.12}$$

This result corresponds to Newton's law of viscosity (Eq. 1.2), with the viscosity given by

$$\eta = \frac{2}{3\pi^{3/2}} \frac{(m\kappa_B T)^{1/2}}{d^2}.$$ (1.13)

A significant conclusion from the above argument is that the viscosity of a gas is independent of pressure and depends only on temperature. This conclusion is in good agreement with experimental data up to about ten atmospheres. However, the temperature dependency is only qualitatively correct, in that η does increase with increasing temperatures; but, quantitatively, the temperature dependency is not satisfactory. Data for real gases indicate that η varies with T^n, with n between 0.6 and unity, rather than 0.5, as in Eq. (1.13).

The more up-to-date kinetic theories replace the billiard-ball model with a more realistic molecular force field by considering the force of attraction and repulsion between molecules. These theories, reviewed by Hirschfelder, Curtiss, and Bird,[1] make use of the potential energy of interaction between a pair of molecules in the gas. This function—often referred to as the *Lennard-Jones potential*—displays the behavior of molecular interactions by exhibiting weak attraction at large separation distances and strong repulsion at small separations, as shown in Fig. 1.6.

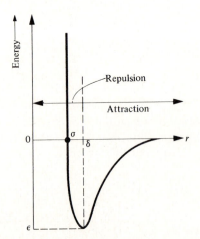

Fig. 1.6 Lennard-Jones potential function describing the interaction of two nonpolar molecules.

The equilibrium position of the molecules is located at δ, where the potential energy is a minimum at $-\varepsilon$; ε is called the *characteristic energy parameter*. Using

[1] J. O. Hirschfelder, C. F. Curtiss, and R. B. Bird, *Molecular Theory of Gases and Liquids*, Wiley, New York, 1954.

the Lennard-Jones potential, Chapman and Enskog have developed the following equation for the viscosity of nonpolar gases at low pressures:

$$\eta = 2.67 \times 10^{-5} \frac{\sqrt{MT}}{\sigma^2 \Omega_\eta}. \tag{1.14}$$

Here M is the gram-molecular weight, η is in poises, T in °K, and σ is a characteristic diameter of the molecule in Å (see Fig. 1.6). The quantity Ω_η is the *collision integral* of the Chapman-Enskog theory, which is a function of a dimensionless temperature parameter $\kappa_B T/\varepsilon$. In order to use Eq. (1.14), we need values of σ and ε/κ_B; these parameters are known for many substances. A partial list of them is given in Table 1.1. We can then determine the collision integral by using Table 1.2.

Table 1.1 Intermolecular force parameters and critical properties

Substance	Molecular weight, M	Lennard-Jones parameters*		Critical constants†	
		σ, Å	ε/κ_B, °K	T_c, °K	$\hat{V}_c$, cm³/g-mol
Light elements					
H_2	2.016	2.915	38.0	33.3	65.0
He	4.003	2.576	10.2	5.26	57.8
Noble gases					
Ne	20.183	2.789	35.7	44.5	41.7
Ar	39.944	3.418	124.	151.	75.2
Kr	83.80	3.498	225.	209.4	92.2
Xe	131.3	4.055	229.	289.8	118.8
Simple polyatomic substances					
Air	28.97	3.617	97.0	132.	86.6
N_2	28.02	3.681	91.5	126.2	90.1
O_2	32.00	3.433	113.	154.4	74.4
CO	28.01	3.590	110.	133.	93.1
CO_2	44.01	3.996	190.	304.2	94.0
SO_2	64.07	4.290	252.	430.7	122.
F_2	38.00	3.653	112.	—	—
Cl_2	70.91	4.115	357.	417.	124.
Br_2	159.83	4.268	520.	584.	144.
CH_4	16.04	3.822	137.	190.7	99.3

* J. O. Hirschfelder, C. F. Curtiss, and R. B. Bird, *Molecular Theory of Gases and Liquids*, Wiley, New York, 1954.
† K. A. Kobe and R. E. Lynn, Jr., *Chem. Rev.* **52**, 117–236 (1952), and *Amer. Petroleum Inst. Research Proj.* **44**, edited by F. D. Rossini, Carnegie Inst. of Technology, 1952.

Table 1.2 Values of Ω-integral for viscosity and of the viscosity function $f(\kappa_B T/\varepsilon)$, based on the Lennard-Jones potential*

$\kappa_B T/\varepsilon$	Ω_η	$f(\kappa_B T/\varepsilon)$
0.3	2.785	0.1969
0.4	2.492	0.2540
0.5	2.257	0.3134
0.6	2.065	0.3751
0.7	1.908	0.4384
0.8	1.780	0.5025
0.9	1.675	0.5666
1.0	1.587	0.6302
2.0	1.175	1.2048
4.0	0.9700	2.0719
6.0	0.8963	2.751
8.0	0.8538	3.337
10	0.8242	3.866
20	0.7432	6.063
40	0.6718	9.488
60	0.6335	12.324
80	0.6076	14.839
100	0.5882	17.137
200	0.5320	26.80
400	0.4811	41.90

* J. O. Hirschfelder, C. F. Curtiss, and R. B. Bird, *Molecular Theory of Gases and Liquids*, Wiley, New York, 1954.

Example 1.2   Compute the viscosity of hydrogen at 1 atm and 2000°F (1364°K

Solution.   From Table 1.1, we find that

$$\varepsilon/\kappa_B = 38.0\,°K, \qquad \sigma = 2.915\,Å.$$

From Table 1.2,

$$\Omega_\eta \cong 0.69.$$

Substituting appropriate values into Eq. (1.14), we have

$$\eta = 2.67 \times 10^{-5}\,\frac{\sqrt{(2)(1364)}}{(2.915)^2(0.69)} = 2.48 \times 10^{-4}\,\text{poise.}$$

(Observed viscosity is 2.44×10^{-4} poise.)

Using Eq. (1.14) to calculate η requires knowing σ and ε/κ_B. When values o

σ are not available, one may use a modified form of Eq. (1.14), as presented by Bromley and Wilke[2]:

$$\eta = 3.33 \times 10^{-5} \frac{\sqrt{M T_c}}{\hat{V}_c^{2/3}} f\left(\frac{\kappa_B T}{\varepsilon}\right) \tag{1.15}$$

where η is in poises, T_c (°K) and $\hat{V}_c$ (cm^3/g mol) are the critical temperature and volume, respectively, and $f(\kappa_B T/\varepsilon)$ is an empirical function also to be found in Table 1.2.

The Chapman-Enskog theory has been extended to include multicomponent gas mixtures at low density. For most purposes, the semiempirical formula of Wilke[3] is quite adequate:

$$\eta_{\text{mix}} = \sum_{i=1}^{n} \frac{x_i \eta_i}{\sum_{j=1}^{n} x_j \Phi_{ij}}, \tag{1.16}$$

in which

$$\Phi_{ij} = \frac{1}{\sqrt{8}}\left(1 + \frac{M_i}{M_j}\right)^{-1/2}\left[1 + \left(\frac{\eta_i}{\eta_j}\right)^{1/2}\left(\frac{M_j}{M_i}\right)^{1/4}\right]^2.$$

Here n is the number of chemical species in the mixture; x_i and x_j are the mole fractions of species i and j; η_i and η_j are the viscosities of species i and j at the system temperature and pressure; and M_i and M_j are the corresponding molecular weights. Note that Φ_{ij} is dimensionless, and, when $i = j$, $\Phi_{ij} = 1$.

To summarize, Eqs. (1.14), (1.15), and (1.16) are useful equations for computing viscosities of nonpolar gases and gas mixtures at low density from tabulated values of the intermolecular force parameters σ and ε. They cannot, however, be applied with confidence to gases consisting of polar or highly elongated molecules such as H_2O, NH_3, CH_3OH, and $NOCl$. A further limitation is that for the most part these equations have been tested only over the temperature range 100°K to 1500°K.

The data on viscosity of several gases as a function of temperature are given in Fig. 1.7. Keep in mind that the data are valid for pressures up to about 10 atmospheres. Note that (1) the viscosity of all gases increases with temperature, and (2) the magnitude of viscosity does not solely depend on the molecular weight of the gas. For example, the data for helium fall in the range of much heavier gases than hydrogen. Sports commentators in particular should heed some of the data and stop propagating the myth that baseballs can be hit farther in dry air than under humid weather conditions.

1.4 VISCOSITY OF LIQUIDS

In dealing with transport processes in liquids, we are always faced with the problem

---

[2] L. R. Bromley and C. R. Wilke, *Ind. Eng. Chem.* **43**, 1641 (1951).
[3] C. R. Wilke, *J. Chem. Phys.* **18**, 517–519 (1950).

that much less is known about the structure of liquids than about the structures of solids or gases. However, there is more similarity between liquids and solids than between liquids and gases. This similarity is based on the small fractional increase in volume on melting (3 to 5% for metals), and the fact that the heat of fusion is quite a bit less than the heat of vaporization. Furthermore, x-ray data tell us that there is at least some degree of short-range order in a liquid. That is, at a short distance from a central atom, the arrangement of nearest neighbors is reasonably predictable. However, as the distance increases, the predictability of atom positions decreases rapidly, unlike in solids.

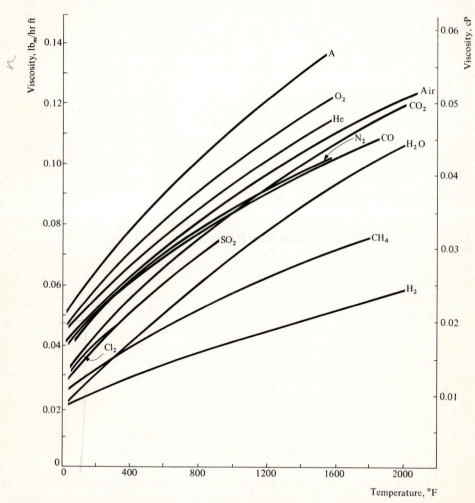

Fig. 1.7 The viscosity of common gases at 1 atm. (Curves drawn from data given in A. Schack, *Industrial Heat Transfer*, sixth edition, Wiley, New York, 1965, and *Handbook of Chemistry and Physics*, 52nd edition, The Chemical Rubber Co., Cleveland, 1971.)

Several theories have been postulated to account for the observed properties of liquids, none without some serious difficulties. The oldest, in terms of the structural picture involved, is the *hole theory*, which postulates that a liquid contains many holes or vacant areas, distributed throughout the liquid, with these holes having some distribution of sizes about an average. Although this theory does not explain all observations of the changes of properties of materials upon fusion, it has been useful in deriving a relatively simple approach to predicting the temperature dependence of the viscosity of liquids. Other models, based on differing assumptions of where the extra volume associated with a liquid is assumed to reside, will be discussed in Chapter 13.

Because liquids near their melting points still represent a dense phase, the concept of transfer of momentum from atom to atom, as utilized in the kinetic theory approach to gas-phase viscosity, is invalid, since the momentum of each atom varies rapidly with the vibration of the atoms within the pseudo-lattice of the liquid. As Frenkel[4] points out, the fact to be explained in the case of liquids is not their resistance to shearing stress, but rather their capability of yielding to stress.

Without recourse to any specific model of a liquid, but assuming only that viscous flow takes place by movement of particles past other particles, we can start by considering the mobility of an individual particle. Einstein has shown that the mobility B of a particle under the influence of an external force is related to the diffusion coefficient D by the relationship

$$D = B\kappa_B T; \tag{1.17}$$

B is the mean velocity divided by the force acting on the particle. Since diffusion appears to be an activated process, i.e., a minimum activation energy $\Delta G^\ddagger$ must be supplied to the particle to move it from one stable position to the next, then the fluidity, which is proportional to the capability of atoms to move, just as diffusion is, must also be thermally activated.

However, fluidity is the inverse of viscosity, so that, although D is proportional to $\exp[-\Delta G^\ddagger/RT]$, η must be proportional to $\exp[+\Delta G^\ddagger/RT]$. That is, the viscosity of *liquids decreases* with increasing temperature. Recall that the viscosity of *gases increases* with temperature. The temperature dependence of η may then be described by an equation of the form

$$\eta = A \exp\left[\frac{\Delta G^\ddagger_{vis}}{RT}\right], \tag{1.18}$$

where η = viscosity, poise, A = constant, poise, T = absolute temperature, °K, R = gas constant, cal/deg mol, and $\Delta G^\ddagger_{vis}$ = activation energy of viscosity, cal/mol.

The constant A is the object of much of the theoretical work done on the structure of liquids. None of the theories to date give satisfactory equations, based

---

[4] J. Frenkel, *Kinetic Theory of Liquids*, New York, Dover, 1955.

on fundamental parameters, which can be used to accurately predict values of A. The closest is *Eyring's theory*, which predicts A according to the equation

$$A \cong \frac{N_0 h}{\hat{V}}, \tag{1.19}$$

in which $N_0 =$ Avogadro's number, $\hat{V} =$ molar volume, and $h =$ Planck's constant.

For *molecular liquids* in which the bonding force is of a van der Waals type, we can predict the activation energy of viscosity from the vaporization energy ΔE_{vap}:

$$\Delta G^{\ddagger} \cong 0.41 \, \Delta E_{vap}. \tag{1.20}$$

Unfortunately, Eqs. (1.19) and (1.20) are *not valid* for liquid metals, nor are they valid for polymers or other chainlike molecules, and should not be used except as a last resort.

It is surprising that the viscosities of many diverse liquids, in terms of bonding nature in the solid state, are very similar. To illustrate this point, Table 1.3 lists groups of various materials under general ranges of viscosity. The viscosities used are those of the material in the normal temperature range of interest.

Table 1.3 Viscosity ranges for various liquids

Viscosity range, poise	Materials
1–100	CaO-Al$_2$O$_3$-SiO$_2$ slags 50% NaOH, 50% H$_2$O Linseed oil
0.1–1.0	H$_2$SO$_4$
0.01–0.1	Molten salts Heavy metals (Pb, Au, Zn, etc.) Alkaline earth metals (Ca, Mg) Transition metals (Fe, Ni, Co, etc.) Water (70°F) Kerosene (70°F)
0.001–0.01	Acetone Alkali metals

Figure 1.8 gives a nomograph for the viscosity of common liquids. More specific aspects of several classes of liquids of particular interest to metallurgists are taken up in the following sections.

1.4.1 Viscosity of liquid metals and alloys

As you are undoubtedly aware, metals are not molecular in nature and neither the

Viscosities of liquids (coordinates for use with Fig. 1.8)

Liquid	x	y	Liquid	x	y
Acetone	14.5	7.2	Nitric acid, 95%	12.8	13.8
Brine, 25% NaCl	10.2	16.6	Nitric acid, 60%	10.8	17.0
Carbon tetrachloride	12.7	13.1	Sodium hydroxide, 50%	3.2	25.8
Fuel oil	6.0	33.7	Sulfuric acid, 100%	8.0	25.1
Hydrochloric acid, 31.5%	13.0	16.6	Titanium tetrachloride	14.4	12.3
Kerosene	10.2	16.9	Water	10.2	13.0
Linseed oil, raw	7.5	27.2			

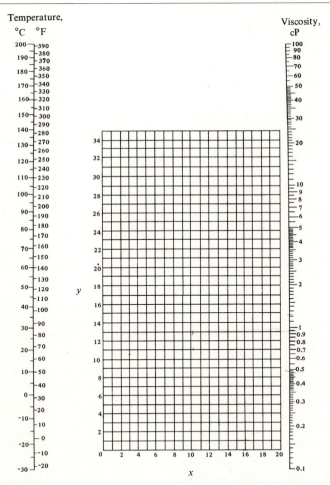

Fig. 1.8 Viscosities of liquids at 1 atm. For coordinates see the table above. (From J. H. Perry (editor), *Chemical Engineers' Handbook*, fourth edition, McGraw-Hill, New York, 1963.)

constants A nor $\Delta G^{\ddagger}_{vis}$ can be predicted by using the simplified equations presented above. Metals do, however, show activated behavior. Figure 1.9 plots the viscosities for many of the common metals as $\log \eta$ versus $1/T$.

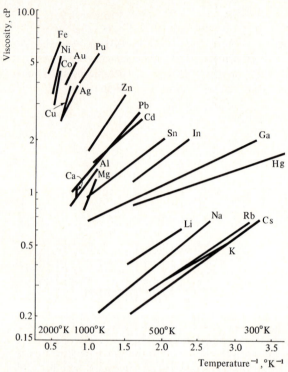

Fig. 1.9 The viscosities of liquid metals and their dependence on temperature. (From Chapman, *ibid.*)

Chapman[5] has analyzed these data in the light of the theory of liquids developed by Kirkwood[6] and by Born and Green.[7] He arrived at a generalized model for the viscosity of liquid metals which involves no assumptions of the structure other than that the atoms are spherical and that the potential between atoms can be expressed by some function $\phi(r)$ of the distance between atoms and an energy parameter, such as is done in the Lennard-Jones picture of the potential energy well between atoms, in which

$$\phi(r) = 4\varepsilon \left[\left(\frac{\delta}{r} \right)^{12} - \left(\frac{\delta}{r} \right)^{6} \right].$$

[5] T. Chapman, *AIChE Journal* **12**, 395 (1966).

[6] J. G. Kirkwood, *J. Chem. Phys.* **14**, 180 (1946).

[7] M. Born and H. S. Green, *A General Kinetic Theory of Liquids*, University Press, Cambridge, 1949.

By obtaining a suitable expression for the time average of the interactions between atoms, when their normal molecular motion is disturbed by imposing a velocity gradient on the liquid, and by attributing virtually all the momentum flux to intermolecular forces (that is, neglecting the very small contribution from the kinetic motion of the atoms), Chapman deduced a relationship between the viscosity, an energy parameter ε, and a separation distance δ. Then, by further assuming that all liquid metals obey the same function $\phi(r)$, he concluded that all substances with this $\phi(r)$ should have a reduced viscosity η^*, which is a function of the reduced temperature T^* and volume V^*, where the functional relationship is given by

$$\eta^*(V^*)^2 = f(T^*), \tag{1.21}$$

and

$$\eta^* = \frac{\eta\delta^2 N_0}{\sqrt{MRT}}, \tag{1.22}$$

$$T^* = \frac{\kappa_B T}{\varepsilon}, \tag{1.23}$$

and

$$V^* = \frac{1}{n\delta^3}. \tag{1.24}$$

The variables used are:

δ = interatomic distance in the close-packed crystal at $0°$K, Å,
ε = energy parameter characteristic of specific metal,
N_0 = Avogadro's number,
M = atomic weight,
R = gas constant,
T = absolute temperature, $°$K,
κ_B = Boltzmann's constant, and
n = number of atoms per unit volume.

The parameter δ is taken as the interatomic spacing for the close-packed crystal at $0°$K. The energy parameters present the largest difficulty, and have been derived in the following manner. The effective Lennard-Jones parameters for sodium and potassium have been determined.[8] Using these two values, one can plot the reduced viscosity data for sodium and potassium as a function of reduced temperature. The data points fall on a smooth curve, as predicted by Eq. (1.21). Then, assuming that all the rest of the metals in Fig. 1.9 obey the same functional relationship, the viscosity–temperature data for the remainder of the pure metals

---

[8] R. C. Ling, *J. Chem. Phys.* **25**, 609 (1956).

are correlated by empirically adjusting the parameter ε/κ_B until all the data for a given metal fall on one point on the curve, given in Fig. 1.10. Table 1.4 shows the resulting values of ε/κ_B starting with the known (measured) values for sodium and potassium.

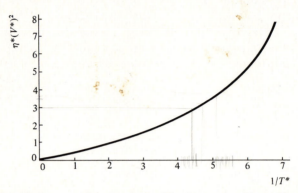

Fig. 1.10 Correlation curve for viscosities of liquid metals. (From Chapman, *ibid.*)

Table 1.4 Empirically determined values of ε/κ_B (from Chapman, *ibid.*)

Metals	δ, Å	ε/κ_B, °K
Na	3.84	1,970
K	4.76	1,760
Li	3.14	2,350
Mg	3.20	4,300
Al	2.86	4,250
Ca	4.02	5,250
Fe	2.52	10,900
Co	2.32	9,550
Ni	2.50	9,750
Cu	2.56	6,600
Zn	2.74	4,700
Rb	5.04	1,600
Ag	2.88	6,400
Cd	3.04	3,300
In	3.14	2,500
Sn	3.16	2,650
Cs	5.40	1,550
Au	2.88	6,750
Hg	3.10	1,250
Pb	3.50	2,800
Pu	3.10	5,550

The fact that, by adjusting an unmeasured parameter, it is possible to correlate all these liquid-metal viscosities on a single curve, might not be taken as significant, except for the fact that one theory of melting indicates that the melting point should be proportional to ε, and that this relationship has been observed for other classes of materials. The empirically determined values of ε/κ_B from Table 1.4 have been plotted as a function of the absolute melting temperature, and an excellent correlation has been observed, leading to the equation

$$\frac{\varepsilon}{\kappa_B} = 5.20 T_{\text{melting}}, {}^{\circ}\text{K}. \tag{1.25}$$

For metals with high melting points, this may be used to predict viscosities, as no further data are available.

Example 1.3   Estimate the viscosity of liquid titanium at 1850°C. The following data are available: $T_m = 1800°C$, $M = 47.9$ g/mol, sp. gr. $= 4.50$ g/cm^3, and $\delta = 2.89$ Å.

Solution.   Using Eq. (1.25), one can estimate ε/κ_B to be

$$\varepsilon/\kappa_B = (5.20)(2073) = 10,780°\text{K}.$$

Then T^* is given by Eq. (1.23), and the product $\eta^*(V^*)^2$ is found from Fig. 1.10:

$$T^* = \frac{1}{10,780} \underset{= 1850 + 273}{(2123)} = 0.197,$$

and $\eta^*(V^*)^2 = 3.6$.

From the given data, one can calculate V^*:

$$V^* = \frac{1}{\left(\dfrac{6.02 \times 10^{23} \text{ atoms}}{47.9 \text{ g}}\right)\left(\dfrac{4.5 \text{ g}}{\text{cm}^3}\right)(2.89 \times 10^{-8} \text{ cm})^3} = 0.733.$$

Then

$$\eta^* = \frac{3.6}{(0.733)^2} = 6.63.$$

Solving for η, we have

$$\eta = \frac{\eta^*(MRT)^{1/2}}{\delta^2 N_0} = \frac{6.63\,[(47.9)(8.314 \times 10^7)(2123)]^{1/2}}{(2.89 \times 10^{-8})^2(6.02 \times 10^{23})}$$

$$= 3.83 \times 10^{-2} \text{ poises,}$$

$$\eta = 3.83 \text{ cP.}$$

At the present time, few data are available on viscosities of molten alloys, and no prediction equations have been developed for them.

In Figs. 1.11 and 1.12, the viscosities of two important binary systems, Al-Si and Fe-C, are superimposed on their respective phase diagrams to illustrate the large effect an alloying element may have on viscosity. Similar anomalies in the regions above structural features on the phase diagram (especially apparent in Fig. 1.11) have been observed in other alloy systems as well.

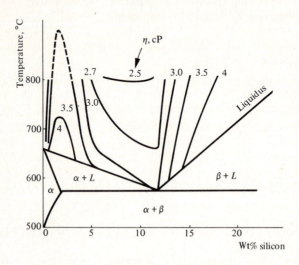

Fig. 1.11 Viscosities of aluminum–silicon alloys. (From W. R. D. Jones and W. L. Bartlett, *J. Inst. Metals* **81**, 145 (1952).)

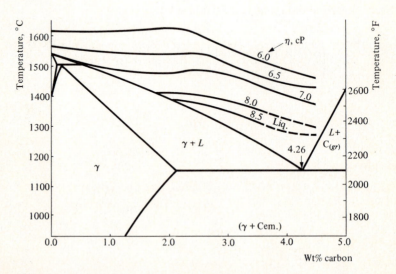

Fig. 1.12 Viscosities of iron–carbon alloys. (From R. N. Barfield and J. A. Kitchener, *J. Iron and Steel Inst.* **180**, 324 (1955).)

1.4.2 Viscosity of molten slags and salts

The structures of molten slag systems have been studied from many aspects. At this point we will not go into the detail of the various structures, but rather discuss briefly their basic aspects which affect viscosity.

In general, slags consist of cations and anions resulting from ionization of basic and acidic constituents in molten oxide solution. We may consider an acidic component to be an oxide which, when dissolved in the slag, acquires additional oxygen ions to form a complex anion, while a basic oxide contributes an oxygen ion to the melt; the cation then remains dissociated from any other ions, and moves about freely. The most common acidic component is SiO_2, and Al_2O_3 behaves in a similar manner. Starting with pure SiO_2, in which the bonding is both strong and highly directional, and in which viscous flow occurs only by breaking bonds, let us examine what happens if we add a basic oxide, such as CaO, to it.

We presume that the structure of pure liquid silica is quite similar to that of solid silica, in which each Si^{4+} ion shares one electron with each of four O^{2-} ions which form a tetrahedron about the Si^{4+} ion. In the solid state, electroneutrality is maintained by each oxygen ion sharing its other electron between two tetrahedra,

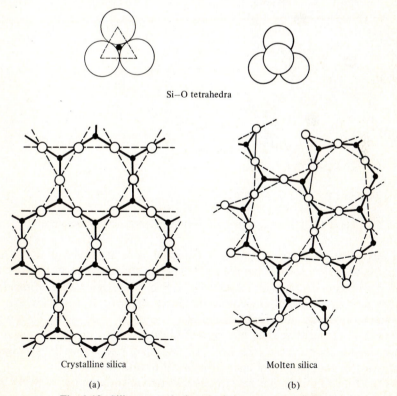

Si–O tetrahedra

Crystalline silica   Molten silica

(a)   (b)

Fig. 1.13 Silicate tetrahedron and the structure of silica.

or Si^{4+} ions, and the structure built up is a regular crystalline array of SiO$_4^{4-}$ groups. This is illustrated in Fig. 1.13(a). When this substance is melted, the arrangement presumably continues, but the long-range order is destroyed, as indicated in Fig. 1.13(b). However, the same Si–O bonds are present and these high energy bonds need to be broken so that viscous flow could take place. The activation energy for this process is quite high (135 kcal) and the viscosity of pure liquid SiO$_2$ at 1940°C is 1.5×10^5 poises.

When CaO, or another similar divalent basic oxide, is added to the melt, the Ca^{2+} ions are accommodated in the interstices of the silicate structure and the O^{2-} ions enter into the network (Fig. 1.14). The O^{2-} ions cause two of the tetrahedra to separate since each corner of these two can now have an oxygen ion of its own. Increasing additions of base results in a progressive breakdown of the original three-dimensional network.

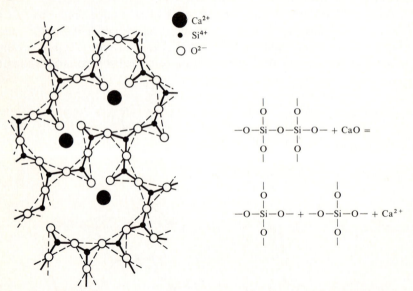

Fig. 1.14 The solution of a divalent metal oxide in molten silica.

Table 1.5 shows this progression as a function of total oxygen atoms to silicon atoms. As the breakdown progresses, we may consider the silicate network to be made up of progressively smaller discrete ions. From the original three-dimensional network, we progress to discrete molecules, such as (Si$_6$O$_{15}$)$^{6-}$ and (Si$_3$O$_9$)$^{6-}$, as more and more oxygen ions are contributed by the basic oxide. Actually, it is more probable that rather than only one type of silicate ion existing at any given ratio of base to acid, there exists a statistical distribution of sizes about the average one, as indicated by the number of links broken per tetrahedron.

As the three-dimensional network breaks down, the number of Si–O bonds that need to be broken during viscous flow decreases. The shear process becomes

Table 1.5 Structural relationships in basic oxide–silicate melts

Total oxygen atoms / Silicon atoms	Corresponding binary molecular formula	Structure	Equivalent silicate ion
2:1	SiO_2	All corners of tetrahedra shared	Infinite network
5:2	$MO \cdot 2SiO_2$	One broken link per tetrahedron	$(Si_6O_{15})^{6-}$ or $(Si_8O_{20})^{8-}$
3:1	$MO \cdot SiO_2$	Two broken links per tetrahedron (ring)	$(Si_3O_9)^{6-}$ or $(Si_4O_{12})^{8-}$
7:2	$3MO \cdot 2SiO_2$	Three broken links per tetrahedron (chain)	$(Si_2O_7)^{6-}$
4:1	$2MO \cdot SiO_2$ (orthosilicate)	All links broken	Discrete $(SiO_4)^{4-}$ tetrahedra

easier as the size of the anions decreases (Fig. 1.15), so that the activation energy $\Delta G_{vis}^{\ddagger}$ falls off continuously as basic oxide is added (Fig. 1.15), up to the orthosilicate composition, beyond which no further significant decrease is expected. At that point the network of $(SiO_4)^{4-}$ tetrahedra is completely broken down. For compositions more basic than the orthosilicate, the viscosity changes that do occur are more likely due to changes in the liquidus temperature.

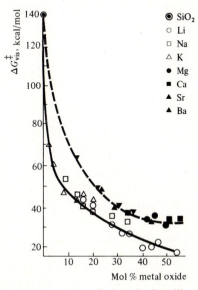

Fig. 1.15 Decrease of the activation energy of viscosity for silica as basic oxides are added. (From Bills, *ibid.* **201**.)

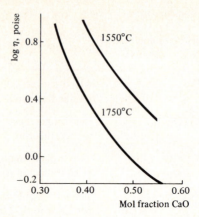

Fig. 1.16 Viscosity of the SiO_2–CaO solutions. (From J. O. M. Bockris, J. A. Kitchener, and J. Mackenzie, *Trans. Faraday Soc.* **51**, 1734 (1955).)

Alumina, which appears to exist in molten oxide solutions as $(AlO_3)^{3-}$ anions, behaves in much the same way as silica. However, silica and alumina are not equivalent on a molar basis, since the basic building block of alumina is $(AlO_3)^{3-}$, and two Al^{3+} ions can replace two Si^{4+} ions only if one Ca^{2+} ion is available to maintain electrical neutrality. Thus, as far as its effect on viscosity is concerned, alumina has a *silica equivalence* X_a, which depends on the Al_2O_3/CaO ratio and on the total Al_2O_3 content, as shown in Fig. 1.17. Turkdogan and Bills,[9] using these data in correlating the viscosity and composition of CaO-MgO-Al_2O_3-SiO_2 slags, have found that, for a given temperature, a smooth curve correlates all compositions in the studied range (Fig. 1.18). To utilize this fact, we convert the

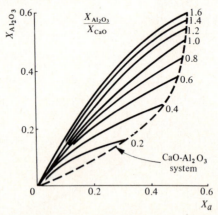

Fig. 1.17 Silica equivalence of alumina related to alumina molar concentrations and alumina : lime molar ratios. (From Bills, *ibid.* **201**.)

---

[9] E. T. Turkdogan and P. M. Bills, *Amer. Ceramic Soc. Bull.* **39**, 682 (1960).

slag analysis to mole fractions, and determine X_a from Fig. 1.17. Magnesium oxide (MgO) is equivalent to CaO, up to about 10 mole percent MgO, and their mole fractions are added together to obtain X_{CaO}. Thus, when we add X_{SiO_2} and X_a, and know the temperature, we can find the viscosity from Fig. 1.18. This acidic behavior of Al_2O_3 is further illustrated by Fig. 1.19 which shows the viscosity of $CaO\text{-}Al_2O_3\text{-}SiO_2$ slags at 1500°C. Note that the isoviscosity lines are essentially parallel to the $Al_2O_3\text{-}SiO_2$ binary system.

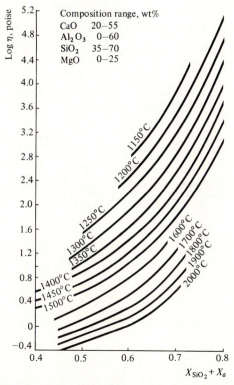

Fig. 1.18 Viscosity of $CaO\text{–}Al_2O_3\text{–}SiO_2\text{–}MgO$ melts. Viscosity is expressed in poises. (From Bills, *ibid.* **201**.)

In another study, Bills[10] determines the effect of FeO on the viscosities of $CaO\text{-}MgO\text{-}Al_2O_3\text{-}SiO_2$ slags. Up to a concentration of 16 mole percent, we may consider FeO to be identical to CaO, as far as their effect on viscosity is concerned. Above 16%, FeO decreases viscosity more than CaO or MgO, except at lower temperatures (Fig. 1.20). In this case X_{CaO} is taken as $X_{CaO} + X_{MgO} + X_{FeO}$ (Fig. 1.17).

Finally, we should note the ability of the well-known flux CaF_2 (fluorspar) to

---

[10] P. M. Bills, *J. Iron and Steel Inst.* **201**, 133 (1963).

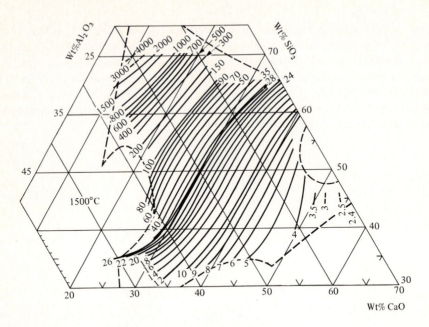

Fig. 1.19 Isoviscosity lines (poise) in the CaO-Al$_2$O$_3$-SiO$_2$ system at 1500°C. (From J. S. Machin and T. B. Yee, *J. Am. Cer. Soc.* **31**, 200 (1948).)

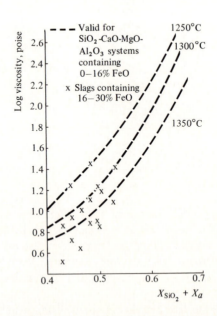

Fig. 1.20 Effect of FeO on viscosity of SiO$_2$-CaO-MgO-Al$_2$O$_3$-FeO systems. (From Bills, *ibid.* **201**.)

decrease the viscosity of oxide slags. Although we do not quite understand the reason, it may be due to the ability of CaF^+ cations to move between silicate anions that are electrostatically bound together, by mutual attraction to a Ca^{2+} cation, thereby decreasing the electrostatic bond and thus lowering the viscosity. The effect is much more pronounced at low rather than at elevated temperatures, and in acidic slags more than in basic slags. Figures 1.21 and 1.22 illustrate the effect for basic and acidic slags, respectively.

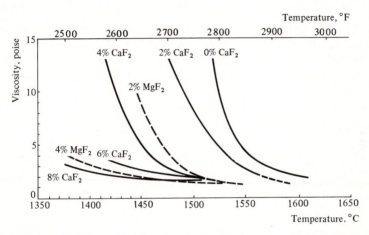

Fig. 1.21 Influence of CaF_2 and MgF_2 additions on the viscosity of a basic slag. Slag composition: CaO—51.7%, MgO—3.2%, Al_2O_3—12.7%, SiO_2—32.4%. (From Elliott, *ibid.*)

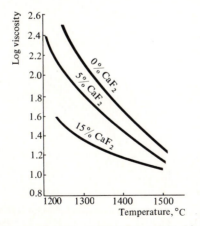

Fig. 1.22 Influence of CaF_2 on viscosity of 50% SiO_2–15% Al_2O_3–35% CaO melt (acid slag). (From Bills, *ibid.* **201**.)

Example 1.4 Calculate the viscosity of a slag with composition 40% CaO, 40% SiO_2, 8% MgO, 12% Al_2O_3 at 1600°C and at 1500°C.

Solution. Converting to mole fractions, we have $X_{CaO} = 0.422$, $X_{MgO} = 0.117$, $X_{SiO_2} = 0.390$, $X_{Al_2O_3} = 0.069$.

For our purposes, we lump X_{CaO} and X_{MgO} together, so the effective $X_{CaO} = 0.539$.

Calculating the ratio $X_{Al_2O_3}/X_{CaO}$ ($=0.128$), enter Fig. 1.17 at $X_{Al_2O_3} = 0.069$, move slightly to the right of $X_{Al_2O_3}/X_{CaO} = 0.2$, and go down to find $X_a \simeq 0.10$. After calculating $X_{SiO_2} + X_a = 0.490$, move to Fig. 1.18, and go up to the curves for 1600°C and 1500°C. The viscosity values are 1.8 and 2.9 poises, respectively.

The simplest molten salts are those in which we take the bonding to be essentially electrostatic; for example, we consider that molten NaCl consists entirely of Na^+ and Cl^- ions. Since there appear to be no large anion or cation complexes analogous to the $(SiO_4)^{4-}$ or $(AlO_3)^{3-}$ ions in molten oxide systems, and since there is a relatively large fraction of nondirectional electrostatic bonding, flow takes place relatively easily in these materials. Figure 1.23 shows viscosities of many molten salts. Note that their viscosities are about 1/100th of those of the molten oxides.

Because of the nondirectionality of the bond forces, the activation energies of viscosity for salts are also lower than those of the slags.

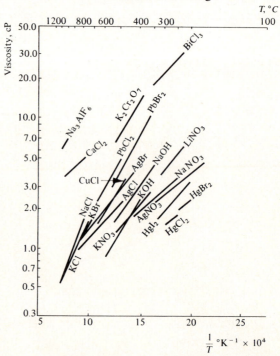

Fig. 1.23 Viscosities of molten salts. (From C. J. Smithells, *Metals Reference Book*, fourth edition, Vol. I, Plenum Press, New York, 1967.)

1.5 NON-NEWTONIAN FLUIDS

According to Newton's law of viscosity (Eq. 1.3) the shear stress τ_{yx}, plotted versus the velocity gradient $-dv_x/dy$, should yield a straight line running through the origin. Experimentally, this has been proved true for all gases and for all single-phase nonpolymeric liquids. Fluids that behave in this manner—and most fluids do—are termed *Newtonian fluids*. However, Eq. (1.3) does not describe the behavior of a large number of fluids which are called *non-Newtonian fluids* and include substances such as molten plastics, slurries, and certain slags.

The study of Newtonian flow is part of a larger discipline of science known as *rheology*. Rheology encompasses the mechanical behavior of gases, liquids, and solids, including Newtonian gases and liquids on the one hand and Hookean behavior of solids on the other. The spectrum of substances between these two extremes concerns the engineers and scientists dealing with non-Newtonian fluids.

A large number of non-Newtonian fluid follows one of the behavioral patterns represented in Fig. 1.24, where the shear stress τ_{yx} is plotted versus the rate of strain $\dot{\gamma}$ ($\dot{\gamma}$ equals $-dv_x/dy$ at steady state). In contrast, the curve of stress–strain rate is also plotted for Newtonian fluids, demonstrating the obvious fact that $\tau_{yx}/\dot{\gamma}$ is independent of $\dot{\gamma}$. On the other hand, non-Newtonian fluids are those in which $\tau_{yx}/\dot{\gamma}$ is a function of $\dot{\gamma}$.

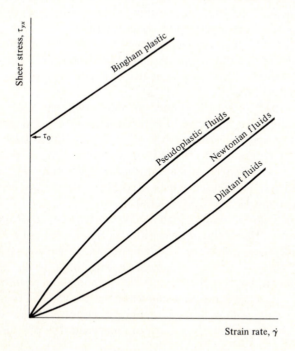

Fig. 1.24 Stress–strain rate curves for time-independent fluids.

The fluids classified as *Bingham plastics* require a finite shear stress τ_0 (yield stress) to initiate flow. In other words, the fluid remains rigid when the shear stress is less than τ_0, but flows when the shear stress exceeds τ_0. An example of a fluid exhibiting Bingham plastic behavior is an aqueous slurry of fine, powdered, coal. The following relationship gives the rate of shear stress and shear strain:

$$\tau_{yx} = \pm\tau_0 + \eta_P\dot{\gamma}, \qquad \tau_{yx} > \tau_0. \tag{1.26}$$

Here η_P is a plastic viscosity or coefficient of rigidity. We use the plus sign when $\dot{\gamma}$ is positive, and the minus sign when $\dot{\gamma}$ is negative.

Pseudoplastic fluids, which are characterized by a decreasing slope of the τ_{yx} versus $\dot{\gamma}$ curve as the stress increases, need, like Newtonian fluids, no yield stress for flow, but, unlike in Newtonian fluids, $\tau_{yx}/\dot{\gamma}$ does depend on τ_{yx}. *Dilatant fluids* differ from pseudoplastics in that the τ_{yx} versus $\dot{\gamma}$ curve has an increasing slope as stress increases. For both pseudoplastic and dilatant fluids we can sometimes use the *Ostwald power law* to describe the relationship of the stress–strain rate:

$$\tau_{yx} = k\dot{\gamma}^n. \tag{1.27}$$

Here, k is a measure of the fluid's consistency, and n is a measure of the fluid's departure from Newtonian behavior. For $n = 1$, Eq. (1.27) reduces to Newton's law of viscosity with $k = \eta$. For $n < 1$, the behavior is pseudoplastic, whereas if $n > 1$ it is dilitant. Aqueous suspensions of clay, lime, and cement rock are examples of various fluids described by the power law.

Since the Ostwald power law is the simplest model to describe dilatant and pseudoplastic fluids, it is often applied. However, when using it, we should keep in mind that this model does not give an accurate portrayal of the low-shear-rate region (i.e., low flow rates).[11]

In addition to these models, numerous other empirical equations have been proposed to express the relationship between shear stress and strain rate. The models we have been discussing are called *generalized Newtonian fluids* (GNF). They should not be applied to those fluids or conditions in which viscoelastic effects are present and play an important role, or to the fluids which exhibit *time-dependent properties*. The fluids in which viscoelastic effects are of great importance under certain conditions are *high polymers*.

Viscoelastic fluids are fluids which exhibit elastic recovery from deformation, that is, they *recoil*. This is in contrast with the behavior of GNF's which do not recoil. This difference is illustrated by the experiment depicted in Fig. 1.25. We observe the behavior of a black line made by injecting a charcoal slurry into a transparent tube. In part (a), the line has just been introduced into the tube with the fluid at rest. The fluid starts flowing when the pressure is lowered at one end of the tube; both fluids flow in the direction of the lower pressure, as shown in parts (b) and (c). An instant after part (c) is reached, the pressure difference is

[11] R. B. Bird, *Can. J. of Chem. Eng.* **43**, 161 (Aug. 1965).

removed and we can see that the GNF moves no further while the viscoelastic fluid goes backward. It does not, however, return to its original position, recoiling only partly.

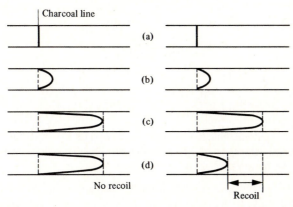

Fig. 1.25 Recoil effect in a viscoelastic fluid. Fluid on the left is a GNF; that on the right is a viscoelastic fluid.

Different conditions create different types of time-dependent fluid behavior. For example, *thixotropic fluids* have a structure that breaks down with time under shear. At a constant shear–strain rate, the viscosity decreases with time and approaches an asymptotic value. Under steady-state flow conditions, when the asymptotic value is maintained, thixotropic fluids may be treated as GNF's. *Rheopectic fluids*, in which viscosity increases with time, behave again quite differently from thixotropic fluids.

The purpose of this brief introduction to non-Newtonian fluids has been to make the reader aware of the fact that there is a large number of fluids that do not necessarily obey Newton's law of viscosity, and that a direct application to such fluids of data and concepts in fluid mechanics dealing with Newtonian fluids could lead to serious engineering blunders. In the remainder of the text, we will deal exclusively with Newtonian fluids.

PROBLEMS

1.1 Compute the steady-state momentum flux in lb_f/ft^2 when a plate moves across another plate, which is stationary, with a velocity of 2 ft/sec. The distance between the plates is $\frac{1}{16}$ in. and the fluid between the plates has a viscosity of 2 cP. What is the direction of the momentum flux? The shear stress?

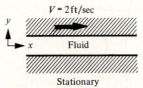

1.2 Water at 100°F flows over a flat plate.

a) If the velocity profile at a point $x = x_1$ is given by $v_x = 3y - y^3$, find the shear stress at the wall at that point.

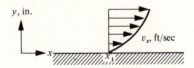

Properties of water at 100°F

$\rho = 62.0 \text{ lb}_m/\text{ft}^3$

$v = 0.027 \text{ ft}^2/\text{hr}$

b) What is the rate of momentum transport at $y = \frac{1}{32}$ in. and $x = x_1$ in the y-direction?

c) Is there momentum transport in the x-direction at $y = \frac{1}{32}$ in. and $x = x_1$? What is it (per unit area normal to flow)?

1.3 a) For problem 1.2 compute the rate of momentum transfer at $y = 1$ in. and $x = x_1$ in the y-direction.

b) Compute the momentum transport in the x-direction at the same point.

c) Compare results with those of Problem 1.2.

1.4 a) Consider air as a one component gas with the parameters given in Table 1.1 and estimate its viscosity at 100°F. Repeat, only now consider it to be 79% N_2 and 21% O_2 and evaluate η_{mix}.

b) Obtain the same results at 1500°F.

c) Compare your answers in (a) and (b) to the experimental values of 0.046 and 0.106 lb_m/hr ft at 100 and 1500°F, respectively.

1.5 Consider that the binary gas $A - B$ is such that at a given temperature $\eta_A = \eta_B$. Plot η_{mix}/η_A versus X_B if M_A/M_B is

a) 100,

b) 10,

c) 1.

What do the results mean to you?

1.6 What is the viscosity of chromium at 2000°C? Data are as follows:

Melting point, 1898°C

Atomic wt, 52.01

Specific gravity, 7.1

δ, 2.72 Å

1.7 Using Bills' method, estimate the viscosity of a 50 wt% SiO_2, 30 wt% CaO, 20 wt% Al_2O_3 slag at 1500°C. Does this agree with Fig. 1.19?

1.8 At 1000°C a melt of Cu–40% Zn has a viscosity 5 cP; at 950°C the same alloy has a viscosity of 6 cP. Estimate the viscosity of the alloy at 1100°C.

1.9 When oxides such as MgO and CaO are added to molten silica, the activation energy of viscosity is reduced from 135 kcal/mol for pure SiO_2 to about 39 kcal/mol for 0.5 mole fraction

SiO_2. We can observe even more dramatic effects when oxides such as Li_2O and Na_2O are added to silica; for example, for 0.5 mole fraction SiO_2, the activation energy is only 23 kcal/mol. Explain with the aid of sketches, why this is so.

$$1 \, lb_f = \frac{32.2 \, lbm \cdot ft}{s^2}$$

UNITS: $\tau_{yx} \left(\frac{lb_f}{ft^2} \right)$

momentum
transport $= mv^2$ DUE TO FLUID MOTION

FLOW OF A FALLING FILM *

VELOCITY DISTRIBUTION OR VELOCITY PROFILE
$$V_z = \frac{-\rho g \cos B \, x^2}{2n} + \frac{\rho g \cos B \, \delta^2}{2n}$$

$$V_z^{MAX} = \frac{\rho g \cos B \, \delta^2}{2n}$$
@ x=0

δ = height
W = width

FOR LAMINAR FLOW

$$\bar{V}_z = \frac{\rho g \, \delta^2 \cos B}{3n}$$
average velocity

$$Q = \frac{\rho g W \delta^3 \cos B}{3n}$$
VOLUME FLOW RATE

FLOW THROUGH A CIRCULAR PIPE * PG 44, 45

(1) SET UP MOMENTUM BALANCE EQUATION
(2) DIVIDE THROUGH BY THE AREA CONSTANTS $2\pi L$ δr PG 44.
(3) TAKE LIMIT, INTEGRATE INDEFINITELY.   PG 44
(4) APPLY B.C. @ $r=0$, $\tau_{rz} = 0$
(5) SUBSTITUTE NEWTON'S LAW VISCOSITY
(6) APPLY B.C. #2 $V_z = 0$ at $r = R$
(7) PG 45. FOR $V_{MAX}, \bar{V}_z, Q$

LAMINAR FLOW AND MOMENTUM BALANCES

In discussing Newton's law of viscosity, we have described fluid motion as flowing parallel layers which, due to viscosity, establish a velocity gradient dependent upon the shear stress applied to the fluid. This velocity gradient has been regarded as a potential or a "reason" for momentum transport from layer to layer.

In this chapter, we shall first derive simple differential equations of momentum for special cases of flow, for example, flow of a falling film, flow between two parallel plates, and flow through tubes. To make it possible for the student to participate in developing hairy formulas, these derivations make use of the concept of a momentum balance and the definition of viscosity. These classic examples of viscous flow patterns certainly apply to rather simplified and idealized conditions. You may be tempted, therefore, to disregard the importance of thoroughly understanding these examples; however, we should point out that, despite the simplicity of the following calculations, you will gain an appreciation of the variables involved. Also, you will obtain a basic tool for analyzing engineering problems: the ability to arrive at pertinent differential equations.

2.1 MOMENTUM BALANCE*

A momentum balance is applied to a small control volume of fluid to develop a differential equation. The differential equations, when their solutions comply with the physical restrictions (boundary conditions), yield the algebraic relationship which can be used to determine the engineering characteristics of the system. The solutions give the velocity distributions from which other characteristics, including the shear stress at the fluid–solid interface, are developed. As we shall see in Chapter 3, the shear stress at the fluid–solid interface is very important in analyzing the disposition of energy flowing through a system. For steady state flow, the momentum balance is

$$\begin{pmatrix} \text{rate of} \\ \text{momentum in} \end{pmatrix} - \begin{pmatrix} \text{rate of} \\ \text{momentum out} \end{pmatrix} + \begin{pmatrix} \text{sum of forces} \\ \text{acting on system} \end{pmatrix} = 0. \quad (2.1)$$

---

* The general aspects of the developments in Sections 2.1–2.6 are similar to those found in Chapters 2–3 in R. B. Bird, W. E. Stewart, and E. N. Lightfoot, *Transport Phenomena*, Wiley, New York, 1960.

Momentum in (or out) may enter a system by momentum transfer according to Newton's equation of viscosity (if the fluid is Newtonian; otherwise various equations for non-Newtonian fluids are used), or it may enter due to the overall fluid motion. The forces applied to the balance are pressure forces and/or gravity forces.

The momentum balance is actually a force balance because we are concerned with the *rate of momentum* that enters and leaves the unit volume. Units of momentum are ML/T (M = mass, L = length, T = time), whereas a rate of momentum is ML/T^2. Classical physics states clearly that forces ($F = ma$) are involved when we consider momentum rates. Thus, if the term momentum balance confuses the reader, he is reassured that a force balance is being applied.

2.2 FLOW OF A FALLING FILM

Consider the flow of a liquid at steady state along an inclined plane (Fig. 2.1). The liquid is at a constant temperature, and therefore its density and viscosity are constant. Furthermore, we consider only that portion of the plane where the entrance and exit of the liquid to the plane are sufficiently removed so as not to influence the velocity v_z. In this situation, v_z is not a function of z but obviously a function of x.* Figure 2.1 also depicts the unit volume as a "shell" with a thickness Δx and length L; the width of the shell extends a distance W, perpendicular to the page. The terms used in Eq. (2.1) are as follows.

Rate of momentum in
across surface at x $(LW)(\tau_{xz})|_x$
(moment transport due to viscosity)

Rate of momentum out
across surface at $x + \Delta x$ $(LW)(\tau_{xz})|_{x+\Delta x}$
(due to viscosity)

Rate of momentum in
across surface at $z = 0$ $(W\Delta x v_z)(\rho v_z)|_{z=0}$
(due to fluid motion)

Rate of momentum out
across surface at $z = L$ $(W\Delta x v_z)(\rho v_z)|_{z=L}$
(due to fluid motion)

Gravity force acting on fluid $(LW\Delta x)(\rho g \cos \beta)$

In this particular problem, the pressure forces are irrelevant because the pressure is equal throughout the liquid. Also note that all terms in the list, including the first two, are z-directed forces. Figure 2.1 shows that momentum in by viscous transport is x-directed, but if we think of interpreting τ_{xz} in an alternative way—

---

* In the region where $v_z = f(x)$ and $v_z \neq f(z)$, we say that the flow is *fully developed*.

namely, as a shear stress—we certainly realize that we are dealing with a z-directed force.

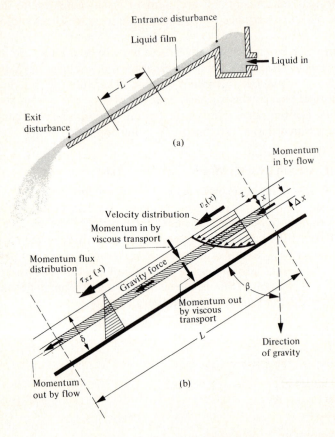

Fig. 2.1   Flow of a falling film.

When all these terms are substituted into the momentum balance, we get

$$LW\tau_{xz}|_x - LW\tau_{xz}|_{x+\Delta x} + W\Delta x\rho v_z^2|_{z=0} - W\Delta x\rho v_z^2|_{z=L} + LW\Delta x\rho g \cos \beta = 0. \tag{2.2}$$

Because we are restricted to that part of the inclined plane which does not feel the effects of the exit and entrance, v_z is independent of z. Therefore, the third and fourth terms cancel one another out. Equation (2.2) is now divided by $LW\Delta x$ and, if Δx is allowed to be infinitely small, we obtain

$$\lim_{\Delta x \to 0} \frac{\tau_{xz}|_{x+\Delta x} - \tau_{xz}|_x}{\Delta x} = \rho g \cos \beta. \tag{2.3}$$

We have now recognized the definition of the first derivative of τ_{xz} with respect to x, and have thus developed the differential equation pertinent to our system:

$$\frac{d\tau_{xz}}{dx} = \rho g \cos \beta. \tag{2.4}$$

This equation is integrated to yield

$$\tau_{xz} = \rho g x \cos \beta + C_1. \quad \text{SHEAR STRESS DISTRIBUTION} \tag{2.5}$$

Equation (2.5) describes the momentum flux (or alternatively the shear-stress distribution), but contains an integration constant C_1. This constant is evaluated by recognizing that the shear stress in the liquid is very nearly zero at a liquid–gas interface. In other words, the gas phase, in this instance, offers little resistance to liquid flow, which results in a realistic *boundary condition*:

$$\text{B.C. 1} \qquad \text{at } x = 0, \qquad \tau_{xz} = 0. \tag{2.6}$$

Substitution of this boundary condition into Eq. (2.5) requires that $C_1 = 0$; hence the momentum flux is

$$\tau_{xz} = \rho g x \cos \beta. \tag{2.7}$$

If the fluid is Newtonian, then we know that the momentum flux is related to velocity gradient according to

$$\tau_{xz} = -\eta \frac{dv_z}{dx}. \tag{2.8}$$

Substituting this expression for τ_{xz} in Eq. (2.7) gives the distribution of the velocity gradient:

$$\frac{dv_z}{dx} = -\frac{\rho g \cos \beta}{\eta} x. \tag{2.9}$$

Integrating Eq. (2.9), we have

$$v_z = -\left(\frac{\rho g \cos \beta}{2\eta}\right) x^2 + C_2. \tag{2.10}$$

Another integration constant has evolved which is evaluated by examining the other boundary condition, namely, that at the fluid–solid interface the fluid clings to the wall; that is,

$$\text{B.C. 2} \qquad \text{at } x = \delta, \qquad v_z = 0. \tag{2.11}$$

Substituting this into Eq. (2.10), we determine the constant of integration; $C_2 = (\rho g \cos \beta / 2\eta)\delta^2$. Therefore the velocity distribution is

$$v_z = \frac{\rho g \delta^2 \cos \beta}{2\eta}\left[1 - \left(\frac{x}{\delta}\right)^2\right], \tag{2.12}$$

and is parabolic. Once the velocity profile has been found, a number of quantities may be calculated:

i) The *maximum velocity*, V_z^{max}, is that velocity at $x = 0$:

$$V_z^{max} = \frac{\rho g \, \delta^2 \cos \beta}{2\eta}.$$ (2.13)

ii) The *average velocity*, $\bar{v}_z$, is simply

$$\bar{v}_z = \frac{1}{\delta} \int_0^\delta v_z \, dx = \frac{\rho g \, \delta^2 \cos \beta}{2\eta} \int_0^\delta \left[1 - \left(\frac{x}{\delta} \right)^2 \right] dx = \frac{\rho g \, \delta^2 \cos \beta}{3\eta}.$$ (2.14)

iii) The *volume flow rate*, Q, is given by the product of the average velocity and the cross section of flow:

$$Q = \bar{v}_z(W \, \delta) = \frac{\rho g W \, \delta^3 \cos \beta}{3\eta}.$$ (2.15)

The foregoing analytical results are valid only when the film is falling in laminar flow (straight streamlines). This condition can easily be satisfied for the slow flow of viscous films, but experimentally, it has been found that as the film velocity increases, the film thickness increases (according to Eq. 2.14) to a critical value, depending on the liquid's kinematic viscosity, where turbulence replaces laminar flow. Of course, when turbulent flow develops, Eqs. (2.12)–(2.15) are no longer valid.

Example 2.1   Molten slag is passed over a matte in the smelting of copper in order to recover most of the copper in the slag. The operation is carried out in a reverberatory furnace 80 ft long and 30 ft wide. Assuming that the matte is stationary (in reality this is oversimplified) and that the slag flows continuously (80 ft³/hr) over the matte, determine: (a) the equation for the velocity distribution in the slag layer; (b) the fraction of material which stays in the furnace for a period twice the mean residence time and longer. The average depth of the slag may be taken as 2 ft.

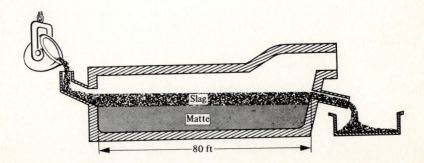

Solution. If the flow is laminar, then Eq. (2.15) gives the volume flow rate, or re-arranged reads

$$\frac{\rho g\, \delta^2 \cos \beta}{2\eta} = \frac{3Q}{2W\,\delta}.$$

a) We write the velocity profile by substituting this expression into Eq. (2.12)

$$v_z = \frac{3Q}{2W\,\delta}\left[1 - \left(\frac{x}{\delta}\right)^2\right] = \frac{(3)(80\ \text{ft}^3/\text{hr})}{(2)(30\ \text{ft})(2\ \text{ft})}\left[1 - \frac{x^2}{4}\right] = 2 - \frac{x^2}{2},$$

where x = depth into slag measured from top of slag, ft, and v_z = velocity of slag, ft/hr.

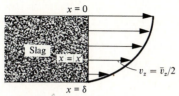

$x = 0$

Slag

$x = x$

$v_z = \bar{v}_z/2$

$x = \delta$

difference between longest time + shortest time

b) The mean residence time would be

$$\Theta = \frac{L}{\bar{v}_z},$$

where L = length of furnace, ft, and Θ = mean residence time, hr. Thus, the fraction of slag that remains for 2Θ is that slag between $x = x'$ and $x = \delta$, where x' is the value of x such that $v_z = \bar{v}_z/2$. Since $\bar{v}_z = Q/W\,\delta$,

$$v_z = \frac{1}{2}\frac{(80)}{(30)(2)} = \tfrac{2}{3}\ \text{ft/hr} \qquad \text{at } x = x'.$$

Then

$$x' = \sqrt{\tfrac{8}{3}}\ \text{ft} = 1.63\ \text{ft},$$

and the fraction of the slag with residence time 2Θ, or greater, is

$$\frac{2 - 1.63}{2} = 0.185.$$

2.3 FLOW BETWEEN PARALLEL PLATES

Consider the flow of fluid between parallel plates in Fig. 2.2. The velocity at the entrance is uniform and, as the flow progresses, velocity gradients must form because the fluid clings to the wall. At some distance downstream from the entrance, the velocity profile becomes independent of the distance from the entrance, and the flow is then fully developed. Let this region of fully developed flow start at $x = 0$ and consider the unit volume in Fig. 2.2 with a thickness Δy, width W, and length L.

Rate of momentum in
across surface at y $(LW)(\tau_{yx})|_y$
(momentum transport due to viscosity)

Rate of momentum out
across surface at $y + \Delta y$ $(LW)(\tau_{yx})|_{y+\Delta y}$
(due to viscosity)

Rate of momentum in
across surface at $x = 0$ $(W \Delta y v_x)(\rho v_x)|_{x=0}$
(due to fluid motion)

Rate of momentum out
across surface at $x = L$ $(W \Delta y v_x)(\rho v_x)|_{x=L}$
(due to fluid motion)

Pressure force on liquid at $x = 0$ $\Delta y W[P(x = 0)] = P_0 \Delta y W$

Pressure force on liquid at $x = L$ $-\Delta y W[P(x = L)] = -P_L \Delta y W$

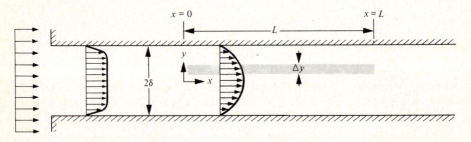

Fig. 2.2 Flow between parallel plates.

Again, the momentum in and out of the system due to fluid motion are equal. We are left with

$$(LW)(\tau_{yx})|_y - (LW)(\tau_{yx})|_{y+\Delta y} + (P_0 - P_L)\Delta y W = 0. \qquad (2.16)$$

Dividing through by $LW \Delta y$ and letting Δy approach zero, we develop the differential equation

$$\frac{d(\tau_{yx})}{dy} = \frac{(P_0 - P_L)}{L}. \qquad (2.17)$$

The boundary conditions are described at the centerline ($y = 0$) and at the solid wall ($y = \delta$) as follows:

$$\begin{array}{llll} \text{B.C. 1} & \text{at } y = 0, & \tau_{yx} = 0; \\ \text{B.C. 2} & \text{at } y = \delta, & v_x = 0. \end{array} \qquad (2.18)$$

It is left as an exercise for the reader to show that the shear stress distribution is given by

$$\tau_{yx} = \frac{(P_0 - P_L)}{L} y, \tag{2.19}$$

and the velocity distribution (for a Newtonian fluid) by

$$v_x = \frac{1}{2\eta} (\delta^2 - y^2) \left[\frac{(P_0 - P_L)}{L} \cdot + \rho g \quad + \frac{P_s - P_L}{L} \right] \tag{2.20}$$

We determine other characteristics of the system by the method shown in Section 2.2. These are:

i) The maximum velocity

$$V_x{}^{max} = \frac{1}{2\eta} \delta^2 \frac{(P_0 - P_L)}{L}. \tag{2.21}$$

ii) The average velocity

$$\bar{v}_x = \frac{1}{\delta} \int_0^\delta v_x \, dy = \frac{\delta^2}{3\eta} \frac{(P_0 - P_L)}{L}. \tag{2.22}$$

iii) The volume flow rate

$$Q = \frac{2}{3} \frac{W \delta^3}{\eta} \left[\frac{P_0 - P_L}{L} + \rho g \right] \tag{2.23}$$

On looking back through this example, we note that in this instance the fluid flows because of the pressure drop $(P_0 - P_L)$. For horizontal flow, such a pressure drop would be necessary to make the fluid flow, in contrast to the flow down an inclined plane (Section 2.2) on which gravity exerts the necessary force for fluid motion.

2.4 FLOW THROUGH A CIRCULAR TUBE

Consider the fully developed flow of a fluid in a long tube of length L and radius R; we specify fully developed flow so that end effects are negligible. Since we are dealing with a pipe, it is convenient to work with cylindrical coordinates. Therefore the shell in Fig. 2.3 is cylindrical, of thickness Δr and length L.

Rate of momentum in
across surface at r $(2\pi r L \tau_{rz})|_r$
(due to viscosity)

Note that here we include the area factor $(2\pi r L)$ in parentheses. This is because the area as well as the shear stress is a function of r.

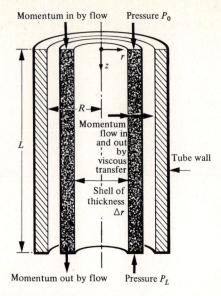

Fig. 2.3 Cylindrical shell chosen for momentum balance in tubes.

Rate of momentum out across surface at $r + \Delta r$ (due to viscosity)	$(2\pi r L \tau_{rz})\vert_{r+\Delta r}$

Since we are considering fully developed flow, the momentum fluxes due to flow are equal; hence these terms are omitted.

Gravity force acting on the cylindrical shell	$(2\pi r \, \Delta r L)\rho g$
Pressure force acting on surface at $z = 0$	$(2\pi r \, \Delta r)P_0$
Pressure force acting on surface at $z = L$	$-(2\pi r \, \Delta r)P_L$

We now add up the contributions to the momentum balance:

$$(2\pi r L \tau_{rz})\vert_r - (2\pi r L \tau_{rz})\vert_{r+\Delta r} + 2\pi r \, \Delta r L \rho g + 2\pi r \, \Delta r (P_0 - P_L) = 0. \quad (2.24)$$

Note that all terms contain the factor r; however, since r is a variable, it should not be used as a common divisor. By dividing through by $2\pi L \, \Delta r$ and taking the limit as Δr goes to zero, we develop the differential equation

$$\frac{d}{dr}(r\tau_{rz}) = \left(\frac{P_0 - P_L}{L} + \rho g\right)r. \quad (2.25)$$

Integration yields

$$\tau_{rz} = \left(\frac{P_0 - P_L}{L} + \rho g\right)\frac{r}{2} + \frac{C_1}{r}. \quad (2.26)$$

At $r = 0$, the velocity gradient (hence, the shear stress) equals zero; this can be realized because of the symmetry of flow.

Thus for this case,

$$\text{B.C. 1} \qquad \text{at } r = 0, \qquad \tau_{rz} = 0. \tag{2.27}$$

Therefore $C_1 = 0$, and the momentum flux is given by

$$\tau_{rz} = \left(\frac{P_0 - P_L}{L} + \rho g\right)\frac{r}{2}. \tag{2.28}$$

Substituting Newton's law of viscosity

$$\tau_{rz} = -\eta\frac{dv_z}{dr}, \tag{2.29}$$

and noting

$$\text{B.C. 2} \qquad \text{at } r = R, \qquad v_z = 0, \tag{2.30}$$

we obtain the solution for the velocity distribution:

$$v_z = \left(\frac{P_0 - P_L}{L} + \rho g\right)\left(\frac{R^2}{4\eta}\right)\left[1 - \left(\frac{r}{R}\right)^2\right]. \tag{2.31}$$

As before:

i) The maximum velocity is at $r = 0$, and is given by

$$V_z^{\,max} = \left(\frac{P_0 - P_L}{L} + \rho g\right)\frac{R^2}{4\eta}. \tag{2.32}$$

ii) The average velocity is

$$\bar{v}_z = \frac{1}{\pi R^2}\int_0^{2\pi}\int_0^R v_z r\, dr\, d\theta = \left(\frac{P_0 - P_L}{L} + \rho g\right)\frac{R^2}{8\eta}. \tag{2.33}$$

iii) Volume flow rate is

o if horizontal pipe

$$Q = \left(\frac{P_0 - P_L}{L} + \rho g\right)\left(\frac{\pi R^4}{8\eta}\right). \tag{2.34}$$

This latter result, which is commonly referred to as the *Hagen-Poiseuille law*, is valid for laminar steady-state flow of incompressible fluids in tubes having sufficient length to make end effects negligible. An entrance length given by $L_e = 0.035\, D\text{Re}$ is required before we can establish fully developed parabolic velocity distribution.

Example 2.2 Water at 60°F flows through a horizontal tube of diameter 1/16 in with a pressure drop of 0.04 psi/ft. Find the mass flow rate through the tube.

Solution. In this situation, the force of gravity does not act on the fluid in the direction of flow, so according to Eq. (2.34) the volume flow rate is

$$Q = \left(\frac{P_0 - P_L}{L}\right)\frac{\pi R^4}{8\eta}.$$

Substituting in values, we obtain the following terms:

$$\frac{P_0 - P_L}{L} = \frac{0.04 \, lb_f}{ft \, in^2}\left|\frac{144 \, in^2}{ft^2}\right. = 5.76 \, lb_f/ft^3,$$

$$\eta = \frac{1.14 \, cP \left|2.09 \times 10^{-5} \, lb_f \, sec/ft^2\right.}{1 \, cP} = 2.38 \times 10^{-5} \, lb_f \, sec/ft^2,$$

$$R = \frac{1}{(2)(16)(12)} \, ft = \frac{1}{384} \, ft.$$

Therefore

$$Q = (5.76)\left(\frac{\pi}{8}\right)\left(\frac{1}{384}\right)^4\left(\frac{1}{2.38 \times 10^{-5}}\right) = 4.31 \times 10^{-6} \, ft^3/sec.$$

Thus we see that the mass flow rate

$$\rho Q = (62.4 \, lb_m/ft^3)(4.31 \times 10^{-6} \, ft^3/sec)$$

$$= 2.74 \times 10^{-4} \, lb_m/sec,$$

or

$$\rho Q = 0.980 \, lb_m/h.$$

We should then check if the flow is laminar, by evaluating the Reynolds number. As mentioned in Chapter 1, the criterion is Re < 2100:

$$Re = \frac{D\bar{V}}{\nu} = \frac{D\bar{V}\rho}{\eta}.$$

Also

$$\rho\bar{V} = \frac{\rho Q}{\pi D^2/4},$$

so that the Reynolds number may be written in the alternative form

$$Re = \frac{D\rho Q}{(\pi D^2/4)\eta} = \frac{4\rho Q}{\pi D\eta},$$

$$\therefore \qquad Re\left(\begin{array}{c}\text{consistent units} \\ \text{must be used}\end{array}\right) = \frac{(4)(62.4)(4.31 \times 10^{-6} \times 3600)}{(\pi)(1/(12 \times 16))(1.14 \times 2.42)}.$$

$$= 86.$$

Since Re < 2100, the flow is laminar.

2.5 GENERAL MOMENTUM EQUATIONS

In the previous sections of this chapter, we determined velocity distributions for some simple flow systems by applying differential momentum balances. The balances for these systems served to illustrate the application of the principle of conservation of momentum. In general, when dealing with isothermal fluid systems which do not involve changes in compositions, we can solve problems by starting with general expressions. This method is better than developing formulations peculiar to the specific problem at hand. The general momentum balance is called the *equation of motion* or the *Navier-Stokes' equation*; in addition the *equation of continuity* is frequently used in conjunction with the equation of motion.

The continuity equation is developed simply by applying the law of conservation of mass to a small volume element within a flowing fluid. The principle of conservation of mass is quite simple to apply and we assume that the reader has used it in developing material balances. We develop the equation of motion by applying the law of conservation of momentum which, in its general form, is an extension of Eq. (2.1). With the aid of these two equations, we can mathematically describe the problems encountered in the previous section, as well as more complicated problems. However, as we shall see, these expressions are rather cumbersome, and exact solutions can be found only in very limited cases. Hence these equations are used primarily as starting points for solving problems. The equations of continuity and motion are simplified to fit the problem at hand. Although theoretically these equations are valid for both laminar and turbulent flows, in practice they are applied only to laminar flow.

2.5.1 Equation of continuity

Consider the stationary volume element within a fluid moving with a velocity having the components v_x, v_y, and v_z, as shown in Fig. 2.4. We begin with the basic representation of the conservation of mass:

$$\begin{pmatrix} \text{rate of mass} \\ \text{accumulation} \end{pmatrix} = \begin{pmatrix} \text{rate of} \\ \text{mass in} \end{pmatrix} - \begin{pmatrix} \text{rate of} \\ \text{mass out} \end{pmatrix}. \qquad (2.35)$$

First, look at the faces perpendicular to the x-axis. The volume flow rate of fluid in across the face at x is simply the product of the velocity (x-component) and the cross-sectional area, yielding $\Delta y\, \Delta z v_x|_x$. The rate of mass in through the face at x is then $\Delta y\, \Delta z (\rho v_x)|_x$. Similarly, the rate of mass out through the face at $x + \Delta x$ is $\Delta y\, \Delta z (\rho v_x)|_{x+\Delta x}$. We may write analogous expressions for the other two pairs of faces, and then enter all the terms that account for the fluid entering and leaving the system into the mass balance, and leave the accumulation term to be developed.

The *accumulation* is the rate of change of mass within the control volume

$$\Delta x\, \Delta y\, \Delta z \frac{\partial \rho}{\partial t}.$$

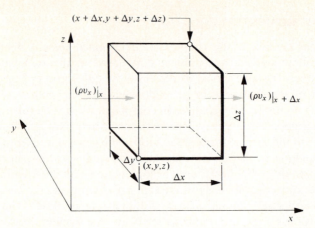

Fig. 2.4 Volume element fixed in space with fluid flowing through it.

The mass balance thus becomes

$$\Delta x\, \Delta y\, \Delta z\, \frac{\partial \rho}{\partial t} = \Delta y\, \Delta z[\rho v_x|_x - \rho v_x|_{x+\Delta x}]$$

$$+ \Delta x\, \Delta z[\rho v_y|_y - \rho v_y|_{y+\Delta y}]$$

$$+ \Delta x\, \Delta y[\rho v_z|_z - \rho v_z|_{z+\Delta z}]. \qquad (2.36)$$

Then, dividing through by $\Delta x\, \Delta y\, \Delta z$, and taking the limit as these dimensions approach zero, we get the *equation of continuity*:

$$\frac{\partial \rho}{\partial t} = -\left(\frac{\partial}{\partial x}\, \rho v_x + \frac{\partial}{\partial y}\, \rho v_y + \frac{\partial}{\partial z}\, \rho v_z \right). \qquad (2.37)$$

A very important form of Eq. (2.37) is the form that applies to a fluid of constant density. For this case, which frequently occurs in engineering problems, the continuity equation reduces to

$$\frac{\partial v_x}{\partial x} + \frac{\partial v_y}{\partial y} + \frac{\partial v_z}{\partial z} = 0. \qquad (2.38)$$

2.5.2 Conservation of momentum

When Eq. (2.1) is extended to include unsteady-state systems, the momentum balance takes the form:

$$\begin{pmatrix} \text{rate of} \\ \text{momentum} \\ \text{accumulation} \end{pmatrix} = \begin{pmatrix} \text{rate of} \\ \text{momentum} \\ \text{in} \end{pmatrix} - \begin{pmatrix} \text{rate of} \\ \text{momentum} \\ \text{out} \end{pmatrix} + \begin{pmatrix} \text{sum of} \\ \text{forces acting} \\ \text{on system} \end{pmatrix}. \qquad (2.39)$$

For simplicity, we begin by considering only the x-component of each term in Eq. (2.39); the y- and z-components may be handled in the same manner.

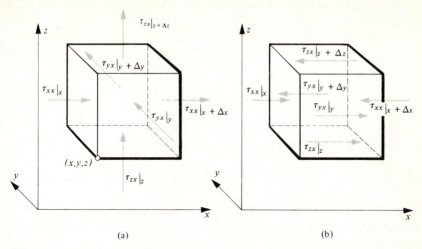

(a)   (b)

Fig. 2.5 Momentum transport (x-component) due to viscosity into the volume element. (a) Directions of viscous momentum transport. (b) Directions of forces.

Figure 2.5(a) shows the x-components of τ as if they were made up of viscous momentum fluxes rather than shear stresses. On the other hand Fig. 2.5(b) shows the x-components of τ as stresses. Note the appearance of τ_{xx}, which by the scheme of subscripts represents the transport of x-momentum in the x-direction. Alternatively, we view τ_{xx} as the x-directed *normal* stress on the x-face, in contrast to τ_{yx} which we view as the x-directed *shear* stress on the y-face.

Let us now develop the terms that enter into Eq. (2.39). First, the net rate at which the x-component of the *convective* momentum enters the unit volume, is

$$\Delta y\, \Delta z(\rho v_x v_x|_x - \rho v_x v_x|_{x+\Delta x}) + \Delta x\, \Delta z(\rho v_y v_x|_y - \rho v_y v_x|_{y+\Delta y})$$

$$+ \Delta x\, \Delta y(\rho v_z v_x|_z - v_z v_x|_{z+\Delta z}). \tag{2.40}$$

Similarly, the net rate of *viscous* momentum flow into the unit volume across the six faces is

$$\Delta y\, \Delta z(\tau_{xx}|_x - \tau_{xx}|_{x+\Delta x}) + \Delta x\, \Delta z(\tau_{yx}|_y - \tau_{yx}|_{y+\Delta y}) + \Delta x\, \Delta y(\tau_{zx}|_z - \tau_{zx}|_{z+\Delta z}). \tag{2.41}$$

The reader who has not come in contact with this development before might find a brief explanation of the meaning of $\rho v_y v_x$ and $\rho v_z v_x$ useful. Remember that we are applying the law of conservation of momentum to the x-component of momentum. Thus v_x represents the x-velocity, and the rate at which mass enters the system through the y-face is given by $\Delta x\, \Delta z \rho v_y|_y$. Hence the rate at which x-momentum enters through the y-face is simply the product of mass-flow rate and velocity:

$$\Delta x\, \Delta z \rho v_y v_x|_y.$$

In most cases, the forces acting on the system are those arising from the pressure P and the gravitational force per unit mass g. In the x-direction, these forces are

$$\Delta y \, \Delta z (P|_x - P|_{x+\Delta x}),\tag{2.42}$$

and

$$\rho g_x \, \Delta x \, \Delta y \, \Delta z,\tag{2.43}$$

respectively. Here g_x is the x-component of the gravitational force. Finally, the rate of accumulation of x-momentum within the element is

$$\Delta x \, \Delta y \, \Delta z \left(\frac{\partial}{\partial t}\, \rho v_x\right).\tag{2.44}$$

Entering Eqs. (2.40)–(2.44) into the momentum balance, dividing through by $\Delta x \, \Delta y \, \Delta z$, and taking the limit as all three approach zero, we obtain the x-component of the momentum conservation equation:

$$\frac{\partial}{\partial t}\, \rho v_x = -\left(\frac{\partial}{\partial x}\, \rho v_x v_x + \frac{\partial}{\partial y}\, \rho v_y v_x + \frac{\partial}{\partial z}\, \rho v_z v_x\right)$$

$$-\left(\frac{\partial}{\partial x}\, \tau_{xx} + \frac{\partial}{\partial y}\, \tau_{yx} + \frac{\partial}{\partial z}\, \tau_{zx}\right)$$

$$-\frac{\partial P}{\partial x} + \rho g_x.\tag{2.45}$$

The y- and z-components, which we obtain in a similar manner, are

$$\frac{\partial}{\partial t}\, \rho v_y = -\left(\frac{\partial}{\partial x}\, \rho v_x v_y + \frac{\partial}{\partial y}\, \rho v_y v_y + \frac{\partial}{\partial z}\, \rho v_z v_y\right)$$

$$-\left(\frac{\partial}{\partial x}\, \tau_{xy} + \frac{\partial}{\partial y}\, \tau_{yy} + \frac{\partial}{\partial z}\, \tau_{zy}\right)$$

$$-\frac{\partial P}{\partial y} + \rho g_y,\tag{2.46}$$

and

$$\frac{\partial}{\partial t}\, \rho v_z = -\left(\frac{\partial}{\partial x}\, \rho v_x v_z + \frac{\partial}{\partial y}\, \rho v_y v_z + \frac{\partial}{\partial z}\, \rho v_z v_z\right)$$

$$-\left(\frac{\partial}{\partial x}\, \tau_{xz} + \frac{\partial}{\partial y}\, \tau_{yz} + \frac{\partial}{\partial z}\, \tau_{zz}\right) -\frac{\partial P}{\partial z} + \rho g_z.\tag{2.47}$$

To describe the general case, all three Equations (2.45), (2.46), and (2.47 are needed. Vector notation can reduce these to one equation which is just as meaningful as all three. The quantities ρv_x, ρv_y, and ρv_z are the components of the mass velocity ρv; similarly g_x, g_y, and g_z are the components of g. Vectorial representation of a velocity and an acceleration is familiar to most readers. However, the terms $\partial P/\partial x$, $\partial P/\partial y$, and $\partial P/\partial z$ all represent pressure *gradients*. By itself, pressure is a scalar quantity, but the gradient of pressure is a vector, denoted, in general by ∇P (sometimes written grad P).

As a simple example to illustrate the necessity of thinking about pressure gradients as vectors, take a tank of water. At any level L, measured from the surface of the water, the pressure is $\rho g L$. If the z-coordinate is that perpendicular to the surface of the water, then the pressure difference in the water *in the z-direction* is $\rho g L$. Note that we have to specify the *direction* in which the difference in pressure is measured; it would not be enough to say simply that the difference in pressure is $\rho g L$ because one could compare the pressure at two different points at the same level; hence the difference would be zero. To be specific, the pressure gradient at the level L must be denoted by $\partial P/\partial x = 0$, $\partial P/\partial y = 0$, and $\partial P/\partial z = \rho g$; being more general, the pressure gradient is ∇P. Therefore

$$\nabla P = \frac{\partial}{\partial x}P + \frac{\partial}{\partial y}P + \frac{\partial}{\partial z}P,$$

and ∇ can be thought to be an operator, such that

$$\nabla = \frac{\partial}{\partial x} + \frac{\partial}{\partial y} + \frac{\partial}{\partial z}.$$

The terms $\rho v_x v_x$, $\rho v_x v_y$, $\rho v_x v_z$, $\rho v_y v_z$, etc., are the nine components of the convective momentum flux ρvv, which is the *dyadic product* of ρv and v. Also τ_{xx}, τ_{xy}, etc., are the nine components of τ.

The vector equation representing Eqs. (2.45)–(2.47) is finally written:

$$\frac{\partial}{\partial t}\rho v = -[\nabla \cdot \rho vv] - \nabla P - [\nabla \cdot \tau] + \rho g. \tag{2.48}$$

Note here that ∇P is the product of a vector (∇) and a scalar (P), yielding a vector. To interpret the mathematical nature of $\nabla \cdot \rho vv$ and $\nabla \cdot \tau$ in physical terms is more difficult. However, for sufficient understanding of this text it is enough if the reader accepts them as mathematical shorthands of the appropriate terms in Eqs. (2.45)–(2.47).

So far we have developed a general expression, namely, Eq. (2.48), for the law of conservation of momentum. However, in order to use this equation for the determination of velocity distributions, it is necessary to insert expressions for the various stresses in terms of velocity gradients and fluid properties. The following equations are presented without proof because the arguments involved are quite lengthy. For *Newtonian fluids*, the nine components of τ are written as follows.[1]

---

[1] V. L. Streeter, *Fluid Dynamics*, McGraw-Hill, New York, 1948, Chapter 10.

$$\text{Normal stresses} \begin{cases} \tau_{xx} = -2\eta \dfrac{\partial v_x}{\partial x} + \tfrac{2}{3}\eta(\mathbf{V} \cdot \mathbf{v}) & (2.49) \\[3mm] \tau_{yy} = -2\eta \dfrac{\partial v_y}{\partial y} + \tfrac{2}{3}\eta(\mathbf{V} \cdot \mathbf{v}) & (2.50) \\[3mm] \tau_{zz} = -2\eta \dfrac{\partial v_z}{\partial z} + \tfrac{2}{3}\eta(\mathbf{V} \cdot \mathbf{v}) & (2.51) \end{cases}$$

$$\text{Shear stresses} \begin{cases} \tau_{xy} = \tau_{yx} = -\eta\left(\dfrac{\partial v_x}{\partial y} + \dfrac{\partial v_y}{\partial x}\right) & (2.52) \\[3mm] \tau_{yz} = \tau_{zy} = -\eta\left(\dfrac{\partial v_y}{\partial z} + \dfrac{\partial v_z}{\partial y}\right) & (2.53) \\[3mm] \tau_{zx} = \tau_{xz} = -\eta\left(\dfrac{\partial v_z}{\partial x} + \dfrac{\partial v_x}{\partial z}\right) & (2.54) \end{cases}$$

These equations constitute a more general statement of Newton's law of viscosity than that given in Eq. (1.2), and apply to complex flow situations. When the fluid flows between two parallel plates in the x-direction so that v_x is a function of y alone, where the y-direction is perpendicular to the plates' surfaces (Fig. 1.4), then Eqs. (2.49)–(2.54) yield

$$\tau_{xx} = \tau_{yy} = \tau_{zz} = \tau_{yz} = \tau_{xz} = 0 \qquad \text{and} \qquad \tau_{yx} = -\eta(\partial v_x/\partial y),$$

which is the same as the simple relationship previously used to describe Newton's law of viscosity. Also in many other problems of physical significance in which v_x is recognized as a function of both y and x, we find that $\partial v_x/\partial y \gg \partial v_x/\partial x$, and the simple rate Eq. (1.2) can be used for τ_{yx} as an example with a high degree of accuracy rather than Eq. (2.52).

2.5.3 Navier-Stokes' equation, constant ρ and η

The continuity equation for constant density is given by Eq. (2.38) or in vector notation,

$$\mathbf{V} \cdot \mathbf{v} = 0. \tag{2.55}$$

Regarding the conservation of momentum, we can write Eqs. (2.45)–(2.47) with constant ρ and η:*

$$\rho\left[\frac{\partial v_x}{\partial t} + v_x\frac{\partial v_x}{\partial x} + v_y\frac{\partial v_x}{\partial y} + v_z\frac{\partial v_x}{\partial z}\right] = -\frac{\partial P}{\partial x} + \eta\left[\frac{\partial^2 v_x}{\partial x^2} + \frac{\partial^2 v_x}{\partial y^2} + \frac{\partial^2 v_x}{\partial z^2}\right] + \rho g_x, \quad (2.56)$$

$$\rho\left[\frac{\partial v_y}{\partial t} + v_y\frac{\partial v_y}{\partial y} + v_x\frac{\partial v_y}{\partial x} + v_z\frac{\partial v_y}{\partial z}\right] = -\frac{\partial P}{\partial y} + \eta\left[\frac{\partial^2 v_y}{\partial x^2} + \frac{\partial^2 v_y}{\partial y^2} + \frac{\partial^2 v_y}{\partial z^2}\right] + \rho g_y, \quad (2.57)$$

---

* This development is the subject of Problem 2.7.

$$\rho\left[\frac{\partial v_z}{\partial t} + v_z\frac{\partial v_z}{\partial z} + v_x\frac{\partial v_z}{\partial x} + v_y\frac{\partial v_z}{\partial y}\right] = -\frac{\partial P}{\partial z} + \eta\left[\frac{\partial^2 v_z}{\partial x^2} + \frac{\partial^2 v_z}{\partial y^2} + \frac{\partial^2 v_z}{\partial z^2}\right] + \rho g_z. \quad (2.58)$$

The bracketed terms on the left side of these equations merit attention. Consider a control volume of fluid moving in space with no mass flow across its surface. The change in the x-component of its velocity with time and position is given by

$$\Delta v_x = \frac{\partial v_x}{\partial t}\Delta t + \frac{\partial v_x}{\partial x}\Delta x + \frac{\partial v_x}{\partial y}\Delta y + \frac{\partial v_x}{\partial z}\Delta z, \quad (2.59)$$

and since the x-component of acceleration is defined as

$$a_x = \lim_{\Delta t\to 0}\frac{\Delta v_x}{\Delta t} = \lim_{\Delta t\to 0}\left\{\frac{\partial v_x}{\partial t}\frac{\Delta t}{\Delta t} + \frac{\partial v_x}{\partial x}\frac{\Delta x}{\Delta t} + \frac{\partial v_x}{\partial y}\frac{\Delta y}{\Delta t} + \frac{\partial v_x}{\partial z}\frac{\Delta z}{\Delta t}\right\}, \quad (2.60)$$

we obtain

$$a_x = \frac{\partial v_x}{\partial t} + v_x\frac{\partial v_x}{\partial x} + v_y\frac{\partial v_x}{\partial y} + v_z\frac{\partial v_x}{\partial z} = \frac{Dv_x}{Dt}. \quad (2.61)$$

This is the acceleration one would feel if riding with the control volume of fluid. We also refer to this time derivative of velocity, Dv_x/Dt, as the *substantial* derivative. Analogous expressions exist for the y- and z-directions. In general, one notation can represent all three substantial derivatives, so that Eqs. (2.56)–(2.58) become

$$\rho\frac{Dv}{Dt} = -\nabla P + \eta\nabla^2 v + \rho g. \quad (2.62)$$

Equation (2.62), or Eqs. (2.56)–(2.58) which taken together represent the expansion of Eq. (2.62), is often referred to as the *Navier-Stokes' equation*. In the form of Eq. (2.62), we can recognize it as a statement of Newton's law in the form *mass* (ρ) × *acceleration* (Dv/Dt) equals the *sum of forces*, namely, the pressure force $(-\nabla P)$, the viscous force $(\eta\nabla^2 v)$, and the gravity or body force ρg.

2.6 THE CONSERVATION OF MOMENTUM EQUATION IN CURVILINEAR COORDINATES

In many instances rectangular coordinates are not useful for analyzing problems. For example, in the Hagen-Poiseuille problem discussed in Section 2.4, we described the axial velocity v_z as a function of only a single variable r by employing cylindrical coordinates. If rectangular coordinates had been used instead, v_z would have been a very complicated function of x and y. Similarly, it would have been difficult to describe and apply the boundary condition at the tube wall.

The equations of continuity and motion in Section 2.5 have been given in rectangular coordinates; spherical or cylindrical coordinates are presented in Tables 2.1–2.7.

Table 2.1 The continuity equation in different coordinate systems*

Rectangular coordinates (x, y, z):

$$\frac{\partial \rho}{\partial t} + \frac{\partial}{\partial x}(\rho v_x) + \frac{\partial}{\partial y}(\rho v_y) + \frac{\partial}{\partial z}(\rho v_z) = 0 \tag{A}$$

Cylindrical coordinates (r, θ, z):

$$\frac{\partial \rho}{\partial t} + \frac{1}{r}\frac{\partial}{\partial r}(\rho r v_r) + \frac{1}{r}\frac{\partial}{\partial \theta}(\rho v_\theta) + \frac{\partial}{\partial z}(\rho v_z) = 0 \tag{B}$$

Spherical coordinates (r, θ, ϕ):

$$\frac{\partial \rho}{\partial t} + \frac{1}{r^2}\frac{\partial}{\partial r}(\rho r^2 v_r) + \frac{1}{r\sin\theta}\frac{\partial}{\partial \theta}(\rho v_\theta \sin\theta) + \frac{1}{r\sin\theta}\frac{\partial}{\partial \phi}(\rho v_\phi) = 0 \tag{C}$$

* Tables 2.1–2.7 are from R. B. Bird, W. E. Stewart, and E. N. Lightfoot, *Transport Phenomena*, Wiley, New York, 1960, pages 83–91. Reprinted by permission.

Table 2.2 The conservation of momentum in rectangular coordinates (x, y, z)

In terms of τ:

x-component $\rho\left(\dfrac{\partial v_x}{\partial t} + v_x\dfrac{\partial v_x}{\partial x} + v_y\dfrac{\partial v_x}{\partial y} + v_z\dfrac{\partial v_x}{\partial z}\right) = -\dfrac{\partial P}{\partial x}$

$$-\left(\dfrac{\partial \tau_{xx}}{\partial x} + \dfrac{\partial \tau_{yx}}{\partial y} + \dfrac{\partial \tau_{zx}}{\partial z}\right) + \rho g_x \qquad \text{(A)}$$

y-component $\rho\left(\dfrac{\partial v_y}{\partial t} + v_x\dfrac{\partial v_y}{\partial x} + v_y\dfrac{\partial v_y}{\partial y} + v_z\dfrac{\partial v_y}{\partial z}\right) = -\dfrac{\partial P}{\partial y}$

$$-\left(\dfrac{\partial \tau_{xy}}{\partial x} + \dfrac{\partial \tau_{yy}}{\partial y} + \dfrac{\partial \tau_{zy}}{\partial z}\right) + \rho g_y \qquad \text{(B)}$$

z-component $\rho\left(\dfrac{\partial v_z}{\partial t} + v_x\dfrac{\partial v_z}{\partial x} + v_y\dfrac{\partial v_z}{\partial y} + v_z\dfrac{\partial v_z}{\partial z}\right) = -\dfrac{\partial P}{\partial z}$

$$-\left(\dfrac{\partial \tau_{xz}}{\partial x} + \dfrac{\partial \tau_{yz}}{\partial y} + \dfrac{\partial \tau_{zz}}{\partial z}\right) + \rho g_z \qquad \text{(C)}$$

In terms of velocity gradients for a Newtonian fluid with constant ρ and η:

x-component $\rho\left(\dfrac{\partial v_x}{\partial t} + v_x\dfrac{\partial v_x}{\partial x} + v_y\dfrac{\partial v_x}{\partial y} + v_z\dfrac{\partial v_x}{\partial z}\right) = -\dfrac{\partial P}{\partial x}$

$$+\eta\left(\dfrac{\partial^2 v_x}{\partial x^2} + \dfrac{\partial^2 v_x}{\partial y^2} + \dfrac{\partial^2 v_x}{\partial z^2}\right) + \rho g_x \qquad \text{(D)}$$

y-component $\rho\left(\dfrac{\partial v_y}{\partial t} + v_x\dfrac{\partial v_y}{\partial x} + v_y\dfrac{\partial v_y}{\partial y} + v_z\dfrac{\partial v_y}{\partial z}\right) = -\dfrac{\partial P}{\partial y}$

$$+\eta\left(\dfrac{\partial^2 v_y}{\partial x^2} + \dfrac{\partial^2 v_y}{\partial y^2} + \dfrac{\partial^2 v_y}{\partial z^2}\right) + \rho g_y \qquad \text{(E)}$$

z-component $\rho\left(\dfrac{\partial v_z}{\partial t} + v_x\dfrac{\partial v_z}{\partial x} + v_y\dfrac{\partial v_z}{\partial y} + v_z\dfrac{\partial v_z}{\partial z}\right) = -\dfrac{\partial P}{\partial z}$

$$+\eta\left(\dfrac{\partial^2 v_z}{\partial x^2} + \dfrac{\partial^2 v_z}{\partial y^2} + \dfrac{\partial^2 v_z}{\partial z^2}\right) + \rho g_z \qquad \text{(F)}$$

Table 2.3 The conservation of momentum in cylindrical coordinates (r, θ, z)

In terms of τ:

r-component* $\quad \rho\left(\dfrac{\partial v_r}{\partial t} + v_r\dfrac{\partial v_r}{\partial r} + \dfrac{v_\theta}{r}\dfrac{\partial v_r}{\partial \theta} - \dfrac{v_\theta^2}{r} + v_z\dfrac{\partial v_r}{\partial z}\right) = -\dfrac{\partial P}{\partial r}$

$$-\left(\dfrac{1}{r}\dfrac{\partial}{\partial r}(r\tau_{rr}) + \dfrac{1}{r}\dfrac{\partial \tau_{r\theta}}{\partial \theta} - \dfrac{\tau_{\theta\theta}}{r} + \dfrac{\partial \tau_{rz}}{\partial z}\right) + \rho g_r \qquad \text{(A)}$$

θ-component $\quad \rho\left(\dfrac{\partial v_\theta}{\partial t} + v_r\dfrac{\partial v_\theta}{\partial r} + \dfrac{v_\theta}{r}\dfrac{\partial v_\theta}{\partial \theta} + \dfrac{v_r v_\theta}{r} + v_z\dfrac{\partial v_\theta}{\partial z}\right) = -\dfrac{1}{r}\dfrac{\partial P}{\partial \theta}$

$$-\left(\dfrac{1}{r^2}\dfrac{\partial}{\partial r}(r^2\tau_{r\theta}) + \dfrac{1}{r}\dfrac{\partial \tau_{\theta\theta}}{\partial \theta} + \dfrac{\partial \tau_{\theta z}}{\partial z}\right) + \rho g_\theta \qquad \text{(B)}$$

z-component $\quad \rho\left(\dfrac{\partial v_z}{\partial t} + v_r\dfrac{\partial v_z}{\partial r} + \dfrac{v_\theta}{r}\dfrac{\partial v_z}{\partial \theta} + v_z\dfrac{\partial v_z}{\partial z}\right) = -\dfrac{\partial P}{\partial z}$

$$-\left(\dfrac{1}{r}\dfrac{\partial}{\partial r}(r\tau_{rz}) + \dfrac{1}{r}\dfrac{\partial \tau_{\theta z}}{\partial \theta} + \dfrac{\partial \tau_{zz}}{\partial z}\right) + \rho g_z \qquad \text{(C)}$$

In terms of velocity gradients for a Newtonian fluid with constant ρ and η:

r-component* $\quad \rho\left(\dfrac{\partial v_r}{\partial t} + v_r\dfrac{\partial v_r}{\partial r} + \dfrac{v_\theta}{r}\dfrac{\partial v_r}{\partial \theta} - \dfrac{v_\theta^2}{r} + v_z\dfrac{\partial v_r}{\partial z}\right) = -\dfrac{\partial P}{\partial r}$

$$+ \eta\left[\dfrac{\partial}{\partial r}\left(\dfrac{1}{r}\dfrac{\partial}{\partial r}(rv_r)\right) + \dfrac{1}{r^2}\dfrac{\partial^2 v_r}{\partial \theta^2} - \dfrac{2}{r^2}\dfrac{\partial v_\theta}{\partial \theta} + \dfrac{\partial^2 v_r}{\partial z^2}\right] + \rho g_r \qquad \text{(D)}$$

θ-component $\quad \rho\left(\dfrac{\partial v_\theta}{\partial t} + v_r\dfrac{\partial v_\theta}{\partial r} + \dfrac{v_\theta}{r}\dfrac{\partial v_\theta}{\partial \theta} + \dfrac{v_r v_\theta}{r} + v_z\dfrac{\partial v_\theta}{\partial z}\right) = -\dfrac{1}{r}\dfrac{\partial P}{\partial \theta}$

$$+ \eta\left[\dfrac{\partial}{\partial r}\left(\dfrac{1}{r}\dfrac{\partial}{\partial r}(rv_\theta)\right) + \dfrac{1}{r^2}\dfrac{\partial^2 v_\theta}{\partial \theta^2} + \dfrac{2}{r^2}\dfrac{\partial v_r}{\partial \theta} + \dfrac{\partial^2 v_\theta}{\partial z^2}\right] + \rho g_\theta \qquad \text{(E)}$$

z-component $\quad \rho\left(\dfrac{\partial v_z}{\partial t} + v_r\dfrac{\partial v_z}{\partial r} + \dfrac{v_\theta}{r}\dfrac{\partial v_z}{\partial \theta} + v_z\dfrac{\partial v_z}{\partial z}\right) = -\dfrac{\partial P}{\partial z}$

$$+ \eta\left[\dfrac{1}{r}\dfrac{\partial}{\partial r}\left(r\dfrac{\partial v_z}{\partial r}\right) + \dfrac{1}{r^2}\dfrac{\partial^2 v_z}{\partial \theta^2} + \dfrac{\partial^2 v_z}{\partial z^2}\right] + \rho g_z \qquad \text{(F)}$$

---

* The term $\rho v_\theta^2/r$ is the *centrifugal force*. It gives the effective force in the r-direction resulting from fluid motion in the θ-direction. This term arises automatically on transformation from rectangular to cylindrical coordinates; it does not have to be added on physical grounds.

Table 2.4 The conservation of momentum in spherical coordinates (r, θ, ϕ)

In terms of τ:

r-component $\quad \rho\left(\dfrac{\partial v_r}{\partial t} + v_r\dfrac{\partial v_r}{\partial r} + \dfrac{v_\theta}{r}\dfrac{\partial v_r}{\partial \theta} + \dfrac{v_\phi}{r\sin\theta}\dfrac{\partial v_r}{\partial \phi} - \dfrac{v_\theta^2 + v_\phi^2}{r}\right)$

$$= -\frac{\partial P}{\partial r} - \left(\frac{1}{r^2}\frac{\partial}{\partial r}(r^2\tau_{rr}) + \frac{1}{r\sin\theta}\frac{\partial}{\partial \theta}(\tau_{r\theta}\sin\theta)\right.$$

$$\left. + \frac{1}{r\sin\theta}\frac{\partial \tau_{r\phi}}{\partial \phi} - \frac{\tau_{\theta\theta} + \tau_{\phi\phi}}{r}\right) + \rho g_r \tag{A}$$

θ-component $\quad \rho\left(\dfrac{\partial v_\theta}{\partial t} + v_r\dfrac{\partial v_\theta}{\partial r} + \dfrac{v_\theta}{r}\dfrac{\partial v_\theta}{\partial \theta} + \dfrac{v_\phi}{r\sin\theta}\dfrac{\partial v_\theta}{\partial \phi} + \dfrac{v_r v_\theta}{r} - \dfrac{v_\phi^2\cot\theta}{r}\right)$

$$= -\frac{1}{r}\frac{\partial P}{\partial \theta} - \left(\frac{1}{r^2}\frac{\partial}{\partial r}(r^2\tau_{r\theta}) + \frac{1}{r\sin\theta}\frac{\partial}{\partial \theta}(\tau_{\theta\theta}\sin\theta) + \frac{1}{r\sin\theta}\frac{\partial \tau_{\theta\phi}}{\partial \phi}\right.$$

$$\left. + \frac{\tau_{r\theta}}{r} - \frac{\cot\theta}{r}\tau_{\phi\phi}\right) + \rho g_\theta \tag{B}$$

ϕ-component $\quad \rho\left(\dfrac{\partial v_\phi}{\partial t} + v_r\dfrac{\partial v_\phi}{\partial r} + \dfrac{v_\theta}{r}\dfrac{\partial v_\phi}{\partial \theta} + \dfrac{v_\phi}{r\sin\theta}\dfrac{\partial v_\phi}{\partial \phi} + \dfrac{v_\phi v_r}{r} + \dfrac{v_\theta v_\phi}{r}\cot\theta\right)$

$$= -\frac{1}{r\sin\theta}\frac{\partial P}{\partial \phi} - \left(\frac{1}{r^2}\frac{\partial}{\partial r}(r^2\tau_{r\phi}) + \frac{1}{r}\frac{\partial \tau_{\theta\phi}}{\partial \theta} + \frac{1}{r\sin\theta}\frac{\partial \tau_{\phi\phi}}{\partial \phi}\right.$$

$$\left. + \frac{\tau_{r\phi}}{r} + \frac{2\cot\theta}{r}\tau_{\theta\phi}\right) + \rho g_\phi \tag{C}$$

In terms of velocity gradients for a Newtonian fluid with constant ρ and η:

r-component $\quad \rho\left(\dfrac{\partial v_r}{\partial t} + v_r\dfrac{\partial v_r}{\partial r} + \dfrac{v_\theta}{r}\dfrac{\partial v_r}{\partial \theta} + \dfrac{v_\phi}{r\sin\theta}\dfrac{\partial v_r}{\partial \phi} - \dfrac{v_\theta^2 + v_\phi^2}{r}\right)$

$$= -\frac{\partial P}{\partial r} + \eta\left(\nabla^2 v_r - \frac{2}{r^2}v_r - \frac{2}{r^2}\frac{\partial v_\theta}{\partial \theta} - \frac{2}{r^2}v_\theta\cot\theta\right.$$

$$\left. - \frac{2}{r^2\sin\theta}\frac{\partial v_\phi}{\partial \phi}\right) + \rho g_r \tag{D}$$

θ-component $\quad \rho\left(\dfrac{\partial v_\theta}{\partial t} + v_r\dfrac{\partial v_\theta}{\partial r} + \dfrac{v_\theta}{r}\dfrac{\partial v_\theta}{\partial \theta} + \dfrac{v_\phi}{r\sin\theta}\dfrac{\partial v_\theta}{\partial \phi} + \dfrac{v_r v_\theta}{r} - \dfrac{v_\phi^2\cot\theta}{r}\right)$

$$= -\frac{1}{r}\frac{\partial P}{\partial \theta} + \eta\left(\nabla^2 v_\theta + \frac{2}{r^2}\frac{\partial v_r}{\partial \theta} - \frac{v_\theta}{r^2\sin^2\theta} - \frac{2\cos\theta}{r^2\sin^2\theta}\frac{\partial v_\phi}{\partial \phi}\right) + \rho g_\theta \tag{E}$$

ϕ-component $\quad \rho\left(\dfrac{\partial v_\phi}{\partial t} + v_r\dfrac{\partial v_\phi}{\partial r} + \dfrac{v_\theta}{r}\dfrac{\partial v_\phi}{\partial \theta} + \dfrac{v_\phi}{r\sin\theta}\dfrac{\partial v_\phi}{\partial \phi} + \dfrac{v_\phi v_r}{r} + \dfrac{v_\theta v_\phi}{r}\cot\theta\right)$

$$= -\frac{1}{r\sin\theta}\frac{\partial P}{\partial \phi} + \eta\left(\nabla^2 v_\phi - \frac{v_\phi}{r^2\sin^2\theta} + \frac{2}{r^2\sin\theta}\frac{\partial v_r}{\partial \phi}\right.$$

$$\left. + \frac{2\cos\theta}{r^2\sin^2\theta}\frac{\partial v_\theta}{\partial \phi}\right) + \rho g_\phi \tag{F}$$

Table 2.5 Components of the stress tensor in rectangular coordinates (x, y, z)

$$\tau_{xx} = -\eta \left[2 \frac{\partial v_x}{\partial x} - \tfrac{2}{3}(\nabla \cdot v) \right] \tag{A}$$

$$\tau_{yy} = -\eta \left[2 \frac{\partial v_y}{\partial y} - \tfrac{2}{3}(\nabla \cdot v) \right] \tag{B}$$

$$\tau_{zz} = -\eta \left[2 \frac{\partial v_z}{\partial z} - \tfrac{2}{3}(\nabla \cdot v) \right] \tag{C}$$

$$\tau_{xy} = \tau_{yx} = -\eta \left[\frac{\partial v_x}{\partial y} + \frac{\partial v_y}{\partial x} \right] \tag{D}$$

$$\tau_{yz} = \tau_{zy} = -\eta \left[\frac{\partial v_y}{\partial z} + \frac{\partial v_z}{\partial y} \right] \tag{E}$$

$$\tau_{zx} = \tau_{xz} = -\eta \left[\frac{\partial v_z}{\partial x} + \frac{\partial v_x}{\partial z} \right] \tag{F}$$

$$(\nabla \cdot v) = \frac{\partial v_x}{\partial x} + \frac{\partial v_y}{\partial y} + \frac{\partial v_z}{\partial z} \tag{G}$$

Table 2.6 Components of the stress tensor in cylindrical coordinates (r, θ, z)

$$\tau_{rr} = -\eta \left[2 \frac{\partial v_r}{\partial r} - \tfrac{2}{3}(\nabla \cdot v) \right] \tag{A}$$

$$\tau_{\theta\theta} = -\eta \left[2 \left(\frac{1}{r} \frac{\partial v_\theta}{\partial \theta} + \frac{v_r}{r} \right) - \tfrac{2}{3}(\nabla \cdot v) \right] \tag{B}$$

$$\tau_{zz} = -\eta \left[2 \frac{\partial v_z}{\partial z} - \tfrac{2}{3}(\nabla \cdot v) \right] \tag{C}$$

$$\tau_{r\theta} = \tau_{\theta r} = -\eta \left[r \frac{\partial}{\partial r} \left(\frac{v_\theta}{r} \right) + \frac{1}{r} \frac{\partial v_r}{\partial \theta} \right] \tag{D}$$

$$\tau_{\theta z} = \tau_{z\theta} = -\eta \left[\frac{\partial v_\theta}{\partial z} + \frac{1}{r} \frac{\partial v_z}{\partial \theta} \right] \tag{E}$$

$$\tau_{zr} = \tau_{rz} = -\eta \left[\frac{\partial v_z}{\partial r} + \frac{\partial v_r}{\partial z} \right] \tag{F}$$

$$(\nabla \cdot v) = \frac{1}{r} \frac{\partial}{\partial r}(r v_r) + \frac{1}{r} \frac{\partial v_\theta}{\partial \theta} + \frac{\partial v_z}{\partial z} \tag{G}$$

Table 2.7 Components of the stress tensor in spherical coordinates (r, θ, ϕ)

$$\tau_{rr} = -\eta\left[2\frac{\partial v_r}{\partial r} - \tfrac{2}{3}(\nabla \cdot v)\right] \tag{A}$$

$$\tau_{\theta\theta} = -\eta\left[2\left(\frac{1}{r}\frac{\partial v_\theta}{\partial \theta} + \frac{v_r}{r}\right) - \tfrac{2}{3}(\nabla \cdot v)\right] \tag{B}$$

$$\tau_{\phi\phi} = -\eta\left[2\left(\frac{1}{r\sin\theta}\frac{\partial v_\phi}{\partial \phi} + \frac{v_r}{r} + \frac{v_\theta\cot\theta}{r}\right) - \tfrac{2}{3}(\nabla \cdot v)\right] \tag{C}$$

$$\tau_{r\theta} = \tau_{\theta r} = -\eta\left[r\frac{\partial}{\partial r}\left(\frac{v_\theta}{r}\right) + \frac{1}{r}\frac{\partial v_r}{\partial \theta}\right] \tag{D}$$

$$\tau_{\theta\phi} = \tau_{\phi\theta} = -\eta\left[\frac{\sin\theta}{r}\frac{\partial}{\partial\theta}\left(\frac{v_\phi}{\sin\theta}\right) + \frac{1}{r\sin\theta}\frac{\partial v_\theta}{\partial\phi}\right] \tag{E}$$

$$\tau_{\phi r} = \tau_{r\phi} = -\eta\left[\frac{1}{r\sin\theta}\frac{\partial v_r}{\partial\phi} + r\frac{\partial}{\partial r}\left(\frac{v_\phi}{r}\right)\right] \tag{F}$$

$$(\nabla \cdot v) = \frac{1}{r^2}\frac{\partial}{\partial r}(r^2 v_r) + \frac{1}{r\sin\theta}\frac{\partial}{\partial\theta}(v_\theta\sin\theta) + \frac{1}{r\sin\theta}\frac{\partial v_\phi}{\partial\phi} \tag{G}$$

2.7 APPLICATION OF NAVIER-STOKES' EQUATION

In this section, we show how to set up problems of viscous flow, by selecting the appropriate equation of motion that applies to the problem at hand and by simplifying it to manageable proportions so that it still relates to the given problem, and yet is not oversimplified. We do so by discarding those terms which are zero, and then recognizing those terms which can be neglected. To decide this, is, to a certain extent, a matter of experience, but in most instances even the novice can make intelligent decisions by making an order-of-magnitude estimate. For this purpose, we shall discuss below an order-of-magnitude technique that can be used to arrive at a more simplified, but still relevant, equation of motion.

We also introduce other topics, such as the *boundary layer*, the *integral method* of solving problems, and *drag forces* exerted by fluids on solids.

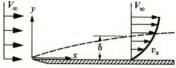

Fig. 2.6 Velocity profile and momentum boundary layer of flow parallel to a flat plate.

2.7.1 Flow over a flat plate—exact solution

Figure 2.6 depicts the velocity profile of a fluid flowing parallel to a flat plate. Before it meets the leading edge of the plate, we assume that the fluid has a uniform velocity V_∞. At any point x downstream from the leading edge of the plate, we observe that the velocity increases from zero at the wall to very near V_∞ at a very short distance δ from the plate. The loci of positions where $v_x/V_\infty = 0.99$, is δ, and it is defined as the *boundary layer*. At the leading edge of the plate ($x = 0$), δ is zero, and it progressively grows as flow proceeds down the plate.

Whenever problems of this type are encountered, namely, in the flow of fluid past a stationary solid, the viscous effects are felt only within the fluid near the solid, that is, $y < \delta$. Of course, this is exactly where the behavior of the fluid should be analyzed, because for $y > \delta$, the happenings from the point of view of this discussion are essentially uneventful, due to the fact that in this region v_x is essentially uniform and constant, being equal to V_∞. Since v_x is uniform and constant for $y > \delta$, Eq. (A) in Table 2.2 reveals that the pressure gradient $\partial P/\partial x$ is zero or, stated differently, pressure everywhere in the bulk stream is uniform. In turn, the pressure within the boundary layer is equal to the pressure in the bulk stream, so that $\partial P/\partial x$ is also zero within the boundary layer. We are now almost ready to pick out the appropriate equation of motion for the flow pattern in Fig. 2.6, but before we do so, let us examine which velocity components are relevant.

As discussed above, v_x is a function of y, and the determination of this functional relationship is indeed a major part of describing the flow and how the fluid and solid surfaces interact. Also note that v_x depends on x. This results from the fact that, as the fluid progresses down the plate, it is retarded more and more by the drag at the plate's surface. Thus $\partial v_x/\partial x$ is not zero, and the equation of continuity for steady two-dimensional flow of fluid with constant ρ and η is

$$\frac{\partial v_x}{\partial x} + \frac{\partial v_y}{\partial y} = 0. \tag{2.63}$$

Thus v_y exists and we should consider both the x- and y-components in Table 2.2.

For the steady-state case with constant density and viscosity, Eqs. (D) and (E) in Table 2.2 reduce to

$$v_x \frac{\partial v_x}{\partial x} + v_y \frac{\partial v_x}{\partial y} = \nu\left(\frac{\partial^2 v_x}{\partial x^2} + \frac{\partial^2 v_x}{\partial y^2}\right), \tag{2.64}$$

and

$$v_x \frac{\partial v_y}{\partial x} + v_y \frac{\partial v_y}{\partial y} = \nu\left(\frac{\partial^2 v_y}{\partial x^2} + \frac{\partial^2 v_y}{\partial y^2}\right) + g_y. \tag{2.65}$$

When we remember that we are primarily interested in the region $y \leq \delta$, at this point it is convenient to define some dimensionless parameters:†

---

† The *Reynolds number* Re_L which was briefly introduced in Chapter 1 reappears here again. Also the *Froude number* Fr_L is introduced. These two dimensionless groups of variables, which so often occur in engineering studies, have been given names in honor of those two early workers in fluid mechanics.

$$u^* = \frac{v_x}{V_\infty}, \qquad x^* = \frac{x}{L}, \qquad \delta^* = \frac{\delta}{L},$$

$$v^* = \frac{v_y}{V_\infty}, \qquad y^* = \frac{y}{L}, \qquad \mathrm{Re}_L = \frac{V_\infty L}{\nu},$$

$$\mathrm{Fr}_L = \frac{V_\infty^2}{g_y L}.$$

Substituting these parameters into Eqs. (2.64), (2.65), and the continuity equation, we get

Continuity:

$$\frac{\partial u^*}{\partial x^*} + \frac{\partial v^*}{\partial y^*} = 0, \tag{2.66}$$

Momentum:

$$u^* \frac{\partial u^*}{\partial x^*} + v^* \frac{\partial u^*}{\partial y^*} = \frac{1}{\mathrm{Re}_L}\left[\frac{\partial^2 u^*}{\partial (x^*)^2} + \frac{\partial^2 u^*}{\partial (y^*)^2}\right]; \tag{2.67}$$

$$u^* \frac{\partial v^*}{\partial x^*} + v^* \frac{\partial v^*}{\partial y^*} = \frac{1}{\mathrm{Re}_L}\left[\frac{\partial^2 v^*}{\partial (x^*)^2} + \frac{\partial^2 v^*}{\partial (y^*)^2}\right] + \frac{1}{\mathrm{Fr}_L}. \tag{2.68}$$

The next step is to make order-of-magnitude estimates of the terms in Eqs. (2.67) and (2.68), realizing that we are interested only in the happenings within the boundary layer.

Order of magnitudes:

$$\Delta x^* \cong 0 \quad \text{to} \quad 1, \tag{2.69}$$

$$\Delta y^* \cong \delta^*, \tag{2.70}$$

$$\Delta u^* \cong 1. \tag{2.71}$$

From (2.69) and (2.71), we get

$$\frac{\partial u^*}{\partial x^*} \cong \frac{1}{1} = 1, \tag{2.72}$$

and from continuity,

$$\frac{\partial v^*}{\partial y^*} \cong 1. \tag{2.73}$$

From (2.70) and (2.71), we obtain

$$\frac{\partial u^*}{\partial y^*} \cong \frac{1}{\delta^*}, \tag{2.74}$$

and from (2.72) and (2.73),

$$\frac{\partial v^*}{\partial x^*} \cong \frac{\delta^*}{1} = \delta^*. \tag{2.75}$$

Now we can estimate all the second derivatives:

$$\frac{\partial^2 u^*}{\partial (x^*)^2} = \frac{\partial}{\partial x^*}\left(\frac{\partial u^*}{\partial x^*}\right) \cong \frac{1}{1} = 1, \tag{2.76}$$

$$\frac{\partial^2 u^*}{\partial (y^*)^2} = \frac{\partial}{\partial y^*}\left(\frac{\partial u^*}{\partial y^*}\right) \cong \frac{1}{\delta^*}\left(\frac{1}{\delta^*}\right) = \frac{1}{\delta^{*2}}, \tag{2.77}$$

$$\frac{\partial^2 v^*}{\partial (x^*)^2} \cong \delta^*, \tag{2.78}$$

$$\frac{\partial^2 v^*}{\partial (y^*)^2} \cong \frac{1}{\delta^*}. \tag{2.79}$$

Insertion of the various magnitudes into Eqs. (2.67) and (2.68) reveals two important facts: $\partial^2 u^*/\partial(y^*)^2 \gg \partial^2 u^*/\partial(x^*)^2$, and the equation involving the x-component of the velocity has much larger terms than that for v_y. Hence we deal only with

$$v_x \frac{\partial v_x}{\partial x} + v_y \frac{\partial v_x}{\partial y} = v \frac{\partial^2 v_x}{\partial y^2}, \tag{2.80}$$

which is the *boundary layer* equation for a flat plate with zero pressure gradient. We now proceed to solve Eq. (2.80) for the boundary conditions

$$\text{B.C. 1} \qquad \text{at } y = 0, \qquad v_x = 0, \qquad v_y = 0; \tag{2.81}$$

$$\text{B.C. 2} \qquad \text{at } y = \infty, \qquad v_x = V_\infty. \tag{2.82}$$

In order to simplify Eq. (2.80), we define the *stream function* ψ as†

$$v_x \equiv \frac{\partial \psi}{\partial y} \qquad \text{and} \qquad v_y \equiv -\frac{\partial \psi}{\partial x}. \tag{2.83}$$

The use of the stream functions simplifies Eq. (2.80) and automatically satisfies continuity (Eq. 2.63). Substituting Eq. (2.83) into Eq. (2.80) yields:

$$\frac{\partial \psi}{\partial y} \frac{\partial^2 \psi}{\partial x \, \partial y} - \frac{\partial \psi}{\partial x} \frac{\partial^2 \psi}{\partial y^2} = v \frac{\partial^3 \psi}{\partial y^3}. \tag{2.84}$$

---

† The *stream function* is not discussed in this text, but it is an important tool often used in fluid dynamics. A discussion of the stream function can be found in V. L. Streeter, *Fluid Dynamics*, McGraw-Hill, New York, 1948, pages 38–41.

A *similarity argument*‡ shows that the stream function may be expressed as

$$\psi \equiv \sqrt{V_\infty \, vx} \, f(\beta), \qquad (2.85)$$

where

$$\beta = \frac{y}{\sqrt{x}} \sqrt{\frac{V_\infty}{v}}. \qquad (2.86)$$

From Eqs. (2.83), (2.85), and (2.86),

$$v_x = \frac{\partial \psi}{\partial y} = \frac{\partial \psi}{\partial \beta} \frac{\partial \beta}{\partial y} = V_\infty \frac{df}{d\beta}, \qquad (2.87)$$

$$v_y = -\frac{\partial \psi}{\partial x} = \frac{1}{2} \sqrt{\frac{v V_\infty}{x}} \left(\beta \frac{df}{d\beta} - f \right). \qquad (2.88)$$

Then Eq. (2.84) becomes

$$2 \frac{d^3 f}{d\beta^3} + f \frac{d^2 f}{d\beta^2} = 0. \qquad (2.89)$$

Mathematically, the use of ψ and β has reduced a partial differential equation to an ordinary differential equation with the boundary conditions also taking equivalent forms:

$$\text{B.C. 1} \qquad \text{at } \beta = 0, \qquad f = 0, \qquad \frac{df}{d\beta} = 0; \qquad (2.90)$$

$$\text{B.C. 2} \qquad \text{at } \beta = \infty, \qquad \frac{df}{d\beta} = 1. \qquad (2.91)$$

Equation (2.89) may be solved by expressing $f(\beta)$ in a power series, that is, $f = \sum_0^\infty a_k \beta^k$. The technique is too involved to develop here, but the solution conforming to the boundary conditions becomes

$$f = \frac{\alpha \beta^2}{2!} - \frac{1}{2} \frac{\alpha^2 \beta^5}{5!} + \frac{11}{4} \frac{\alpha^3 \beta^8}{8!} + \cdots, \qquad (2.92)$$

where $\alpha = 0.332$. Then Eqs. (2.87) and (2.88) give expressions for v_x and v_y; the solution for v_x is shown graphically in Fig. 2.7. The position, where $v_x/V_\infty = 0.99$, is located at $\beta = 5.0$; thus the boundary-layer thickness δ is

$$\delta = 5.0 \sqrt{\frac{vx}{V_\infty}}. \qquad (2.93)$$

---

‡ Blasius showed that Eq. (2.84) could be solved in this manner. (See H. Blasius, *NACA Tech. Mem.*, 1949, page 1217.)

Note that if we divide Eq. (2.93) by x, both sides become dimensionless:

$$\frac{\delta}{x} = 5.0\sqrt{\frac{v}{xV_\infty}} = \frac{5.0}{\sqrt{Re_x}}. \tag{2.94}$$

Note also that as a result of the analysis, the Reynolds number ($Re_x = xV_\infty/v$) has evolved; in this instance we give Re the subscript x in order to emphasize that it is a local value with the *characteristic dimension* x. We can also calculate the *drag force*, which is exerted by the fluid on the plate's surface. If the plate has a length L and width W, the drag force F_K is

$$F_K = \int_0^W \int_0^L \left(\eta \frac{\partial v_x}{\partial y} \right)_{y=0} dx\, dz. \tag{2.95}$$

In other words, the shear stress at the solid surface is integrated over the entire surface. From Fig. 2.7 we find that

$$\left[\frac{\partial(v_x/V_\infty)}{\partial \beta} \right]_{\beta=0} = 0.332. \tag{2.96}$$

Knowing the integrand in Eq. (2.95), we can now perform the integration. The result is

$$F_K = 0.664\sqrt{\rho\eta L W^2 V_\infty^3}. \tag{2.97}$$

This is the drag force exerted by the fluid on one surface only.

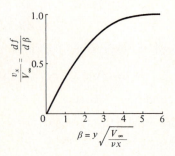

Fig. 2.7 Solution for the velocity distribution in the boundary layer over a flat plate. (From L. Howarth, *Proc. Roy. Soc.*, London **A164**, 547 (1938).)

2.7.2 Flow over a flat plate—approximate integral method

The exact solution method yields the velocity profile, boundary layer, and the drag force on the plate. In this section, we give an alternative solution which provides only the boundary layer thickness and drag force. Consider the portion of the developing boundary layer between x and $x + \Delta x$ in Fig. 2.8.

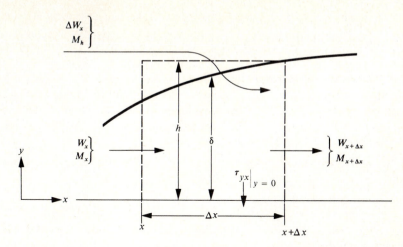

Fig. 2.8 Integrated mass and momentum quantities in the boundary layer over a flat plate.

For unit thickness of fluid, the mass flow rate at x is

$$W_x = \int_0^h \rho v_x \, dy, \qquad (2.98)$$

and the momentum flow is

$$M_x = \int_0^h \rho v_x v_x \, dy. \qquad (2.99)$$

The net change in the mass flow rate between x and Δx is

$$\Delta W_x = W_{x+\Delta x} - W_x = \frac{\partial}{\partial x}(W_x) \cdot \Delta x = \frac{\partial}{\partial x}\left[\int_0^h \rho v_x \, dy\right] \cdot \Delta x. \qquad (2.100)$$

Then in order to satisfy continuity, an amount of fluid ΔW_x must enter the control element through its top surface, $y = h$.† The momentum associated with this fluid is therefore

$$M_h = \Delta W_x \cdot V_\infty. \qquad (2.101)$$

We can now write a momentum balance that also includes the viscous momentum leaving at $y = 0$:

$$(M|_{x+\Delta x} - M_x) - M_h = (\tau_{yx})_{y=0} \cdot \Delta x, \qquad (2.102)$$

---

† The reader should satisfy himself that fluid does enter rather than leave the control element through its top surface.

$$\frac{d}{dx}\left[\int_0^h \rho v_x^2 \, dy\right]\Delta x - V_\infty \frac{d}{dx}\left[\int_0^h \rho v_x \, dy\right]\Delta x = -\eta\left(\frac{\partial v_x}{\partial y}\right)_{y=0}\Delta x.$$

Both integrals are split into two, that is,

$$\int_0^h = \int_0^\delta + \int_\delta^h,$$

and when we realize that between δ and h, $v_x = V_\infty$, then Eq. (2.102) reduces to

$$\frac{d}{dx}\left[\int_0^\delta (V_\infty - v_x)v_x \, dy\right] = \nu\left(\frac{\partial v_x}{\partial y}\right)_{y=0} \tag{2.103}$$

for constant ρ and η.

Equation (2.103) is known as the *von Karman integral relation* for zero pressure gradient. It is solved by assuming a reasonable velocity distribution. Even with an assumed distribution that departs from the actual distribution, the results for the drag force and boundary-layer thickness are very good.

The velocity distribution is *assumed* to have the following form:

$$v_x = ay + by^3. \tag{2.104}$$

There is nothing sacred about Eq. (2.104), except that it does represent a velocity profile that at least looks like what we would expect. The constants a and b may be evaluated by utilizing the conditions

$$\text{at } y = 0, v_x = 0; \qquad \text{at } y = \delta, v_x = V_\infty, \text{ and } \partial v_x/\partial y = 0$$

with the following result:

$$\frac{v_x}{V_\infty} = \frac{3}{2}\left(\frac{y}{\delta}\right) - \frac{1}{2}\left(\frac{y}{\delta}\right)^3. \tag{2.105}$$

When this is substituted into Eq. (2.103) and the indicated operations are performed, a differential equation for the boundary layer results:

$$\delta \, d\delta = \frac{140}{13}\frac{\nu}{V_\infty}\, dx. \tag{2.106}$$

Since $\delta = 0$ at $x = 0$, we obtain

$$\frac{\delta}{x} = 4.64\sqrt{\frac{\nu}{xV_\infty}}. \tag{2.107}$$

Thus Eqs. (2.105) and (2.107) together give the velocity distribution as a function of both y and x. The drag force is then evaluated by using Eq. (2.95) with the result

$$F_K = 0.646\sqrt{\rho\eta LW^2V_\infty^3}. \tag{2.108}$$

Equations (2.107) and (2.108) compare very favorably with the exact results given by Eqs. (2.94) and (2.97), respectively. This is just one example where an integral technique is found to be useful. It is also a powerful tool for solving many heat transfer problems in which exact solutions are not always available. Therefore, later in the text, we will discuss this method again.

2.7.3   Flow in inlet of circular tubes

In Section 2.4, we considered the flow of fluid in a long tube so that end effects were negligible. Now we wish to study the flow conditions at the inlet where the flow is not fully developed. The fluid enters the tube with a uniform velocity V_0 in the z-direction. The important component for this system is the z-component, just as the x-component is most important for the flow past a flat plate. According to Eq. (F) in Table 2.3, the momentum equation with $v_\theta = 0$, $\partial v_z/\partial t = 0$, and $g_z = 0$, reduces to

$$v_r \frac{\partial v_z}{\partial r} + v_z \frac{\partial v_z}{\partial z} = -\frac{1}{\rho}\frac{\partial P}{\partial z} + v\left[\frac{1}{r}\frac{\partial}{\partial r}\left(r\frac{\partial v_z}{\partial r}\right) + \frac{\partial^2 v_z}{\partial z^2}\right].$$

Again, the viscous effect in the direction parallel to flow is negligible, so that $\partial^2 v_z/\partial z^2 \cong 0$, and we are left with

$$v_r \frac{\partial v_z}{\partial r} + v_z \frac{\partial v_z}{\partial z} = -\frac{1}{\rho}\frac{\partial P}{\partial z} + \frac{v}{r}\frac{\partial}{\partial r}\left(r\frac{\partial v_z}{\partial r}\right). \tag{2.109}$$

We can deduce the equation of continuity from Eq. (B) in Table 2.1:

$$\frac{1}{r}\frac{\partial}{\partial r}(rv_r) + \frac{\partial v_z}{\partial z} = 0. \tag{2.110}$$

The method for the solution of Eq. (2.109) is not given here, but one should understand why Eq. (2.109) is the starting point. Langhaar[2] has developed the solution

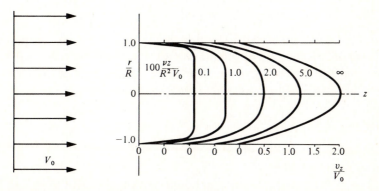

Fig. 2.9   Velocity distribution for laminar flow in the inlet section of a tube.

---

[2] H. L. Langhaar, *J. Appl. Mech.* **9**, A55–58 (1942).

for this problem, as described in Fig. 2.9. His analysis shows that a fully developed flow is not established until $vz/R^2 V_0 \cong 0.07$. Thus an entrance length $(z = L_e)$ of approximately $0.035\, (D^2 V_0 \rho)/\eta$ is required for buildup to the parabolic profile of the fully developed flow.

2.7.4 Creeping flow around a solid sphere

Consider the flow of an incompressible fluid about a solid sphere (Fig. 2.10). The fluid approaches the sphere upward along the z-axis with a uniform velocity V_∞.

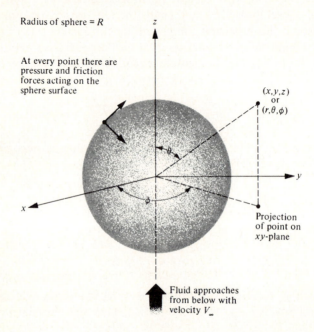

Radius of sphere = R

At every point there are pressure and friction forces acting on the sphere surface

(x,y,z) or (r,θ,ϕ)

Projection of point on xy-plane

Fluid approaches from below with velocity V_∞

Fig. 2.10 Coordinate system used in describing the flow of a fluid about a rigid sphere.

Clearly, the momentum equation for this situation does not involve the ϕ-component. In addition, if the flow is slow enough, the acceleration terms in Navier-Stokes' equation can be ignored. Therefore, in spherical coordinates, from Eqs. (D) and (E) in Table 2.4, we obtain for the r-component:

$$-\frac{\partial P}{\partial r} + \eta\left(\nabla^2 v_r - \frac{2}{r^2} v_r - \frac{2}{r^2}\frac{\partial v_\theta}{\partial \theta} - \frac{2}{r^2} v_\theta \cot\theta\right) + \rho g_r = 0, \qquad (2.111)$$

and for the θ-component:

$$-\frac{1}{r}\frac{\partial P}{\partial \theta} + \eta\left(\nabla^2 v_\theta + \frac{2}{r^2}\frac{\partial v_r}{\partial \theta} - \frac{v_\theta}{r^2 \sin^2\theta}\right) + \rho g_\theta = 0. \qquad (2.112)$$

The continuity equation (Eq. (C) in Table 2.1) is

$$\frac{1}{r^2}\frac{\partial}{\partial r}(r^2 v_r) + \frac{1}{r\sin\theta}\frac{\partial}{\partial\theta}(v_\theta\sin\theta) = 0. \tag{2.113}$$

The momentum flux-distribution, pressure distribution, and velocity components have been found analytically:[3]

$$\tau_{r\theta} = \frac{3}{2}\frac{\eta V_\infty}{R}\left(\frac{R}{r}\right)^4\sin\theta, \tag{2.114}$$

$$P = P_0 - \rho gz - \frac{3}{2}\frac{\eta V_\infty}{R}\left(\frac{R}{r}\right)^2\cos\theta, \tag{2.115}$$

$$v_r = V_\infty\left[1 - \frac{3}{2}\left(\frac{R}{r}\right) + \frac{1}{2}\left(\frac{R}{r}\right)^3\right]\cos\theta, \tag{2.116}$$

$$v_\theta = -V_\infty\left[1 - \frac{3}{4}\left(\frac{R}{r}\right) - \frac{1}{4}\left(\frac{R}{r}\right)^3\right]\sin\theta. \tag{2.117}$$

We can check the validity of the results by showing that Eqs. (2.111)–(2.113) and the following conditions are satisfied:

　　　　B.C. 1　　　at $r = R$,　　　$v_r = 0 = v_\theta$;

　　　　B.C. 2　　　at $r = \infty$,　　　$v_z = V_\infty$.

In Eq. (2.115), the quantity P_0 is the pressure in the plane $z = 0$ far away from the sphere, $-\rho gz$ is simply the hydrostatic effect, and the term containing V_∞ results from the fluid flow around the sphere. These equations are valid for a Reynolds number (DV_∞/ν) less than approximately unity.

With these results, we can calculate the net force which is exerted by the fluid on the sphere. This force is computed by integrating the normal force and tangential force over the sphere surface as follows.

The *normal force* acting on the solid surface is due to the pressure given by Eq. (2.115) with $r = R$ and $z = R\cos\theta$. Thus

$$P(r = R) = P_0 - \rho gR\cos\theta - \frac{3}{2}\frac{\eta V_\infty}{R}\cos\theta.$$

The z-component of this pressure multiplied by the surface area on which it acts, $R^2\sin\theta\,d\theta\,d\phi$, is integrated over the surface of the sphere to yield the net force due to the pressure difference:

$$F_n = \int_0^{2\pi}\int_0^\pi\left[P_0 - \rho gR\cos\theta - \frac{3}{2}\frac{\eta V_\infty}{R}\cos\theta\right]R^2\sin\theta\,d\theta\,d\phi. \tag{2.118}$$

---

[3] V. L. Streeter, *Fluid Dynamics*, McGraw-Hill, New York, 1948, pages 235–240.

Equation (2.118), integrated, yields two terms:

$$F_n = \tfrac{4}{3}\,\pi R^3 \rho g + 2\pi \eta R V_\infty, \tag{2.119}$$

the *buoyant force* and *form drag*, respectively.

At each point on the surface, there is also a shear stress acting tangentially, $-\tau_{r\theta}$, which is the force acting on a unit surface area. The z-component of this force is $(-\tau_{r\theta})(-\sin\theta)$; again, integration over the sphere's surface yields

$$F_t = \int_0^{2\pi}\int_0^\pi (\tau_{r\theta}|_{r=R}\,\sin\theta)R^2\,\sin\theta\,d\theta\,d\phi.$$

From Eq. (2.114), we get

$$\tau_{r\theta}|_{r=R} = \frac{3}{2}\,\frac{\eta V_\infty}{R}\,\sin\theta,$$

so that the *friction drag* results:

$$F_t = 4\pi\eta R V_\infty. \tag{2.120}$$

The total force F of the fluid on the sphere is given by the sum of Eqs. (2.119) and (2.120):

$$F = \tfrac{4}{3}\,\pi R^3 \rho g + 6\pi\eta R V_\infty. \tag{2.121}$$

These two terms are designated as F_s (the force exerted even if the fluid is stationary) and F_K (the force associated with fluid movement). Thus these forces are

$$F_s = \tfrac{4}{3}\,\pi R^3 \rho g, \tag{2.122}$$

$$F_K = 6\pi\eta R V_\infty. \tag{2.123}$$

We use Eq. (2.121), known as *Stokes' law*, primarily for determining the terminal velocity, V_t, of small spherical particles moving through fluid media. The fluid media are stagnant; the spherical particle moves through the fluid, and V_∞ is viewed as V_t. With this in mind, we may use Stokes' law as the basis of a *falling-sphere* viscometer, in which the liquid is placed in a tall transparent cylinder and a sphere of known mass and diameter is dropped into it. The terminal velocity of the falling sphere can be measured, and this in turn relates to the fluid's viscosity.

Example 2.3 Apply Stokes' law to the falling sphere viscometer and write an expression for the viscosity of the liquid in the viscometer.

Solution. A force balance on the sphere, as it falls through the liquid, is made according to the diagram:

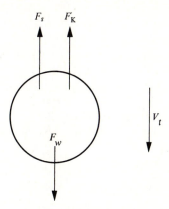

Here F_s is the buoyant force exerted by the liquid and is therefore directed upward; F_K is often called the *drag force* and as such always acts in the opposite direction to that of motion and is therefore directed upward. The only force in the downward direction is the weight of the sphere. The terminal velocity is reached when the force system is in equilibrium. Therefore $F_s + F_K = F_w$, and by substituting in expressions for each of these forces, we have

$$\tfrac{4}{3}\pi R^3 \rho g + 6\pi\eta R V_t = \tfrac{4}{3}\pi R^3 \rho_s g, \tag{2.124}$$

where ρ_s is the sphere's density.

By solving Eq. (2.124) for η, we arrive at

$$\eta = \frac{2R^2(\rho_s - \rho)g}{9V_t}. \tag{2.125}$$

This result is valid only if $2RV_t/v$ is less than approximately unity.

PROBLEMS

2.1 A continuous sheet of metal is cold-rolled by passing vertically between rolls. Before entering the rolls, the sheet passes through a tank of lubricating oil equipped with a squeegee device that coats both sides of the sheet uniformly as it exits. The amount of oil that is carried through can be controlled by adjusting the squeegee device. Prepare a control chart that can be used to determine the thickness of oil (in inches) on the plate just before it enters the roll as a function of the volume rate of oil (in lb_m per hour). Values of interest for the thickness of the oil film range from 0–0.024 in.

Data:
Oil density, 60 lb_m/ft^3
Oil viscosity, 10 lb_m/hr ft
Width of sheet, 5 ft
Velocity of sheet, 1 ft/sec

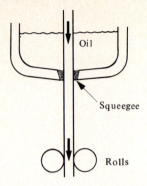

2.2 Develop expressions for the flow of a fluid between vertical parallel plates. The plates are separated by a distance of 2δ. Consider fully developed flow and determine

a) the velocity distribution,

b) the volume flow rate.

   Compare your expressions with Eqs. (2.20) and (2.23).

2.3 Repeat Problem 2.2, but now orient the plates at an angle β to the direction of gravity and obtain expressions for

a) the velocity distributions,

b) the volume flow rate.

   Compare your expressions with the results of Problem 2.2 and Eqs. (2.20) and (2.23), and try to grasp the "madness behind the question."

2.4 A liquid is flowing through a vertical tube 1 ft long and 0.1 in. I.D. The density of the liquid is 1.26 g/cm^3 and the mass flow rate is 0.005 lb/min.

a) What is the viscosity in cP? lb$_m$/ft-hr?

b) Check on the validity of your results.

2.5 A liquid flows upward through a tube, overflows, and then flows downward as a film on the outside.

a) Develop the pertinent momentum balance that applies to the falling film, for steady-state laminar flow, neglecting end effects.

b) Develop an expression for the velocity distribution.

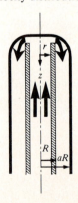

2.6 A wire is cooled after a heat treating operation by being pulled through the center of an open-ended, oil-filled tube which is immersed in a tank. In a region in the tube where end effects are negligible, obtain an expression for the velocity profile assuming steady state and all physical properties constant.

Tube inner radius: R
Wire radius: KR
Wire velocity: U

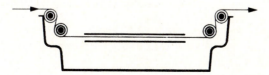

2.7 Starting with the x-component of the momentum equation (Eq. 2.45), develop the x-component for the Navier-Stokes' equation (constant ρ and η, (Eq. 2.56)).

2.8 a) Consider a very large flat plate bounding a liquid that extends to $y = +\infty$. Initially, the liquid and the plate are at rest; then suddenly the plate is set into motion with velocity V_0 as shown in the figure below. Write (1) the pertinent differential equation in terms of velocity, for constant properties, that applies from the instant the plate moves, and (2) the appropriate boundary and initial conditions. The solution to these equations will be discussed in Chapter 9.

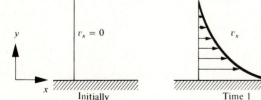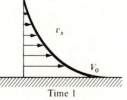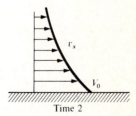

b) A liquid flows upward through a long vertical conduit with a square cross section. With the aid of a clearly labeled sketch, write (1) a pertinent differential equation that describes the flow for constant properties, and (2) the appropriate boundary conditions. Consider only that portion of the conduit where flow is fully developed and be sure that your sketch and equations correspond to one another.

2.9 a) For flow parallel to a flat plate, develop an expression for the momentum boundary layer thickness assuming that $v_x = a \sin by$, where a and b are constants that must be evaluated. Apply the approximate integral technique.

b) Apply the approximate integral technique for flow parallel to a flat plate and develop an expression for the momentum boundary layer thickness, assuming $v_x = a + by$, where a and b are constants that must be evaluated. Also, write an expression for the drag force exerted by the fluid on the plate's surface for this velocity distribution.

2.10 Liquid steel (0.75 % C) at 2900°F is deoxidized by the addition of aluminum which forms alumina (Al_2O_3). We can obtain a better quality of steel if the alumina particles are allowed to float to the surface. Determine the size of the smallest particles that reach the surface two minutes after the steel is deoxidized if the melt is 5 ft deep.

Data: ρ (steel) $= 0.30 \, \text{lb/in}^3$

   ρ (Al$_2$O$_3$) $= 0.12 \, \text{lb/in}^3$

2.11 Molten aluminum (1300°F) is degassed by gently bubbling 75% N$_2$-25% Cl$_2$ gas through it. The gas passes through a graphite tube ($\frac{1}{16}$-in I.D. $\times$ 3 ft long) at a desired rate of 240 in^3/min (1 atm, 1300°F). Calculate the pressure that should be maintained at the tube entrance if the pressure over the bath is 1 atm.

Data: ρ (Al) $= 160 \, \text{lb}_m/\text{ft}^3$

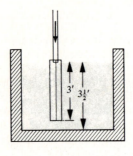

$3'$ $3\frac{1}{2}'$

3

TURBULENT FLOW AND
EXPERIMENTAL RESULTS

In Chapter 2 we discussed only laminar flow problems where we knew differential equations describing the flow, and could calculate the velocity distribution for simple systems. But more often than not, the flow is turbulent, and then experimental information must be sought. For example, it is practically impossible for laminar flow to persist in a pipe at values of Re greater than about 2100. This value of Re is subject to variations in that laminar flow has been maintained up to values of Re as high as 50,000. However, in such cases, the flow is extremely unstable, and the least disturbance transforms it instantly into turbulent flow. Also, the transition Re is higher in a converging pipe and lower in a diverging pipe than in a straight pipe, and even depends to some degree on the inside-surface roughness of the pipe.

If we accept 2100 as the maximum value of Re for laminar flow in pipes with normal roughness, we can easily show that turbulent flow is the usual case. Consider water at 60°F which has a kinematic viscosity of 1.22×10^{-5} ft²/sec. In a pipe of 1-in. diam. the average velocity that still corresponds to laminar flow is

$$V_{\text{crit}} = \frac{v}{D} \text{Re}_D = \frac{1.22 \times 10^{-5} \text{ ft}^2/\text{sec}}{8.33 \times 10^{-2} \text{ ft}} \times 2100$$

$$= 0.308 \text{ ft/sec.} \quad in= \frac{1}{12} ft = 8.33 \times 10^{-2} ft$$

Velocities as small as these (in 1-in. pipes) are not often encountered in practical engineering, so that most problems of engineering importance occur in the region of turbulent flow.

In this chapter we shall examine some semiempirical information available for various systems. We may classify the flow systems into two groups: flow through closed conduits and flow past submerged bodies. In the former group, pressure-drop data are usually desired and correlations given in terms of a *friction factor*; in the latter group, forces acting on submerged bodies are usually sought, and correlations reported in terms of a *drag coefficient*.

3.1 FRICTION FACTORS FOR FLOW IN TUBES

As an example of a flow system, consider a length of horizontal pipe between

$z = 0$ and $z = L$. We presume that in this length of pipe the fluid is flowing with an average velocity independent of time, and that the flow is fully developed. For flow in such a system, we may write the force of the fluid on the inner wall due to kinetic behavior as

$$F_K = 2\pi R L \tau_0. \tag{3.1}$$

Here τ_0 is defined as the shear stress at the wall, or, in terms of previous notation, $\tau_0 \equiv \tau_{rz}(r = R)$. According to Eq. (2.28), we can express τ_0 in terms of the pressure drop:

$$\tau_0 = \left(\frac{P_0 - P_L}{L}\right)\frac{R}{2}. \tag{3.2}$$

Equation (3.2) is also valid for turbulent flow as it specifies conditions only at the wall. Using Eq. (3.2), we may write F_K in an alternative form:

$$F_K = \pi R^2 (P_0 - P_L). \tag{3.3}$$

Equation (3.3) focuses our attention on what should be considered if we ask the following question: What pressure drop is necessary to deliver a given volume of fluid through a tube? Thus we learn from Eq. (3.3) that F_K must be determined; for laminar flow, this can be calculated because the velocity distribution is amenable to analysis, and pressure drops can be determined *a priori*. For turbulent flow, the uncertainties involved in parallel analysis have led engineers to take an experimental approach to the problem.

In turbulent flow we may think that the flow pattern starts with a *laminar boundary layer*, in which the flow can be described by Newton's law of viscosity, followed by a transition region in which the degree of turbulence steadily increases and laminar effects diminish, until finally the region of fully developed turbulence is reached. These regions are illustrated in Fig. 3.1. In turbulent flow, therefore, the fluid still clings to the solid wall. Thus Eq. (3.1) is applicable, but, in general, the value of τ_0 is not determined by analytical means. More often we employ an empirical technique and express the force F_K as the product of a *characteristic area* A, a *characteristic kinetic energy* K (per unit volume), and a dimensionless quantity f, known as the *friction factor*:

$$F_K = AKf. \tag{3.4}$$

This is a useful definition because f is a function only of the dimensionless Reynolds number for a given geometrical shape. This is shown in a following discussion, but first let us present how f is related to τ_0.

For flow in conduits, A is taken to be the wet surface $2\pi RL$, and K is taken to be the kinetic energy based on the average velocity, that is, $\frac{1}{2}\rho \bar{V}^2$. Thus for flow in a filled pipe of length L, we have

$$F_K = f(2\pi RL)(\tfrac{1}{2}\rho \bar{V}^2). \tag{3.5}$$

Combining Eqs. (3.1) and (3.5), we see that

$$f = \frac{\tau_0}{(\frac{1}{2}\rho \overline{V}^2)}. \qquad (3.6)$$

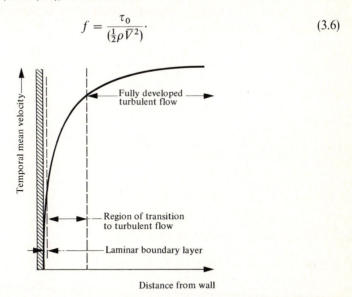

Fig. 3.1 Velocity distribution for turbulent flow in tubes in the region near the wall.

3.1.1 Dimensional analysis for friction factor

Now we resort to *dimensional analysis* which is a method of deducing logical groupings of the variables involved in a process. One of these methods, called the *similarity technique*, is applied to systems which are geometrically similar. For example, in two circular tubes, all comparable lengths have identical ratios. Thus we write

$$K_D = \frac{z_1}{z_2} = \frac{r_1}{r_2} = \frac{D_1}{D_2}, \qquad (3.7)$$

where z_1 and z_2 are comparable distances along the lengths of tube; r_1 and r_2 are comparable radii in tubes 1 and 2, respectively.

For the flow system in either tube, the differential equation for the conservation of momentum that applies in the steady state is

$$\eta \left[\frac{1}{r} \frac{d}{dr} \left(r \frac{dv_z}{dr} \right) \right] - \frac{dP}{dz} = 0, \qquad (3.8)$$

and a boundary condition is

$$\tau_{rz} \bigg|_{r=R} \equiv \tau_0 = -\eta \left(\frac{dv_z}{dr} \right)_{r=R}. \qquad (3.9)$$

The differential equation and the boundary condition require certain relationships between velocities, dP/dz, and fluid properties, at corresponding points in the two systems. Thus, in addition to the ratios of Eq. (3.7), we define the ratios for kinematic and dynamic similarity:

$$K_v \equiv \frac{v_{z1}}{v_{z2}} = \frac{\overline{V}_1}{\overline{V}_2}, \qquad K_P \equiv \frac{dP_1/dz_1}{dP_2/dz_2},$$

$$K_\eta \equiv \frac{\eta_1}{\eta_2}, \qquad K_\rho \equiv \frac{\rho_1}{\rho_2}. \tag{3.10}$$

To investigate the conditions which must be satisfied by these ratios, we write Eq. (3.8) specifically for system 1:

$$\eta_1 \left[\frac{1}{r_1} \frac{d}{dr_1} \left(r_1 \frac{dv_{z1}}{dr_1} \right) \right] - \frac{dP_1}{dz_1} = 0. \tag{3.11}$$

Now if we replace r_1, v_{z1}, etc., by their equivalents in Eqs. (3.7) and (3.10), then we can write Eq. (3.11) for system 2:

$$\left(\frac{K_\eta K_v}{K_D^2} \right) \eta_2 \left[\frac{1}{r_2} \frac{d}{dr_2} \left(r_2 \frac{dv_{z2}}{dr_2} \right) \right] - K_P \frac{dP_2}{dz_2} = 0. \tag{3.12}$$

Multiplying Eq. (3.12) by $K_D/K_v^2 K_\rho$, we finally obtain:

$$\left(\frac{K_\eta}{K_D K_v K_\rho} \right) \eta_2 \left[\frac{1}{r_2} \frac{d}{dr_2} \left(r_2 \frac{dv_{z2}}{dr_2} \right) \right] - \left(\frac{K_P K_D}{K_\rho K_v^2} \right) \frac{dP_2}{dz_2} = 0. \tag{3.13}$$

It is apparent that Eq. (3.13) written with subscripts 1 would also be valid, and because both systems must obey the original differential equation, then

$$\frac{K_\eta}{K_D K_v K_\rho} = 1 \qquad \text{and} \qquad \frac{K_P K_D}{K_\rho K_v^2} = 1,$$

or

$$\frac{D_1 \overline{V}_1 \rho_1}{\eta_1} = \frac{D_2 \overline{V}_2 \rho_2}{\eta_2}, \tag{3.14}$$

and

$$\frac{D_1 (dP_1/dz_1)}{\rho_1 \overline{V}_1^2} = \frac{D_2 (dP_2/dz_2)}{\rho_2 \overline{V}_2^2}. \tag{3.15}$$

The group in Eq. (3.14) is the Reynolds number which, of course, is dimensionless. Once again it appears—in fact, in all cases of forced convection, the Reynolds number is a significant dimensionless group. The group in Eq. (3.15) is another dimensionless group and, as shown in the following paragraph, is actually twice the friction factor.

If we define a new ratio as

$$K_\tau = \tau_{01}/\tau_{02},$$

then the boundary condition, (Eq. 3.9), applied to both systems, yields

$$\frac{K_\tau K_D}{K_\eta K_v} = 1 \qquad \text{or} \qquad \frac{\tau_{01} D_1}{\eta_1 \bar{V}_1} = \frac{\tau_{02} D_2}{\eta_2 \bar{V}_2}.$$

Dividing by the Reynolds number, we obtain

$$\frac{\tau_{01}}{\rho_1 \bar{V}_1^2} = \frac{\tau_{02}}{\rho_2 \bar{V}_2^2}, \tag{3.16}$$

which are equivalent to half the friction factor (Eq. 3.6). Now the dimensional analysis is complete; it has been shown that for flow in the two arbitrary systems having geometric similarity, the friction factors are equal when the Reynolds numbers are equal. Thus for flow in tubes, the friction factor f may be correlated as a function of the Reynolds number.

3.1.2 Experimental results for friction factor

Experimentally, f can be measured by noting that if F_K is eliminated between Eqs. (3.3) and (3.5), then

$$f = \frac{1}{4}\left(\frac{D}{L}\right)\left(\frac{P_0 - P_L}{\frac{1}{2}\rho \bar{V}^2}\right), \tag{3.17}$$

and the connection with Eq. (3.15) is made apparent. Also, as we have shown, the friction factor is only a function of the Reynolds number: $f = f(\text{Re}_D)$. In most designs the pressure drop is not usually known *a priori*; we therefore use Re_D to evaluate the friction factor and then the pressure drop developed due to friction. Since f is only a function of Re_D, it is sufficient to plot only a single curve of f against $D\bar{V}/v$ rather than determine how f varies for separate values of D, $\bar{V}$, ρ, and η. The lower curve in Fig. 3.2 gives a plot of f versus Re_D for smooth tubes. This curve reflects the laminar and turbulent behavior of fluids in long, smooth, circular tubes. For the laminar region,

$$f = \frac{16}{\text{Re}}, \tag{3.18}$$

which can be derived by using the Hagen-Poiseuille law developed in Chapter 2.

The turbulent region has been established solely by experimental data. The entire turbulent curve closely approximates

$$1/\sqrt{f} = 4.0 \log(\text{Re}\sqrt{f}) - 0.40, \tag{3.19}$$

and a simpler expression exists for $2.1 \times 10^3 < \text{Re} < 10^5$, namely,

$$f = 0.0791 \, \text{Re}^{-1/4}. \tag{3.20}$$

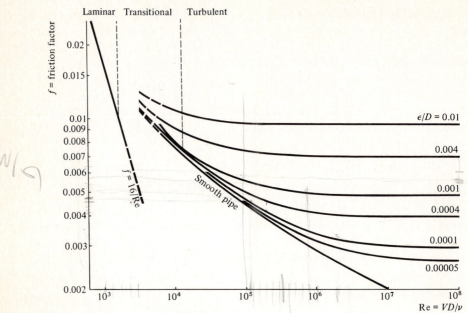

Fig. 3.2 Friction factors for flow in tubes. (Adapted from L. F. Moody, *Trans. ASME* **66**, 671 (1944), and *Mech. Eng.* **69**, 1005 (1947).)

Table 3.1 Values of absolute roughness ε for new pipes

	Ft	In.
Drawn tubing, brass, lead, glass, centrifugally spun cement, bituminous lining, transite...	0.000005	0.00006
Commercial steel or wrought iron..............	0.00015	0.0018
Welded-steel pipe......................	0.00015	0.0018
Asphalt-dipped cast iron....................	0.0004	0.0048
Galvanized iron........................	0.0005	0.006
Cast iron, average......................	0.00085	0.0102
Wood stave................................	0.0006 to 0.003	0.0072 to 0.036
Concrete..................................	0.001 to 0.01	0.012 to 0.12
Riveted steel.............................	0.003 to 0.03	0.036 to 0.36

$$Note: \frac{\varepsilon}{D} = \frac{\varepsilon \text{ in ft}}{D \text{ in ft}} = \frac{\varepsilon \text{ in in.}}{\text{diam. in in.}}.$$

If the tubes are rough, then in turbulent flow the friction factor is higher than that indicated for smooth tubes. The *relative roughness*, ε/D, enters the correlation where ε is the height of protuberances on the tube wall. For use with Fig. 3.2, we can obtain values of ε from Table 3.1. However, the reader should note that these values apply for new, clean pipes, and even then the values may vary. For old pipes, they may be much higher and certainly ε varies greatly with age, depending on the fluid being transported. In addition, in small pipes, deposits may substantially reduce the inside diameter. Therefore we must carefully use our judgment in estimating a value of ε and consequently of f.

Example 3.1 Determine the mass flow rate (lb_m/hr) of water at 80°F through 600 ft of horizontal pipe, having I.D. of 5 in., under a pressure drop of 5 lb/in². The relative roughness ε/D of the pipe is estimated as 0.001.

Solution. The solution to this problem can be found by using Eq. (3.17) and solving for the average velocity $\bar{V}$. However, $\bar{V}$ cannot be determined unless f is known and, because $f = f(\text{Re}_D)$, a trial and error solution is in order.

First approximation. In Fig. 3.2, $f = 0.005$ when $\varepsilon/D = 0.001$ and $\text{Re} = 9 \times 10^5$. For this friction factor, the velocity is

$$\bar{V} = \sqrt{\left(\frac{1}{2f}\right)\left(\frac{D}{L}\right)\left(\frac{\Delta P}{\rho}\right)};$$

$$\frac{D}{L} = \frac{\frac{5}{12}\,\text{ft}}{600\,\text{ft}} = \frac{1}{1440},$$

$$\Delta P = 5\,\text{lb/in}^2 = (5)(4633)\frac{\text{lb}_m}{\text{ft sec}^2},$$

$$\frac{\Delta P}{\rho} = \frac{(5)(4633)\,\text{lb}_m}{\text{ft sec}^2}\left|\frac{\text{ft}^3}{62.4\,\text{lb}_m}\right|\frac{3600^2\,\text{sec}^2}{1\,\text{hr}^2} = 4.81 \times 10^9\,\text{ft}^2/\text{hr}^2.$$

Then

$$\bar{V} = \sqrt{\left(\frac{1}{0.01}\right)\left(\frac{1}{1440}\right)(4.81 \times 10^9)} = 1.83 \times 10^4\,\text{ft/hr}.$$

This velocity gives the following Reynolds number:

$$\text{Re}_D = \frac{D\bar{V}\rho}{\eta} = \frac{5\,\text{ft}}{12}\left|\frac{1.83 \times 10^4\,\text{ft}}{\text{hr}}\right|\frac{62.4\,\text{lb}_m}{\text{ft}^3}\left|\frac{\text{hr ft}}{2.25\,\text{lb}_m}\right.$$

$$= 2.11 \times 10^5.$$

Second approximation. For $Re_D = 2.11 \times 10^5$, $f = 0.0052$. Thus, with this f, $\bar{V}$ is slightly corrected:

$$\bar{V} = \left(\frac{0.0050}{0.0052}\right)^{1/2}(1.83 \times 10^4) = 1.82 \times 10^4 \text{ ft/hr.}$$

Therefore the mass flow rate W is

$$W = \left(\frac{\pi D^2}{4}\right)\rho\bar{V} = \left(\frac{25\pi}{4 \times 144}\right)(62.4)(1.82 \times 10^4)$$

$$= 1.55 \times 10^5 \text{ lb}_m/\text{hr.}$$

3.2 FLOW IN NONCIRCULAR CONDUITS

If the pipes are not circular and turbulent flow exists, then an *equivalent diameter* D_e* replaces D in the Reynolds number:

$$D_e = \frac{4 \times \text{flow area}}{\text{wetted perimeter}} = \frac{4A}{P_W}. \tag{3.21}$$

With this modification, we can determine the friction factor from Fig. 3.2 where ε/D is replaced by ε/D_e. Similarly in Eq. (3.17), we replace D by D_e, and calculate the pressure drop due to friction in noncircular conduits. This approach gives good results for turbulent flow, but for laminar flow the results are poor. For example, in a thin annulus in which the spacing z is very much less than the width, laminar flow has a parabolic distribution perpendicular to the walls. This situation closely approximates flow between parallel flat plates, and we can show the friction factor to be

$$f = \frac{24}{Re}. \tag{3.22}$$

This expression, of course, differs from Eq. (3.18) which applies to laminar flow in circular pipes.

For laminar flow in a rectangular duct of dimensions $z_1 \times z_2$, the friction factor is

$$f = \frac{16}{\phi Re}. \tag{3.23}$$

Here we evaluate the Reynolds number using D_e, which for rectangular ducts is

$$D_e = \frac{2(z_1 z_2)}{(z_1 + z_2)},$$

---

* Many other texts refer to the hydraulic radius R_h rather than equivalent diameter. In such instances, D is replaced by $4R_h$, where $R_h = A/P_W$; the end result is the same whether D_e or R_h is employed.

and ϕ is given by Fig. 3.3. The laminar-to-turbulent transition for noncircular conduits is still at a Reynolds number of approximately 2100.

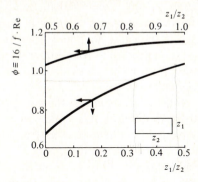

Fig. 3.3 Values of ϕ for laminar flow in rectangular ducts. (From W. M. Rohsenow and H. Y. Choi, *Heat, Mass, and Momentum Transfer*, Prentice-Hall, Englewood Cliffs, New Jersey, 1961, page 63.)

3.3 FLOW PAST SUBMERGED BODIES

3.3.1 Turbulent boundary layer on flat plate

In Chapter 2, laminar flow over a flat plate was analyzed. The force exerted by the fluid on one side of the plate (drag force F_K) was calculated for laminar flow as

$$F_K = 0.664\sqrt{\rho\eta LW^2V_\infty^3}. \tag{2.97}$$

From Eq. (3.4),

$$f = \frac{F_K}{AK}, \tag{3.4a}$$

where, for this case, the characteristic area is conveniently defined $A = LW$ and the characteristic kinetic energy $K = \frac{1}{2}\rho V_\infty^2$. The friction factor for laminar flow is evaluated by combining Eqs. (2.97) and (3.4a) for laminar flow:

$$f = 1.328(V_\infty L/\nu)^{-1/2} = 1.328\,\mathrm{Re}_L^{-1/2}. \tag{3.24}$$

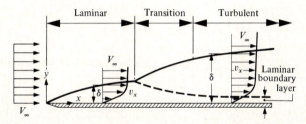

Fig. 3.4 Laminar and turbulent boundary-layers next to a flat plate (the vertical scale is greatly exaggerated).

We again see that the friction factor is a function only of a Reynolds number.* These results hold so long as the boundary layer itself remains laminar. However, at a value of Re_x between 300,000 and 500,000 the layer becomes turbulent, increasing significantly in thickness, and displaying a marked change in velocity distribution. We depict this transition in Fig. 3.4 which shows a much steeper gradient near the wall and flatter gradient throughout the remainder of the boundary layer for the turbulent zone. As a result, the shear stress at the wall is greater in the turbulent layer than in the laminar layer.

Using the applicable differential equation, the geometric similarity argument would again indicate that the friction factor is a function of the same dimensionless group, Re. Again, experimental data have been gathered to correlate the friction factor and the Reynolds number. The results in Fig. 3.5 show that the laminar region may extend up to a Reynolds number of about 10^6. However, this condition is only achieved on smooth surfaces, and if the system is protected against external disturbances, such as minute vibrations caused by marching protestors, passing trucks, or bomb explosions on other parts of the campus. The results can be represented as a simple function for a limited range:

$$f = 0.072\left(\frac{v}{V_\infty L}\right)^{1/5}, \qquad 5 \times 10^5 < Re < 10^7. \tag{3.25}$$

A somewhat more complicated function[1] exists for the entire graph:

$$f = \frac{0.455}{(\log Re)^{2.58}}. \tag{3.26}$$

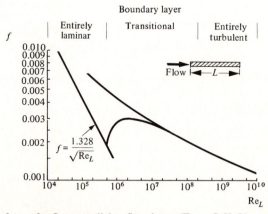

Fig. 3.5 Friction factor for flow parallel to flat plates. (From J. K. Vennard, *Elementary Fluid Mechanics*, Wiley, New York, 1954, page 384.)

---

* Most texts on fluid mechanics use the term *drag coefficient* C_D rather than friction factor f, when considering flow past submerged bodies. When the form drag is entirely friction drag, an even more specific term, *skin friction coefficient* C_f is sometimes used.

[1] H. Schlichting, *Boundary Layer Theory*, McGraw-Hill, New York, 1955, page 439.

These equations are valid for a flat plate at zero incidence to the flow, providing separation does not occur.

Example 3.2   In order to study the mixing action in a molten bath of aluminum contained in an induction furnace, we use a steel *flag*. The flag is held in a vertical position, and placed in the central part of the furnace where the metal flows upward, as depicted in the figure below.

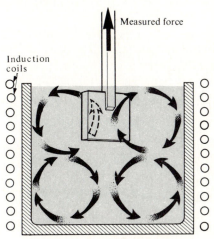

Arrows indicate the melt's motion
due to the electromagnetic forces.

If provisions are made to measure the added force exerted on the flag due to fluid motion, prepare a control chart that relates this force to the velocity of the molten aluminum in the central portion of the furnace. Assume that turbulent flow encompasses the conditions of interest.

Data: Viscosity of aluminum, 1 cP,
             Density of aluminum, 160 lb_m/ft^3,
             Flag dimensions, 1 ft × 1 ft × 0.01 ft.

Solution. In this problem it is convenient to choose values of Re_L and then calculate the corresponding sets of F_K and V_∞. The velocity of the aluminum is found by:

$$V_\infty = \frac{\eta}{L\rho} Re_L,$$

and then we can write the corresponding force exerted by the moving aluminum on both sides of the flag:

$$F_K = 2f(LW)(\tfrac{1}{2}\rho V_\infty^2)$$

$$= fLW\rho\left(\frac{\eta^2}{L^2\rho^2} Re_L^2\right)$$

$$= f\left(\frac{W}{L}\right)\left(\frac{1}{\rho}\right)(\eta^2 \, \text{Re}_L^2) \cdot$$

For $\text{Re}_L = 10^6$, Fig. 3.5 gives $f = 0.0045$.

$$V_\infty = \frac{2.42 \, \text{lb}_m}{\text{hr ft}} \left|\frac{\text{ft}^3}{1 \, \text{ft}}\right| \frac{1 \, \text{hr}}{160 \, \text{lb}_m} \left|\frac{10^6}{3600 \, \text{sec}}\right| = 4.20 \, \text{ft/sec},$$

and

$$F_K = \frac{0.0045}{1 \, \text{ft}} \left|\frac{1 \, \text{ft}}{160 \, \text{lb}_m}\right| \frac{\text{ft}^3}{\text{hr}^2 \, \text{ft}^2} \left|\frac{2.42^2 \, \text{lb}_m^2}{3600^2 \, \text{sec}^2}\right| \frac{1 \, \text{hr}^2}{32.2 \, \text{ft lb}_m} \left|\frac{\text{sec}^2 \, \text{lb}_f}{}\right| \, 10^{12}$$

$$= 0.382 \, \text{lb}_f.$$

In a similar manner, the corresponding values of V_∞ and F_K can be found for other values of Re_L:

Re_L	V_∞, ft/sec	F_K, lb$_f$
10^5	0.420	0.006
5×10^5	2.10	0.108
10^6	4.20	0.382
2×10^6	8.40	1.36
4×10^6	16.8	4.89
5×10^6	21.0	7.22

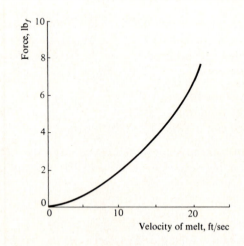

3.3.2 Flow past a sphere and other submerged objects

In Chapter 2, we analyzed laminar flow past a sphere, and gave the kinetic force of the fluid exerted on the sphere:

$$F_K = 6\pi\eta R V_\infty. \tag{2.123}$$

We have found that this force, often termed the *form drag*, is composed of two parts: the *friction drag* and the *pressure drag*. In a well-streamlined body (flow parallel to a flat plate, airplane wing, hull of a boat, etc.), the friction drag not only forms the major part of the total drag, but may even comprise it entirely. In the case of a body with sharp corners, or a sphere, or a cylinder oriented normally to flow, separation occurs, and a turbulent wake forms.

With a plate oriented perpendicularly to flow, the separation always occurs at the same point, and the wake extends across the full projected width of the body, as shown in Fig. 3.6; this results in almost all pressure drag comprising the form drag. If the body has curved sides, the location of the separation point is determined according to whether the leading boundary layer is laminar or turbulent, as depicted in Fig. 3.7. In turn, this location determines the size of the wake and the amount of the pressure drag. After this brief introduction, let us now consider some of the available data.

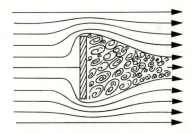

Fig. 3.6 Wake formation for flow normal to flat plate.

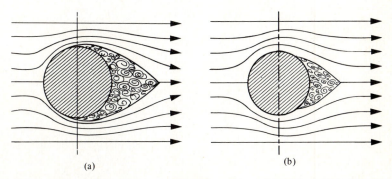

(a)   (b)

Fig. 3.7 Separation of boundary layer and wake formation with a cylinder or sphere immersed in a flowing fluid when (a) the leading boundary layer is laminar, and (b) the leading boundary layer is turbulent.

For the presentation of data, we use Eq. (3.4). As we come across the friction factor again, let us repeat the expression

$$F_K = fAK;$$ (3.4)

K still remains $\frac{1}{2}\rho V_\infty^2$, but for shapes such as spheres, cylinders, ellipsoids, etc., we choose the area A as the projected area normal to the flow. For example, for flow past a sphere, F_K takes the form

$$F_K = (\pi R^2)(\tfrac{1}{2}\rho V_\infty^2)f. \tag{3.27}$$

From Stokes' law, which is the result of an analytical solution for laminar flow around a sphere, we can write the following equation by rearranging Eq. (2.123):

$$F_K = (\pi R^2)(\tfrac{1}{2}\rho V_\infty^2)\left(\frac{24}{DV_\infty/\nu}\right). \tag{3.28}$$

By comparing Eqs. (3.27) and (3.28), we obtain the friction factor (drag coefficient) for laminar flow past a sphere:

$$f = \frac{24}{\mathrm{Re}}. \tag{3.29}$$

Once again, the friction factor is a function of the Reynolds number. This is also true for turbulent flow but, as you may suspect, Eq. (3.29) does not apply for all velocities, and experimental correlation is sought in the form of $f = f(\mathrm{Re})$. Figure 3.8 shows such a correlation for spheres, as well as for some other geometrical shapes. We also include Fig. 3.9 to indicate other possibilities. Note that Stokes' law applies up to $\mathrm{Re} \cong 1$ for spheres.

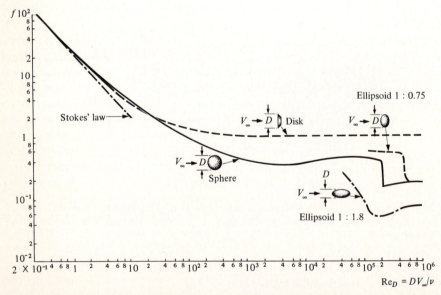

Fig. 3.8 Friction factors for submerged bodies. (Adapted from F. Eisner, *Proc. 3rd Intern. Congr. Appl. Mech.*, 1930, page 32.)

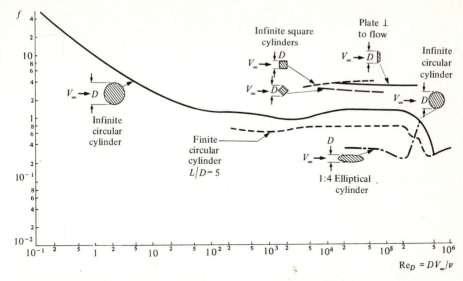

Fig. 3.9 Friction factors for submerged bodies. (Adapted from F. Eisner, *Proc. 3rd Intern. Congr. Appl. Mech.*, 1930, page 32.)

The phenomenon of boundary-layer separation has a major effect on the form drag. This effect is noticeable at large Re where the form drag comprises most of the total drag. At this high Re, the transition to turbulent flow occurs before separation, and the separation point is moved farther downstream, as illustrated in Fig. 3.7, thereby reducing the form drag.

In metallurgical processes, the discussed principles of flow past submerged bodies mostly apply to situations in which the body moves in a fluid. The friction factors are determined experimentally, and it makes no difference whether the body moves in a still fluid or is held stationary while the fluid flows past it. However, when the body is irregularly shaped, friction factors which are measured while the body is stationary may not be applicable when the body moves, because during free motion the orientation of the body may change rapidly and repeatedly.

It is often desirable to separate particulate mixtures of solids. These mixtures may consist of particles of the same material which have different sizes (homogeneous mixtures), or they may consist of particles of different materials and various sizes (heterogeneous mixtures). Homogeneous mixtures can be separated by screening, but for separating mixtures of fine sizes, free settling or centrifugal methods are preferred. In general, screening does not work for heterogeneous mixtures, and therefore separation is gained by free settling or centrifugal methods.

Example 3.3 In powder metallurgy operations, it is important to separate homogeneous mixtures in the range of particle sizes from 1 to 50 μ. For this purpose we apply the principle of air elutriation, which is illustrated by the device on page 90.

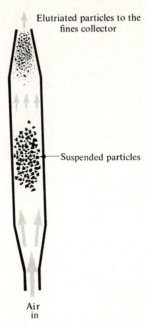

Elutriated particles to the
fines collector

—Suspended particles

Air
in

The force exerted by the air stream is great enough to suspend particles of a given diameter, and all particles of smaller diameter are carried upward and to the collector of fines. Larger particles fall back against the air stream and down into a settling chamber. For the effective operation of this device it is necessary to know the size of particles suspended at a given flow rate.

Assuming that a spherical particle, with a specific gravity of 4.0, applies to a homogeneous mixture of iron powder, draw a graph that relates the diameter of suspended particles to the velocity of the air in the expanded portion of the tube.

Solution. To obtain the solution, recognize that particles which have terminal velocities equal to the upward air velocities are those which remain suspended. Further, examine the criterion that Stokes' law applies for $Re_D < 1$ and that a simple expression for the diameter may be appropriate.

For air at 80°F, the fluid properties are $\eta = 1.80 \times 10^{-4}$ g/cm sec and $\rho = 0.00118$ g/cm^3. For $Re_D = 1$ and $D = 5 \times 10^{-3}$ cm (the largest particle size of interest), the corresponding velocity is

$$V_\infty = Re_D \frac{\eta}{D\rho}.$$

Using consistent units, we have

$$V_\infty = 1 \frac{1.80 \times 10^{-4}}{(5 \times 10^{-3})(1.18 \times 10^{-3})} = 30.5 \text{ cm/sec.}$$

This is a very high velocity; thus, for all situations, $Re_D < 1$, so that Stokes' law does apply. According to Eq. (2.125),

$$D = \sqrt{\frac{18V_\infty\eta}{(\rho_s - \rho)g}}.$$

Here we may ignore the fluid density ρ.

For $V_\infty = 10 \text{ cm/sec}$,

$$D = \sqrt{\frac{(18)(10)(1.80 \times 10^{-4})}{(4.0)(980)}} = 0.00287 \text{ cm} = 28.7 \, \mu.$$

We can calculate values of D for other air velocities in a similar manner; the results are presented in the graph below.

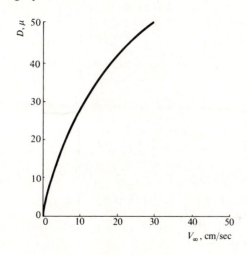

3.4 FLOW THROUGH PACKED BEDS OF SOLIDS

Packed beds of granular solids or agglomerates of fine particles occur in many metallurgical processing systems, ranging from sinter plants for agglomerating iron ores to the production of intricate parts via powder metallurgy. The flow of fluids through porous aggregates is not simple, especially due to the effect of the properties of the aggregate on the flow. But under certain conditions we may predict the flow reasonably well.

3.4.1 D'Arcy's law

If the flow occurs under low pressure conditions, that is, it is slow enough, the rate of flow is essentially proportional to the pressure drop per unit length of packing, $\Delta P/L$:

$$Q = \frac{k_D A \Delta P}{L}, \tag{3.30}$$

where Q = volume of fluid flowing per unit time, cm^3/sec, A = cross-sectional area, cm^2, and k_D = permeability coefficient, a constant depending on the fluid, temperature, and packing characteristics, $cm^4/dyn\text{-}sec$. Equation (3.30), which is known as *D'Arcy's law*, has been applied to the problem of the permeation of gases through foundry molding sands, and the filtration of water through filter cake.

The k_D in Eq. (3.30) is often called the *permeability*, which is satisfactory as long as we carry out the test with the same fluid at the same temperature, but, in general, it is more common and more desirable to define a *specific permeability* $\mathscr{P}$ by means of

$$k_D = \frac{\mathscr{P}}{\eta}, \tag{3.31}$$

where η is the viscosity of the fluid. This allows $\mathscr{P}$ to be specific to the packing only, and therefore allows the flow of other fluids, or the same fluid at other temperatures, to be predicted. The units of specific permeability $\mathscr{P}$ are

$$\mathscr{P} = \left(\frac{cm^4}{dyn\ sec}\right)\left(\frac{dyn\ sec}{cm^2}\right) = cm^2,$$

where the unit of specific permeability, the D'Arcy, is now defined as

$$1\ \text{D'Arcy} = 1 \times 10^{-8}\ cm^2.$$

Unfortunately, this value has not been standardized and various other definitions of the unit have been proposed. The British Standard Specifications define the D'Arcy as expressed with viscosity in centipoises and the pressure drop in atmospheres. This D'Arcy differs from the above definition by 1.3%, since

$$1\ \text{D'Arcy (BSS)} = 0.987 \times 10^{-8}\ cm^2.$$

3.4.2 Tube-bundle theory and Ergun's equation

D'Arcy's law is an empirically observed law. However, a semitheoretical approach to the problem yields D'Arcy's law with more insight into the effect of the packing on $\mathscr{P}$. The theory, which is known as the *tube-bundle theory*, regards the bed as a bundle of tangled tubes with weird cross-sections. It is assumed that the packing is uniform without isolated porosity, that there is no *channeling* of flow, and that the diameter of the column is much greater than that of the particles. To start with, we look to the Hagen-Poiseuille formula for laminar flow in Chapter 2 (Eq. 2.33). By analogy, we say that

$$\bar{V} = K_1 \frac{\Delta P R_h^2}{L\eta}, \tag{3.32}$$

where K_1 = constant of proportionality, and $\bar{V}$ = average velocity in the interstices of the bed.

Engineers usually know V_0 rather than $\bar{V}$; V_0 is called the *superficial velocity* and is defined by

$$V_0 = Q/A, \tag{3.33}$$

and since

$$V_0 = \bar{V}\omega, \tag{3.34}$$

then

$$V_0 = K_1 \frac{\Delta P R_h^2 \omega}{\eta L}, \tag{3.35}$$

where ω = void fraction.* The concept of the *hydraulic radius* R_h was introduced with the equivalent diameter (Eq. 3.21).

For packed beds, R_h is defined as

$$R_h = \frac{\text{average cross section available for flow}}{\text{average wetted perimeter}} = \frac{A_h}{P_w}. \tag{3.36}$$

By the average wetted perimeter, we mean the average total boundary line between the fluid and the packing, viewed by a slice through the bed normal to the axis of the bed. Thus, in a bed of length L,

$$R_h = \frac{\text{volume available for flow}}{\text{total wetted surface}} = \frac{A_h L}{P_w L}. \tag{3.37}$$

Here $A_h L$ is also the volume of voids in the packing, and if both the numerator and denominator of Eq. (3.37) are normalized by the bed volume, we obtain

$$R_h = \frac{A_h L/V}{P_w L/V} = \frac{\omega}{S}, \tag{3.38}$$

where ω = void fraction or voidage and $S = S_0(1 - \omega)$.

The factor S_0 is the total surface area of particles per unit volume of *particles*, while S is the total surface area per unit volume of *bed*. From this,

$$R_h = \frac{\omega}{S_0(1 - \omega)}. \tag{3.39}$$

We may substitute Eq. (3.39) into Eq. (3.35) and obtain

$$V_0 = K_1 \frac{\Delta P}{L\eta S_0^2} \frac{\omega^3}{(1 - \omega)^2}. \tag{3.40}$$

Equation (3.40) which is the form of the pressure-drop relationship for flow

---

* We use the symbol ω here to denote the void fraction, instead of the more common ε or P, in order to avoid confusion with the emissivity ε, which is introduced in later chapters, and with the pressure P used in this chapter.

through packed beds, is valid for the lower range of Reynolds number (laminar flow region), where K_1^{-1} has been found to equal 4.2.[2] Insertion of this value into Eq. (3.40) gives

$$V_0 = \frac{1}{4.2} \frac{\Delta P}{L\eta S_0^2} \frac{\omega^3}{(1-\omega)^2}, \tag{3.41}$$

which is the *Blake-Kozeny* equation. Note that this is the same as the D'Arcy equation, where

$$k_D = \frac{\omega^3}{4.2\eta S_0^2 (1-\omega)^2}, \tag{3.42}$$

or

$$\mathscr{P} = \frac{\omega^3}{4.2 S_0^2 (1-\omega)^2}. \tag{3.43}$$

These equations emphasize the fact that k_D depends on the properties of both the fluid and solid, while $\mathscr{P}$ depends only on the properties of the solid phase.

According to the flow behavior in all forced convection systems, a fluid velocity is eventually reached beyond which laminar flow no longer prevails. Under these conditions we again resort to the use of a friction factor which can be correlated solely as a function of a Reynolds number. For packed beds, the modified friction factor may be measured by

$$f_c = \frac{\Delta P \omega^3}{L S_0 \rho V_0^2 (1-\omega)}. \tag{3.44}$$

This is analogous to Eq. (3.17). We utilize a modified Reynolds number for packed beds in the correlation

$$\mathrm{Re}_c = \frac{\rho V_0}{\eta(1-\omega)S_0}. \tag{3.45}$$

When the flow exceeds that for $\mathrm{Re}_c \cong 2$, then Eq. (3.41) no longer applies and we use Fig. 3.10. The equation

$$\frac{\Delta P}{L} = \frac{4.2\eta V_0 S_0^2 (1-\omega)^2}{\omega^3} + \frac{0.292 \rho V_0^2 S_0 (1-\omega)}{\omega^3} \tag{3.46}$$

describes the entire curve analytically; the second term on the right describes the pressure drop under highly turbulent flow conditions. In dimensionless groups,

$$f_c = \frac{4.2}{\mathrm{Re}_c} + 0.292. \tag{3.47}$$

---

[2] S. Ergun, *Chem. Eng. Prog.* **48**, 93 (1952). The constant 4.2 is not universally selected; some believe the value to be as high as 5.0, based on a paper by P. C. Carman, *Trans. Inst. of Chem. Eng.* **15**, 150–166 (1937).

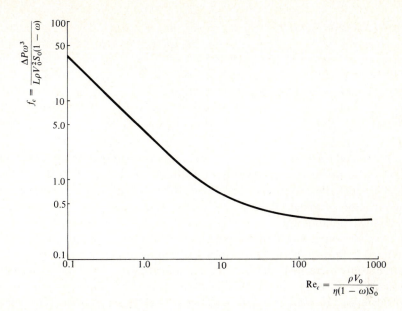

Fig. 3.10 Pressure-drop correlation for flow through packed beds. (Based on S. Ergun, *Chem. Eng. Prog.* **48**, 93 (1952).)

When applying this expression to gases, we take the density of the gas as the arithmetic average of the end densities. For large pressure drops, however, it is probably best to apply Eq. (3.46), to express pressure gradient in the differential form, and to integrate over the bed thickness, taking into account the variations in density, viscosity, and superficial velocity.

Thus far, we have defined the modified Reynolds number and friction factor in terms of S_0, the particle surface area per unit particle volume. If the particles in the bed are spheres of uniform size, S_0 may be easily related to the diameter, D_P, by

$$S_0 = \frac{\pi D_P^2}{(\pi D_P^3/6)} = \frac{6}{D_P}.$$ (3.48)

Then Eq. (3.46) becomes

$$\frac{\Delta P}{L} = \frac{150\eta V_0(1 - \omega)^2}{D_P^2\omega^3} + \frac{1.75\rho V_0^2(1 - \omega)}{D_P\omega^3},$$ (3.49)

which is known as *Ergun's equation*. It is then useful to define a new friction factor which is often associated with Ergun's work:

$$f_E = 6f_c = \frac{D_P \Delta P\omega^3}{L\rho V_0^2(1 - \omega)},$$ (3.50)

and a new Reynolds number

$$\mathrm{Re_E} = 6\,\mathrm{Re}_c = \frac{D_P \rho V_0}{\eta(1 - \omega)}. \tag{3.51}$$

Now, in dimensionless form, Eq. (3.49) becomes

$$f_E = \frac{150}{\mathrm{Re_E}} + 1.75. \tag{3.52}$$

When a bed contains a mixture of different-size particles, the question arises as to what should be used for D_P in Eqs. (3.49)–(3.51). To this day, the question is still unanswered, but a suggestion commonly put forward (although not adequately tested) is to use the *volume–surface mean diameter*, $\bar{D}_{vs}$. This parameter is discussed in the following paragraph.

For a bed containing a mixture of different-size *spherical particles*, S_0 may be determined from the specific surface of the mixture, S_W (cm^2/mass of particles). If the material is screened, and the diameter of each size fraction, $\bar{D}_{P_i}$, is taken as

Table 3.2 Shape factors for screened particles

Substance	Type of measurement	Shape factor, λ
Sand (jagged)	Permeability	1.68[1]
Sand (nearly spherical)	Permeability	1.15[1]
Sand (angular)	Permeability	1.49[1]
Sand (flakes)	Permeability	2.54[1]
Sand (rounded)	Metallographic	1.24[1]
Crushed glass (jagged)	Metallographic	1.54[1]
Coal (pulverized)	Metallographic	1.37[1]
Coal dust (up to $\frac{3}{8}$ in.)	Metallographic	1.54[1]
Tungsten powder	Metallographic	1.12[1]
Cu shot	Settling techniques	1.00[2]
Sand	Settling techniques	1.59[2]
Coal	Settling techniques	1.72[2]
Silimanite	Settling techniques	1.72[2]
Limestone	Settling techniques	2.20[2]
Flake graphite	Settling techniques	7.96[2]
Crushed and screened ores		1.75[3]

[1] Values calculated from data presented by P. C. Carman, *Trans. Inst. of Chem. Eng.* **15**, 155 (1937).

[2] Derived from data in F. A. Zenz and D. F. Othmer, *Fluidization and Fluid Particle Systems*, Reinhold, New York, 1960, pages 184, 213.

[3] Typical value suggested by A. M. Gaudin, *Principles of Mineral Dressing*, McGraw-Hill, New York, 1939, page 132.

the arithmetic mean of the openings of the two screens defining the corresponding mass fraction, $\Delta\phi_i$, collected between them, then

$$S_W = \frac{6}{\rho_P} \sum_{i=1}^{n} \frac{\Delta\phi_i}{\bar{D}_{P_i}}, \tag{3.53}$$

where n is the number of screens used, and ρ_P is the density of the particulate material. The average particle diameter for this mixture is called the *volume–surface mean diameter*, defined as

$$\bar{D}_{vs} = \frac{6}{S_W \rho_P}. \tag{3.54}$$

Noting that $S_0 = S_W \rho_P$, then

$$S_0 = \frac{6}{\bar{D}_{vs}}, \tag{3.55}$$

and, by comparing Eq. (3.55) with Eq. (3.48), the proper value of D_P to use is $\bar{D}_{vs}$.

Finally, the situation arises that we have a bed of nonspherical particles. In this case, we define a shape factor λ, which does not depend on particle size and which is a function of the shape of the particles only, by:

$$S_0 = \frac{6\lambda}{D_P}. \tag{3.56}$$

Here D_P is a characteristic dimension of the particle, and thus serves to define λ. For a cube or a sphere, the simplest choice for D_P is the edge length or the diameter, respectively, and for both, λ is then unity. For screened irregular particles, D_P corresponds to $\bar{D}_{vs}$, in which $\bar{D}_{P_i}$ is still the arithmetic mean of the screen openings. With this definition of D_P, corresponding values of λ for screened materials are given in Table 3.2.

Now, replace D_P in Eq. (3.49) by D_P/λ or $\bar{D}_{vs}/\lambda$, and other forms of Ergun's equation evolve:

Uniform size particles

$$\frac{\Delta P}{L} = \frac{150\eta V_0 \lambda^2}{D_P^2} \frac{(1-\omega)^2}{\omega^3} + \frac{1.75\rho V_0^2 \lambda}{D_P} \frac{(1-\omega)}{\omega^3}, \tag{3.57a}$$

Nonuniform size particles

$$\frac{\Delta P}{L} = \frac{150\eta V_0 \lambda^2}{\bar{D}_{vs}^2} \frac{(1-\omega)^2}{\omega^3} + \frac{1.75\rho V_0^2 \lambda}{\bar{D}_{vs}} \frac{(1-\omega)}{\omega^3}. \tag{3.57b}$$

We should caution the reader that the literature in this field abounds with confusion concerning the definition of D_P for nonspherical particles, and the shape factor λ. In general, values of λ are very difficult to obtain, and therefore pressure drop equations utilizing D_P, such as Eq. (3.49), are not of much use unless the particles are spherical. It is usually more satisfactory to use Eq. (3.46), determining S_0 by means of permeability tests and Eq. (3.43).

Example 3.4 Sintering of iron ore is an important metallurgical process in which gases must penetrate through a bed of solids. In this process, loosely-packed fine particles of ore are sintered into larger particles by passing air through the bed, which in turn reacts with admixed coal to develop very high temperatures in the sinter. It is necessary that large amounts of air can pass through the bed without creating large pressure drops, which would require unduly large fans.

Calculate the pressure drop, prior to ignition, across a bed of sinter 12 in. deep ($\omega = 0.39$) for air flowing at 60°F, and with $V_0 = 25$ cm/sec. The surface area S measures 81 cm^2/cm^3.

$$\rho_{air} = 1.23 \times 10^{-3} \text{ g/cm}^3 \text{ (STP)},$$

$$\eta_{air} = 178 \times 10^{-6} \text{ poise (60°F)}.$$

Solution

$$\text{Re}_c = \frac{\rho V_0}{\eta S} = \frac{(1.23 \times 10^{-3})(25)}{(178 \times 10^{-6})(81)} = 2.13.$$

Using Eq. (3.46), we get

$$\frac{\Delta P}{L} = \frac{\rho V_0^2 S}{\omega^3}\left(\frac{4.2}{\text{Re}_c} + 0.292\right) = 2385 \text{ (dyn/cm}^2)/\text{cm}$$

$$= 2.34 \text{ in. H}_2\text{O/in. bed};$$

$$\Delta P = 2.34 \times 12 = 28.08 \text{ in. H}_2\text{O}.$$

The measured value in the case studied[3] was 25 in.

3.4.3 Wall effect

It is probably obvious to the reader that the container diameter must be a good deal larger than the mean particle diameter of the packing, in order for the above equations to be valid, since the void fraction at the wall will be larger than the bulk value of ω. This *wall effect* is demonstrated in Fig. 3.11 which shows that the container diameter should be approximately 20 times the particle diameter in order that Eq. (3.46) could predict the pressure drop to within 10% for Re_c between 2 and 150. If necessary, we may first compute the pressure drop using Eq. (3.46), and then use Fig. 3.11 for correction.

3.4.4 Applications

In practice, several tests have been developed for control of the permeability of aggregates. In foundries, a standard test subjects a cylindrical sample of sand 2 in.

---

[3] D. W. Mitchell, *J. Iron & Steel Inst.* **198**, 358 (1961).

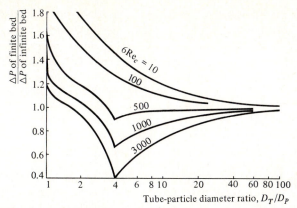

Fig. 3.11 Correction factor for wall effect. (Derived by H. E. Rose and A. M. Rizk, *Proc. Inst. Mech. Engrs.*, London, **60**, 493 (1950).)

in diameter ($A = 20.268$ cm^2), by 2 in. in length ($L = 5.08$ cm), to a standard pressure differential of 10 cm water (ΔP), and determines the time (t, sec) for 2000 cm^3 of air at room temperature to pass through the specimen. In this case, we use D'Arcy's law (Eq. 3.30) in its simplest form, and the permeability k_D is obtained from

$$k_D = \frac{3007.2}{t}\left(\frac{\text{cm}^2}{\text{min-cm H}_2\text{O}}\right). \tag{3.58}$$

The units are common only to this test. Sands with high permeability to gases have k_D values in the range 300–500; a typical value would be 60–120 cm^2/min-cm H$_2$O.

Voice, Brooks, and Gledhill[4] studied the relationship between pressure drop across a sinter bed and the rate of flow of air and found that

$$V_0 = \frac{Q}{A} = k_D\left(\frac{\Delta P}{L}\right)^{-n}, \tag{3.59}$$

where k_D is the permeability constant, and n is a variable which depends on the stage of the sintering process, but which averages 0.60 for the process as a whole, as long as extra-fine ore is not used. For the cooling of hot sinter, values of 0.53 have been found for n. If one considers Eq. (3.46) and assumes turbulent flow, that is, only the second term on the right-hand side is significant, then one finds

$$V_0 = \left(\frac{\omega^3}{0.292\rho S_0(1-\omega)}\right)^{0.5}\left(\frac{\Delta P}{L}\right)^{0.5} = k_D\left(\frac{\Delta P}{L}\right)^{0.5},$$

which is in good agreement with the observed values, considering the assumptions made.

---

[4] E. W. Voice, S. H. Brooks, and P. K. Gledhill, *I.S.I. Spec. Report* **53** (1955).

In powder metallurgy, the permeability of a sintered metal compact is of importance because it is a measure of the relationship between processing variables and the pore structure developed during sintering, and because, in some instances, the flow of a fluid through a porous sintered metal part is important. Self-lubricating bearings and porous metal filters for air, water, etc. can serve as examples.

Morgan[5] has proposed an equation for steady flow conditions through porous sintered metal:

$$\frac{\Delta P}{L} = \frac{\eta V_0}{\phi} + \frac{\rho V_0^2}{\Phi},$$
(3.60)

where ϕ is called the viscous permeability, and Φ the inertial permeability; V_0, η, and ρ are defined as before. We should compare this equation with Eq. (3.46), from which it appears that

$$\phi = \frac{\omega^3}{4.2 S_0^2 (1 - \omega)^2} \equiv \mathscr{P} \quad \text{and} \quad \Phi = \frac{\omega^3}{0.292 S_0 (1 - \omega)}.$$

Now, at low values of ω, S_0 is the surface area *seen* by the fluid (since there may be internal porosity which is not connected to any flow channels, and which contributes to the ω determined *in situ*, but whose ω surface area does not contribute to the S_0 which affects the flow). Because ω is then too large, S_0 measured in this manner will be larger than the true value of S_0. The ω used when the isolated pore volume becomes significant should only be the interconnected void volume fraction.

3.4.5 The void fraction

An important parameter in the above equations for flow through packed beds is the voidage ω, which is often difficult to know or predict under industrial conditions. We define the voidage, or void fraction, by

$$\omega = \frac{\text{volume of voids}}{\text{total bed volume}},$$

$$\omega = \frac{\text{volume of voids}}{\text{volume of voids} + \text{volume of solids}},$$
(3.61)

or

$$\omega = 1 - \frac{\text{bulk density of the bed or compact}}{\text{true density of the solid material}}.$$
(3.62)

The closest possible packing of equal-size spheres gives $\omega = 0.259$. This is rarely achieved in practice, since most materials are at least somewhat irregular

[5] V. T. Morgan, *Symposium sur la Metallurgie des Poudres*, Editions Metaux (sponsored by Société Française de Metallurgie), Paris, June, 1964, page 419.

in shape, and exhibit a relatively high degree of friction between particles. Typical values of ω lie between 0.35 and 0.5, except, of course, where pressure is applied in order to force the particles together. In most instances, we must measure ω *in situ*, using Eq. (3.62).

Since there are voids in the packing of equal-size spheres, small particles may enter without changing the overall volume of the bed, so clearly, the size distribution of the particles is going to have an effect on the bulk density of the bed. Furnas[6] has made some classical studies on the void fractions in packed beds, using binary mixtures of particles (two different sizes) in various proportions. He started, in each case, with materials with the same initial (single component) voidages ω^0, mixed them, and measured ω for the mixture. Figure 3.12 shows his results for $\omega^0 = 0.5$. It is clear that, as the difference in the particle size increases, lower and lower ω values are obtainable, with minimum voidages occurring in the range 55–67 wt % of the larger-size material. Essentially, the same range for minimum voidages is found when $\omega^0 = 0.6$ and 0.4.

Furnas has also made calculations of the *minimum* void fractions for three- and four-component mixtures of particles in which each component *alone* exhibits the same voidage. If the coarse and fine particles in a binary mixture are of *equal particle density* and *equal initial voidage*, then when the voids, ω, in the coarse material are saturated with fines, the volume fraction of coarse material is $1/(1 + \omega)$, and the amount of fine material is $1 - [1/(1 + \omega)]$. A third, still smaller, component can then be added to the binary mixture, filling the interstices of the second component, etc. The *total volume fraction* of solids in the mixture is then

$$(1 - \omega_m) = \frac{1}{1 + \omega} + \left(1 - \frac{1}{1 + \omega}\right) + \left(1 - \frac{1}{1 + \omega}\right) \frac{\left(1 - \dfrac{1}{1 + \omega}\right)}{\left(\dfrac{1}{1 + \omega}\right)} + \cdots,$$

which simplifies to

$$(1 - \omega_m) = \frac{1}{1 + \omega} + \frac{\omega}{1 + \omega} + \frac{\omega^2}{1 + \omega} + \cdots + \frac{\omega^n}{1 + \omega}. \qquad (3.63)$$

When each term in the series (Eq. 3.63) is multiplied by $100/(1 - \omega_m)$, the result is the percentage of each component in a mixture which will produce the minimum voids. Figure 3.13 illustrates the results obtained by Furnas for two samples of mixtures with all initial components having $\omega = 0.4$ and 0.6, respectively. It should be emphasized that Figs. 3.12 and 3.13 refer only to minimum voids.

White and Walton,[7] who used geometric considerations and *assumed close packing of spheres*, computed the number and size of particles needed to fill interstices in the packing with each addition of a smaller component. This has led

[6] C. C. Furnas, *Ind. & Eng. Chem.* **23**, 1052 (1931).

[7] H. White and S. Walton, *J. Am. Ceramic Soc.* **20**, 155 (1937).

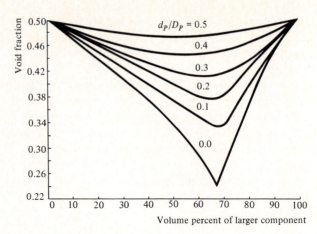

Fig. 3.12 Experimental voidage of two-component particle mixtures, both having initial void fractions of 0.5. The numbers on the curves refer to the ratio of the particle diameters. (From Furnas, *ibid.*)

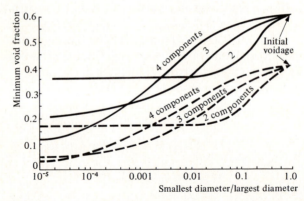

Fig. 3.13 Calculated minimum voidage in two-, three-, and four-component particle mixtures. (From Furnas, *ibid.*)

to a reduction in the overall void fraction as smaller particles are added. With an ideal five-component mixture, the voidage would be 0.149, decreased from the initial 0.259. Table 3.3 indicates some of the results of their calculations. Experience with foundry sands indicates that this approach, although idealized, is useful. For details of the calculations, the reader should consult the original paper.

3.5 FLUIDIZED BEDS

For an up-draft packed bed system, an upper limit exists for the fluid flow rate. Let us consider what happens if the flow rate through a *static* bed steadily increases. The pressure drop across the bed increases steadily with increasing flow rate until,

Table 3.3 Effect of size gradations on properties of a rhombohedral packing. (From White and Walton, *ibid.*)

| Property | Primary | Mixture with: | | | |
		Secondary	Tertiary	Quaternary	Quinary
Diameter	d	$0.414d$	$0.225d$	$0.177d$	$0.116d$
Relative number	1	1	2	8	8
Volume of space	$0.524d^3$	$0.037d^3$	$0.006d^3$	$0.0026d^3$	$0.0008d^3$
Volume of spheres added	$0.524d^3$	$0.037d^3$	$0.012d^3$	$0.021d^3$	$0.0064d^3$
Total solid volume of spheres added	$0.524d^3$	$0.561d^3$	$0.573d^3$	$0.595d^3$	$0.602d^3$
Fractional voids in mixtures	0.2595	0.207	0.190	0.158	0.149
Weight of spheres in final mixture, %	77.08	5.47	1.75	3.31	0.97
Total surface area of spheres in mixture		$3.14d^2$	$3.68d^2$	$4.00d^2$	$4.77d^2$	$5.11d^2$

at a certain point, the bed expands slightly, and the individual particles become supported in the fluid stream with freedom to move relative to each other. At this point, the bed is no longer static, and is said to be *fluidized*.

We relate the pressure drop for the start of fluidization (incipient fluidization) to the weight of solid particles supported in the fluid stream by

$$\frac{\Delta P}{L} = (\rho_s - \rho)g(1 - \omega), \tag{3.64}$$

where ρ_s = density of solid particle, and ρ = density of fluid. Up to the point of fluidization, however, we give the pressure drop by a relationship such as Eq. (3.57a), and when (3.57a) and (3.64) are equated, we obtain, after rearranging,

$$\frac{15\lambda^2(1 - \omega)}{\omega^3}\left(\frac{\rho V_0 D_P}{\eta}\right) + \frac{1.75\lambda}{\omega^3}\left(\frac{\rho V_0 D_P}{\eta}\right)^2 = \frac{D_P^3(\rho_s - \rho)\rho g}{\eta^2}. \tag{3.65}$$

By analogy with flow past submerged spheres, we can define a Reynolds number in fluidized beds:

$$\text{Re}_D' = \frac{D_P V_0 \rho}{\eta}, \tag{3.66}$$

$$\frac{150\lambda^2(1 - \omega)}{\omega^3}\text{Re}_D' + \frac{1.75\lambda}{\omega^3}\text{Re}_D'^2 = \frac{D_P^3(\rho_s - \rho)\rho g}{\eta^2}. \tag{3.67}$$

Wen and Yu[8] have shown that there are two empirical relationships between the shape factor λ and the void fraction at minimum fluidization ω_{mf}:

---

[8] C. Y. Wen and Y. H. Yu, Fluid Particle Technology, *Chem. Eng. Prog. Symposium Series*, No. 62, *AIChE*, New York, 1966.

$$\frac{\lambda^2(1 - \omega_{mf})}{\omega_{mf}^3} \cong 11 \qquad \text{and} \qquad \frac{\lambda}{\omega_{mf}^3} \cong 14.$$

When these are put into Eq. (3.67), we derive an expression, which is independent of λ or ω_{mf}, for the velocity at minimum fluidization:

$$\text{Re}'_{D_{mf}} = \sqrt{(33.7)^2 + 0.0408\ \text{Ga}} - 33.7, \tag{3.68}$$

where

$$\text{Ga} = \frac{D_P^3(\rho_s - \rho)\rho g}{\eta^2}$$

is called the Galileo number. Figure 3.14 is a plot of Eq. (3.68). In a bed with a narrow distribution of sizes, the particle diameter $\bar{D}_{vs}$, as defined by Eq. (3.54), should replace D_P. If the size distribution is wide, at least as far as the significant portions of the material are concerned, then the minimum fluidization velocity must be evaluated for the largest significant size present. Wen and Yu suggest

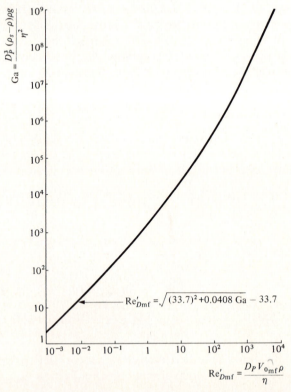

Fig. 3.14 Generalized correlation for minimum fluidization velocity. (Taken from Wen and Yu, *ibid.*)

that a ratio of the largest to the smallest significant particle diameters of about 1.3 is the dividing point between the two methods of defining which D_P to use.

The void fraction at this point, ω_{mf}, corresponds to the loosest possible packing of the material in a fixed-bed configuration, and this is usually greater than the initial fixed-bed voidage, so that some internal rearrangement of the bed takes place prior to the onset of fluidization, as shown in Fig. 3.15. With increasing velocity, the bed expands, or ω increases. If this expansion is uniform, with the interparticle spacing increasing uniformly, then we speak of *particulate fluidization*. Bed expansion continues until the velocity required to balance or support a single particle acting alone is reached, which is the terminal or free-fall velocity V_t. This point is reached when ω approaches 1, and from that point on, the pressure drop is determined by the equation for flow through an empty tube. At velocities above V_t, the particles are entrained and blow out of the reactor. Figure 3.15 shows the various stages of fluidization schematically.

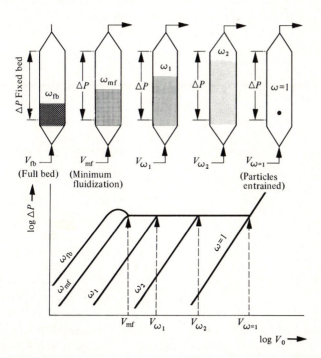

Fig. 3.15 Schematic representation of the relationship between the void fraction in the fluidized bed, the superficial velocity, and the pressure drop across the entire reactor, for particulate fluidization.

In Section 3.3.2 we discussed friction factors for flow past individual particles. In a fluidized bed, the particles, although suspended by the gas stream, exhibit a varying friction factor as the voidage varies, until finally at $\omega = 1$ the friction factor is the same as that for free-fall conditions. Wen and Yu found the relationships between the friction factor and ω, and expressed their results analytically:

$$\omega^{4.7}\, Ga = 18\, Re'_D + 2.70\, Re'^{1.687}_D. \tag{3.69}$$

This is shown in Fig. 3.16. Thus, for a given Re'_D value, one may calculate the bed voidage, or for a given bed expansion and particle size, one may calculate the required superficial velocity. Finally, at $\omega = 1$, the terminal velocity is reached:

$$V_t = \sqrt{\frac{4}{3}\, \frac{D_P(\rho_s - \rho)g}{f\rho}}, \tag{3.70}$$

where f is the friction factor for flow past the particles.

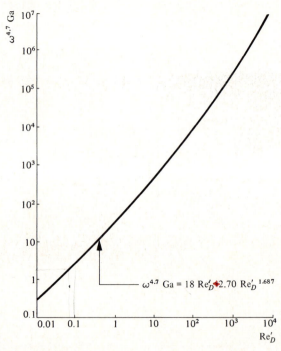

Fig. 3.16 Generalized correlation between the particle, bed, and fluid characteristics for a fluidized bed. (Taken from Wen and Yu, *ibid.*)

Fluidized beds allow more uniform heating and fluid–solid contact than fixed beds. They also offer the advantages of continuous operation since they may be continuously fed and the product withdrawn *ad infinitum*. For operation in this condition, the superficial velocity has to lie between that given by Eq. (3.68) and Eq. (3.70).

Applications of fluidized beds to metallurgical processes include the roasting of sulfide ores, the reduction of oxides, halides, and sulfides to metals, the drying of coal and sand, the heat treating of small parts, and the coating of pipe by fluidizing the organic coating material around the pipe.

Example 3.5 In developing a method for the recovery of uranium from spent nuclear fuel elements, engineers have conducted research on the feasibility of fluorinating U_3O_8 to UF_6 in a fluidized bed reactor.[9] The U_3O_8 particles are mixed with Al_2O_3 particles (90% by volume) and fluidized with N_2 at 750°K, to which a small amount of F_2 has been added. In order to start an experiment of this type, we must make some estimates.

First, we estimate the minimum fluidizing velocity. Since most of the particles are Al_2O_3, we make appropriate calculations for the Al_2O_3, and then see whether the velocities are satisfactory for the U_3O_8 particles.

The appropriate constants and data are

	ρ(g/cc)	η	D_P
U_3O_8	8.3	—	6×10^{-4} cm (6μ)
Al_2O_3	3.9	—	1.2×10^{-2} cm (120 mesh)
N_2 (750°K)	4.8×10^{-4}	3.35×10^{-4} P	—

Solution. In order to calculate $V_{0_{mf}}$, we need the Galileo number Ga:

$$Ga^{(Al_2O_3)} = \frac{(1.2 \times 10^{-2})^3(3.9)(4.8 \times 10^{-4})(981)}{(3.35 \times 10^{-4})^2} = 28.2.$$

From Fig. 3.14,

$$Re''^{(Al_2O_3)}_{D_{mf}} = 2 \times 10^{-2} = \frac{(4.8 \times 10^{-4})(1.2 \times 10^{-2})V_{0_{mf}}}{(3.35 \times 10^{-4})},$$

$$V_{0_{mf}} = 1.17 \text{ cm/sec}.$$

Check to see if this is satisfactory for the U_3O_8 particles:

$$Ga^{(U_3O_8)} = \frac{(6 \times 10^{-4})^3(8.3)(4.8 \times 10^{-4})(981)}{(3.35 \times 10^{-4})^2} = 7.53 \times 10^{-3}.$$

Using $V_{0_{mf}} = 1.17$ cm/sec, we obtain

$$Re'_{D_{mf}} = \frac{(6 \times 10^{-4})(1.17)(4.8 \times 10^{-4})}{3.35 \times 10^{-4}} = 1.00 \times 10^{-3},$$

which is greater than that for minimum fluidization if Ga $= 7.53 \times 10^{-3}$. Therefore, 1.17 cm/sec is the minimum fluidizing velocity.

---

[9] *Argonne National Laboratory Report 6763, Oct. 1963.*

Next, calculate the velocity for the operating condition when the bed voidage is 0.8:

$$ (\epsilon)^{4.7}\, Ga^{(Al_2O_3)} = (0.8)^{4.7}(28.2) = 9.9. $$

From Fig. 3.16,

$$ Re_D^{\prime(Al_2O_3)} = 0.5 = \frac{(1.2 \times 10^{-2})(4.8 \times 10^{-4})V_0}{3.35 \times 10^{-4}}, $$

$$ V_0 = 29.2 \text{ cm/sec}. $$

Under these conditions, there is elutriation of the U_3O_8, but when the U_3O_8 particles are fed into the bed from the bottom, the residence time within the bed is long enough for the reaction to take place, and the U_3O_8 is converted to UF_6 before it reaches the top of the bed. Whatever particles reach the top of the bed, are necessarily blown out, since

$$ V_t(6\mu\text{-}U_3O_8) = \frac{(6 \times 10^{-4})^2(8.3)(981)}{(18)(3.35 \times 10^{-4})} = 0.485 \text{ cm/sec}. $$

Here, the terminal velocity has been determined by substituting $f = 24\eta/D_P V_t \rho$ into Eq. (3.70). Therefore, if the reactor is operated at 29.2 cm/sec superficial velocity, the Al_2O_3 is fluidized, and the U_3O_8 is entrained when it reaches the top of the bed. The critical factor which determines whether the method works or not is the time it takes a U_3O_8 particle to work its way up through the Al_2O_3 bed.

PROBLEMS

3.1 Water at 80°F flows through a brass tube 100 ft long × $\frac{1}{2}$-in. I.D. at a rate of 50 gpm. Calculate the pressure drop (in psi) that accompanies this flow.

3.2 Evaluate the pressure drop in a horizontal 100 ft length of galvanized rectangular duct of dimensions 1 in. × 3 in. for

a) an average air flow velocity of 1.5 ft/sec at 100°F and 15 psia;

b) repeat for a flow velocity of 15 ft/sec.

3.3 Show that for flow through a slit with a spacing much less than the width, Eq. (3.22) gives the friction factor for laminar flow.

3.4 Refer to Problem 2.10 and calculate the maximum diameter of spherical alumina particles that would be expected to obey Stokes' law.

3.5 A steel ball (radius = 0.291 ft) is dropped through molten slag in order to determine its viscosity. The steel's density is twice that of the slag and a terminal velocity of 5 ft/sec is experimentally determined. Calculate the *kinematic* viscosity of the slag in ft²/sec.

3.6 A tower 50 ft high and 20 ft in diam. is packed with spherical pebbles (0.1 ft in diam.). Gas enters the top of the bed at 1000°F and leaves at the same temperature. The pressure at the bottom of the tower is 20 lb_f/in^2 and the bed porosity is 0.40. The gas has a viscosity of 0.1 lb_m/hr ft at 1000°F and atmosphere pressure; the density of the gas at 1000° is approximated by

$$\rho(lb_m/ft^3) = 0.05\ P\ (atm).$$

If the gas flows through at a rate of 753,600 lb_m/hr, what is the inlet pressure?

3.7 A bed of solids is 60 ft high and 15 ft in diam., with material A forming a central column with diameter 10 ft and material B filling the annulus between A and the outside. The pressure at the top of the bed is 10 psig and that at the bottom is 25 psig. For the following characteristics, calculate the fraction of the gas that passes through the A material.

$$A: \omega = 0.40,\ D_P = 3\ in.$$
$$B: \omega = 0.25,\ D_P = 0.75\ in.$$

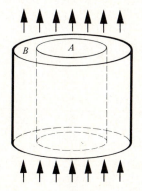

Assume that the temperature of the bed is uniform across the bed at any height L, that is, ρ_{gas} is the same in both columns. Assume that turbulent flow conditions prevail.

3.8 Gas flows through a packed bed 10 ft × 10 ft square cross section × 46.3 ft long. An inlet pressure of 15.1 psia is measured when 200 lb_m/hr of the gas flow through the bed at 200°F discharging at 15.0 psia. All we know about the bed porosity is that $0 < \omega < 0.6$. Evaluate the bed porosity from the above conditions and the data below:

$$D_P = 0.1\ ft$$
$$\eta = 0.05\ lb_m/hr\ ft$$
$$\rho = 0.0075\ lb_m/ft^3\ at\ 200°F,\ 15\ psia.$$

3.9 In the production of titanium, rutile (TiO_2) is fluidized with gaseous chlorine, and the reaction

$$TiO_2 + 2Cl_2 \rightarrow TiCl_4(g) + O_2$$

occurs with the oxygen being continuously removed by reaction with coke particles,

$$C + O_2 \rightarrow CO_2. \qquad \rho = 4.26\ g/cc$$

The rutile ore has a screen analysis of

$$-60 + 80M \qquad 5.0\%$$

$180\mu M$

$150 \ \mu m$

$-80 + 100M$ 106 6.2%

$-100 + 140M$ 77.4% $228 \ \mu p$

$-140 + 200M$ 75 11.4% $250 \degree C$

Calculate the possible range of flow of the chlorine at 950°C which is needed to fluidize the rutile in a bed 4 ft in diam. Give your results in terms of lb Cl_2/min, and in terms of velocity, ft/sec.

3.10 During the compaction of metal powders into sheet material (powder rolling), as shown in the figure below, the entrapped air is expelled from the loose powder. This expulsion occurs at the line AB. Below AB, the powder is coherent, that is, the particles are locked together, but above AB, the powder is loose, i.e., a normal packed bed.

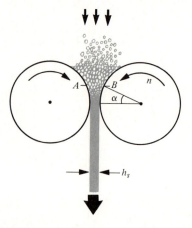

If the velocity of expulsion exceeds the minimum fluidization velocity, the powder does not feed properly into the roll gap and the sheet product is not satisfactory. This, in fact, limits the production rate of the process.

The following equation gives the superficial velocity V_0 (cm/sec), of gas expulsion from the coherent zone:

$$V_0 = \frac{2\pi R^2 n(\alpha - \sin\alpha\cos\alpha)}{\alpha[2R(1 - \cos\alpha) + h_s]},$$

where R = roll radius, cm, n = rolling speed, rev/sec, α = roll-coherent powder contact angle, rad, and h_s = roll gap, cm.

For copper powder, the angle α has been found to be 6°.

a) Calculate the rolling speed at which copper powder, with properties given below, will just begin to be fluidized at the plane AB, for strip thickness 0.05 cm and roll diameter 20 cm.

b) Calculate the corresponding strip speed.

c) Calculate the maximum allowable rolling speed for this powder (elutriation conditions).

d) Discuss the effect that order-of-magnitude changes in particle size and roll radius would have on production rates. Data are given as follows:
$\eta_{air} = 1.8 \times 10^{-4}$ poise, $\rho_{air} = 1.3 \times 10^{-3}$ g/cm^3, $D_{p_{Cu}} = 0.004$ cm, $\rho_{Cu} = 6.9$ g/cm^3.

4

ENERGY BALANCE APPLICATIONS IN FLUID FLOW

One of the most powerful analytical tools available for quantitative problems in engineering is the principle of the conservation of energy. In fluid flow, this principle is most often applied to open systems in the form of the *mechanical energy balance* which is also commonly called *Bernoulli's theorem*. The approach is macroscopic, since we examine the entire flow system with an entrance and an exit. This method is different from the one presented in Chapter 2, where we analyzed the microscopic volume element through which fluid flows as the starting point for solving problems.

4.1 CONSERVATION OF ENERGY

The conservation of energy can be applied to the flow of fluids such as that depicted in Fig. 4.1. With the aid of the symbols in Table 4.1, the general energy balance can be written

$$\frac{d}{dt}(E_{\text{tot}}) = -\Delta[(H + E_P + E_K)W] + Q - M + S_R. \tag{4.1}$$

In this form, the macroscopic energy balance is written for unsteady-state conditions in which the left side of the equation represents the accumulation of total energy in the system. The total energy, E_{tot}, is defined as the sum of internal,

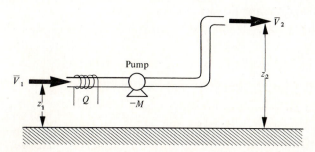

Fig. 4.1 Flow system for the application of the energy balance.

111

Table 4.1 Symbols used in the general energy balance

	Entering the system	Leaving the system
Mass flow rate	W_1	W_2
Pressure	P_1	P_2
Volume per unit mass	$1/\rho_1$	$1/\rho_2$
Energy per unit mass		
Potential	E_{P1}	E_{P2}
Kinetic	E_{K1}	E_{K2}
Internal	U_1	U_2
Enthalpy	$H_1 = U_1 + P_1(1/\rho_1)$	$H_2 = U_2 + P_2(1/\rho_2)$

Mechanical work done by the system, M

Net energy generation within the system from chemical reactions or other sources, S_R

Net heat input, Q

potential, and kinetic energy. The operator Δ signifies exit minus entrance quantities, and note that the expression $\Delta H = \Delta U + \Delta(P/\rho)$ has been employed in the usual manner for open systems.

The terms P_1/ρ_1 and P_2/ρ_2, which are parts of H_1 and H_2, respectively, are often called the *flow work* or *PV work*. They are recognized as the work required to put a unit of mass into the system at "1" in Fig. 4.1 and the work done by the system on a unit mass leaving at 2, respectively. After this brief discussion explaining the basis of the problem at hand, namely, the flow of fluids, we now proceed to focus attention on some of the specific terms in Eq. (4.1) for steady-state conditions,† namely,

$$\frac{d}{dt}(E_{\text{tot}}) = 0, \qquad W_1 = W_2 = W, \qquad \text{and} \qquad S_R = 0.$$

4.1.1 Evaluation of the kinetic energy terms

The rate of kinetic energy entering the system through area A_1 (normal to flow) is given by

$$W_1 E_{K1} = \frac{1}{2}\left[\int_0^{A_1} (\rho_1 v_1)v_1^2 \, dA_1\right]. \tag{4.2}$$

---

† For a less specific but still concise presentation of the general energy balance, refer to D. M. Himmelblau, *Basic Principles and Calculations in Chemical Engineering*, Prentice-Hall, Englewood Cliffs, New Jersey, 1967, pages 261–281 and 397–401.

Examine Eq. (4.2) for meaning; one may be tempted at this point to simply write $\frac{1}{2}(\rho A_1 \bar{V}_1)\bar{V}_1^2$ to represent the kinetic energy entering the system. However, because the velocity varies in some manner across the cross-sectional area, the integration indicated by Eq. (4.2) is in order.

To account for the velocity distribution, the kinetic energy term can be written as:

$$W_1 E_{K1} = \frac{1}{2\beta_1} W_1 \bar{V}_1^2, \tag{4.3}$$

where β_1 is defined by

$$\frac{1}{\beta_1} \equiv \frac{1}{A_1} \int_0^{A_1} \left(\frac{v_1}{\bar{V}_1}\right)^3 dA_1. \tag{4.4}$$

We can also use a similar expression for the rate at which kinetic energy leaves the system; so we write the difference between leaving and entering the system

$$W \Delta E_K = W\left(\frac{\bar{V}_2^2}{2\beta_2} - \frac{\bar{V}_1^2}{2\beta_1}\right). \tag{4.5}$$

For laminar flow in a circular tube, $\beta = 0.5$, and for turbulent flow, β is nearly unity, and is usually approximated as such.

4.1.2 Evaluation of the potential energy terms

Potential energy is defined relative to some arbitrary reference plane for the fluid both leaving and entering the system. Therefore, the difference in potential energy of the fluid leaving and entering is simply

$$W\Delta E_P = Wg(z_2 - z_1),$$

or

$$W \Delta E_P = Wg \, \Delta z. \tag{4.6}$$

4.1.3 Mechanical energy balance, Bernoulli's theorem

Substituting Eqs. (4.5) and (4.6) into Eq. (4.1) for steady-state conditions yields

$$\Delta H + \left(\frac{\bar{V}_2^2}{2\beta_2} - \frac{\bar{V}_1^2}{2\beta_1}\right) + g \, \Delta z + M^* - Q^* = 0. \tag{4.7}$$

Here $M^* = M/W$, and $Q^* = Q/W$. In this form, all the terms are now expressed per unit mass of fluid flowing through the system. Further, if we express H by $U + P/\rho$, we obtain *the overall energy balance* form of Bernoulli's theorem:

$$\Delta U + \Delta\left(\frac{P}{\rho}\right) + \left(\frac{\bar{V}_2^2}{2\beta_2} - \frac{\bar{V}_1^2}{2\beta_1}\right) + g \, \Delta z + M^* - Q^* = 0. \tag{4.8}$$

The equation is also commonly found in its differential form for a differential segment of the system:

$$dU + d\left(\frac{P}{\rho}\right) + d\left(\frac{\bar{V}^2}{2\beta}\right) + g\,dz - \delta Q^* = 0. \tag{4.9}$$

Here δ prefacing Q^* emphasizes that it is not a true differential but that it merely represents the heat which is transferred across the system boundary.

A more common form of the energy balance for applications to fluid flow is a variation referred to as the *mechanical energy balance* form of Bernoulli's theorem. The change in the internal energy for the unit mass of fluid as it passes through a short segment of the system is

$$dU = \delta Q^* - P d(1/\rho) + \delta E_f, \tag{4.10}$$

where δE_f is the *mechanical energy per unit mass lost as frictional conversion into heat*. Substituting Eq. (4.10) into (4.9) and expanding $d(P/\rho)$ into

$$[(1/\rho)\,dP + P d(1/\rho)]$$

yields

$$\frac{1}{\rho}\,dP + d\left(\frac{\bar{V}^2}{2\beta}\right) + g\,dz + \delta E_f = 0.$$

The integral of this equation, which applies to the whole system, finally yields (with M^* reinserted) the mechanical energy balance, or, Bernoulli's theorem in the form which can be applied directly to most problems of fluid flow:

$$\int_{P_1}^{P_2} \frac{dP}{\rho} + \left[\frac{\bar{V}_2^2}{2\beta_2} - \frac{\bar{V}_1^2}{2\beta_1}\right] + g\,\Delta z + M^* + E_f = 0. \tag{4.11}$$

Actually, Bernoulli's equation only applies to the flow of ideal fluids, where $M^* = E_f = 0$, but Eq. (4.11) having been used so frequently in engineering, has taken on the connotation of Bernoulli's equation as well. Also note that it is written in terms of unit mass of material flowing.

4.2 FRICTION LOSSES IN STRAIGHT CONDUITS

Due to the utility of Eq. (4.11) in engineering calculations, considerable effort has been made to develop methods for estimating the friction loss E_f in various flow systems.

If we consider the friction loss in a straight conduit with steady flow of constant-density fluid, we can determine E_f from the friction factor f. If a conduit has an arbitrary but constant cross-sectional area A and a length L, and if the fluid flows only due to a pressure difference of $P_2 - P_1$, then Eq. (3.3) gives us the force exerted by the fluid on the wall.

$$F_K = (P_1 - P_2)A. \tag{4.12}$$

Since we are discussing a horizontal conduit with no mechanical work supplied to or withdrawn from the system, Eq. (4.11) reduces to

$$E_f = \frac{1}{\rho}(P_1 - P_2). \tag{4.13}$$

By comparing Eqs. (4.12) and (4.13), we see that

$$E_f = \frac{F_K}{\rho A}. \tag{4.14}$$

For circular tubes, $A = \pi R^2$ and $F_K = 2\pi R L(\frac{1}{2}\rho \bar{V}^2)f$. Thus,

$$E_f = 2f\left(\frac{L}{D}\right)\bar{V}^2. \tag{4.15}$$

Again, for noncircular conduits, D may be replaced by the equivalent diameter D_e for turbulent flow. Equation (4.15) is a form of the well-known *Fanning equation*. Caution must be exercised by the reader in the exact definition of the friction factor; in much of the literature the friction factor is defined such that it is four times the friction factor employed here. Therefore, the Fanning equation may be found in a different form from that in Eq. (4.15).

Example 4.1 A fan draws air at rest, and sends it through a straight rectangular duct 8 in. × 12 in. and 180 ft long. The air enters at 60°F and 750 mm Hg pressure at a rate of 1000 ft³/min. What theoretical horsepower is required of the fan if the air discharges at 750 mm Hg?

Solution. The problem does not state whether the duct is horizontal, vertical, or at an angle; thus, the potential energy change of the gas as it passes through the system cannot be defined. However, since the density of a gas is quite small, we may safely ignore the potential energy term in Eq. (4.11). Consider the system below.

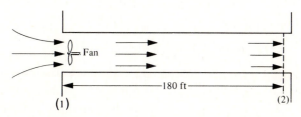

With the fan included within the system, Bernoulli's equation is written to include M^*.

Since $P_1 \cong P_2 = 750$ mm Hg, then $\int_{P_1}^{P_2} dP = 0.$† Also $\bar{V}_1 = 0$ and $\Delta z = 0$. Under these conditions, Eq. (4.11) reduces to

$$\frac{\bar{V}^2}{2\beta_2} + M^* + E_f = 0.$$

Now

$$D_e = \frac{2(\frac{2}{3} \times 1) \text{ ft}^2}{(\frac{2}{3} + 1) \text{ ft}} = 0.80 \text{ ft}.$$

At 1 atm and 60°F, the density of air is 0.0763 lb_m/ft^3. Therefore,

$$\rho \cong \frac{750 \text{ mm Hg}}{760 \text{ mm Hg}} \left| \frac{0.0763 \text{ lb}_m}{\text{ft}^3} \right| = 0.0753 \text{ lb}_m/\text{ft}^3.$$

(Why is the approximate sign rather than the equal sign employed here?) Also, $\eta = 0.043$ $\text{lb}_m/\text{hr ft}$,

$$\bar{V}_2 = \frac{1000 \text{ ft}^3}{\text{min}} \left| \frac{1}{\frac{2}{3} \text{ ft}^2} \right| \frac{60 \text{ min}}{1 \text{ hr}} = 90,000 \text{ ft/hr}.$$

The Reynolds number is now determined:

$$\text{Re} = \frac{D_e \bar{V}_2 \rho}{\eta} = \frac{(0.8)(90,000)(0.0753)}{(0.043)}$$

$$= 1.26 \times 10^5.$$

From Fig. 3.2, $f = 0.0042$ (smooth). Since

$$E_f = 2\left(\frac{L}{D_e}\right)\bar{V}^2 f,$$

the energy balance can be written as

$$M^* + \left[\frac{1}{2\beta_2} + 2f\left(\frac{L}{D_e}\right)\right]\bar{V}^2 = 0.$$

For $\beta_2 \cong 1$ (turbulent flow),

$$M^* = -\left[\frac{1}{2} + (2)(0.0042)\left(\frac{180}{0.80}\right)\right](9 \times 10^4)^2 \text{ ft}^2/\text{hr}^2$$

$$= -\frac{1.94 \times 10^{10} \text{ ft}^2}{\text{hr}^2} \left| \frac{1 \text{ hr}^2}{3600^2 \text{ sec}^2} \right| \frac{\text{sec}^2\text{-lb}_f}{32.2 \text{ ft-lb}_m}$$

---

† By stating that $P_2 = 750$ mm Hg, we ignore the *exit loss* which we shall discuss in Section 4.3.

$$= -46.5 \frac{\text{ft-lb}_f}{\text{lb}_m}.$$

Then,

$$\text{hp} = \frac{46.5 \text{ ft-lb}_f}{\text{lb}_m} \left| \frac{0.0753 \text{ lb}_m}{\text{ft}^3} \right| \frac{1000 \text{ ft}^3}{\text{min}} \left| \frac{60 \text{ min}}{1 \text{ hr}} \right| \frac{5.05 \times 10^{-7} \text{ hp}}{1 \text{ ft-lb}_f/\text{hr}}$$

$$= 0.106.$$

For straight conduits, Eq. (4.15) gives the friction loss E_f. However, in most flow systems there are fittings, bends, changes in cross sections, valves, etc. Therefore additional resistances must be included in E_f for the best application of Eq. (4.11). The following sections review some of the methods used to calculate such resistances.

4.3 ENLARGEMENT AND CONTRACTION

The friction loss associated with a *sudden enlargement* or a *sudden contraction* (Fig. 4.2) is calculated by using a *friction-loss factor* e_f. The friction loss for the particular geometry that upsets flow in conduits is evaluated in the following manner:

$$E_f = \tfrac{1}{2}\bar{V}^2 e_f, \tag{4.16}$$

where $\bar{V}$ is the velocity of fluid in the *smaller* cross section. (Since other conventions are also used, the reader should always be certain of which velocity is chosen to define e_f.) By analogy with Eq. (4.15), e_f corresponds to a friction factor with *built-in* geometry effects. Hence, e_f depends on a Reynolds number and a geometric ratio of the system. We present values of e_f for enlargements and contractions in circular cross sections in Figs. 4.3 and 4.4, respectively. For both these figures, D is the smaller diameter, Re is calculated using the same D, and L is the length of the smaller pipe considered.

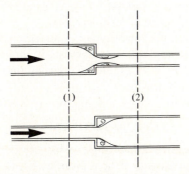

Fig. 4.2 Sudden contraction and enlargement.

The results for e_f in the case of enlargements apply equally well to all exit shapes (except for gradual expansions discussed in the following paragraphs), as

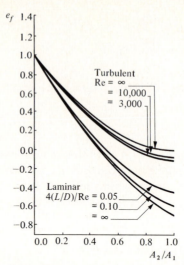

Fig. 4.3 Friction-loss factor for sudden expansion. (From W. M. Kays and A. L. London, *Trans. ASME* **74**, 1179 (1952).)

can be seen by viewing the flow in Fig. 4.2. The extent of the region of vortices after the expansion does not depend on whether the corners are rounded. However, e_f for contraction (entrances) can be modified significantly. Figure 4.5 shows the ratio of e_f for various inlet shapes to that for a sharp-edged entry.

In flow through *gradual enlargements*, the energy losses are significantly reduced due to the elimination of the vortices shown in Fig. 4.2. Experiments on tapered enlargements show that the friction-loss factor e_f depends on both the taper angle β and the area ratio A_1/A_2 (Fig. 4.6). In reality, e_f should also depend on a Reynolds number, but this is not given in the figure, since, in turbulent flow, the dependence of e_f on the Reynolds number is small, as can be seen in Figs. 4.3 and 4.4. Consequently, data presented for e_f are usually assumed to apply to all Reynolds numbers in the regime of turbulent flow.

In *gradual contractions*, the converging boundaries have a tendency to steady the flow, and if the included angle of convergence is 30° or less, the energy loss factor can be approximated as 0.05–0.10. Again, when using these data, one must realize that they apply only to turbulent flow.

4.4 FLOW THROUGH VALVES AND FITTINGS

To evaluate flow through valves and fittings, we usually assign an equivalent length to the fixture such that the pressure drop is given by Eq. (4.15), where L is replaced by L_e. Therefore

$$E_f = 2f \frac{L_e}{D} \bar{V}^2, \tag{4.17}$$

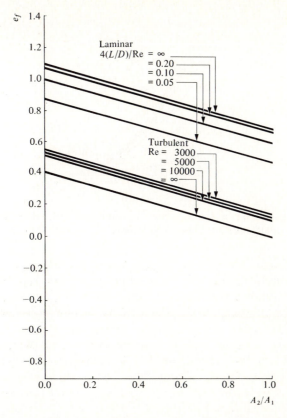

Fig. 4.4 Friction-loss factor for sudden contraction. (From Kays and London, *ibid.*)

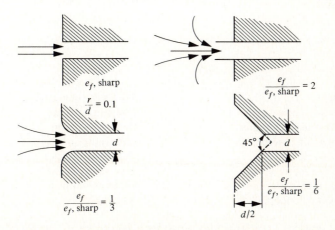

Fig. 4.5 Entrance-loss coefficients.

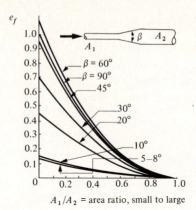

Fig. 4.6 Friction-loss factor for turbulent flow through gradual enlargements.   (From *Steam—Its Generation and Use*, Babcock & Wilcox, New York, 1963, page 8–17.)

where f is evaluated for Re with the pipe diameter equal to the fitting diameter, and L_e/D for turbulent flow of various fittings is given in Table 4.2. Thus, when evaluating the friction loss through a system with fittings, the sum of the equivalent lengths of all the fittings is added to the length of the pipe; E_f is then determined by using Eq. (4.17).

For example, if a pipe network contains 20 ft of $\frac{3}{4}$-in. pipe (I.D. = 0.824 in.), two elbows (90°, medium radius), and a wide open gate valve, then the L/D ratio to be substituted into Eq. (4.15) would be

$$\left(\frac{L}{D}\right) = \left(\frac{L}{D}\right)_{\text{pipe}} + 2\left(\frac{L_e}{D}\right)_{\text{elbows}} + \left(\frac{L_e}{D}\right)_{\text{valve}}$$

$$= \frac{20}{0.824/12} + (2)(26) + 7 = 351.$$

Table 4.2 Equivalent length for various fixtures, turbulent flow*

Fitting	L_e/D	Fitting	L_e/D
45° elbow	15	Tee (as el, entering branch)	90
90° elbow, standard radius ..	31	Couplings, unions ....Negligible	
90° elbow, medium radius...	26	Gate valve, open............	7
90° elbow, long sweep.........	20	Gate valve, $\frac{1}{4}$ closed.........	40
90° square elbow..............	65	Gate valve, $\frac{1}{2}$ closed.........	190
180° close return bend......	75	Gate valve, $\frac{3}{4}$ closed.........	840
Swing check-valve, open...	77	Globe valve, open............	340
Tee (as el, entering run)......	65	Angle valve, open............	170

* W. M. Rohsenow and H. Y. Choi, *Heat, Mass, and Momentum Transfer*, Prentice-Hall, Englewood Cliffs, New Jersey, 1961, p. 64.

4.5 FLOW THROUGH SMOOTH BENDS AND COILS

In curved pipes, the friction loss may rise considerably above the values for straight lengths of pipe, and the transition Reynolds number may be much higher than 2100. The maximum Reynolds number for laminar flow, or the critical Reynolds number, Re_c, is given as a function of the coil diameter D_c and pipe diameter D:[1]

$$Re_c = 20{,}000\left(\frac{D}{D_c}\right)^{0.32}, \qquad 15 < \frac{D_c}{D} < 860. \tag{4.18}$$

This result is shown in Fig. 4.7.

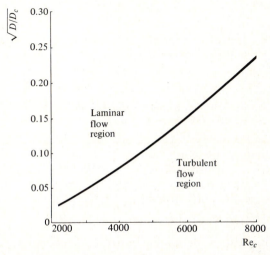

Fig. 4.7 Transition Reynolds number for flow through smooth curved pipe or coil. (Eq. 4.18 from H. Ito, *Trans. ASME* **82**, Series D, 123–34 (1959).)

The friction loss for laminar flow in a curved pipe can be expressed in terms of the ratio of the friction factor for the curved pipe to that for the same length of straight pipe. This ratio is a function of the Dean number, $Re(D/D_c)^{1/2}$, as shown in Fig. 4.8. Note by referring to Fig. 4.8 that the effect of curvature is negligible for $Re(D/D_c)^{1/2} < 10$.

We can compute the friction loss for turbulent flow by using Eq. (4.15) and applying the ratio of curved-to-straight pipe friction factors, f_c/f_s, given by[2]

$$\frac{f_c}{f_s} = \left[Re\left(\frac{D}{D_c}\right)^2\right]^{1/20} \tag{4.19}$$

for $Re(D/D_c)^2 > 6$.

---

[1] H. Ito, *Trans. ASME* **82**, Series D, 123–34 (1959).
[2] H. Ito, *ibid.*

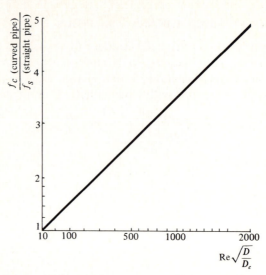

Fig. 4.8 Effect of curvature on f in laminar flow. (From *Steam—Its Generation and Use*, Babcock & Wilcox, New York, 1963, page 8–17.)

Example 4.2 An induction melting furnace has a 14-turn coil of $\frac{1}{2}$ in. I.D. copper-tubing wound with a coil diameter of 16 in. For adequate cooling, the manu-facturer's specifications call for water at 4 gpm. Determine the water pressure that should be available in the water supply line in order to deliver the 4 gpm through the coil if the water discharges at 1 atm.

Solution. For this situation, Bernoulli's equation reduces to:

$$E_f = \frac{P_2 - P_1}{\rho} = \frac{\Delta P}{\rho},$$

neglecting the potential energy contribution. The task at hand, then, is to evaluate the friction loss E_f:

$$\text{Re} = \frac{D\bar{V}\rho}{\eta} = \frac{4W}{\pi D \eta},$$

where W is the mass-flow rate (lb_m/hr).

$$\text{Re} = \frac{4}{\pi} \left| \frac{4 \text{ gal}}{\text{min}} \right| \frac{8.33 \text{ lb}_m}{1 \text{ gal}} \left| \frac{60 \text{ min}}{1 \text{ hr}} \right| \frac{24}{1 \text{ ft}} \left| \frac{\text{ft hr}}{2.42 \text{ lb}_m} \right.$$

$$= 2.52 \times 10^4.$$

$$\left(\frac{D}{D_c}\right)^{1/2} = \left(\frac{1/2}{16}\right)^{1/2} = 0.177.$$

From Fig. 4.7, the flow is shown to be turbulent, and so we apply Eq. (4.19):

$$\frac{f_c}{f_s} = \left[2.52 \times 10^4 \left(\frac{1}{32} \right)^2 \right]^{1/20} = 1.17.$$

In order to determine f, we refer to Fig. 3.2 which yields $f = f_s = 0.0075$, so that $f_c = 0.0088$.

The length of pipe in the coil is $14 \times \pi D_c$, so that we write the friction loss, using Eq. (4.15), as

$$E_f = 2f_c \left(\frac{14\pi D_c}{D} \right) \bar{V}^2.$$

The pressure drop is then

$$\Delta P = 2f_c \left(\frac{14\pi D_c}{D} \right) \rho \bar{V}^2.$$

Since

$$\rho \bar{V}^2 = \mathrm{Re}^2 \left(\frac{\eta}{D} \right)^2 \frac{1}{\rho} = 3.44 \times 10^{10} \frac{\mathrm{lb}_m}{\mathrm{ft \, hr}^2}.$$

$$\Delta P = \frac{(2)(0.0088) \left| 14\pi \times 16 \text{ in.} \right| 3.44 \times 10^{10} \text{ lb}_m \left| 1 \text{ hr}^2 \right| \sec^2 \text{ lb}_f \left| 1 \text{ ft}^2 \right.}{\left| \frac{1}{2} \text{ in.} \right| \quad \mathrm{ft \, hr}^2 \left| 3600^2 \sec^2 \right| 32.2 \text{ ft lb}_m \left| 144 \text{ in}^2 \right.}$$

$$= 14.2 \text{ lb}_f/\text{in}^2.$$

Thus the line pressure must be at least 14.2 psig.

4.6   FLOW MEASUREMENT

Efficient operation and control of engineering processes and experimental set-ups often require information concerning the quantities of flowing fluids. For measuring flow in closed conduits, there is a variety of measuring devices available; for example, velocity meters, head meters, or area meters. Although all of them can be used to determine the mass-flow rate, which is the information usually needed, none can measure the flow *directly*, so that, particularly in the case of compressible fluids, the meters require careful calibration.

4.6.1 Velocity meters

A commonly encountered velocity meter is the pitot-static tube which measures local velocities. The pitot tube consists essentially of a tube with an open end facing the stream, as shown in Fig. 4.9. The velocity of the fluid along the streamline x–y decreases to zero at y which corresponds to the tip of the pitot-tube opening, and is called the stagnation point. Applying Bernoulli's equation between planes (1) and (2), we have

$$\frac{P_2}{\rho} + 0 = \frac{P_1}{\rho} + \frac{v_1^2}{2},$$ (4.20)

where P_1 denotes the upstream pressure, and v_1 the approach velocity. The pressure P_2, which is called the stagnation pressure, is made up of two parts: (1) the pressure to be measured if the fluid were not moving—the static pressure, and (2) the pressure resulting from the sudden cessation of the streamline's kinetic energy—the velocity pressure. In other words,

$$P_2 = P_1 + \frac{\rho v_1^2}{2},$$

which is, in fact, a rearrangement of Eq. (4.20). The construction of the pitot tube in Fig. 4.9 shows the static holes; at these locations, static pressure is measured, while the pressure differential $P_2 - P_1$ is measured by the complete assembly. Thus, after including a calibration constant C_p which takes care of nonideal behavior in the system, we rewrite Eq. (4.20) to yield the velocity of the fluid:

$$v_1 = C_p \sqrt{\frac{2}{\rho}(P_2 - P_1)};$$ (4.21)

C_p, which is sometimes called the pitot-tube coefficient, usually has a value between 0.98 and 1.00.

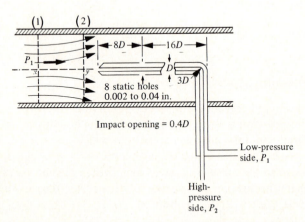

Fig. 4.9 Pitot-static tube and recommended dimensional relationships.

Because the pitot tube measures only local velocities, a traverse of the conduit must be made in order to obtain the complete profile, and from it the average velocity $\bar{V}$. To obtain the density ρ, we usually measure the temperature upstream from the probe tip. In the flow in a straight circular pipe at points preceded by a run of at least 50 diameters without obstructions, the ratio of V_{max} (the velocity measured at the center line) to $\bar{V}$ (the average velocity) can be found by relating the ratio to the Reynolds number. For laminar flow, of course,

$$\frac{\bar{V}}{V_{max}} = \frac{1}{2}, \qquad 0 < \text{Re} < 2.1 \times 10^3, \tag{4.22}$$

and for a large region of the turbulent range, namely, $10^4 < \text{Re} < 10^7$,

$$\frac{\bar{V}}{V_{max}} = 0.62 + 0.04 \log\left(\frac{DV_{max}\rho}{\eta}\right). \tag{4.23}$$

In the range $2.1 \times 10^3 < \text{Re} < 10^4$, a traverse must be made because $\bar{V}/V_{max}$ is not predictable.

For gases at low flow rates (< 200 ft/sec) and isothermal flow conditions, we may use Eq. (4.21) as it is. However, if there is a substantial temperature rise (high velocity), then an energy balance must be made between the upstream point, at which a temperature is measured, and the tip of the pitot tube. For this case, Bernoulli's equation should be written

$$\int_{P_1}^{P_2} \frac{dP}{\rho} = \frac{v_1^2}{2}. \tag{4.24}$$

If the fluid is a compressible gas, we may assume that the flow is adiabatic and the gas ideal; thus the equation of state is

$$P\left(\frac{1}{\rho}\right)^\gamma = \text{constant}, \tag{4.25}$$

where γ is the ratio of the heat capacity at constant pressure to that at constant volume. Integrating Eq. (4.24) by using Eq. (4.25) yields

$$v_1 = C_p \sqrt{\left(\frac{2\gamma}{\gamma - 1}\right)\left(\frac{P_1}{\rho_1}\right)\left[1 - \left(\frac{P_1}{P_2}\right)^{(\gamma-1)/\gamma}\right]}. \tag{4.26}$$

Example 4.3 A pitot-static tube is installed with its impact opening along the center line of a long pipe with an I.D. of 1 ft. Air at 150°F and 12 psig passes through the pipe; the barometric pressure is 745 mm Hg. The pressure difference measured by the pitot-static tube is 0.42 in. water. Calculate the total flow rate in the pipe (lb_m/hr).

Solution. Here we apply Eq. (4.21) and, in the absence of data, assume that $C_p = 1$. The absolute pressure in the pipe where the measurement is made is

$$P_1 = \frac{12}{14.7}\text{ atm} + \frac{745}{760}\text{ atm} = 1.80\text{ atm}.$$

Then we calculate the gas density:

$$\rho = \frac{1 \text{ lb-mol}}{359 \text{ ft}^3} \bigg| \frac{1.80 \text{ atm}}{1 \text{ atm}} \bigg| \frac{492°R}{610°R} \bigg| \frac{28.8 \text{ lb}_m}{1 \text{ lb-mol}} = 0.116 \text{ lb}_m/\text{ft}^3.$$

We convert the measured pressure difference into more convenient units,

$$P_2 - P_1 = \frac{0.42 \text{ in. H}_2\text{O}}{} \bigg| \frac{5.20 \text{ lb}_f/\text{ft}^2}{1 \text{ in. H}_2\text{O}} \bigg| \frac{32.2 \text{ lb}_m\text{-ft}}{\sec^2\text{-lb}_f} = 70.4 \text{ lb}_m/\text{ft sec}^2.$$

Now we utilize Eq. (4.21) to determine $V_{\max}$:

$$v_1 = V_{\max} = \sqrt{\frac{2}{0.116}(70.4)\frac{\text{ft}^2}{\sec^2}} = 34.8 \text{ ft/sec}.$$

(At this velocity and pressure drop, the use of the more complicated expression in Eq. (4.26) is not necessary.)

To find the average velocity, we utilize Eq. (4.23):

$$\frac{\bar{V}}{V_{\max}} = 0.62 + 0.04 \log\left[\frac{(1)(34.8)(3600)(0.116)}{(0.0196)(2.42)}\right] = 0.84.$$

Therefore

$$\bar{V} = (0.84)(34.8 \text{ ft/sec}) = 29.2 \text{ ft}^2/\text{sec},$$

and the mass-flow rate W can finally be determined:

$$W = \frac{29.2 \text{ ft}}{\sec} \bigg| \frac{3600 \text{ sec}}{1 \text{ hr}} \bigg| \frac{\pi \text{ ft}^2}{4} \bigg| \frac{0.116 \text{ lb}_m}{\text{ft}^3}$$

$$= 9600 \text{ lb}_m/\text{hr}.$$

4.6.2 Head meters

There are essentially three types of head meters. They are called so not because of their use in lavatories, but because all of them place some sort of restriction in

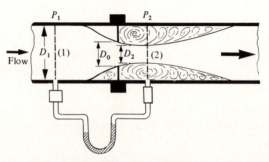

Fig. 4.10 Orifice meter.

the flow line, causing a local increase in the velocity of the fluid and a corresponding decrease in the pressure *head*.* The simplest is perhaps the *orifice plate* illustrated in Fig. 4.10; however, the *venturi meter* and *flow nozzle* (Figs. 4.11 and 4.12) are based on the same principle, and the same group of equations apply to all of them.

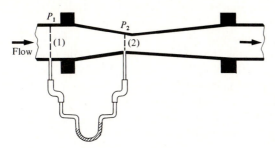

Fig. 4.11   Venturi meter.

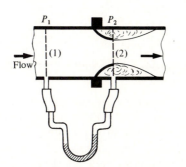

Fig. 4.12   Flow nozzle.

Specifically, consider the orifice meter in Fig. 4.10, which is of very simple construction. A thin plate with a centrally located circular hole is inserted into the duct, and, as indicated, the flow contracts before the hole, and continues to contract for a short distance downstream from the orifice to the location of the smallest flow section, known as the *vena contracta*. If we apply Bernoulli's equation to the system between (1) and (2), and neglect friction loss for the time being, then the mechanical energy balance (for incompressible fluids) with $\beta = 1$ is

$$\frac{P_2 - P_1}{\rho} + \frac{\bar{V}_2^2}{2} - \frac{\bar{V}_1^2}{2} = 0. \tag{4.27}$$

Also, for incompressible fluids (by continuity),

$$\bar{V}_1 = \left(\frac{D_2}{D_1}\right)^2 \bar{V}_2, \tag{4.28}$$

* We shall discuss the term *head* in Chapter 5.

so that when Eq. (4.27) is solved for $\bar{V}_2$ in terms of the difference in pressure between the taps, we obtain

$$\bar{V}_2 = \sqrt{\frac{2(P_1 - P_2)}{\rho \left(1 - \left(\dfrac{D_2}{D_1}\right)^4\right)}}.$$ (4.29)

This is the theoretical velocity at the *vena contracta*, which disregards frictional energy losses and can never be achieved in practice. Therefore we utilize a *discharge coefficient* C_D which accounts for such losses and an additional geometric factor. The largest pressure drop is measured at the *vena contracta*, but it is more convenient to relate the velocity at the orifice plate $\bar{V}_0$ to the pressure drop, $(P_1 - P_2)$, and at the same time use the diameter of the plate opening D_0 instead of the diameter of the *vena contracta*. Thus we get the final form of the expression

$$\bar{V}_0 = C_D \sqrt{\frac{2}{\rho} \frac{(P_1 - P_2)}{(1 - B^4)}},$$ (4.30)

in which $B \equiv D_0/D_1$, or

$$\bar{V}_0 = K \sqrt{\frac{2}{\rho}(P_1 - P_2)},$$ (4.31)

where K is the *flow coefficient*, defined as $C_D/\sqrt{1 - B^4}$.

Equations (4.30) and (4.31) also apply to venturi and nozzle meters as well as to orifice plates, but, as will be discussed below, C_D and K are quite different. In addition, for both venturi and nozzle meters, as shown in Figs. 4.11 and 4.12, the minimum flow contraction corresponds to the position where P_2 is measured; hence $D_2 = D_0$, and $\bar{V}_2 = \bar{V}_0$.

For liquids (incompressible fluids), the mass flow rate is

$$W = KA_0\sqrt{2\rho(P_1 - P_2)}.$$ (4.32)

For gases, we can modify this equation to a more general form, utilizing the *expansion factor Y*,

$$W = KYA_0\sqrt{2\rho(P_1 - P_2)}.$$ (4.33)

Table 4.3 Expansion factors Y for flow meters*

Meter type	Y
Venturi and nozzle	$\sqrt{\dfrac{\gamma(1 - r^{1 - 1/\gamma})(1 - B^4)}{(\gamma - 1)(1 - r)(r^{-2/\gamma} - B^4)}}$
Orifice (square-edged)	$1 - \left(\dfrac{1 - r}{\gamma}\right)\left(0.41 + 0.35B^4\right)$

* In these expressions for Y, $r = P_2/P_1$ and $\gamma = C_p/C_v$.

Expansion factors for gases are different for each type of head meter; these are summarized in Table 4.3.

For convenience, Fig. 4.13 shows values of Y for each of the above meter types. For liquids, Y is taken to be unity, as it is also for gases when the pressure difference is small compared with the total pressure.

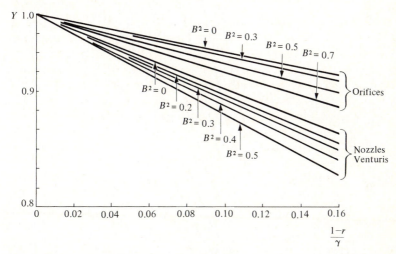

Fig. 4.13 Expansion factors for gases flowing through various head meters. (From Perry, *ibid.*, page 403.)

For the orifice plate, we have to choose: (1) whether to use a square-edged hole, or one with a 45°-taper and a knife-edge inlet, and (2) where to locate the pressure taps with respect to the plate. Figure 4.14 illustrates the effect of various locations of taps on the flow coefficient, and shows how the stream is narrowed as it leaves the orifice plate, reaching a minimum at the *vena contracta*. In most cases, the upstream tap is located from one to two pipe diameters ahead of the plate, and the downstream tap either one half of a pipe diameter downstream or exactly at the *vena contracta*, which would have to be found experimentally. As specified in the legend, the data of Fig. 4.14 apply specifically to square-edged circular orifices. Generally, such data depend to some extent on the exact design of the orifice plate and on the location of the pressure taps so that, to be safe, one should rely on manufacturer's information and/or self-calibration. However, there are data available in the general literature for various designs that may be consulted.[3,4,5]

---

[3] P. S. Barna, *Fluid Mechanics for Engineers*, second edition, Butterworth, Washington D.C., 1964, pages 103–104.

[4] *Flow Meters: Their Theory and Application*, fifth edition, *ASME*, New York, 1959.

[5] J. H. Perry, *Chemical Engineers' Handbook*, third edition, McGraw-Hill, New York, 1950, pages 405–406.

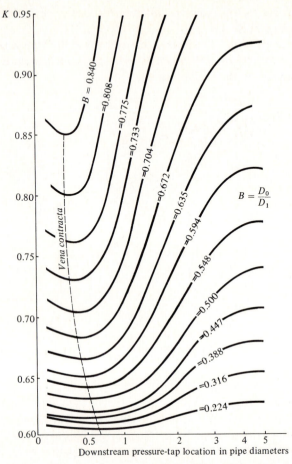

Fig. 4.14 Effect of tap locations on the flow coefficient. Applies to square-edged circular orifices with Re (at the orifice) greater than 30,000. (From Perry, *ibid.*, page 405.)

The square-edged orifice plate is most commonly used, but we use the sharp-edged plate to ensure reproducible coefficients when the plates must be often replaced, and recalibration is to be avoided each time. This is a problem, for example, in systems where the gas is dirty and the velocities high, which results in rapid erosion of the opening in the plate.

Orifice plates are the simplest and cheapest types of head meters, but they also cause the largest *permanent* pressure drop in the system. This refers to the fact that the pressure difference, $P_1 - P_2$, across the orifice is not entirely lost, and a partial pressure recovery is experienced downstream. The results of experiments indicate that the permanent loss of pressure for orifice plates is given by

$$\Delta P_{\text{loss}} = (1 - B^2)(P_1 - P_2). \tag{4.34}$$

The design of the venturi meter is such that a gradual restriction in flow area precedes the throat, which is a short, straight section; then the flow area gradually returns to the original area. On the upstream side, the included angle of convergence is about 25°, and on the downstream side, we employ an included angle of divergence between 5 and 7°. A venturi tube is nearly frictionless under turbulent flow conditions, so that the typical values of C_D are between 0.98 and 1.0. However, in laminar flow, C_D drops rapidly with decreasing Reynolds number and calibration data should be consulted.* For venturi meters, the permanent pressure drop is much lower than for orifice plates and may be approximated as ten percent of the measured differential.

Nozzles are similar to orifices in some respects, but are designed so that the discharge is preceded by a smooth contracting passage. As a result, nozzles have less eddying upstream and the measured pressure drop corresponds more closely to that for a venturi rather than an orifice meter. However, downstream from the nozzle discharge, the flow behavior is more similar to that of an orifice; thus the permanent pressure drop corresponds more closely to the orifice than to the venturi.

4.6.3 Area meters

These meters are also based on the principle of placing a restriction on the flowing stream, creating a pressure drop and a corresponding change in flow velocity through the restricted flow area. However, in this case the pressure drop stays constant and the flow area changes as the velocity changes, rather than *vice versa*. The most common type of area meter—called a rotameter—is illustrated in Fig. 4.15. The flow is read by measuring the height of a float in the *slightly tapered* column.

A force balance applied to the float determines the equilibrium position. When a fluid of density ρ moves past the float and maintains it in suspension, we can use the same force balance that was used several times in Chapters 2 and 3 for particle dynamics. The net buoyant weight of the float is balanced by the upward force created by the moving fluid. This is expressed as

$$(\rho_f - \rho)\frac{m_f}{\rho_f}g = F_K, \qquad (4.35)$$

where m_f is the mass of the float, and ρ_f is the float density.

For a given meter through which a given fluid flows, the left side of Eq. (4.35) is constant and independent of flow rate. Accordingly, F_K is constant when the float is at equilibrium, and, if the flow rate changes, the float counters the effect of this change by taking on a new equilibrium position. For example, if the float is at some equilibrium position corresponding to some mass flow rate and then the mass flow rate increases, F_K becomes larger, and the float rises. However, as the float

---

* For example, see Perry, *ibid.*, page 407.

rises, the tapered tube presents a larger cross-sectional area for flow, and the velocity of the fluid between the float and the tube wall decreases, so that a new equilibrium position is eventually reached, where F_K returns to the value expressed by Eq. (4.35).

The variety in designs of rotameters is so great that there does not exist one relationship valid for all types of rotameters to describe how the mass flow rate varies with height. However, the manufacturers usually supply calibration data for their devices, each set of data being appropriate for a specific fluid. Thus, if a gas mixture such as He plus 10% O_2 is being passed through a rotameter calibrated for He alone, then the user is fooling himself. The rotameter must be recalibrated for the He plus 10% O_2 mixture, or a dimensional analysis of the system that would clearly indicate how the physical parameters interact must be carried out.

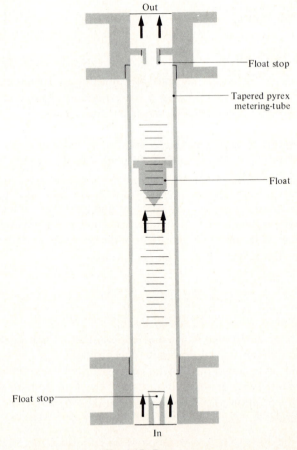

Fig. 4.15 Rotameter.

When we apply a dimensional analysis to a float of a given geometry, the following functional relationship between dimensionless groups evolves[6]

$$\frac{W}{D_f\sqrt{m_f g \rho(1 - \rho/\rho_f)}} = f\left(\frac{\sqrt{m_f g \rho(1 - \rho/\rho_f)}}{\eta} \quad, \quad \frac{D_t}{D_f}\right),\qquad (4.36)$$

where W = mass flow rate, D_f = characteristic diameter of float, and D_t = diameter of the tube.

The ratio D_t/D_f, of course, is directly related to the meter reading h, so that Eq. (4.36) does show the general form $W = W(h)$. Equation (4.36) is important because it indicates how the same set of curves generated to fit its functional relationship can be used for all fluids. Thus, if one intends to utilize a rotameter for many different fluids (for example, as a laboratory item), one should know enough characteristics of the rotameter to be able to make it completely versatile.

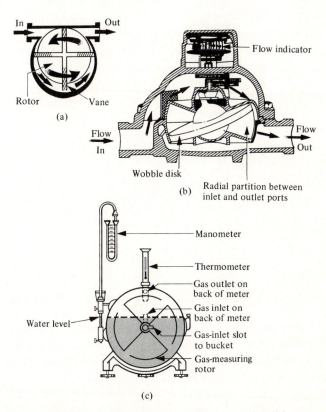

Fig. 4.16 Various flow totalizers. (a) Rotary vane meter, (b) rotating disk meter, and (c) liquid-sealed gas meter. In each case, the shaft feeds a mechanical counter.

---

[6] W. L. McCabe and J. C. Smith, *Unit Operations of Chemical Engineering*, McGraw-Hill, New York, 1956, pages 120–121.

4.6.4 Flow totalizers

In some cases, for example, in pilot or bench-scale research work, the total flow through a line is required. There are many flow totalizers available, depending on the magnitude of the flow being studied. A common type is a volumetric meter, called the *rotary vane meter* (see Fig. 4.16a), which is applicable for either liquids or gases. Such meters measure flows from a few cfm to several thousand cfm with an accuracy of better than half of a percent.

For metering and totalizing water or other liquids, the *rotating disk meter* is used (Fig. 4.16b). It operates over a range from 1 to 1000 gpm with an accuracy of one percent.

For totalizing gas flow, the *liquid-sealed gas meter* is employed (Fig. 4.16c). This type is designed for the range from a fraction of a cfm to many thousand cfm, and has an accuracy of about half of a percent.

4.7 FLOW FROM LADLES

In metallurgical operations, molten metal is contained in refractory lined vessels (ladles). These ladles transport the metal to various points within a plant where molten metal is required. It is desirable to know the time required to completely or partially empty a ladle.

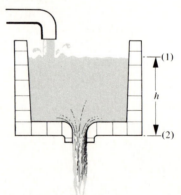

Fig. 4.17 Flow from a vessel in which h can be maintained constant.

As a means of applying Bernoulli's equation, consider the vessel depicted in Fig. 4.17. The liquid level is obviously constant if the mass flow rates of the incoming and discharging streams are equal. For this steady-state system, $P_1 = P_2$, $M^* = 0$, and $\bar{V}_2^2 \gg \bar{V}_1^2$; thus Eq. (4.11) reduces to

$$\frac{\bar{V}_2^2}{2\beta_2} + g\,\Delta z + E_f = 0. \tag{4.37}$$

The friction loss E_f for this system is entirely due to the contraction nozzle, and we can evaluate it by the method applied in Section 4.3:

$$E_f = \tfrac{1}{2}e_f \bar{V}_2^2. \tag{4.38}$$

Combining Eqs. (4.37) and (4.38) yields

$$\frac{\bar{V}_2^2}{2}\left(\frac{1}{\beta_2} + e_f\right) + g\,\Delta z = 0. \tag{4.39}$$

Let h be the height of liquid; then

$$\bar{V}_2 = \left(\frac{1}{\beta_2} + e_f\right)^{-1/2}\sqrt{2gh}. \tag{4.40}$$

For a given nozzle, we can evaluate the term in parentheses, which relates to the flow. As a vessel empties, β_2 may change as well as e_f which is a function of the Reynolds number. However, it is common engineering practice to recognize that $\bar{V}_2 = \sqrt{2gh}$ represents the maximum possible velocity of the exit stream, and then to assign a discharge coefficient C_D to a real situation:

$$\bar{V}_2 = C_D\sqrt{2gh}. \tag{4.41}$$

By comparing Eqs. (4.40) and (4.41), we can compute the discharge coefficients for ladles:

$$C_D = \left(\frac{1}{\beta_2} + e_f\right)^{-1/2}. \tag{4.42}$$

Now consider the case when the mass flow rate of the incoming stream is reduced so that the liquid level gradually diminishes. Then the velocity of the exit stream changes with time, but the velocity at any instant is still given by Eq. (4.41), where h is the liquid level at that instant. Finally, if the vessel contains a liquid and

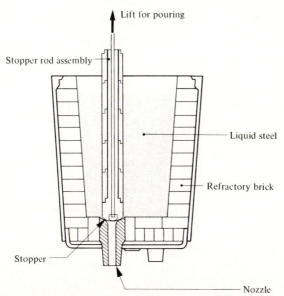

Fig. 4.18 Single-nozzle bottom-pour ladle for steel.

no liquid enters at the top, and then a bottom plug is removed, Eq. (4.41) still applies, but h must be recognized to be a function of time. This is the situation that applies to ladles; Fig. 4.18 shows a typical ladle design for molten steel.

The mass flow rate discharging from the ladle with a nozzle of area A_N is given by

$$\frac{dw}{dt} = A_N \rho \bar{V}_2, \tag{4.43}$$

so that

$$\frac{dw}{dt} = A_N \rho C_D \sqrt{2gh}. \tag{4.44}$$

The integration limits are at time $t = 0$, $h = h_0$, and $w = w_0$.

$$\int_{w_0, h_0}^{w, h} \frac{dw}{\sqrt{2gh}} = A_N \rho \int_0^t C_D \, dt. \tag{4.45}$$

It is usually assumed that C_D is constant; also, many ladles have a constant circular cross section of $\pi D_L^2 / 4$, so that

$$dw = -\rho \frac{\pi D_L^2}{4} dh. \tag{4.46}$$

Thus we obtain the time t_f to completely empty the ladle by integrating Eq. (4.45):

$$t_f = \frac{\pi D_L^2}{2 A_N C_D} \left(\frac{h_0}{2g} \right)^{1/2}. \tag{4.47}$$

Example 4.4 A ladle having an I.D. of 3 ft with a capacity height of 4 ft has a nozzle that tapers (45°) to a 3-in. exit diam.: a) Calculate the time it takes to empty the ladle if it is filled with Al–7 % Si alloy at 1300°F. b) Calculate the discharge rate (lb_m/sec) (1) initially, and (2) when the ladle is 75 % empty.

Data: $\eta = 2.75 \text{ cP} = 6.59 \text{ lb}_m/\text{ft hr},$
            $\rho = 150 \text{ lb}_m/\text{ft}^3.$

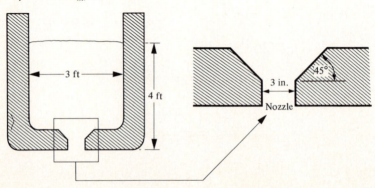

Solution

Equation (4.46) may be utilized if it is reasonable to treat C_D as a constant. Since e_f and β_2 both depend on the Reynolds number through the nozzle, C_D is actually a function of the height. Let us first examine the manner in which C_D varies. This involves calculating the friction-loss factor for flow through a contraction; hence, we use Figs. 4.4 and 4.5.

Completely full

Reynolds number is estimated by assuming $C_D = 1$.

$$\bar{V}_2 = \sqrt{2gh} = \sqrt{2 \times 32.2 \times 4} = 16.04 \text{ ft/sec},$$

$$\text{Re} = \frac{(\frac{1}{4})(16.04)(3600)(150)}{(6.59)} = 3.29 \times 10^5,$$

$$e_f = (0.45)(\tfrac{1}{6}) = 0.075, \qquad \beta_2 \cong 1,$$

$$C_D = (1 + 0.075)^{-1/2} = 0.96.$$

Nearly empty

$$\text{Re} \to 0, \qquad e_f \cong 0.8, \qquad \beta_2 \cong \tfrac{1}{2} \text{ (laminar)},$$

$$C_D \cong [2.8]^{-1/2} \cong 0.60.$$

1-in. level

$$\bar{V}_2 = \sqrt{2 \times 32.2 \times 1/12} = 2.32 \text{ ft/sec},$$

$$\text{Re} = 4.75 \times 10^4, \qquad C_D = 0.96.$$

Calculations at all levels show that C_D is constant at 0.96 except for the very last bit. Hence, it is a good assumption to treat C_D as a constant.

a) $t_f = \dfrac{\pi D^2}{2A_N C_D}\left(\dfrac{h_0}{2g}\right)^{1/2}$

$$= \frac{(\pi)(9)}{2\pi(\frac{1}{4})^2(\frac{1}{4})(0.96)}\left(\frac{4}{2 \times 32.2}\right)^{1/2} = 74.9 \text{ sec}.$$

b) Initial discharge-rate (4-ft level)

$$\frac{dw}{dt} = \bar{V}_2 A_N \rho = (0.96)(16.04)(\pi)(\tfrac{1}{4})^2(\tfrac{1}{4})(150)$$

$$= 113.3 \text{ lb}_m/\text{sec}.$$

Discharge rate at 75% empty (1-ft level)

$$\frac{dw}{dt} = 113.3\sqrt{\tfrac{1}{4}} = 56.7 \text{ lb}_m/\text{sec}.$$

PROBLEMS

4.1 Hot-rolled steel sheet is quenched by passing under nine water sprays as depicted below. Each spray requires 15 gpm of water at 70°F, and the pressure drop across each nozzle at this flow rate is 25 psi. Calculate the theoretical horsepower required by the pump.

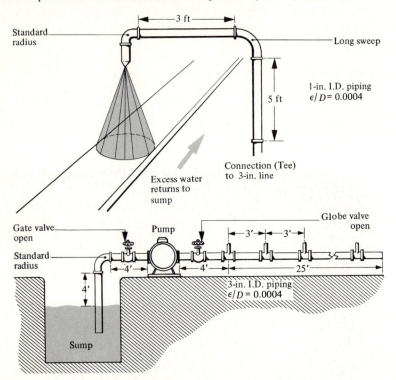

4.2 A fan draws air at rest and sends it through a straight duct 500 ft long. The diam. of the duct is 2 ft, and a pitot-static tube is installed with its impact opening along the center line. The air enters at 80°F and 1 atm and discharges at 1.2 atm. Calculate the theoretical work (in ft-lb$_f$/lb$_m$) of the fan if the pitot-static tube measures a pressure difference of 1.0 in. of water.

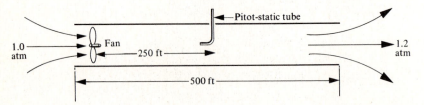

4.3 Compressed air at 100 psia and 100°F flows through an orifice plate meter installed in a 3 in. I.D. pipe. The orifice has a 1.00 in. hole and the downstream pressure tap location is 1.5 in. from the plate. When the manometer reading is 14.1 in. Hg,

a) What is the flow rate of air (lb$_m$/hr)?

b) What is the permanent pressure drop?

4.4 A venturi meter is installed in an air duct of circular cross section 1.5 ft in diameter which carries up to a maximum of 2500 ft³/min (STP) of air at 80°F and 1 atm. The throat diameter is 0.75 ft.

a) Determine the maximum pressure drop that a manometer must be able to handle, i.e., what range of pressure drops will be encountered? Express results in in. of water.

b) Instead of the venturi meter, an orifice meter is proposed and the maximum pressure drop to be measured is 2 in. of water. Calculate what diameter of sharp edged orifice should be installed in order to obtain the full scale reading at maximum flow.

c) Estimate the permanent pressure drop for the devices in parts (a) and (b).

d) If the air is supplied by a blower operating at 50% efficiency, what is the power consumption associated with each installation?

4.5 In practice, when metal is poured from ladles, the nozzles erode. The erosion can be related to the quantity of metal that passes through by this simple expression:

$$D_N - D_N^0 = kW,$$

in which D_N^0 = initial nozzle diameter, D_N = nozzle diameter during pouring that varies, W = mass of metal passed through, and k = proportionality constant.

Develop an expression for the time to empty a ladle that accounts for nozzle erosion.

4.6 In order to *slush* cast seamless stainless-steel pipe, a *pressure pouring* technique is utilized, as depicted below.

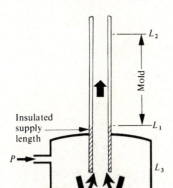

The mold must be filled rapidly up to level L_2 so that no solidification takes place until the entire mold is filled. Determine an expression that we can use to give the time it takes to fill the mold *only*. Consider the mold to be between L_1 and L_2 and open to atmospheric pressure at the top. Neglect the change of metal height L_3 in the ladle; you may also neglect the friction loss associated with the supply tube and mold tube walls, but the entrance loss may *not* be neglected.

4.7 Calculate the time to fill the mold, as depicted below, with molten metal if the metal level at plane A is maintained constant and the time to fill the runner system (entering *piping*) is ignored.

Data (all may be taken as constant):

$\eta = 4 \text{ lb}_m/\text{hr-ft}$

$\rho = 400 \text{ lb}_m/\text{ft}^3$

$f = 0.0025 \text{ (runner)}$

$e_f(\text{contraction}) = 0.1$

$e_f(\text{enlargement for liquid levels below } B) = 0$

$e_f(\text{enlargement for liquid levels above } B) = 1.0$

$L/D \text{ (90}° \text{ turn)} = 25$

$\beta = 1.0$

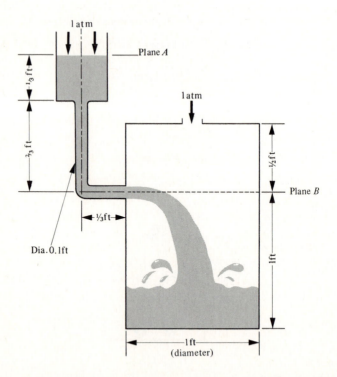

4.8 Molten steel is tapped into a ladle. The full ladle is locked into a pressurized chamber. Air pressure in the chamber is increased to 45 psig forcing the molten steel in the ladle up through the ceramic pouring tube and into the graphite mold which has been secured to the top of the chamber. Determine the time to fill the mold. Data and dimensions are as follows:

Inside dimensions	Diameter, ft	Height or length, ft
Ladle	3.0	3.0
Mold	2.12	3.0
Tube	0.33	4.0

The tube extends 2 ft below the initial melt surface. Density of the steel is 0.272 lb_m/in^3 and its viscosity is 6 cP.

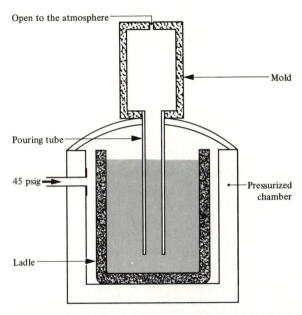

4.9 In planning continuous casting, we use fluid flow analysis. Consider the illustrated configuration of the equipment, which includes in-line vacuum degassing.

a) Determine the tundish and degasser nozzle sizes which are necessary to operate the system at a rate of 25 t/hr per strand. Suppose that for operational reasons it is desirable to maintain tundish and degasser bath depths of 30 and 72 in., respectively.

b) If only $\frac{1}{2}$-in. diameter degasser nozzles are available, how would their use affect the casting operation?

Inside dimensions:

Tundish, $8' \times 8' \times 4'$
Degasser, $4' \times 4' \times 8'$
Wall thickness = 6 in.
Liquid-steel density = 0.272 lb_m/in^3
Discharge coefficients for tundish and vacuum degasser nozzles: $C_D = [1/\beta + e_f]^{-1/2} = 0.8$
Vacuum pressure = 10^{-3} atm

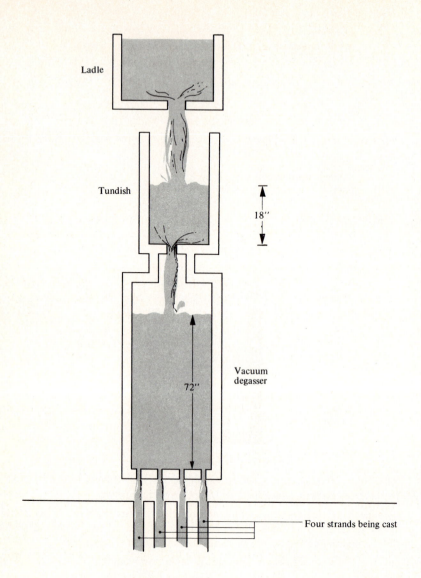

5

FLOW AND VACUUM PRODUCTION

The metallurgist must know not only how to measure flow, but also how to produce it. Some methods of producing flow and vacuum, such as gravity flow and suction lift, are well known, but there are many less obvious types of fans, pumps, and flow devices which are of particular interest, since they are often required and used in metallurgical laboratories and plants. In this chapter, we will consider the fundamentals of the behavior and application of such devices, and also the behavior of high-velocity gas jets, which are playing an ever-increasing role in metallurgical processes.

5.1 PUMPS

In general, pumps may be classified as either positive-displacement pumps or centrifugal pumps. We may subdivide positive-displacement pumps into reciprocal or rotary types, and centrifugal pumps into tangential or axial-flow types.

5.1.1 Positive-displacement pumps

Positive-displacement pumps are widely used to pump liquids. Their characteristic feature is that they endeavour to deliver a definite volume of liquid at every stroke or revolution, regardless of the head against which they are working. There are two common types of reciprocating positive-displacement pumps—the double-acting piston pump and the single-acting plunger pump—and a large number of different designs of rotary positive-displacement pumps, the gear pump perhaps being the best known.

In general, positive-displacement pumps of one type or another are employed where delivery of a relatively small volume of liquid against high pressure is required. The principles of fluid dynamics do not enter into their design, since the geometry of the pump determines the volume flow, given by the swept volume per stroke less the leakage rate.

In reciprocating pumps, as the piston withdraws, the fluid discharge ceases, and so the delivery is in pulses. In double-acting pumps, this pulsating characteristic decreases. In Fig. 5.1(a) and (b) we indicate the principle of a double-acting reciprocal pump and a rotary gear pump.

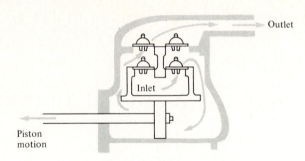

Outlet

Inlet

Piston
motion

(a) Double-acting reciprocating pump

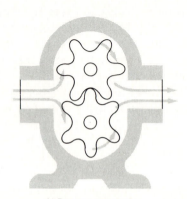

(b) Rotary gear pump

Fig. 5.1 Positive-displacement pumps. In pump (a), reversal of the piston motion results in reversal of the valve positions, but in continuation of flow in the outlet.

Since the flow rate is determined by the size and speed of these pumps, the major specification problem is to calculate the power which the pump requires to carry out a specific job. This is done by applying Bernoulli's equation (4.11) to the system. Consider that the pump and the fluid being pumped (Fig. 5.2) are the system, and write Bernoulli's equation from the suction point (3) to the discharge point (1).

$$\left(\frac{P_1}{\rho_1} - \frac{P_3}{\rho_3}\right) + \left(\frac{\bar{V}_1^2}{2\beta_1} - \frac{\bar{V}_3^2}{2\beta_3}\right) + g(z_1 - z_3) + E_f = -M^*. \tag{5.1}$$

If the frictional loss within the pump itself is accounted for by a mechanical efficiency Γ, then the work done *by the pump* is

$$\Gamma M_P^* = \left(\frac{P_1}{\rho_1} + \frac{\bar{V}_1^2}{2\beta_1} + gz_1\right) - \left(\frac{P_3}{\rho_3} + \frac{\bar{V}_3^2}{2\beta_3} + gz_3\right) + E_f. \tag{5.2}$$

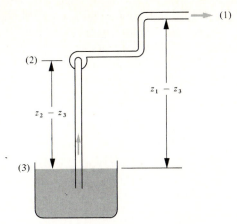

Fig. 5.2 Reference levels for the analysis of pump application.

In the nomenclature associated with pumps, the quantities

$$\left(\frac{P_i}{\rho_i} + \frac{\bar{V}_i^2}{2\beta_i} + gz_i \right)$$

are called *heads, h*, and so

$$M_P^* = \frac{\Delta h}{\Gamma}. \tag{5.3}$$

Since M_P^* is defined in terms of the mass flow rate, then the power required by the pump from a motor (brake horsepower) is given by

$$P_B = \frac{W M_P^*}{550} = \frac{W \Delta h}{550 \Gamma} \text{ (horsepower)}, \tag{5.4}$$

where W is the mass flow rate in lb/sec and the Δh is given in ft-lb$_f$/ lb$_m$. The power absorbed by the fluid (fluid horsepower) is

$$P_f = \frac{W \Delta h}{550}, \tag{5.5}$$

and, therefore,

$$\Gamma = \frac{P_f}{P_B}. \tag{5.6}$$

On many occasions, the flow rate q is given in gallons per minute (gpm) and the liquid density in lb/ft^3. In such cases, this expression is convenient:

$$P_B = \frac{q\rho \, \Delta h}{2.47 \times 10^5 \Gamma}. \tag{5.7}$$

In pumping liquids, there is a limit on the net positive *suction-head* that cannot be exceeded. This is set by the fact that if the dynamic pressure of the liquid $(P + \rho \bar{V}^2/2\beta)$ falls below the vapor pressure of the liquid P_V, then the liquid vaporizes, and no liquid is drawn into the pump. This phenomenon is called *cavitation*. For example, consider the liquid just before it enters the pump at plane (2) in Fig. 5.2. In order to avoid cavitation,

$$\left(\frac{\bar{V}_2^2}{2\beta_2} + \frac{P_2}{\rho} \right) > \frac{P_V}{\rho}. \tag{5.8}$$

The pump lifts the liquid from the reservoir through a pipe, and Bernoulli's equation, written between the reservoir surface (3), which is open to the atmosphere, and the suction entrance of the pump (2), assuming $\bar{V}_3 = 0$, is

$$\left(\frac{P_2}{\rho} - \frac{P_3}{\rho} \right) + \frac{\bar{V}_2^2}{2\beta_2} + g(z_2 - z_3) + E_f = 0. \tag{5.9}$$

From Eqs. (5.8) and (5.9), we obtain the following requirement in terms of the height of lift and the pressure at the reservoir:

$$\frac{P_3}{\rho} - g(z_2 - z_3) - E_f > \frac{P_V}{\rho}.$$

Therefore, the net positive suction-head defined below must indeed be positive:

$$h_{nps} = \left(\frac{P_3}{\rho} - \frac{P_V}{\rho} \right) - g(z_2 - z_3) - E_f > 0. \tag{5.10}$$

5.1.2 Centrifugal pumps

In centrifugal pumps (Fig. 5.3), the fluid enters axially at the suction connection, accelerates radially out along the blades of the impeller, collects in the volute, and leaves the pump tangentially at a high velocity. For an ideal centrifugal pump, we

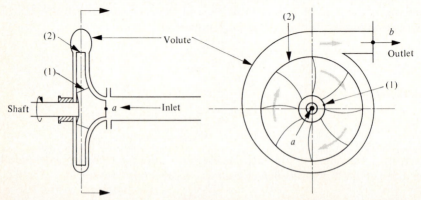

Fig. 5.3 Schematic diagram of centrifugal pumps including the reference points of Bernoulli's equation.

can make a theoretical analysis, giving the relationships between the head developed Δh, and the volumetric flow rate Q:

$$\Delta h = V_p(V_p - Q/A_p \tan \beta), \tag{5.11}$$

where V_p = peripheral velocity, A_p = cross-sectional area of the volute channel around the periphery, Q = flow rate, ft^3/sec, and β = angle between the blade and a tangent to the hub, as shown in Fig. 5.4.

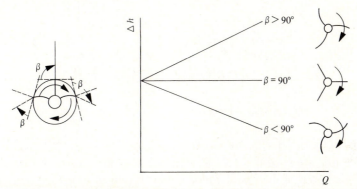

Fig. 5.4 Schematic diagram showing the effect of vane configuration on the shape of the theoretical head versus flow-rate curve.

If β is less than 90°, then the slope of the theoretical Δh–Q curve is negative; if β equals 90°, then the line is horizontal; and if β is greater than 90°, then the slope is positive. Unfortunately, there are losses due to circulatory flow, fluid friction, shock, and mechanical friction, none of which is predictable, except by experience. Figure 5.5 illustrates the effect of these losses on a pump with β less than 90°.

The efficiency of centrifugal pumps, which is given by Eq. (5.6), is a maximum at the design conditions, falling off on either side of them (Fig. 5.6). Pumps are best

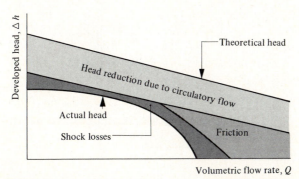

Fig. 5.5 Theoretical head, actual head, and various loss contributions for a centrifug?[1] pump with β less than 90°.

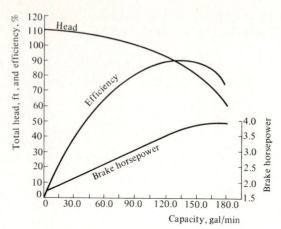

Fig. 5.6 Characteristic curves for a centrifugal pump.

described in terms of the head-flow rate characteristic curve. The head is usually given in terms of feet of the fluid flowing through the pump, and indicates the height of a column of the liquid which could be pulled up through a frictionless pipe to the pump itself. In practice, of course, the friction loss and any fitting or meter losses would have to be subtracted from this height.

Example 5.1 A strip of metal emerging from a set of rolls is to be cooled by means of a spray of water. If 100 gpm are required, and the pressure drop across the spray nozzle is 20 psi at this flow rate, is it possible to install the pump, as depicted below, if the characteristic curves given in Fig. 5.6 apply? Assume 3-in. welded steel-piping up to the pump and 1-in. welded steel-piping to the nozzle, as shown below.

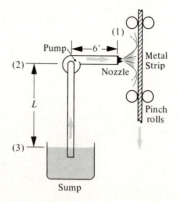

Solution. From Fig. 5.6, we obtain the total head delivered by the pump, $-M^*$, which is 104 ft at 100 gpm. (In reality, the units are ft-lb$_f$/lb$_m$.)

Next we write Bernoulli's equation from planes (3) to (1). In this instance, Eq. (5.1) applies, with $L = z_1 - z_3$ and $\bar{V}_3 = 0$, $\beta_1 \cong 1$, and $\rho_1 = \rho_3 = \rho$.

$$\left(\frac{P_1 - P_3}{\rho}\right) + \frac{\bar{V}_1^2}{2} + gL + E_f = -M^*,$$

$$\left(\frac{P_1 - P_3}{\rho}\right) = \frac{20 \text{ lb}_f}{\text{in}^2} \left| \frac{\text{ft}^3}{62.4 \text{ lb}_m} \right| \frac{144 \text{ in}^2}{\text{ft}^2} = 46.2 \frac{\text{ft-lb}_f}{\text{lb}_m}.$$

At 100 gpm, the velocity in the 1-in. pipe is 32.1 ft/sec and 3.56 ft/sec in the 3-in. pipe. Thus,

$$\frac{\bar{V}_1^2}{2} = \frac{(32.1)^2 \text{ ft}^2}{2 \text{ sec}^2} \left| \frac{\text{sec}^2 \text{ lb}_f}{32.2 \text{ lb}_m\text{-ft}} \right| = 16.0 \frac{\text{ft-lb}_f}{\text{lb}_m}.$$

Now we formulate E_f:

$$E_f = \tfrac{1}{2} e_f \bar{V}_{3-2}^2 + 2f\left(\frac{L}{D}\right)\bar{V}_{3-2}^2 + 2f\left(\frac{l}{D}\right)\bar{V}_1^2.$$

entrance   3-in. pipe   1-in. pipe
loss   loss   loss

From Figs. 4.4 and 4.5, we obtain

$$e_f = (2)(0.4) = 0.8.$$

The respective Reynolds numbers and friction factors are:

	Re	f
1-in. pipe	2.5×10^5	0.0057
3-in. pipe	8.3×10^4	0.0065

Therefore

$$E_f = (\tfrac{1}{2})(0.8)\left(\frac{3.56^2}{32.2}\right) + \frac{(2)(0.0065)(3.56)^2}{(\frac{3}{12})\,(32.2)}\,L + \frac{(2)(0.0057)(6)(32.1^2)}{(\frac{1}{12})\,(32.2)}$$

$$= 0.157 + 0.021L + 26.2$$

$$= 26.4 + 0.021L, \qquad \text{ft-lb}_f/\text{lb}_m.$$

Substituting all available data into the mechanical energy balance, we have

$$26.4 + 0.021L + 46.2 + 16.0 + L = 104.0,$$

or

$$L = \frac{15.4}{1.02} = 15.1 \text{ ft.}$$

Finally, we should examine the net positive suction-head. (We use E_f here only between (3) and (2).) $(P_3 - P_V)/\rho \cong P_3/\rho$ in this case.

$$h_{nps} = \frac{P_3}{\rho} - gL - E_f$$

$$= 34.0 - 15.1 - 0.3 = 18.6 \text{ ft}.$$

Since $h_{nps} > 0$, cavitation will not occur with $L = 15.1$ ft.

5.1.3 Electromagnetic pumps

In the development of nuclear power reactors, heat transfer from the reactor to the turbine propulsion system via liquid Na and Na–K alloys has played an important role. It should be apparent, however, that circulation of the molten metal is quite difficult to achieve because of the corrosive nature of the metal and the temperatures involved. In order to overcome these difficulties, an entirely different sort of pump has been developed—the electromagnetic pump.

The dc electromagnetic pump operates on the same principle as a dc motor. The liquid metal, as a conductor, is subjected to a force when a current and a magnetic field are passed through it at right angles to each other, as shown in Fig. 5.7. The current connection is made directly to the duct, which is thin-walled, so that most of the current passes through the molten metal and not the duct wall. The efficiency of such pumps is in the range 10–50%, in terms of conversion of applied current and magnetic flux to fluid motion. The flow rate is directly proportional to the current input at the pump duct.

5.2 FANS AND BLOWERS

Fans and blowers are used for moving gases. We employ fans when the pressure drop in the system to overcome is not larger than about 50 in. w.c.,* although

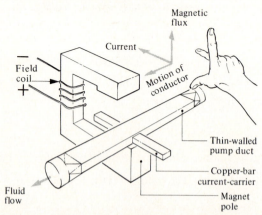

Fig. 5.7 Schematic diagram of an electromagnetic pump for liquid metals.

---

* The notation in. w.c., or inches water-column, is often used instead of in. H_2O, or inches water.

volumes up to several hundred thousand cfm may be involved. Blowers and
turboblowers, on the other hand, are used in situations where larger pressure drops
occur.

5.2.1 Fans

As in the case of pumps, there are many different designs of fans—axial flow,
propeller and cross-flow fans as well as centrifugal types with a variety of blade
configurations. All manufacturers of gas-moving equipment supply characteristic
curves describing the performance of their equipment under the specified operating
conditions at the inlet, as, for example, in Fig. 5.8, which presents curves for both
the static and total pressures and the static and total efficiencies. If one wants to
use a fan for moving a specified amount of gas against a given system resistance,
it is necessary to know the static pressure developed by the fan at that flow rate.

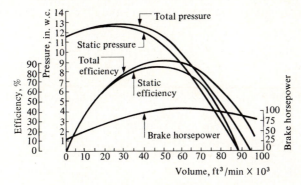

Fig. 5.8 Characteristic curves for a fan.

First, consider how the fan characteristics are measured. Figure 5.9(a) illustrates
an appropriate system provided with a damper, so that the flow throughput may
be varied. The pitot tube measures the total pressure P_t, which is made up of the

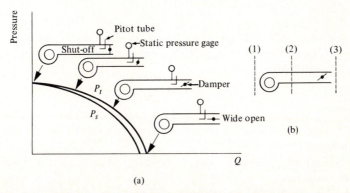

Fig. 5.9 Obtaining characteristic curves for a fan.

static pressure P_s (measured by the static pressure gage), and the dynamic pressure $\frac{1}{2}\rho \bar{V}^2$. By adjusting the damper to various positions, the values of P_t and P_s corresponding to the various flow rates can be generated. Both of these values are relative to the inlet conditions, and are really pressure increases developed by the fan. At any flow rate, a theoretical horsepower may be calculated corresponding to either P_s or P_t, so that the definition of the *static efficiency*, Γ_s, as well as of the total efficiency, Γ, is based on the actual, or so-called brake horsepower.

The above statement can also be demonstrated by use of Bernoulli's equation applied to the system shown in Fig. 5.9(b). Between planes (1) and (3), the energy balance is simply

$$M^* + E_f = 0. \tag{5.12}$$

In this case, the system friction is composed of the resistance offered by the damper and the exit losses, that is, the friction between planes (2) and (3). The mechanical energy balance between planes (2) and (3) is then expressed

$$E_f = \left(\frac{P_2 - P_3}{\rho}\right) + \left(\frac{\bar{V}_2^2}{2}\right) \tag{5.13}$$

assuming $\beta = 1$.

Hence, by combining Eqs. (5.12) and (5.13), and noting $P_1 = P_3$, we arrive at

$$-M^* = \left(\frac{P_2 - P_1}{\rho}\right) + \left(\frac{\bar{V}_2^2}{2}\right), \tag{5.14}$$

or

$$-M^*\rho = (P_2 - P_1) + \left(\frac{\rho\bar{V}_2^2}{2}\right). \tag{5.15}$$

Writing the above expression with the actual or brake horsepower P_B and the fan's efficiency Γ, we obtain

$$\Gamma P_B = -M^*\rho Q = \left[(P_2 - P_1) + \frac{\rho\bar{V}_2^2}{2}\right]Q. \tag{5.16}$$

Thus the total power delivered is based on the sum of the static pressure $(P_2 - P_1)$ and the dynamic pressure $(\frac{1}{2}\rho\bar{V}_2^2)$, or what we call the total pressure, as incorporated in brackets.

5.2.2 Blowers

In order to produce large heads to overcome large pressure drops in systems, either positive displacement or centrifugal blowers are used. Blowers and turboblowers find their typical metallurgical applications in producing the air blast for blast furnaces and cupolas.

Blowers are essentially constant-pressure machines with power consumption

almost directly proportional to the volume delivered. Characteristic performance
curves for a variable speed turboblower are given in Fig. 5.10, and a typical blower
is depicted in Fig. 5.11.

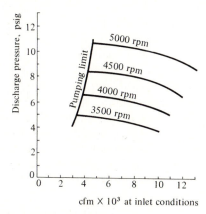

Fig. 5.10 Typical characteristic curves for a variable-speed blower.

Fig. 5.11 A Roots pump or turboblower.

5.2.3 Interactions between fans and systems

Since a fan is part of an overall system, the system as a whole determines the size
of the fan required. For any system, there is a certain curve of volume flow versus
system resistance or pressure drop (see Fig. 5.12). The reader should realize by
now that the resistance usually increases as the square of the volume flow. The
system may contain any combination of duct work, beds of fluidized or packed
solids, dust collectors, flues, etc. There is usually a specific volume throughput Q',
required for proper operation of the process; this fixes the pressure drop resistance
of the system, which in turn must be overcome by the fan or blower. Essentially

then, the fan characteristic curve must intersect the system curve at the desired Q'–ΔP coordinate (the so-called *operating point*). The efficiency and required horsepower are then fixed. Note that a different fan with a different characteristic curve would place the operating point at a different position on the system curve, resulting in a flow different from Q'.

The normal operating range for a fan is to the right of the peak of its pressure–volume curve. The so-called *pumping limit*, defines the furthest possible left-hand operating point. *Pumping* occurs when the fan is slowed down to the point where the static pressure created is less than the static pressure in the discharge line. At that point a flow reversal takes place. After this reversal, a momentary drop in the pressure in the discharge line occurs, and flow starts in a normal direction again. This pattern is repeated very rapidly and results in a pumping-type action, which may cause damage to the fan unless flow below this limit is avoided. It usually turns out that this point is reached at the maximum on the fan *total pressure* curve, which in turn is usually slightly to the right of the maximum of the static pressure curve.

It is not always possible to make desired changes in the operating point of a fan system. Changes in operating temperature or pressure drop require reevaluation of the system resistance curve, and if this changes, the operating point shifts, unless the fan is adjusted to keep Q' constant. Conversely, if it is desired to change Q', then the operating point has to be changed, and consequently the fan characteristic curve must be changed.

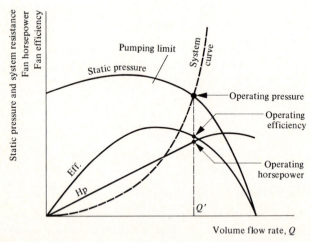

Fig. 5.12 Relationship between the fan curves and system curve.

In order to estimate the effect of deviations from the specified conditions, or the effect of a change in fan size or speed, on the behavior of the fan or the gas, the so-called *fan laws* are used to convert from one set of operating variables to another. The variables are given in Table 5.1.

Table 5.1   Variables for fan laws

Speed of rotation	n
Impeller diameter	D
Gas density	ρ
Static pressure	P_s
Power	P_B
Volume flow	Q

The fan laws are:

$$Q = k_1 D^3 n, \tag{5.17}$$

$$P_s = k_2 D^2 n^2 \rho, \tag{5.18}$$

$$P_B = k_3 D^5 n^3 \rho. \tag{5.19}$$

Here is an example of the use of the fan laws. If the density of the gas varies, and the speed and volume flow remain constant, then Eqs. (5.18) and (5.19) tell us that both the horsepower and the static pressure would be expected to vary *directly* with the density. The proportionality constants k_1, k_2, and k_3 are constants over a limited range, which should not deviate too far from a given point on the curve of pressure versus flow rate.

Changes in the pattern of flow of the gases entering the fan will also change the characteristic curve, since the calibration is performed with gases entering smoothly at right angles to the rotor. If the gases enter otherwise, then their flow through the fan will affect the characteristic curves.

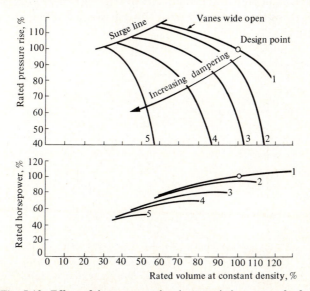

Fig. 5.13 Effect of dampers on the characteristic curves of a fan.

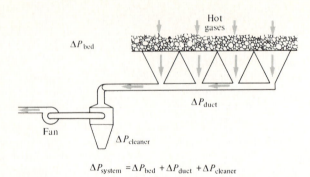

$$\Delta P_{system} = \Delta P_{bed} + \Delta P_{duct} + \Delta P_{cleaner}$$

Fig. 5.14 Schematic diagram of an iron-ore pelletizing plant.

We can accomplish control of the flow by varying the speed, by using inlet guide vanes, or by placing a damper in the system. The effect of guide vanes, or the placing of a damper in the system, on the rated (without vanes) capacity is shown in Fig. 5.13.

It is best, therefore, to overrate fans somewhat on both the pressure and volume specifications, but not too much, since severe dampering, if necessary, decreases the efficiency and wastes the available horsepower. The problems that can arise from errors in matching the system and fan curves are illustrated in the following example.

Example 5.2 Part of an iron-ore pelletizing plant consists of a moving chain grate (carrying a bed of pellets subjected to hot gases) and associated gas cleaning and exhaust equipment, as shown schematically in Fig. 5.14. The total pressure drop due to duct work, at the required flow of 320,000 cfm at 250°F, is 3.5 in. w.c., and,

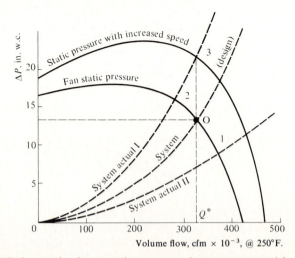

Fig. 5.15 Interaction between the system-resistance curves and fan curves.

when added to the *calculated* design bed-resistance of 9.5 in. w.c. at the design flow, gives a total system *design resistance* of 13.0 in. w.c. The fan to be used in this system is then specified to meet a 320,000 cfm flow at 250°F with a 13.0 in. w.c. static pressure drop (point O in Fig. 5.15).

Suppose, however, that when the plant is built, the *actual* pressure drop across the bed of pellets at the required flow is 21.0 in. w.c. This means that the actual system curve (I) is different from the design system curve. Now the fan operates at point 2 (Fig. 5.15), and it pulls less than the required flow through the system (280,000 cfm). In this case, the only alternatives to regain a flow of 320,000 cfm are either to decrease the system resistance, or to change the characteristic curve by increasing the fan speed. In the first instance, one could decrease the system resistance by decreasing the bed height. If this is done, however, the lateral speed of the bed must be increased in order to keep the production rate (lb solids treated/hr) at the design value. However, this may not be possible, since there is usually a minimum reaction-time for the exposure of the solids to the hot gases, and the original design is usually close to that minimum. Changes in duct work may also be possible, but usually are not practical once the plant is built.

The other alternative—speeding up the fan until the operating condition is at point 3—is easier but may be very expensive. Reference to the fan laws shows us that the power required is proportional to the cube of the flow; that is, for constant fan-size and gas-density, $Q \propto n$, and $P_B \propto n^3$, so that $P_B \propto Q^3$. Therefore, for an increase of flow in the ratio 320,000/280,000 = 1.14, the ratio of the new horsepower required to the old will be $(1.14)^3 = 1.50$; in other words, a motor 50% more powerful is required. This may be quite expensive and could cause many problems in the electrical system. The ultimate solution is to buy a new fan.

If, on the other hand, the actual bed-resistance were, e.g., 7 in. w.c., then the fan would pull too much gas through the system, as at point 1 (Fig. 5.15). This can easily be corrected by increasing the system resistance through the use of dampers, until the system and fan curves coincide at the required flow and pressure drop, point O.

5.3 HIGH-VELOCITY JETS

During the past decade, jets of gas have become important in several metallurgical processes; therefore metallurgists should have a good understanding of the characteristics and behavior of jets, in order to make the best use of them.

5.3.1 Nozzle design

Consider the flow nozzle described in Section 4.6.2. If the area of the nozzle opening is considerably smaller than that of the approach pipe, then the velocity in the pipe is negligible (the gas is stagnant) with respect to the velocity at the nozzle opening, as long as the pressure in the pipe is at or above the *stagnation* pressure. The application of Bernoulli's equation to the adiabatic and frictionless flow of an ideal compressible gas under these conditions yields an equation for the velocity in the nozzle opening $\bar{V}_t$:

$$\bar{V}_t = \sqrt{\frac{2P_0}{\rho_0}\left(\frac{\gamma}{\gamma-1}\right)\left[1 - \left(\frac{P_t}{P_0}\right)^{(\gamma-1)/\gamma}\right]}, \tag{5.20}$$

where reference points are defined in Fig. 5.16, and $\gamma = C_P/C_V$. We may use this equation for a flow nozzle operating at subsonic velocities.

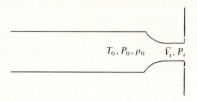

$$T_0, P_0, \rho_0 \qquad \bar{V}_t, P_t$$

Fig. 5.16 Reference points for flow-nozzle equations.

Now the speed at which a compression–expansion wave passes through a medium, that is, the speed of sound V_s in that medium is given by

$$V_s = \sqrt{\left(\frac{\partial P}{\partial \rho}\right)_{\substack{\text{constant}\\ \text{entropy}}}}. \tag{5.21}$$

For an ideal gas,

$$V_s = \sqrt{\frac{\gamma P}{\rho}}. \tag{5.22}$$

The equation of momentum for one-dimensional flow is

$$\frac{dP}{\rho} + \bar{V}\,d\bar{V} = 0. \tag{5.23}$$

Substituting $dP = V_s^2\,d\rho$ into Eq. (5.23), we get

$$V_s^2\frac{d\rho}{\rho} + \bar{V}\,d\bar{V} = 0. \tag{5.24}$$

The continuity equation, which requires that the mass flow rate remain constant at all points in the nozzle,

$$\frac{d\rho}{\rho} + \frac{d\bar{V}}{\bar{V}} + \frac{dA}{A} = 0, \tag{5.25}$$

must be satisfied along with Eq. (5.24), and so

$$\frac{d\bar{V}}{\bar{V}}\left(\frac{\bar{V}^2}{V_s^2} - 1\right) = \frac{dA}{A}. \tag{5.26}$$

The Mach number, M, is defined as $\bar{V}/V_s$, so that finally

$$\frac{d\bar{V}}{\bar{V}}(M^2 - 1) = \frac{dA}{A}.$$ (5.27)

Now, if the velocity at any point in the nozzle is *less than* M = 1, and if the area of the nozzle decreases at that point (dA/A = negative), then the velocity increases at that point ($d\bar{V}/\bar{V}$ = positive).† At the throat, (dA = 0), either M = 1 or $d\bar{V}/\bar{V}$ =

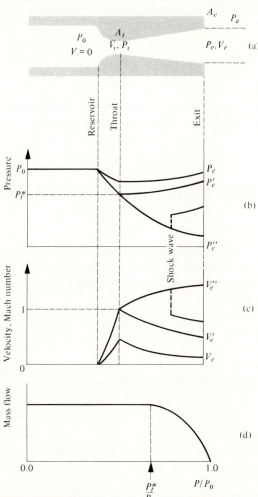

Fig. 5.17 Reference points and schematic internal conditions for various operating pressure ratios for converging-diverging nozzles.

---

† The reader should consider whether or not M > 1 can be achieved in the converging portion of a nozzle.

0. If $M < 1$ at the throat, then $d\bar{V}/\bar{V}$ is zero and no further increase in velocity is possible, i.e., this is the maximum velocity which can be achieved.

If $M = 1$ at the throat, then we have reached the point at which supersonic flow can take place. By adding a diverging section on the end, and referring to Eq. (5.27), we see that if dA/A is positive, then either $d\bar{V}/\bar{V}$ will be positive, and a further increase to higher velocities is achieved in the diverging section, or the flow must come to a stop. The former case occurs as long as $P_t > P_{exit}$. A converging–diverging nozzle is shown in Fig. 5.17. It is usually called a *deLaval nozzle*.

For a converging nozzle, we obtain the conditions required to attain sonic velocity at the throat by using the energy relation for an ideal gas undergoing adiabatic flow:

$$\frac{T_0}{T} = 1 + \frac{\gamma - 1}{2} M^2. \tag{5.28a}$$

When $M = 1$, sonic conditions exist, denoted by an asterisk, and since the flow is also isentropic, substitution of the ideal gas relationship

$$\frac{P}{P_0} = \left(\frac{T}{T_0}\right)^{\gamma/\gamma - 1} = \left(\frac{\rho}{\rho_0}\right)^{\lambda} \tag{5.28b}$$

gives the *critical pressure ratio*

$$\frac{P_t^*}{P_0} = \left(\frac{2}{\gamma + 1}\right)^{\gamma/\gamma - 1}. \tag{5.29}$$

This means that the ratio of reservoir pressure to throat pressure at the sonic flow condition is governed only by the value of γ. For air and oxygen, $P_t^*/P_0 = 0.528$. For pressure ratios between 1.00 and 0.528, the mass flow rate of the gas is $W_t = \rho_t \bar{V}_t A_t$, and using Eq. (5.29b),

$$W_t = A_t \sqrt{\frac{2\rho_0 P_0 \gamma}{\gamma - 1}\left[1 - \left(\frac{P_t}{P_0}\right)^{(\gamma - 1)/\gamma}\right]}\left(\frac{P_t}{P_0}\right)^{1/\gamma}. \tag{5.30}$$

Any further decrease in the ratio P_t/P_0 below that for P_t^*/P_0, caused, for example, by increasing the reservoir pressure P_0, will not cause a further increase in the mass flow rate out of a converging nozzle, since the condition of isentropic flow is no longer satisfied beyond the nozzle. If $(P_t/P_0) \leq (P_t^*/P_0)$, then the nozzle is said to be *choked*, and the mass flow rate is

$$W_t^* = A_t \sqrt{\rho_0 P_0 \gamma \left(\frac{2}{\gamma + 1}\right)^{(\gamma + 1)/(\gamma - 1)}}. \tag{5.31}$$

For the nozzle with the additional diverging section, the mass flow at sonic or supersonic conditions is still given by Eq. (5.31). However, since the velocity increases along the length of the diverging portion, the pressure correspondingly decreases from that at the throat. The problem is to determine how long this portion of the nozzle should be made in order to produce a jet with exit pressure P_e, equal to ambient pressure. Such a condition establishes the most effective condition, as discussed in Section 5.3.2.

Up to this point, it has not been established what the absolute pressure at the throat, relative to the exist pressure, is. Several possible solutions arise, depending on the nozzle design. There are two values of P_t^*/P_e—and therefore two values of P_e, shown in Fig. 5.17(b), which result in isentropic shockless flow in the diverging portion of the nozzle for the conditions where $M = 1$ at the throat. The higher pressure P_e' will cause the flow to become subsonic again immediately at the throat, and will cause a decrease in velocity; the lower pressure P_e'' will allow supersonic flow throughout the nozzle. If the exit pressure is between P_e' and P_e'', and $P_t = P_t^*$, then somewhere in the nozzle, a discontinuous transition from a lower to higher pressure (and simultaneously from supersonic to subsonic flow) occurs, as shown schematically in Fig. 5.17. This produces a *shock wave*.

If the nozzle design is such that $P_e = P_e''$, then we may proceed to calculate the exit velocity, using the equation

$$V_e = \sqrt{\left(\frac{2}{\gamma - 1}\right)\frac{\gamma P_0}{\rho_0}\left[1 - \left(\frac{P_e}{P_0}\right)^{(\gamma - 1)/\gamma}\right]} \tag{5.32}$$

or

$$M_e^2 = \frac{2}{\gamma - 1}\left[\left(\frac{P_0}{P_e}\right)^{(\gamma - 1)/\gamma} - 1\right]. \tag{5.33}$$

The mass flow rate and throat area are still given by Eq. (5.31).

Now that we have the mass flow rate, we may calculate the area of the exit by means of a mass balance, since $W_t^* = W_e$,

$$A_t\sqrt{\rho_0 P_0\gamma\left(\frac{2}{\gamma + 1}\right)^{(\gamma + 1)/(\gamma - 1)}} = A_e\sqrt{\rho_0 P_0\gamma\left(\frac{2}{\gamma - 1}\right)\left[1 - \left(\frac{P_e}{P_0}\right)^{(\gamma - 1)/\gamma}\right]} \cdot \left(\frac{P_e}{P_0}\right)^{1/\gamma}, \tag{5.34}$$

$$\left(\frac{A_t}{A_e}\right)^2 = \left(\frac{2}{\gamma - 1}\right)\left(\frac{\gamma + 1}{2}\right)^{(\gamma + 1)/(\gamma - 1)}\left(\frac{P_e}{P_0}\right)^{2/\gamma}\left[1 - \left(\frac{P_e}{P_0}\right)^{(\gamma - 1)/\gamma}\right]. \tag{5.35}$$

Ideally, we want $P_e = P_{ambient}$, since then the ultimate adiabatic expansion is reached, and the jet issues from the nozzle at atmospheric pressure. If P_e is less than P_a, the ambient atmosphere compresses the flowing jet and collapses it in a series of shock waves. If $P_e > P_a$, the jet continues to expand beyond the nozzle tip. In both non-ideal situations, the efficiency of conversion of the nozzle velocity to jet momentum decreases.

The angle of divergence of the nozzle is usually about $7°$ in order to avoid separation of flow from the nozzle walls.

Example 5.3 In the basic oxygen steelmaking process, we decarburize a bath of molten iron–carbon alloy with gaseous oxygen which is blown into the bath from a lance held above the bath. Determine the dimensions of the converging–diverging nozzle which are required to achieve a velocity of Mach 2 at the exit with an oxygen flow rate of 15,000 scfm.* What is the required driving pressure?

Solution. From Eq. (5.33), we can obtain P_0 by assuming that $P_e = 14.7$ psia to achieve the best behavior of the jet, and noting that γ for O_2 is 1.4:

$$\left(\frac{P_0}{P_e}\right)^{0.286} = \frac{2^2}{\left(\frac{2}{0.4}\right)} + 1.0,$$

and thus

$$\frac{P_0}{P_e} = 7.85.$$

For $P_e = 14.7$ psia, $P_0 = 115.3$ psia. Now we can make use of Eq. (5.35) to find the ratio A_t/A_e:

$$\left(\frac{A_t}{A_e}\right)^2 = \left(\frac{2}{0.4}\right)\left(\frac{2.4}{2}\right)^{2.4/0.4}\left(\frac{14.7}{115.3}\right)^{2/1.4}\left[1 - \left(\frac{14.7}{115.3}\right)^{0.4/1.4}\right]$$

$$\left(\frac{A_t}{A_e}\right) = 0.595.$$

Finally, we can calculate A_t from Eq. (5.31), since the nozzle is choked, as otherwise supersonic velocities could not be reached.

The mass flow is

$$W^* = 15,000 \text{ scfm} \times 0.089 \frac{lb_m}{ft^3} \times \frac{1 \text{ min}}{60 \text{ sec}}$$

$$= 22.2 \text{ lb}_m/\text{sec}.$$

$$22.2 = A_t\sqrt{\left(\frac{0.650 \text{ lb}_m}{ft^3}\right)\left(\frac{115.3 \text{ lb}_f}{in^2}\left|\frac{144 \text{ in}^2}{ft^2}\right.\right)(1.4)\left(\frac{2}{2.4}\right)^6\left(\frac{32.17 \text{ lb}_m\text{-ft}}{lb_f\text{-sec}^2}\right)}$$

$$= A_t\sqrt{16.25 \times 10^4},$$

$$A_t = 0.0551 \text{ ft}^2$$

$$= 7.94 \text{ in}^2.$$

Thus the throat diameter $d_t = 3.19$ in., and the exit diameter $d_e = 4.14$ in.

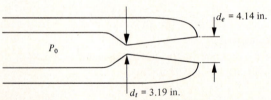

* Standard cubic feet per minute.

5.3.2 Jet behavior

As a jet exits from the nozzle, it entrains adjacent slow-moving air, which in turn acts as a drag, creating turbulence. This slows some of the supersonically flowing gas to sonic and subsonic velocities and the supersonic core of the jet gradually decays, until at some distance from the nozzle the core disappears, and the entire jet is subsonic. Figure 5.18, based on the work of Anderson and Johns,[1] shows the relationship between the exit velocity (in terms of Mach number) and the length of the supersonic core (in terms of either the ratio x/d_t, where x is the distance from the nozzle and d_t is the throat diameter, or the ratio x/d_e, where d_e is the exit diameter).

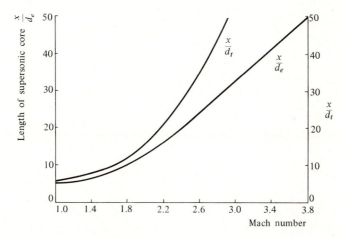

Fig. 5.18 Relationship between the length of supersonic core and exit Mach number. (Adapted from Anderson and Johns by Smith and by Holden and Hogg, *ibid.*)

We measure the overall spreading of the jet either by the ratio of the velocity V_r at a given radius r (measured from the center line of the jet) to the center-line velocity V_c, or by the ratio r_0/r_t, where we define r_0 as the radius at which the velocity is one-half that at the center line. The typical velocity or impact pressure profile is that of a normal distribution about the center line. However, the spreading of the jet is minor until the supersonic core has decayed. Once this point has been reached, the jet expands at an included angle of about 18°. Figure 5.19 indicates the spreading profile of two jets, and Fig. 5.20 gives a dimensionless graph for determining the width of the jet as a function of the exit Mach number and the distance from the nozzle. We then take the *effective jet radius r* as $2r_0$, and can calculate it at any distance from the nozzle with the help of Fig. 5.20.

Because of the entrainment of more gas into the jet, the mass of the jet increases as we get further from the nozzle, but the velocity decreases. For supersonic jets,

---

[1] A. R. Anderson and F. R. Johns, *Jet Propulsion* **25**, 13 (1955).

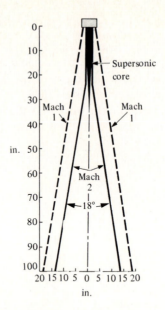

Fig. 5.19 Spreading of Mach 1 and Mach 2 jets with identical flow rates of 6500 scfm O_2. (From Smith, *ibid.*)

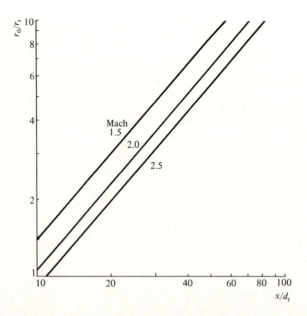

Fig. 5.20 Jet-spreading characteristics as a function of Mach number and distance from the nozzle. (Adapted from Anderson and Johns by Smith, *ibid.*)

the impact pressure at the center line (virtually entirely dynamic pressure) has been correlated as a function of the distance from the nozzle; Fig. 5.21[2] shows this correlation. For *subsonic* jets, the velocity at the center line has been observed as proportional to (d_t/x).[3] Due to the fact that the impact pressure is essentially proportional to the square of the velocity, we would expect

$$\frac{P_c}{P_0 - P_a} = k\left(\frac{x}{d_t}\right)^{-2}.$$
(5.36)

When we test this against the supersonic data in Fig. 5.21, we find that at Mach number 1.5, the relationship still holds, but that for higher Mach numbers, the slope is no longer -2.

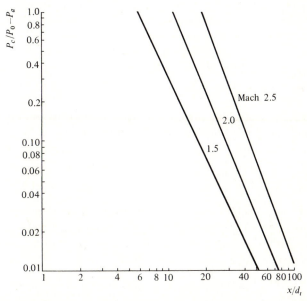

Fig. 5.21 Maximum impact pressure as a function of Mach number and distance from the nozzle. (Adapted from Anderson and Johns by Smith, *ibid.*)

The increase in the mass flow of a subsonic jet due to the entrainment of the surrounding gas has been found to be directly proportional to the distance from the nozzle exit, according to the equation

$$W_x = W_e + k\sqrt{M_e\rho_a}\left(\frac{x}{d}\right),$$
(5.37)

[2] G. C. Smith, *J. Metals*, 846 (July, 1966).

[3] W. G. Davenport, D. H. Wakelin, and A. V. Bradshaw, article in *Heat and Mass Transfer in Process Met.*, Inst. of Mining & Met., London, 1967.

where M_e = jet momentum at nozzle exit ($= W_e V_e$), x/d = distance from the nozzle exit, W_e = jet mass flow rate at nozzle exit, W_x = jet mass flow at x, ρ_a = density of the ambient gas.

In supersonic jets, the entrainment in the region where supersonic flow predominates is less than in the subsonic region, where Eq. (5.37) is satisfactory. A very satisfactory representation of the increased jet mass for supersonic jets with exit velocities between Mach 1 and 2 is given by*

$$\frac{W_x - W_e}{W_e} = 9 \times 10^{-3} \sqrt{\frac{T_0}{T_a A_e}} \left[1 + \frac{(\gamma - 1)M^2}{2}\right]^{-1} \left[\left(\frac{x}{d_e}\right) - \left(\frac{x}{d_e}\right)_{core}\right], \quad (5.38)$$

where T_0 = stagnation temperature, °R, T_a = ambient temperature, °R, and $(x/d_e)_{core}$ = length of the supersonic core determined from Fig. 5.18. This equation does *not* describe the increasing entrainment in the region from the nozzle to a distance of about 10 nozzle diameters downstream, but since this is not the region of usual interest and also the rate of increase is not very strong in this region, the equation is still useful. Note that as the ambient temperature increases, the entrainment decreases.

We should point out that the entrainment of the surrounding gases dilutes the jet gas-concentration. For example, the concentration of oxygen in cold jets of pure oxygen issuing into cold air, at a distance of 6 ft from the nozzle, is 90% in the case of a Mach 3 nozzle versus 80% in the case of a subsonic nozzle with 25 psig driving pressure.

Example 5.4 At what height above the bath should an oxygen lance discharging 15,000 scfm O_2 at Mach 2 be placed in order to have an impact area of 400 in² on a bath of molten iron–carbon alloy? What is the impact pressure at the center line? How long is the supersonic core?

Solution. The effective radius of the jet in order to cover 400 in² is

$$r = \sqrt{400/3.14} = 12.72 \text{ in.}$$

Since $r = 2r_0$,

$$r_0 = 6.37 \text{ in.}$$

$$r_t = 1.595 \text{ in. from Example 5.3.}$$

Thus

$$r_0/r_t = 6.37/1.595 = 3.99.$$

---

* This equation is derived from data presented in the report by J. D. Kapner and Kun Li, *Mixing Phenomena of Turbulent Supersonic Jets*, American Iron and Steel Inst., June 26, 1967.

From Fig. 5.20, at Mach 2.0, $x/d_t = 31$; $x = (31)(3.19) = 99$ in. above the bath. The impact pressure is found from Fig. 5.21:

$$\frac{P_c}{P_0 - P_a} = 0.09,$$

$$P_0 - P_a = 115.3 - 14.7 = 100.6 \text{psi},$$

$$P_c = 9.06 \text{ psig}.$$

The supersonic core is dissipated at a distance $x = 13(d_e) = (13)(4.14 \text{ in.}) = 53.8$ in. from the nozzle.

Supersonic jets in metallurgy have thus far been applied mainly in the basic oxygen steelmaking process, as mentioned in the examples above. The inherent rate of the reaction is so fast that the rate of decarburization is limited only by the rate at which oxygen is supplied to the bath of pig iron. The rate of supply is a maximum in the case of the converging–diverging nozzle operated at supersonic conditions. The jet of oxygen penetrates into the liquid, displacing metal. The distance to which the jet penetrates has been studied,[4,5] but the general correlation is not well enough established to warrant its inclusion at this time.

Subsonic jets of gas are often injected into molten metals, such as in the older practice of oxygen *lancing* of steel, but this is usually done with a straight pipe, and the analysis above is not applicable here. For a discussion of the behavior of such a jet, the reader can consult the review articles by Holden and Hogg[6] and by Davenport *et al.*[5]

5.4 VACUUM PRODUCTION

In recent years many metallurgical processes have been developed in which vacuum plays an important role. These include vacuum annealing, vacuum deposition of coatings, vacuum melting, and vacuum degassing. In addition, vacuum equipment is a standard item in almost every metallurgical laboratory. For these reasons, it is important that the metallurgist be acquainted with the principles and operation of vacuum-producing equipment.

The most important variable in the design and specification of equipment for a vacuum system is the pressure which the pumping system must be able to maintain in the work chamber. In vacuum technology the standard unit of pressure is the torr.* It is defined as the pressure exerted by a column of mercury 1 mm high and

---

[4] R. A. Flinn, R. D. Pehlke, D. R. Glass, and P. O. Hays, *Trans. A.I.M.E.* **239**, 1776 (1967).

[5] W. G. Davenport *et al.*, *ibid.*, pages 5–22.

[6] C. Holden and A. Hogg, *J. Iron and Steel Inst.* **198**, 318 (1960).

* The name torr comes from E. Torricelli, a student of Galileo, who devised the first single stroke pump by inverting a closed tube of mercury into a dish containing mercury, creating a vacuum in the tube.

measured at 0°C, sea level and a latitude of 45°. A vacuum of 1 torr is no longer considered to be a particularly good, or *hard* vacuum, since now we are able to produce much lower vacuums. Figure 5.22 compares the various pressure scales, indicating the ranges of application of various types of vacuum gages and pumps.

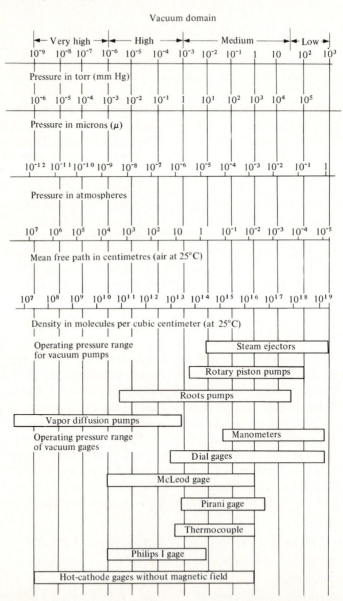

Fig. 5.22 Comparative pressure scales and ranges of application of various vacuum pumps and gages.

Another variable is the speed at which the desired pressure is reached. We define the *speed* S_p, of any type of vacuum pump by

$$S_p = Q/P, \tag{5.39}$$

where P is the pressure at the inlet to the pump, and Q is the *throughput* at that point. Typical units for S_p are liters/sec, for P, torr, and for Q, torr-liters/sec. We may use the same definition to express the pump-down speed of a system, S_s, consisting of a working chamber, connecting duct work, and pumps, but the speed depends on the design of the rest of the system as well as on the pump itself.

A system, initially at atmospheric pressure, is first *roughed out* by either mechanical pumps or the first stage of an ejector system until a pressure is reached where vapor diffusion pumps, or further stages of an ejector system, become effective and can be used for final evacuation to the desired limiting pressure. During the initial roughing out, the gas density will be high enough, so that the mean free path of the gas is very small ($\sim 10^{-5}$ cm) compared with the dimensions of the conduit, and the flow rate of the gas is governed by the viscosity of the gas through the equations developed in previous chapters for turbulent and laminar flows. However, at low pressures, the density eventually reaches a value such that the mean free path is much greater than the conduit dimensions. At this point, viscosity is no longer important in determining the flow rate, and the gas flow becomes *molecular flow*. At an intermediate pressure, a transition flow regime is encountered when the mean free path is of the same order of magnitude as the equipment dimensions.

5.4.1 Molecular flow mechanics

In the *molecular flow* regime, the gas molecules move randomly, with a Maxwell–Boltzmann distribution of velocities, and collisions between molecules are rare compared with collisions with the walls. The only transfer of momentum is between molecules and the wall instead of from molecule to molecule. Net flow results from the statistical effect that the number of molecules leaving a given region is proportional to the number of molecules in the region, and the number reentering will be proportional to the number in the adjacent region. Knudsen developed the

Fig. 5.23 Schematic chamber being evacuated from an internal pressure of P_1 to an external pressure of P_2 through an aperture of area A.

equations which govern the flow of gases through various geometries under molecular flow conditions. Consider a chamber with an aperture of area A, as in Fig. 5.23. The pressure on the downstream side of the aperture is P_2 and that in the chamber is P_1. If the concentration of gas molecules is given by n (molecules/cm³), and the number hitting the wall is

$$Z = \frac{n\bar{V}}{4} \quad \left(\frac{\text{molecules}}{\text{cm}^2\text{-sec}}\right), \tag{5.40}$$

where $\bar{V}$ is the average molecular speed given in Eq. (1.4), then the frequency Z of gas molecules per unit area passing from the high pressure side of the aperture through it, is

$$Z_1 = \frac{n_1}{2\sqrt{\pi}}\left(\frac{2\kappa_B T}{m}\right)^{1/2}, \quad \frac{\text{molecules}}{\text{cm}^2\text{-sec}}. \tag{5.41}$$

The frequency from the low-pressure side back through the opening

$$Z_2 = \frac{n_2}{2\sqrt{\pi}}\left(\frac{2\kappa_B T}{m}\right)^{1/2}, \tag{5.42}$$

yields a net frequency of

$$Z_{\text{net}} = Z_1 - Z_2 = \frac{1}{2\sqrt{\pi}}\left(\frac{2\kappa_B T}{m}\right)^{1/2}(n_1 - n_2). \tag{5.43}$$

For a circular opening, the net rate at which molecules leave the chamber is

$$R = \frac{\pi D^2}{4}Z_{\text{net}} = \frac{\sqrt{\pi}}{8}\left(\frac{2\kappa_B T}{m}\right)^{1/2}D^2(n_1 - n_2). \tag{5.44}$$

Since $n = P/\kappa_B T$,

$$R = \left(\frac{2\pi}{m\kappa_B T}\right)^{1/2}\frac{D^2}{8}(P_1 - P_2). \tag{5.45}$$

On the other hand, Eq. (5.45) simplifies to

$$Q = C(P_2 - P_1), \tag{5.46}$$

where C is the *conductance* of the aperture. The conductance of an entire vacuum system composed of several different components may be approximately calculated by analogy with electrical circuits. Specifically,

$$\frac{1}{C_{\text{system}}} = \frac{1}{C_1} + \frac{1}{C_2} + \frac{1}{C_3} + \cdots \tag{5.47}$$

for a system with components 1, 2, 3, etc., in series, or, if the components are in parallel, then

$$C_{\text{system}} = C_1 + C_2 + C_3 + \cdots \tag{5.48}$$

Table 5.2   Conductances of various geometric shapes for molecular flow*

Shape†	Conductance (liters/sec)	C, for air at 25°C
	$C = 3.64A\left(\dfrac{T}{M}\right)^{1/2}$	$= 11.7A$
	$C = 19.4\dfrac{A^2}{BL}\left(\dfrac{T}{M}\right)^{1/2}$	$= 12.2\dfrac{D^3}{L}$
	$C = 3.81\dfrac{D^3}{L}\left(\dfrac{T}{M}\right)^{1/2}$	$= 12.2\dfrac{(D_2 - D_1)^2(D_2 + D_1)}{L}$
	$C = 9.70\dfrac{b^2c^2}{(b + c)L}\left(\dfrac{T}{M}\right)^{1/2}$	$= 31.1\dfrac{b^2c^2}{(b + c)L}$
	$C = 2.85D^2\left(\dfrac{T}{M}\right)^{1/2}\left(\dfrac{1}{1 + 3L/4D}\right)$	$= 9.14\dfrac{D^2}{1 + 3L/4D}$
	$C = 3.64\left(\dfrac{T}{M}\right)^{1/2}\left(\dfrac{A}{1 - (A/A_t)}\right)$	$= \dfrac{11.7A_0}{1 - A_0/A_t}$
	$C = 3.81\dfrac{D^3}{L}\left(\dfrac{T}{M}\right)^{1/2}\left(\dfrac{1}{1 + \dfrac{4D}{3L}\left(1 - \dfrac{D^2}{D_t^2}\right)}\right)$	$= \dfrac{12.2D^3}{L\left[1 + \dfrac{4D}{3L}\left(1 - \dfrac{D^2}{D_t^2}\right)\right]}$

* This table is taken from J. M. Lafferty, *Techniques of High Vacuum*, General Electric Report No. 64-RL-3791G, 1964.

† The variables and their respective dimensions are

      A = area, cm²,

      D = diameter, cm,

  L, b, c = length dimensions, cm,

      B = perimeter, cm,

      T = absolute temperature, °K,

      M = molecular weight, g.

Table 5.2 presents the conductances of other shapes besides apertures.
Pumping speed has the same units as conductance, but it should be noted that
conductance implies a pressure gradient across a specific geometry. Pumping
speed is simply the volume of gas flowing across any plane in a system per second,
which is measured at the pressure existing at that particular plane.

The pump-down speed of a system depends on both the pump speed and the
conductance of the connections. Refer to the schematic diagram of a laboratory
vacuum melting system in Fig. 5.24. Since $P = Q/S$ at the inlet to the duct,
$P_p = Q/S_p$ at the inlet to the pump, and $(P - P_p) = Q/C$ over the duct length, then

$$\frac{1}{S} - \frac{1}{S_p} = \frac{1}{C}, \tag{5.49}$$

or

$$S = S_p\left(\frac{1}{1 + S_p/C}\right). \tag{5.50}$$

This means that the effective pump speed S of a system being evacuated by a pump
with rated speed S_p, cannot exceed S_p or C, whichever is the smaller. If the conductance of the duct is the same as the pump speed, then $S = S_p/2$. This should
emphasize why it is desirable to make connections between the working chamber
and the pump as short and as wide as possible.

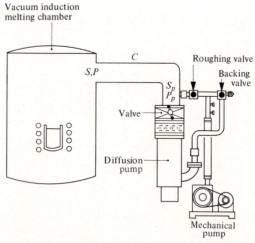

Fig. 5.24 A typical laboratory vacuum melting system.

5.4.2 Mechanical pumps

Most mechanical vacuum pumps are of the positive-displacement, rotary-piston
type with sliding vanes, and sealed with oil (Fig. 5.25). A small quantity of gas from

the system is isolated, compressed, and discharged to the atmosphere with each rotation of the piston. These pumps have *intrinsic speeds*, S_0, ranging in value from 0.5 to 350 liters/sec.

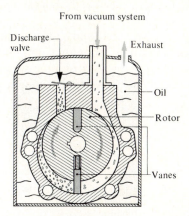

Fig. 5.25 A mechanical pump of the rotary, oil-sealed, vane type.

There is a lower limit to the pressure that a pump may produce, known as the ultimate pressure P_a, at which point the speed drops to zero. This limit is determined by the amount of back leakage of a very small quantity of gas Q_a, which is nearly independent of pressure. At the inlet to the pump,

$$S_p = \frac{Q - Q_a}{P_p} = S_0\left(1 - \frac{Q_a}{Q}\right). \tag{5.51}$$

At the ultimate pressure, $Q_a = Q$ and $S_0 = Q_a/P_a$. Therefore

$$S_p = S_0\left(1 - \frac{P_a}{P_p}\right). \tag{5.52}$$

Substitution of Eq. (5.52) into Eq. (5.50), and elimination of S_p and P_p by the use of $PS = Q = (P - P_p)C$, leads to a more realistic value for the speed of the system

$$S = S_0\left(\frac{1 - P_a/P}{1 + S_0/C}\right), \tag{5.53}$$

or

$$S = S'\left(1 - \frac{P_a}{P}\right), \tag{5.54}$$

where

$$S' = S_0\left(\frac{1}{1 + S_0/C}\right).$$

Single-stage pumps have ultimate pressures in the range from 10^{-2}–10^{-3} torr. Two-stage mechanical pumps can reduce the ultimate pressure to 10^{-4}–10^{-5} torr. In order to reduce the pressure from any water vapor in the system, a trap, either cold or chemical, should be placed ahead of the pump to remove moisture from the incoming gas stream.

If a slightly lower pressure is desired, and if the quantity of gas is large, a Roots pump (shown in Fig. 5.11) may be used in conjunction with an oil-sealed mechanical forepump. This combination may produce an ultimate pressure of 10^{-6} torr.

Typical characteristic curves for various types of mechanical pumps are shown in Fig. 5.26.

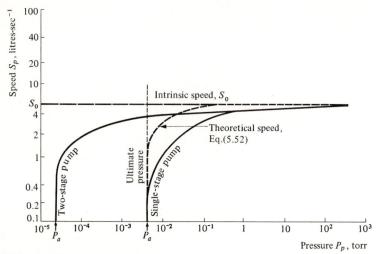

Fig. 5.26 The speed–pressure characteristic curves for a single- and two-stage rotary, oil-sealed, vane-type pump.

5.4.3 Diffusion pumps

We apply the term diffusion pump to a jet pump which utilizes the vapor from low-vapor pressure liquids to impart increased momentum to the gas molecules being removed from the system, eventually forcing them out of the discharge into a mechanical forepump. Figure 5.27 illustrates a typical diffusion pump. The pump fluid is heated in the boiler until its vapor pressure reaches an optimum value of about 1 torr. This vapor is carried to a nozzle from which it is ejected as a high-velocity jet directed away from the incoming gas and towards the wall of the pump. The gas molecules flow into the annular space between the wall and column, by molecular diffusion. Some fraction, H, of the molecules that encounter the jet in the first stage is entrained into the jet and driven downstream with higher velocities than the molecules would normally have. The jet expands and eventually strikes the water-cooled wall, the working fluid condenses out, and flows down the walls back to the boiler. In multiple-jet pumps, the gas molecules being pumped are

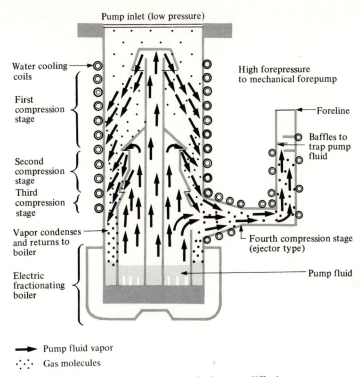

Pump inlet (low pressure)

Water cooling coils

First compression stage

Second compression stage

Third compression stage

Vapor condenses and returns to boiler

Electric fractionating boiler

High forepressure to mechanical forepump

Foreline

Baffles to trap pump fluid

Fourth compression stage (ejector type)

Pump fluid

→ Pump fluid vapor

·.·.· Gas molecules

Fig. 5.27 Cross section of a typical vapor diffusion pump.

caught in a succession of jet spray stages, and eventually ejected from the diffusion pump into the foreline.

Since free molecular flow is needed in order for their successful operation, diffusion pumps usually operate at inlet pressures of 10^{-3} torr, or below. The compression ratios are not great enough to allow direct discharge to the atmosphere, so a relatively low forepressure must be maintained by a mechanical forepump, ultimately discharging to the atmosphere. For any diffusion pump there is a *limiting forepressure*, which is the pressure above which the boundary between the jet of pump vapor molecules and the randomly moving incoming gas molecules does not extend to the cold pump walls. In this situation, there is a direct connection between the high-vacuum and low-vacuum (forepump) sides of the jet, and effective pumping ceases. This pressure is typically of the order of 0.5 torr for multistage diffusion pumps and 0.05 torr for single-stage pumps.

We define the speed of diffusion pumps in the same way as for mechanical pumps. If A is the area of the pumping annulus, then the rate at which gas molecules are entrained by the jet is

$$HZA = \frac{HAn\bar{V}}{4}, \tag{5.55}$$

using the previously defined nomenclature. Then, the intrinsic pump speed S_0 is

$$S_0 = \frac{HZA}{n} = \frac{Q}{P_p} = \frac{HA\bar{V}}{4},$$

(5.56)

and, substituting Eq. (1.4) for $\bar{V}$, we obtain

$$S_0 = \frac{H}{2\sqrt{\pi}}\left(\frac{2\kappa_B T}{m}\right)^{1/2} A,$$

(5.57)

or

$$S_0 = 3.64H\left(\frac{T}{M}\right)^{1/2} A, \quad \text{liters/sec.}$$

(5.58)

Specifically, for air at 20°C,

$$S_0 = 11.6(HA), \quad \text{liters/sec,}$$

(5.59)

where A is measured in cm^2. The coefficient H is a measure of the collection efficiency of the pump and is called the *Ho coefficient* (named after T. L. Ho). For most diffusion pumps the Ho coefficient is about 0.5.

These equations imply that S_0 is independent of the pressure for diffusion pumps, and this is nearly true, as shown in Fig. 5.28.

The ultimate pressures attainable with diffusion pumps depend on their design, including the number of stages, and also on the working fluid used. If we use mercury, a vapor trap, in the form of a baffle system externally cooled to low temperatures, must be placed between the diffusion pump and the system being evacuated in order to prevent back-diffusion of mercury vapor ($p_{Hg} = 1 \times 10^{-3}$ torr at 20°C). In this case, the effective speed of the pump is

$$S_p = S_0\left(\frac{1}{1 + S_0/C_t}\right),$$

(5.60)

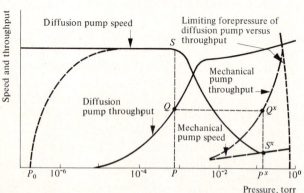

Fig. 5.28 Characteristic curve for a typical vacuum diffusion-pump, with a matching mechanical forepump's curves included.

where C_t is the conductance of the trap. For well-designed traps, $S_p \approx S_0/2$. The ultimate pressures attainable with *well-trapped* mercury diffusion pumps are of the order of 10^{-6}–10^{-7} torr. (A pressure of 10^{-13} torr is reported to have been developed with a liquid nitrogen-trapped, three-stage mercury pump constructed of glass.) More commonly, low-vapor-pressure hydrocarbon oils or silicone oils are used as the working fluids. They are capable of producing vacuums of 10^{-6} torr without traps, since their inherent vapor pressures at room temperature are of the order of 10^{-10} torr.

5.4.4 Pumpdown time

Returning to our laboratory vacuum melting unit, the basic equation relating the change in pressure in the tank to the pumping speed of the system, is

$$PS = -V\frac{dP}{dt} + Q_l, \tag{5.61}$$

where P = pressure measured at a specified point in the system, S = speed at that point, V = system volume, and Q_l = additional gas flow made up of the leak rate, interior surface outgassing, and any process gases. In well-maintained systems with no process gas evolution, Q_l is eventually brought to a negligible level.

Solving Eq. (5.61) for S and equating with Eq. (5.54), we obtain

$$\frac{dP}{(P - P_a)} = -\frac{S'}{V}dt. \tag{5.62}$$

We find the pumpdown time by integrating this equation from the initial tank pressure P_1 to the final pressure P_2

$$t = \frac{2.30\,V}{S'} \log\left(\frac{P_1 - P_a}{P_2 - P_a}\right). \tag{5.63}$$

However, since S' is a function of the conductance, it will also be a function of pressure at the higher pressures where viscous flow occurs. Therefore, in order to use Eq. (5.63), our approach will have to be to add several values of t obtained by incrementing S' until it is no longer a function of pressure. Equation (5.63) is not completely accurate, but it is a good approximation down to a pressure of the order of 1 torr.

Finally, we must consider the interaction and matching of the forepump and diffusion pump. Figure 5.28 includes the performance curves for a diffusion pump and a mechanical pump. In operation, the forepump is turned on, and run alone until a forepump inlet pressure less than the limiting forepressure of the diffusion pump is reached, at which point the lower stages of the diffusion pump become operative. If the pressure at the diffusion pump inlet is P, then the throughput is at point Q, and since the throughput is the same at any instant for both pumps, then the forepump inlet pressure is P^x, its speed is S^x, and its throughput Q^x.

Since the speed and throughput required of mechanical pumps when employed as forepumps are often rather low values, their use in large systems for roughing out would require excessively long pumpdown times. Therefore separate mechanical roughing pumps are often used, and then turned off when the diffusion pump–forepump combination can be applied.

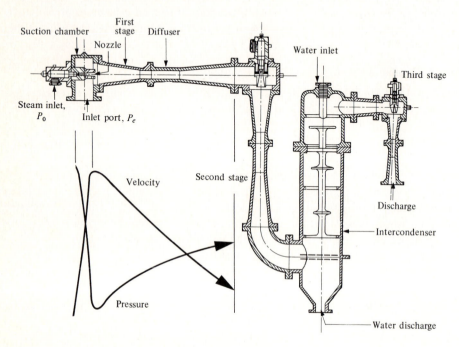

Fig. 5.29 Three-stage steam ejector with intercondenser. Schematic diagram of pressure–velocity relationships in the first-stage ejector is included.

5.4.5 Ejectors

For very large systems, other types of vapor pumps, called *steam ejectors*, are used. Figure 5.29 shows a schematic diagram of such an ejector. The steam is made to pass through a converging–diverging nozzle, designed to reach Mach 2 or higher at the nozzle exit and, correspondingly, a very low pressure P_e. The inlet port design pressure is then this P_e, and the steam jet entrains gas molecules at this pressure, as it leaves the nozzle. Flow into the port is again due to statistical molecular flow, at the lower pressures. Once entrained, the gas and steam slow down and are compressed in the diffuser so that they may exit at an exhaust pressure equal to the atmosphere, in the case of a single-stage pump, or at the design inlet pressure to the next stage in the case of a multiple-stage ejector. In order to lighten the load on the succeeding stages, condensers are often inserted between stages to remove the steam from the preceding stage.

The pumping capacity or throughput of a steam ejector is generally given in terms of pounds of dry air removed per hour. For comparison purposes

$$1 \text{ lb air at } 68°F/hr \equiv 79.5 \text{ torr-liters/sec.}$$

Figure 5.30 illustrates the range of pressures and throughputs obtainable with typical combinations of ejector stages and condensers.

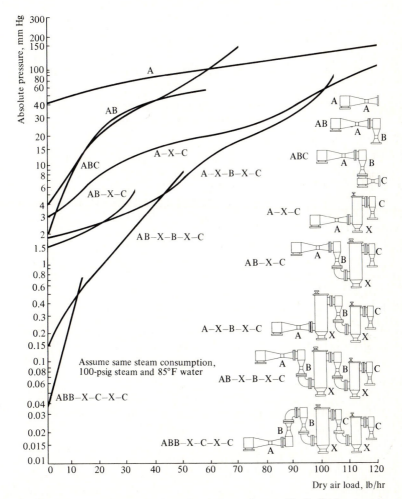

Fig. 5.30 Ranges of application of various steam-ejector (A, B, C)–intercondenser (X) combinations. (From an article by F. Berkeley, *Chem. Eng.*, April, 1957, page 255.)

PROBLEMS

5.1 Design the system and specify the pumps which are needed to evacuate a laboratory melting chamber of 30-ft³ volume to 10μ pressure in 20 min. Discuss how your design would change if the requirements were changed as follows:

- a) final pressure is to be 1 mm,
- b) pumpdown time is to be 5 min, and
- c) pumpdown time is to be 60 min.

5.2 A continuous vapor deposition process utilizes a vacuum to protect easily oxidized metal vapors as they travel between a source and metal strip moving through the chamber, upon which they condense. Discuss the relative importance of the absolute pressure attainable versus the leak-up rate, recalling that a large vacuum pump may be able to attain a very low pressure even with a large throughput.

5.3 Design the nozzles and stipulate the operating pressure for a three-hole supersonic nozzle delivering 10,000 scfm oxygen at Mach 2 with the requirement that the jet boundaries touch at 60 in. from the nozzle.

5.4 Read up on techniques for producing ultralow vacuums. How does a "getter" work? What are the requirements—in terms of physical size, shape, composition, and temperature—for its successful operation?

5.5 A 50-ton heat of steel is to be degassed (H_2 and N_2 removed) from 5 ppm H_2 to 1 ppm H_2 and from 100 ppm N_2 to 75 ppm N_2 in a period of 15 min. The metal is at 1600°C (assumed constant) and the chamber has 300 ft^3 of space occupied by air after the top is closed with the ladle inside. At what pressure would you recommend operating the system? Calculate the required vacuum system capacity and specify a steam ejector to do the job. [Ref.: *Chemical Engineering*, page 255, April 1957.]

PART TWO

ENERGY TRANSPORT

In the processing of materials, a situation almost invariably arises that necessitates a change in the temperature of a material, as, for example, in heat treating processes. We assume that the student is aware of the importance of being able to calculate the heat which is produced and used in such processes, by making heat balances. However, it is also desirable to have an understanding of how heat is transferred into, out of, and within metallurgical processes, because many operations take place in such a way that the *rate* of energy transfer becomes the controlling factor in raising and lowering the temperature. Since we usually consider this energy to be virtually all thermal energy, we speak of *heat transfer* as the controlling factor.

There are two basic types of heat-transfer mechanisms: *conduction* and *radiation*. Quite frequently, however, three mechanisms are set forth, namely, conduction, radiation, and convection. To be more specific, convection is rather a process involving mass movement of fluids, than a real mechanism of heat transfer. In regard to this distinction it is better to speak of "heat transfer with convection" rather than of "heat transfer by convection." The term convection implies fluid motion, and mechanisms of heat transfer anywhere within the fluid are only conduction and radiation. We shall discuss conduction and radiation, and present the relationships used for describing heat transfer by each mechanism in the following chapters. Here, as an introduction, we shall consider them in brief.

Conduction is the transfer of heat by molecular motion which occurs between two parts of the same body, or between two bodies which are in physical contact with each other. In fluids, heat is conducted by molecular collisions; in solids, heat is conducted either by lattice waves in nonconductors or by a combination of lattice waves with the drift of the conduction electrons in conducting materials. The macroscopic theory of conduction is merely *Fourier's law*:

$$q_y = -k \frac{\partial T}{\partial y},$$

where q_y is the heat flux in the y-direction, $\partial T/\partial y$ is the temperature gradient in the y-direction, and k is the thermal conductivity. Note the analogy with Newton's law of viscosity (Eq. 1.3). In both cases, fluxes are proportional to gradients, and the relationships also define the proportionality constants, namely, viscosity by Newton's law and thermal conductivity by Fourier's law.

The nature of heat transfer by radiation is quite different from that by conduction, and consequently the basic rate equation for heat transfer by radiation is in no way similar to Fourier's law. In Chapter 11, we shall analyze radiation problems in detail; here, we state briefly that thermal radiation is part of the electromagnetic spectrum. The energy flux emitted by an ideal radiator is proportional to the fourth power of its absolute temperature

$$e_b = \sigma T^4,$$

where e_b is the emissive power (a special term for thermal energy transferred by radiation) and σ is the Stefan-Boltzmann constant. The processes of conduction and radiation frequently occur simultaneously, even within certain media. In many practical situations, however, one mode is negligible with respect to the other, and may be ignored.

When a moving fluid at one temperature is in contact with a solid at a different temperature, heat exchanges between the solid and the fluid by conduction at a rate given by Fourier's law where k is the conductivity of the fluid, and $\partial T/\partial y$ is the temperature gradient in the fluid normal to the wall at the fluid–solid interface. If the details of the convection are known in a given situation, then we can determine the distribution of temperature within the fluid, and calculate the heat flux at the wall. In many cases, such a detailed analysis is not available; then it is convenient to define the *heat-transfer coefficient*, h, by the equation

$$h = \frac{q_0}{T_s - T_f} = \frac{-k(\partial T/\partial y)_0}{T_s - T_f},$$

where T_s is the surface temperature, T_f is usually taken as some bulk fluid temperature, and q_0 is the heat flux at the wall. The units of h, which is a function of the fluid and the flow pattern of the system, are Btu/hr-ft^2 °F. Much of the research on heat transfer has been devoted to the determination of h, since it enters into all problems of heat transfer with convection.

Convection is usually classified as either a *forced* convection or *free* (*natural*) convection. When a pump or other mechanical device causes the fluid to move, we call this process forced convection; when a fluid moves as a result of density difference, then we speak of a free or natural convection. Thus, when a radiator heats the air which rises, displacing colder air in the upper part of a room, the fluid motion is by natural convection.

In this section of the text, we shall examine the thermal conductivity of materials and the various modes of heat transfer applied to situations which metallurgists are likely to encounter.

6

FOURIER'S LAW AND THERMAL CONDUCTIVITY OF MATERIALS

Thermal conductivity is an intrinsic property of materials. In this chapter we shall consider the thermal conductivity of various materials, such as gases, liquids, and solids, with the emphasis on solids, including not only dense solids but also porous bulk materials.

6.1 FOURIER'S LAW AND THERMAL CONDUCTIVITY

Consider a slab of solid material of area A bounded by two large parallel surfaces a distance Y apart (Fig. 6.1). Initially, the solid material is at a uniform temperature T_0. At some instant, the lower surface is suddenly raised to a temperature T_1, which is maintained. The material beneath the heated surface becomes heated, and the

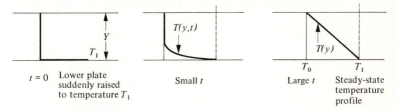

$t = 0$ Lower plate suddenly raised to temperature T_1	Small t	Large t Steady-state temperature profile

Fig. 6.1 Build-up to steady-state temperature profile for a solid slab.

heat is gradually transferred across the solid toward the colder surface, which is kept at T_0. A steady-state temperature distribution is attained when a constant rate of heat flow through the slab is required to maintain the temperature difference $T_1 - T_0$. It is found that for sufficiently small temperature differences, the rate of heat flow per unit area q is proportional to the temperature difference and inversely proportional to the distance between the surfaces; hence,

$$q = k\left(\frac{(T_1 - T_0)}{Y}\right). \tag{6.1}$$

Equation (6.1) also applies to liquids and gases, provided no convection or heat transfer by radiation is allowed to take place in the fluid. As stated, the temperature difference must be sufficiently small, because the thermal conductivity depends not only on the specific material, but also on the temperature. Equation (6.1) is therefore valid for fixed values of T_1 and T_0 for all values of Y.

As the temperature difference and the separation distance approach zero, the differential form of Fourier's law of heat conduction arises:

$$q_y = -k\frac{\partial T}{\partial y}.$$ (6.2)

The flow of heat per unit area (heat flux) in the y-direction is proportional to the temperature gradient in the y-direction.

For a three-dimensional situation in which the temperature varies in all three directions, we write

$$q_x = -k\frac{\partial T}{\partial x}; \qquad q_y = -k\frac{\partial T}{\partial y}; \qquad q_z = -k\frac{\partial T}{\partial z}.$$ (6.3)

These three relations form the components of the single equation

$$q = -k\mathbf{\nabla}T.$$ (6.4)

Note the similarity between Eq. (6.2) for one-dimensional heat flux and Eq. (1.2) for one-dimensional momentum flux. In each case, the flux is proportional to the gradient of a potential variable, temperature, and velocity, respectively. The mathematical similarity ends, however, when we compare the three components for heat flux versus the nine components that express momentum transfer in three dimensions. This is because energy is a scalar quantity whereas momentum is a vector quantity.

The units involved in Fourier's law of heat conduction are

	Engineering units	Scientific units*
q_x, q_y, q_z	Btu/hr-ft^2	cal/sec-cm^2
T	°F or °R	°C or °K
x, y, z	ft	cm
k	Btu/hr-ft °F	cal/sec-cm °C

The thermal conductivity of a material reflects the relative ease or difficulty of the transfer of energy through the material. This, in turn, depends on the bonding and structure of the material. When considering the thermal conductivity of most materials used in engineering, bear in mind that the effective thermal conductivity depends on the interaction between the intrinsic thermal conductivities of the phases present and the mode of energy transfer between them. In the following sections, we shall first consider the thermal conductivity of various pure phases, and then the effective thermal conductivity of some bulk materials.

---

* 1 cal/sec-cm °C = 241.9 Btu/hr-ft °F.

6.2 THERMAL CONDUCTIVITY OF GASES

Conduction of energy in a gas phase is primarily by transfer of translational energy from molecule to molecule as the faster moving (higher-energy) molecules collide with the slower ones.

For a simple, monatomic gas, we can develop an expression for the thermal conductivity in a similar manner as we did in Eq. (1.13) for viscosity. Assume that the gas in Fig. 1.5 is under the influence of a temperature gradient, $\partial T/\partial y$. The temperatures of the molecules at $y - \bar{y}$ and at $y + \bar{y}$ are, respectively,

$$T\Big|_{y-\bar{y}} = T\Big|_{y} - \frac{2}{3}\lambda\frac{\partial T}{\partial y}, \tag{6.5}$$

and

$$T\Big|_{y+\bar{y}} = T\Big|_{y} + \frac{2}{3}\lambda\frac{\partial T}{\partial y}. \tag{6.6}$$

Here the average distance moved in the y-direction between collisions is $\bar{y}$ and $\bar{y} = \frac{2}{3}\lambda$, where λ is the mean free path as defined by Eq. (1.7).

If the heat capacity per molecule is c, then the net flux of thermal energy across the plane at y is given by the net difference between the energy of the molecules crossing the y-plane in the positive and in the negative directions:

$$q_y = ZcT|_{y-\bar{y}} - ZcT|_{y+\bar{y}}$$

$$= Zc(T|_{y-\bar{y}} - T|_{y+\bar{y}}). \tag{6.7}$$

Equation (1.6) gives the expression for Z, the flux of molecules (number/sec-cm^2) crossing the y-plane in either the positive or the negative direction; Eqs. (6.5) and (6.6) give the expressions for the temperatures, so that ultimately

$$q_y = -\frac{nc\bar{V}\lambda}{3}\left(\frac{\partial T}{\partial y}\right). \tag{6.8}$$

Denoting nc as C_v, the heat capacity *per unit volume*, and by comparing Eq. (6.8) to Eq. (6.2), we arrive at

$$k = \frac{C_v\bar{V}\lambda}{3}. \tag{6.9}$$

This equation is basic for an understanding of the thermal conductivity of gases. In some cases, it can be extended to aid in theorizing the thermal conductivity of solids and liquids. For dilute gases with atoms assumed as rigid spheres, Eqs. (1.4) and (1.5) may be substituted for $\bar{V}$ and λ in Eq. (6.9), with the result that

$$k = \frac{1}{d^2}\sqrt{\frac{\kappa_B^3 T}{\pi^3 m}}. \tag{6.10}$$

This result suggests that the thermal conductivity of gases does not depend on pressure, but that it does depend on the square root of the temperature. For pressures to about 10 atm, this lack of dependence on pressure is essentially correct. The thermal conductivity, however, varies with temperature more than predicted by Eq. (6.10).

In the case of polyatomic gases, Eucken[1] developed an equation for the thermal conductivity of these gases at normal pressures:

$$k = \eta\left(C_p + \frac{1.25R}{M}\right),\qquad(6.11)$$

where M is the molecular weight and C_p the heat capacity at constant pressure.

In the case of gas mixtures, we can estimate the thermal conductivity within a few percent by

$$k_{mix} = \frac{\sum_i X_i k_i M_i^{1/3}}{\sum_i X_i M_i^{1/3}},\qquad(6.12)$$

where X_i is the mole fraction of component i having molecular weight M_i and intrinsic thermal conductivity k_i. A comparison[2] of observed conductivities with calculated values, using Eq. (6.12), for binary gas mixtures involving air, CO, CO_2, H_2O, N_2, N_2O, NH_3, CH_4, C_2H_2, He, and Ar, indicates average discrepancies of only 2.7%, over the temperature range 273–353°K. Tests at elevated temperatures are not available, but the errors would probably be no larger. There are some far more complex equations, which reduce the errors to about 1%, but they are not as easy to use.

Figure 6.2 gives the thermal conductivities of several common gases as a function of temperature. The data are valid to at least 10 atm. For corrections at higher pressures, the reader should consult references[3,4] below, and for more detailed data, Tsederberg.[5]

Example 6.1 Calculate the thermal conductivity of a gas containing 40 mol% CH_4 and 60 mol% H_2, at 1.5 atm pressure and 1800°F.

Solution. From Fig. 6.2,

$$k_{H_2} = 0.295 \text{ Btu/hr-ft °F}$$

$$k_{CH_4} = 0.127. \text{ Btu/hr-ft °F}.$$

[1] A. Eucken, *Physik Z.* **14**, 324 (1913).

[2] L. Friend and S. Adler, article in *Transport Properties in Gases*, Northwestern University Press, Evanston, Ill., 1958.

[3] R. Bird, W. Stewart, and E. Lightfoot, *Transport Phenomena*, Wiley, New York.

[4] J. O. Hirschfelder, C. F. Curtiss, and R. B. Bird, *Molecular Theory of Gases and Liquids*, Wiley, New York, 1954.

[5] N. V. Tsederberg, *Thermal Conductivity of Gases and Liquids*, M.I.T. Press, Cambridge, Massachusetts, 1964.

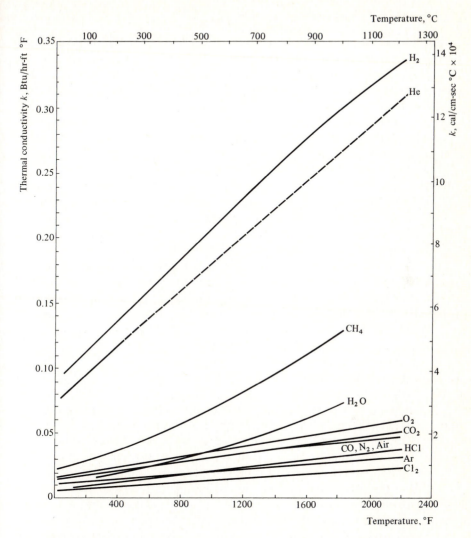

Fig. 6.2   Thermal conductivity of several gases.   Data valid for up to 10 atm.

Using Eq. (6.12) we get

$$k_{\text{mix}} = \frac{(2)^{1/3}(0.295)(0.6) + (16)^{1/3}(0.127)(0.4)}{(0.6)(2)^{1/3} + (0.4)(16)^{1/3}} = 0.199 \; \frac{\text{Btu}}{\text{hr-ft °F}}.$$

The fact that the pressure is 1.5 atm has no significance here, since the thermal conductivity is independent of pressure at this pressure level.

6.3 THERMAL CONDUCTIVITY OF SOLIDS

Solids transmit thermal energy by two modes, either one of which, or both, may operate. In all solids, energy may be transferred by means of elastic vibrations of the lattice moving through the crystal in the form of waves. In some solids, notably metals, free electrons moving through the lattice also carry energy in a manner similar to thermal conduction by a true gas phase.

Recalling the fundamental treatments of bonding in solids, we remember that at ordinary temperatures all solids store thermal energy as vibratory motion of their atoms, and potential energy in the bonding between atoms. Einstein has developed his theory for the heat capacity of solids by assuming that each atom vibrates independently of its neighbors. This, however, would result in no conduction via lattice waves, which is known to exist, being caused by the fact that neighboring atoms actually do interact and therefore do not vibrate independently. Debye, in the process of improving Einstein's model of heat capacity, assumed that the lattice is made up of independent oscillators, and that these oscillators are simply considered to be elastic waves traveling in different directions with different polarizations and wavelengths, not necessarily associated with the atoms themselves. This has led to his theory of specific heat ($C_v \propto T^3$ at low temperatures) which is more in agreement with experiments than Einstein's theory, and also gives a more satisfactory picture of thermal conductivity of insulating solids. Each lattice vibration (there is always a spectrum of vibrations) may be described as a traveling wave carrying energy and obeying the laws of quantum mechanics. By analogy with light theory, the waves in a crystal exhibit the attributes of particles and are called *phonons*.

If the forces between atoms were perfectly harmonic, two phonons moving through the crystal could collide and combine, with the resulting phonon having the same total energy and momentum; energy transport would continue in the direction of the resultant of the original phonons. If this were the only type of phonon–phonon interaction (Normal or N-type), then the only limit to the mean free-path of the phonons would be the dimensions of the piece of the material itself, since the resultant phonon acts in the same manner as the two original phonons as far as energy transport is concerned. Thermal conduction would thus be extremely easy. But the forces between atoms are not in perfect harmony, and there exists another type of collision between phonons, known as an Unklapp (U-process) collision. In this process, energy is conserved, while phonon momentum is not, and the result of such a collision is that a phonon moves in a direction opposite to the resultant of the original phonons. For these processes, the mean free-path of phonons is extremely short, being of the order of the distance between atoms. Since the number of phonons increases with temperature, the number of the U-processes also increases with temperature and the wavelength of the phonons λ_{ph} is proportional to $1/T$. At room temperature and above, $\hat{C}_v$ for most materials is roughly constant, and if we use Eq. (6.9) to describe the thermal conductivity of a solid which conducts energy only by phonons, then k must decrease with increasing

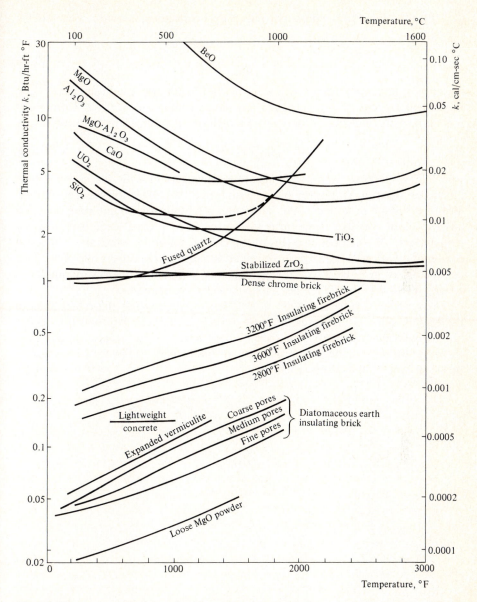

Fig. 6.3 Thermal conductivity of oxides and various insulating materials. (From A. Schack, *Industrial Heat Transfer*, Wiley, New York, 1965.)

temperature. This is what happens in most electrically insulating substances, such as the oxides shown in Fig. 6.3 (but not in the form of porous, bulk materials).

When heat enters a crystal, the crystal expands, and this thermal expansion of the lattice is related to the bulk modulus B by a constant γ, known as Gruneisen's constant

$$\gamma = \frac{\hat{V}}{\hat{C}_v}\left(\frac{\partial P}{\partial T}\right)_v = \frac{3B\hat{V}\alpha}{\hat{C}_v},$$ (6.13)

where $\hat{V}$ = molar volume, α = linear thermal expansion coefficient, and $\hat{C}_v$ = molar heat capacity. For most solids at ordinary temperatures, $\gamma \cong 2$. Combining this result with Lindeman's theory of the melting of solids, Debye showed that

$$\lambda_{\text{ph}} = \frac{20T_m d}{\gamma^2 T},$$ (6.14)

where T_m = melting point, T = absolute temperature, and d = crystal-lattice dimension, resulting in the previously mentioned intuitive relation between phonon mean free path and temperature. Note that a high-melting-point material has a large value of λ at low temperatures, and therefore a large value of k at room temperature. We can observe this in the case of diamond, which has a thermal conductivity at room temperature comparable to that of copper, one of the best conductors known. However, the conductivity of diamond drops very rapidly to a small fraction of that of copper as the temperature increases.

At very low temperatures ($T < {\sim}10°\text{K}$), λ_{ph} approaches the dimensions of the crystal, since U-processes diminish rapidly as the temperature approaches absolute zero; consequently, the thermal conductivity rises rapidly. In this case, the larger the crystal, the higher the thermal conductivity, as illustrated in Fig. 6.4; this is an exception to Fourier's law.

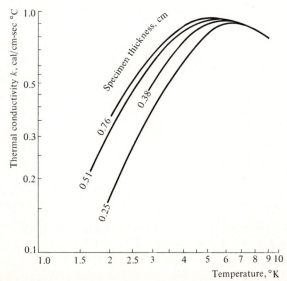

Fig. 6.4 Effect of specimen thickness on thermal conductivity of potassium chloride single crystals. (From W. J. deHaas and T. Biermasz, *Physica* **2**, 673 (1935); **4**, 752 (1937); **5**, 47, 320, and 619 (1938).)

Phonons are also scattered by differences in isotopic masses, chemical impurities, dislocations, and second phases. With these imperfections quantitative prediction of thermal conductivity is exceedingly difficult. The only thing that we can be sure about is that the less perfect the crystal is, the lower will be the thermal conductivity due to phonons.

As we proceed from electrical insulators to conductors, we deal with materials with increasing concentrations of conduction electrons. The conduction electrons in metals form an *electron gas* which obeys the laws of quantum mechanics. The electronic contribution to the total heat capacity (per cm³) of a metal is given by the expression

$$C_{v,\text{el}} = \frac{\pi^2 n_e \kappa_B^2 T}{2\varepsilon_F}.$$
(6.15)

Here ε_F is the Fermi energy of the particular metal and n_e is the number of free electrons per cm³. The Fermi energy is related to the average velocity of the electrons by

$$\bar{V}_F = \sqrt{\frac{2\varepsilon_F}{m_e}},$$
(6.16)

where $\bar{V}_F$ = electron velocity at the Fermi surface, and m_e = electron mass. Using Eq. (6.15) for the heat capacity and Eq. (6.16) for the Fermi energy, Eq. (6.9) becomes

$$k_{\text{el}} = \frac{\pi^2 n_e \kappa_B^2 T \lambda_{\text{el}}}{3 m_e \bar{V}_F},$$
(6.17)

where k_{el} is now the electronic contribution to the thermal conductivity. This predicts that the electronic contribution in metals *increases* with temperature, provided that λ_{el} does not decrease just as strongly with temperature.

To prove the theory that the electrons in a metal carry a major portion of the thermal energy, the thermal and electrical conductivities should be proportionally related. This proportionality is known as the Wiedmann–Franz law and the constant of proportionality as the Lorentz number L:

$$L = \frac{k_{\text{el}}}{\sigma T} = \frac{\pi}{3}\left(\frac{\kappa_B}{e}\right)^2 = 2.45 \times 10^{-8} \text{ watt-ohm/deg}^2,$$
(6.18)

where σ = the electrical conductivity, ohm⁻¹-cm⁻¹, and e = the charge on an electron. When the experimental value of L is close to, or equal, to the theoretical value, then we assume that the electronic contribution to the thermal conductivity predominates. For pure metals near room temperature, experimental values of L range from 2.23×10^{-8} watt-ohm/deg² for copper to 3.04×10^{-8} watt-ohm/deg² for tungsten.

The thermal conductivities of pure metals are shown in Fig. 6.5. Since the

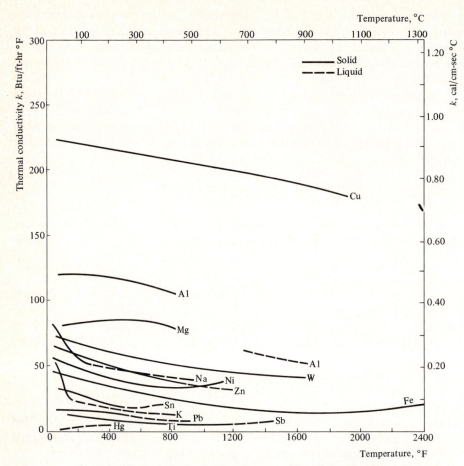

Fig. 6.5 Thermal conductivities of pure, solid, and liquid metals.

thermal conductivity of all crystalline materials is made up of contributions from both k_{ph} and k_{el}, we can only crudely predict the temperature dependence. It is evident that the increase in k with temperature predicted by Eq. (6.17) does not operate in a strong fashion, and that most metals actually show a decrease in k with temperature. In the cases of pure nickel and pure iron, k decreases with temperature at low temperatures; at higher temperatures, the electronic contribution presumably overwhelms the phonon contribution, and k increases with temperature.

Figures 6.6 and 6.7 show the effect of alloying. Substitutional alloying and alloying, resulting in a second phase, both lower the conductivity from that of the pure metal, although a wide variety of ferrous alloys all reach the same limiting value at temperatures at or above the α-γ transition point. Note that the *alloys* in both Figs. 6.6 and 6.7 have absolute values of k that are about the same as those for the oxides in Fig. 6.3, which clearly are not metallic in nature. This implies—and

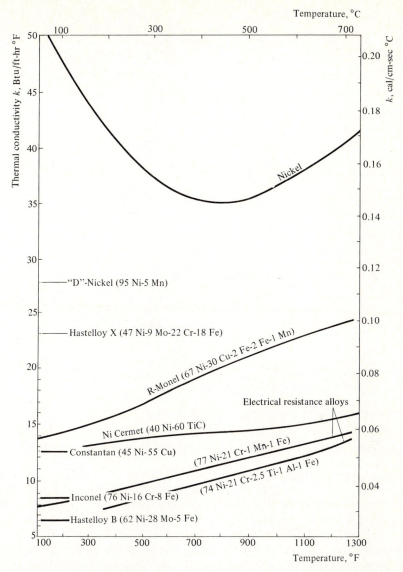

Fig. 6.6 Thermal conductivities of pure nickel and of nickel-base alloys.

research has confirmed—that in alloys a substantial portion of the thermal conduction is via phonons.

In the case of semiconducting compounds, the thermal conductivity may be strongly influenced by phonon mechanisms at low or at moderate temperatures, but as more electrons are excited from valence to conduction states, the electron contribution increases, until k_{el} may predominate. In the intrinsic conduction

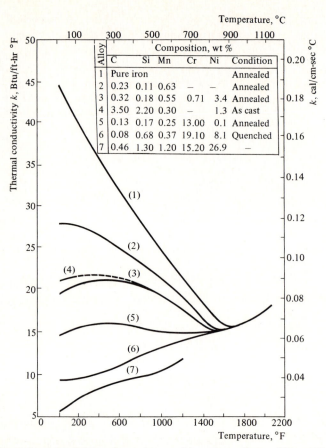

Fig. 6.7 Thermal conductivities of pure iron and iron-base alloys. (From Schack, *ibid.*)

range, the Lorentz number is given by

$$L = 2\left(\frac{\kappa_B}{e}\right)^2 + \left(\frac{\kappa_B}{e}\right)^2 \frac{1}{(\sigma_n + \sigma_p)}\left(4 + \frac{E_g}{RT}\right)^2,\tag{6.19}$$

where σ_n = electrical conductivity via electrons in the conduction band, σ_p = electrical conductivity via holes in the valence band, and E_g = energy gap between bands. It has been shown[6,7] that this contribution is significant in the case of UO_2, and is certainly significant for other semiconducting oxides, such as TiO_2, ZrO_2, and Nb_2O_5. Since this contribution increases with temperature ($\sigma_n + \sigma_p$, σ_n, and σ_p all increase with temperature), it must be superimposed on the phonon contribution in order to totally explain the rise in the thermal conductivity at very high temperatures for semiconductors.

---

[6] J. L. Bates, *Nucleonics* **19**, 83 (1961).

[7] D. R. deHelar, *Nucleonics* **2**, 92 (1963).

Conversely, other oxides, for example, SiO_2, Al_2O_3, and MgO, never show significant electronic conduction, and yet their thermal conductivities also increase with temperature at very high temperatures (see Fig. 6.3). This phenomenon has been explained on the basis of the ability of certain materials to transmit radiant energy (a *photon* contribution to thermal conductivity). This occurs when a material is no longer opaque, but translucent to incident radiation. Without going into the details of the interaction of radiation and solids, we can compute the radiant energy conductivity k_r from

$$k_r = \frac{16}{3} \frac{\sigma v^2 T^3}{a}, \tag{6.20}$$

where $\sigma =$ the Stefan–Boltzmann constant (1.37×10^{-12} cal/cm^2-sec°K^4), $v =$ the index of refraction, and $a =$ the absorption coefficient equal to $1/\lambda_{ph}$, cm^{-1}. When a is large (opaque material), the ability to absorb incident radiation is large; thus radiant energy is not transmitted through the solid as photons, but rather is absorbed at the surface and transferred to energy transmitted by the phonon or electron mechanisms.

When a is small (transparent material), k_r may be quite significant. In general, the contribution of photon or radiant energy becomes significant for dense oxides at temperatures in the neighborhood of 1500°C.

Figure 6.8 presents thermal conductivity data for a variety of nonmetallic crystalline materials used for high-temperature applications. Some of these exhibit semiconducting electrical properties at elevated temperatures, and the transition in temperature dependence is probably caused by an increasing influence of the electronic conductance over the phonon contribution. The absolute value is again in the same range as that for the alloys in Figs. 6.6 and 6.7.

Finally, for amorphous materials such as high polymers and glasses, thermal conduction is mainly via atomic migration (or radiation at high temperatures), since the material is too irregular in structure to support a phonon mechanism and the electron contribution is negligible. This results in very low values of conductivity, as indicated in Table 6.1.

Table 6.1 Thermal conductivities of amorphous or molecular solids

Substance	Temperature, °F	k, Btu/hr-ft °F
Glass	212	0.44
Lead glass	32	0.50
Pyrex glass	212	0.67
Quartz glass	212	0.82
Asphalt	68	0.44
Polystyrene	68	0.07
Polyvinyl chloride	68	0.15

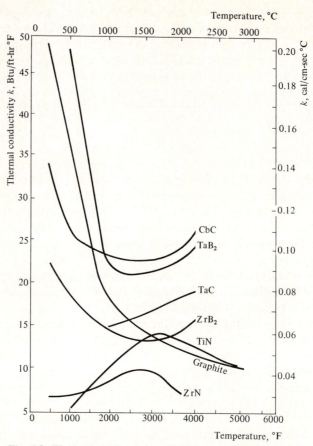

Fig. 6.8 Thermal conductivity of high-temperature materials.

To summarize, we should remember that the value of the thermal conductivity of solids is determined by the sum of several mechanisms, including those of phonons, electrons, photons, and atomic migration, regardless of the materials, and it is due to this fact that the prediction of this property is so difficult.

6.4 THERMAL CONDUCTIVITY OF LIQUIDS

As usual, when dealing with liquids, we are faced with a lack of knowledge of their structure. Using Eq. (6.9), Bird et al.[8] have modified a theory due to Bridgman which results in an expression for the thermal conductivity of liquids at densities away from the critical value:

$$k = 2.8 \kappa_B V_s \left(\frac{N_0}{\hat{V}} \right)^{2/3}, \qquad (6.21)$$

---

[8] Bird, Stewart, and Lightfoot, *ibid.*, page 260.

where $\hat{V}$ = the molar volume, N_0 = Avogadro's number, and V_s = the speed of sound through the liquid, given by

$$V_s = \sqrt{\frac{C_p}{C_v}\left(\frac{1}{\rho\beta}\right)_T},$$

where β = compressibility. This is based on the assumption that the molecules in the liquid are arranged in a cubic lattice and energy transfer is via collisions between molecules. The thermal conductivity of ordinary liquids near room temperature is considerably below that of crystalline solids, as indicated in Table 6.2, emphasizing the fact that energy transfer in ordinary liquids is difficult owing to the lack of both a phonon and a free-electron mechanism. However, liquid metals show much higher conductivities than other liquids, since presumably electronic conduction is still possible, and a survey of the Lorentz numbers for liquid metals[9] indicates close agreement with the theoretical value. Data for liquid metals are included in Fig. 6.5. A notable feature of the available data is that liquid metal conductivities are all within the range 1–50 Btu/ft-hr °F, while the range for the solid metals is 1–225. The thermal conductivity of good conductors drops significantly prior to melting, as the phonon contribution decreases. On the basis of this observation, one might assume that, if no other data are available, the unknown thermal conductivity of any liquid metal or alloy would fall in the range 1–50 Btu/hr-ft °F.

Table 6.2   Thermal conductivities of various liquids

Substance	Temperature, °F	k, Btu/hr-ft °F
Water	60	0.319
Water	100	0.364
Light oil	60	0.077
Light oil	100	0.079
Benzene	177	0.083
Fluoride salts	900	3.2
Slag	2900	2.3

From this discussion, we can see that to predict the thermal conductivity of solids and liquids is an extremely difficult task. One must be able to make intelligent estimates and extrapolations from known data, when the occasion calls for it. This in turn requires an understanding of the role of the various conduction mechanisms we have just discussed, and of the effects that structural and chemical variations may have on these mechanisms. We can only make order-of-magnitude estimates if experimental data are not available. Figure 6.9 gives a summary of the typical ranges of thermal conductivities for various classes of materials.

---

[9] J. R. Wilson, *Structure of Liquid Metals and Alloys*, Institute of Metals Report, 1967, page 482.

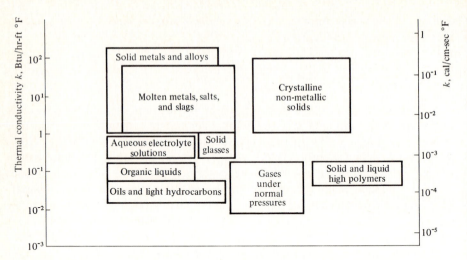

Fig. 6.9 Summary of the ranges of thermal conductivity for various classes of materials.

6.5 THERMAL CONDUCTIVITY OF BULK MATERIALS

So far, we have presented equations, and looked at data describing the thermal conductivity of individual phases. We have seen the difficulties of arriving at exact predictions of thermal conductivity for most solids, particularly complex alloys. In most engineering situations, we are often faced with even more complex materials, for example, porous bulk materials. The thermal conductivity of such materials is certainly not equal to the intrinsic thermal conductivity of the solid involved. The question is: What value should be used?

6.5.1 Two-phase mixtures

In nuclear and aerospace technology, alloys known as cermets have been evolved. These are usually metallic alloys with a dispersion of a ceramic phase within the matrix. The thermal conductivity of such materials is a function of the volume fraction of each phase. For up to one tenth volume fraction, V_d, of a dispersed phase of spherical particles with intrinsic thermal conductivity, k_d, it has been suggested that the thermal conductivity of the mixture, k_{mix}, may be found from the Maxwell-Eucken equation,

$$k_{mix} = k_c \left[\frac{1 + 2V_d\left(\dfrac{1 - k_c/k_d}{2k_c/k_d + 1}\right)}{1 - V_d\left(\dfrac{1 - k_c/k_d}{k_c/k_d + 1}\right)} \right]. \tag{6.22}$$

If the thermal conductivity of the continuous phase, k_c, is much larger than k_d, then

$$k_{mix} \cong k_c \left(\frac{1 - V_d}{1 + V_d/2} \right). \tag{6.23}$$

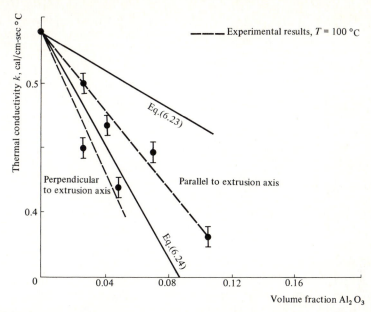

Fig. 6.10 Thermal conductivity of Al–Al_2O_3 (SAP) alloys at 100°C. Similar results are obtained on data available at 500°C. (Data from D. Nobili and M. A. DeBacci, *ibid.*)

Data for the thermal conductivity of SAP Al–Al_2O_3 alloys[10] are plotted in Fig. 6.10, along with k_{mix} calculated by using Eq. (6.23) and the data from Figs. 6.3 and 6.5. The agreement is less than satisfactory.

However, if we assume that the material behaves as if it were a series of plates of Al and Al_2O_3 normal to the direction of heat flow, then they would be equivalent to a series of resistors in an electric circuit, and

$$\frac{1}{k_{mix}} = \frac{V_c}{k_c} + \frac{V_d}{k_d}.$$ (6.24)

Using this equation, we obtain a much better agreement with the experimental results. We should emphasize, however, that we lack the physical basis for this assumption, and that the purpose in including it at this point is only to stress the difficulty of predicting k for two-phase materials.

6.5.2 Porous materials

Most ceramic materials of construction and some powder metallurgy products have some, low, internal porosity ω, as a result of having been sintered from powders. The pores are generally isolated from one another. At temperatures up to and slightly above the room temperature, the porosity has a thermal conductivity

---

[10] D. Nobili and M. A. DeBacci, *J. Nuclear Materials* **18**, 187 (1966).

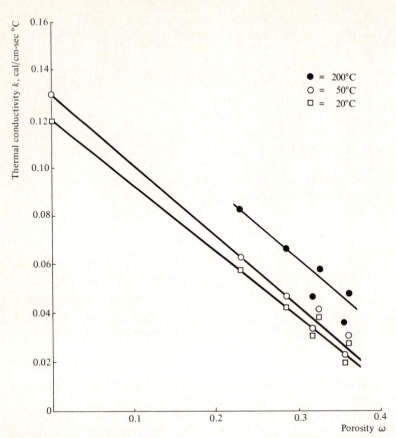

Fig. 6.11 Thermal conductivity of phosphor-bronze powder compacts as a function of porosity ω. (From P. Grootenhuis, R. W. Powell, and R. P. Tye, *Proc. Phys. Soc.* **65**, 502 (1952).)

which is essentially zero. The effective thermal conductivity of the bulk material is often estimated to be

$$k_{\text{eff}} = k_c(1 - \omega). \tag{6.25}$$

At larger porosities, the pores are not simple isolated shapes and the porosity becomes continuous. Conductivity at room temperature is thus even lower than that predicted by Eq. (6.25). Figure 6.11 illustrates the effect of larger amounts of porosity on the thermal conductivity of partially sintered powder metal compacts.

In general, the conductivity of a loose packing, or packed bed, is a function of the thermal conductivity of the gas in the pores, the solid, the void fraction in the bed, and the temperature, because, as the temperature increases, the particle-to-particle thermal radiation also increases. Schotte[11] has developed a technique for predicting the conductivity of packed beds, in which the gas phase is continuous; it is outlined below.

Taking the gas first, we can use the methods outlined in Section 6.2 to obtain k_g, except in the following instance. When the effective pore dimensions are of the same order of magnitude as the mean free path of the gas, the true thermal conductivity of the gas decreases. This occurs when the particle diameter D_p is about 1000 times the mean free path of the gas. (The mean free path of air at room temperature and pressure is $\sim 10^{-5}$ cm. At 2800°F and 1 atm, it is $\sim 10^{-4}$ cm. Thus we are talking about situations where $D_p < 10^{-2}$–10^{-1} cm.) Deissler and Eian[12] have developed a correlation for the *break-away* pressure P_b in lb_f/ft^2, *above* which no correction for this effect is required:

$$P_b = (1.77 \times 10^{-21}) \frac{T}{D_p\, d^2}, \tag{6.26}$$

where $T =$ the absolute temperature, °R, $D_p =$ the average particle diameter, ft, and $d =$ the mean diameter of the gas molecules, ft. If the actual pressure is greater than P_b, no correction to the value of k_g read from Fig. 6.2 is needed, but if the actual pressure is less than P_b calculated by Eq. (6.26), the thermal conductivity of the gas phase should then be calculated according to the equation

$$k_g = \frac{k_g^0}{1 + 2.03 \times 10^{-22} \left(\dfrac{C_p/C_v}{1 + C_p/C_v} \right)_g \left(\dfrac{1 - \omega}{\omega} \right) \left(\dfrac{Tk_g^0}{PD_p\, d^2 C_p \eta} \right)}, \tag{6.27}$$

where k_g^0 is the uncorrected value obtained from Fig. 6.2; all units for the variables are the same as for Eq. (6.26) with C_p in Btu/lb °F and η in lb/ft-hr. It can readily be seen that k_g decreases, at any given temperature, as D_p decreases, and if the pores are small enough, k_g is effectively zero. This effect is illustrated in Fig. 6.3 by the data for the thermal conductivity of porous brick.

Having obtained k_g, either directly or by using Eq. (6.27) if required, we use k_s, the intrinsic thermal conductivity of the solid phase, to calculate the ratio k_s/k_g. Then, using Fig. 6.12 we may find k_b, the *effective thermal conductivity* of the packed bed, from the ratio k_b/k_g. This correlation (Fig. 6.12) has been developed from data on many packed beds, and is accurate as given at temperatures less than 200°C.[13] However, Fig. 6.12 should not be used outside the limits $0.2 < \omega < 0.6$ and $1 < k_s/k_g < 6000$. If k_s/k_g is greater than 6000, then there may be appreciable contact conduction between particles which has not been accounted for in the

[11] W. Schotte, *A.I.Ch.E. Journal* **6**, 63 (1960).

[12] R. G. Deissler and C. S. Eian, NACA, RM E52CO5 (1952).

[13] For another substantial review article, see R. Krupiczka: Analysis of thermal conductivity of granular materials, *Intl. Chem. Engr.* **1**, 122 (Jan. 1967). He arrives at a graph that is virtually the same as that of Deissler and Eian.

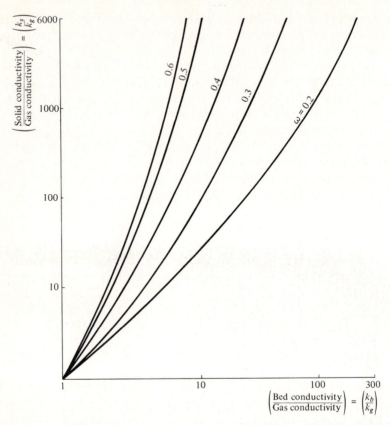

Fig. 6.12 Effect of porosity on packed bed thermal conductivity. (From Deissler and Eian, *ibid.*)

correlation of Fig. 6.12. However, this situation does not often arise, except possibly in compacts of powdered metals.

If the temperature is above 200°C we must add another term to k_b, found in the same manner as above, in order to obtain the true effective thermal conductivity of the bed. At higher temperatures, there is a contribution to the effective thermal conductivity of the bed by radiation heat-transfer from particle to particle. This contribution, for example, is important for 1-mm particles above 400°C and for 0.1-mm particles above 1500°C. Schotte considered the radiation from a plane on one side of a spherical particle to a plane on the other side. First, there is a direct radiation heat-transfer across the void space past the particle, with heat-transfer contribution:

$$Q_1 = -h_r \left(D_p^2 \frac{\pi}{4} \frac{\omega}{1-\omega} \right) \left(D_p \frac{dT}{dx} \right), \tag{6.28}$$

where h_r is the radiation heat-transfer coefficient between a particle and its neighbor, given by

$$h_r = 0.692\varepsilon \frac{T^3}{10^8}, \qquad \text{Btu/hr-ft}^2 \, {}^\circ\text{F}.$$

Here ε is the emissivity of the solid, and T is the absolute temperature, $^\circ$R. Secondly, there is radiation to and from the particle, in series with conduction through the particle. This is written

$$Q_2 = -D_p^2 \frac{\pi}{4}\left(\frac{k_s h_r D_p}{k_s + h_r D_p}\right)\left(\frac{dT}{dx}\right). \tag{6.29}$$

When Q_1 and Q_2 are added, one obtains the total heat flow

$$Q = -k_r\left(D_p^2 \frac{\pi}{4}\frac{1}{1-\omega}\right)\left(\frac{dT}{dx}\right),$$

where k_r is the *radiation contribution to the effective thermal conductivity of the bed*:

$$k_r = \frac{1-\omega}{(1/k_s)+(1/k_r^0)} + \omega k_r^0, \tag{6.30}$$

and

$$k_r^0 = 0.692 \, \varepsilon \, D_p\left(\frac{T^3}{10^8}\right).$$

To evaluate the thermal conductivity of a packed bed, we use the correlation in Fig. 6.12 (if necessary, use Eq. (6.27) to obtain k_g), then add k_r from Eq. (6.30) to obtain the effective thermal conductivity, as illustrated in Example 6.2. The reader will note that even at high temperatures, it is possible for porosity to contribute only slightly to k_r, provided the pore size is quite small. This is generally true of isolated pores as well as of pores in loose materials.

Example 6.2 Bearing in mind the importance of the effective thermal conductivity of molding sand in relation to the solidification rate of castings, let us see if we can predict the thermal conductivity of silica sand molds as a function of temperature, using the method described above.

Solution. Assume that the material is entirely quartz (SiO_2) with thermal conductivity given in Fig. 6.3, and that the atmosphere is dry air. If the bulk porosity is 40%, the grain size D_p is 0.015 in. (AFS 43), and the emissivity of quartz is as given below, then we can proceed as outlined above.

1. Calculate P_b at 40°F (500°R)

$$P_b = \frac{1.77 \times 10^{-21}T}{D_p \, d^2}.$$

Assuming $d = 4\,\text{Å} \equiv 1.31 \times 10^{-9}$ ft,

$$P_b = \frac{(1.77 \times 10^{-21})(500)}{(0.015/12)(1.31 \times 10^{-9})^2} = 423 \; \text{lb}_f/\text{ft}^2.$$

Since atmospheric pressure $= 2120 \, lb_f/ft^2$, then P_b is less than P_{atm}, and therefore we need to make no correction to the value read from Fig. 6.2 at this temperature. Subsequent calculations show that this condition holds true up to 2100°F. At 2340°F (2800°R), k_g is calculated using Eq. (6.27).

2. Using the value of k_g in the table below, we calculate the ratio k_s/k_g at 40°F:

$$k_s/k_g = 5.0/0.013 = 384.$$

Then, from Fig. 6.12, we find that for $\omega = 0.4$, $k_b/k_g \cong 12$, or, since $k_g = 0.013$, $k_b = 0.156$ Btu/hr-ft °F.

This operation is carried out at each temperature:

T, °F	k_g	k_s/k_g	k_b/k_g	k_b Btu/hr-ft °F
40	0.013	384	12.0	0.156
440	0.021	167	9.5	0.200
840	0.029	90	7.9	0.229
1240	0.036	79	7.5	0.248
1540	0.040	67	7.0	0.273
2040	0.044	136	9.0	0.396
2340	0.040	380	11.9	0.452

3. The radiation contribution to k_b is next calculated, specifically at 40°F:

$$k_r^0 = \frac{(0.692)(0.82)(1.25 \times 10^{-3})(500)^3}{10^8} = 0.00089 \text{ Btu/hr-ft °F}.$$

Then,

$$k_r = \frac{(1.0 - 0.4)}{(1/5.0) + (1/0.00089)} + (0.4)(0.00089)$$

$$\cong 0.0009 \text{ Btu/hr-ft °F}.$$

On the other hand, at 2340°F:

$$k_r^0 = \frac{(0.692)(0.27)(1.25 \times 10^{-3})(2800)^3}{10^8} = 0.0514 \text{ Btu/hr-ft °F},$$

$$k_r = \frac{1.0 - 0.4}{(1/15.0) + (1/0.0515)} + (0.4)(0.0515)$$

$$= 0.0512 \text{ Btu/hr-ft °F}.$$

These values are added to the k_b values, previously calculated, to obtain the final result.

$T, °F$	ε_{SiO_2}	k_r^0	k_r	$+$	k_b	$=$	k_{eff} (Btu/hr-ft °F)
40	0.82	0.00089	0.0009		0.156		0.157
440	0.75	0.0047	0.0048		0.200		0.205
840	0.68	0.0130	0.0131		0.229		0.242
1240	0.58	0.0248	0.0245		0.248		0.273
1540	0.49	0.0339	0.0337		0.273		0.307
2040	0.32	0.0400	0.0398		0.396		0.436
2340	0.27	0.0514	0.0512		0.452		0.503

The results of these calculations are plotted in Fig. 6.13. Some data from experimental studies are compared with the calculated values. The results are in surprisingly good agreement, and the calculations are probably as good as the measured values, considering the experimental difficulties involved.

For another interesting application of this method in analyzing a metallurgical

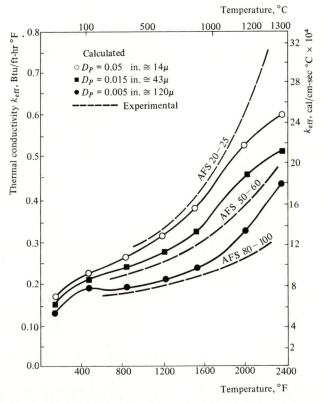

Fig. 6.13 Calculated and experimental thermal conductivity of silica molding sand. (Experimental values from L. F. Lucks, C. L. Linebrink, and K. L. Johnson, *Trans. A.F.S.* **55**, 62 (1947).)

process, we refer the reader to a paper by Downing[14] on the thermal behavior of a ferro-alloy furnace.

PROBLEMS

6.1 In the same system described in Problem 1.2, the temperature profile at $x = x_1$ is given by

$$T = 6 \sin (\pi/2)y, \qquad 0 \le y \le 1,$$

where T is measured in °F and y in ft. Find the heat flux through the wall at that point. (The thermal conductivity of water at 100°F is 0.36 Btu/hr-ft^2 °F, and the specific heat is 1 Btu/lb$_m$-°F).

6.2 Determine the thermal conductivity of a test panel 6 in. × 6 in. and $\frac{1}{2}$ in. thick, if during a two-hour period 80 Btu are conducted through the panel when the temperature of the two faces are 67°F and 79°F.

6.3 At steady state, the temperature profile in a laminated system appears thus:

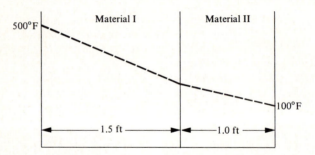

Determine the thermal conductivity of II if the steady-state heat flux is 4000 Btu/hr-ft^2 and the conductivity of I is 30 Btu/hr-ft °F.

6.4

a) The thermal conductivity of helium at 200°F and 1 atm is reported to be 0.098 Btu/hr-ft °F. What is helium's thermal conductivity (in the same units) at 1000°F?

b) Estimate the thermal conductivity (in Btu/hr-ft °F) of carbon dioxide at 2000°F.

6.5 Show that Fourier's law can be written (for constant ρC_p) as

$$q_y = -\alpha \frac{d}{dy} (\rho C_p T)$$

for one-dimensional heat flow. In addition, show that Newton's law, for constant ρ, is

$$\tau_{yx} = -v \frac{d}{dy} (\rho v_x).$$

Discuss the analogies between the fluxes, constants, and gradients as they appear in these equations.

---

[14] J. H. Downing, *Electric Furnace Proceedings*, *A.I.M.E.* **26**, 81 (1968).

HEAT TRANSFER AND THE ENERGY EQUATION

We have designed this chapter to introduce the reader to three interwoven topics. First, we develop differential equations in terms of temperature in space (and with time if transient conditions apply) for several simple physical problems, by writing energy balances for suitable unit volumes. In order to obtain useful solutions to the problems, we integrate the differential equations to ascertain the temperature and arbitrary constants, and then apply boundary and initial conditions to obtain the particular solution. The general procedure is conceptually identical to that followed in Chapter 2 for obtaining the velocity profiles.

Second, several of the examples to be discussed concern heat transfer to and from moving fluids; in this regard, we deal only with laminar convection, but this enables the reader to become involved in the fundamentals of heat transfer with convection.

Third, we bring to the reader's attention more general forms of the equation of energy, leading to Tables 7.2–7.4 which may be used in a manner similar to the general momentum equations given in Chapter 2.

7.1 HEAT TRANSFER WITH FORCED CONVECTION IN A TUBE

Consider a fluid in laminar flow in a circular tube of radius R, as depicted in Fig. 7.1. If the tube and the fluid exchange heat, then clearly the fluid's temperature is a function of both the r- and z-directions. A suitable unit volume is a ring-shaped element, Δr thick and Δz high. Energy enters and leaves this ring by thermal conduction; also, a unit mass of fluid, which enters with an enthalpy corresponding to its temperature (sensible heat), must leave the ring with a different enthalpy. Let us now develop the energy balance for the unit volume.

Rate of energy in by conduction across surface at r $2\pi r\, \Delta z q_r|_r$

Rate of energy out by conduction across surface at $r + \Delta r$ $2\pi(r + \Delta r)\, \Delta z q_r|_{r+\Delta r}$

Rate of energy in by conduction across surface at z $2\pi r\, \Delta r q_z|_z$

Energy out by conduction across surface at $z + \Delta z$ $2\pi r\, \Delta r q_z|_{z+\Delta z}$

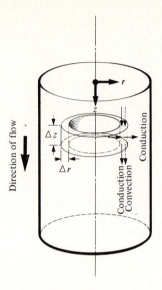

Fig. 7.1 Elemental circular ring used to develop the differential energy balance for laminar tube flow.

Energy in due to fluid flow (sensible heat) across surface at z $\rho v_z 2\pi r \, \Delta r H|_z$

Energy out due to fluid flow across surface at $z + \Delta z$ $\rho v_z 2\pi r \, \Delta r H|_{z+\Delta z}$

Here H is the enthalpy per unit mass, and v_z is the velocity in the z-direction. At steady state, the energy balance requires equal inputs and outputs. If we divide all terms by $2\pi \, \Delta r \, \Delta z$, we obtain

$$\frac{rq_r|_{r+\Delta r} - rq_r|_r}{\Delta r} + r\frac{q_z|_{z+\Delta z} - q_z|_z}{\Delta z} + r\rho v_z \frac{H|_{z+\Delta z} - H|_z}{\Delta z} = 0. \qquad (7.1)$$

Now Δr and Δz are allowed to approach zero,

$$\frac{\partial(rq_r)}{\partial r} + r\frac{\partial q_z}{\partial z} + r\rho v_z \frac{\partial H}{\partial z} = 0. \qquad (7.2)$$

If C_p is the heat capacity, then

$$\frac{\partial H}{\partial z} = C_p \frac{\partial T}{\partial z}. \qquad (7.3)$$

Also

$$q_r = -k(\partial T/\partial r) \quad \text{and} \quad q_z = -k(\partial T/\partial z). \qquad (7.4 \text{ a, b})$$

Substituting Eqs. (7.3) and (7.4) into Eq. (7.2) yields an energy equation written in terms of temperature:

$$v_z \frac{\partial T}{\partial z} = \frac{k}{\rho C_p}\left[\frac{1}{r}\frac{\partial}{\partial r}\left(r\frac{\partial T}{\partial r}\right) + \frac{\partial^2 T}{\partial z^2}\right].$$
(7.5)

We can further simplify the energy balance (Eq. 7.5), since, except for the very slow flow of liquid metals, the term $(k/\rho C_p)\partial^2 T/\partial z^2$ is negligible even though $v_z(\partial T/\partial z)$ is not. With this assumption, Eq. (7.5) reduces to

$$v_z \frac{\partial T}{\partial z} = \frac{k}{\rho C_p}\left[\frac{1}{r}\frac{\partial}{\partial r}\left(r\frac{\partial T}{\partial r}\right)\right].$$
(7.6)

Equation (7.6) contains v_z, the factor that ties together heat transfer and convection. For the purpose of this discussion, consider fully developed laminar flow; the velocity distribution is therefore parabolic, and, as previously derived, is given by Eqs. (2.31) and (2.33):

$$v_z = 2\bar{V}_z\left[1 - \left(\frac{r}{R}\right)^2\right].$$

By including this velocity distribution, Eq. (7.6) becomes

$$2\bar{V}_z\left[1 - \left(\frac{r}{R}\right)^2\right]\frac{\partial T}{\partial z} = \frac{k}{\rho C_p}\left[\frac{1}{r}\frac{\partial}{\partial r}\left(r\frac{\partial T}{\partial r}\right)\right].$$
(7.7)

Here we consider the special case where a *fully developed temperature profile* exists. For any set of boundary conditions, a fully developed temperature profile exists when $(T_R - T)/(T_R - T_m)$ is a unique function of r/R, *independent of z*. Then

$$\frac{T_R - T}{T_R - T_m} = f(r/R),$$
(7.8)

or

$$\frac{\partial}{\partial z}\left(\frac{T_R - T}{T_R - T_m}\right) = 0,$$
(7.9)

where T_R = temperature of fluid at the wall, and T_m = mean temperature of fluid. A fully developed temperature profile is analogous to fully developed flow. This is exemplified by Fig. 7.2, where the liquid flowing in the z-direction encounters the heated section of the tube. Over a finite interval downstream from this point, the temperature profile changes from uniform to fully developed.

For a fully developed temperature profile, an important corollary arises; namely, the heat-transfer coefficient is uniform along the pipe. We realize this by employing the definition of the heat-transfer coefficient based on the mean temperature of the fluid:

$$h \equiv \frac{q_0}{T_R - T_m} = -\frac{k}{R}\frac{\partial}{\partial(r/R)}\left(\frac{T_R - T}{T_R - T_m}\right)_{r=R}.$$
(7.10)

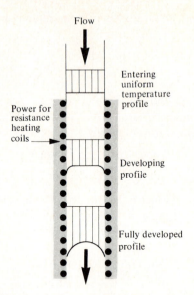

Fig. 7.2 Heating a fluid in a tube showing the development of the temperature profile.

Because the derivative in Eq. (7.10) has a unique value at the wall, independent of z, h is therefore uniform along the pipe under the fully developed temperature conditions.

Now consider the case where q_0 is uniform. This represents a uniform heat flux at the wall, and could be physically obtained by using an electric heater, depicted in Fig. 7.2. Further, since h and q_0 are constant, Eq. (7.10) specifies that $T_R - T_m$ is constant, and

$$\frac{\partial T_R}{\partial z} = \frac{\partial T_m}{\partial z}. \tag{7.11}$$

(Note that T_R and T_m themselves are not constants.) Now expand Eq. (7.9) in a general sense where each quantity varies as follows,

$$\left(\frac{\partial T_R}{\partial z} - \frac{\partial T}{\partial z}\right) - \left(\frac{T_R - T}{T_R - T_m}\right)\left(\frac{\partial T_R}{\partial z} - \frac{\partial T_m}{\partial z}\right) = 0. \tag{7.12}$$

Then Eq. (7.11) shows that

$$\frac{\partial T_R}{\partial z} = \frac{\partial T}{\partial z} = \frac{\partial T_m}{\partial z}. \tag{7.13}$$

Equation (7.13) is important because it allows Eq. (7.7) to be integrated directly using $\partial T/\partial z = \partial T_m/\partial z$:

$$2\bar{V}_z\left(\frac{\partial T_m}{\partial z}\right) \int_{r=0}^{r} r\left[1 - \left(\frac{r}{R}\right)^2\right] dr = \frac{k}{\rho C_p} \int_{\partial T/\partial r=0}^{\partial T/\partial r} d\left(r\frac{\partial T}{\partial r}\right). \tag{7.14}$$

Here the lower integration limits represent a boundary condition, namely,

B.C.1   at $r = 0$, $\partial T / \partial r = 0$.

Integrating, we get

$$2\bar{V}_z \left(\frac{\partial T_m}{\partial z}\right) \frac{r}{2} \left[1 - \frac{1}{2} \left(\frac{r}{R}\right)^2 \right] = \frac{k}{\rho C_p} \frac{\partial T}{\partial r}. \tag{7.15}$$

A second integration with

B.C. 2   at $r = R$, $T = T_R$,

finally results in the temperature distribution

$$T_R - T = \left(\frac{\bar{V}_z \rho C_p}{8R^2 k}\right) \left(\frac{\partial T_m}{\partial z}\right) (3R^4 - 4r^2 R^2 + r^4). \tag{7.16}$$

Having obtained the temperature profile, we can evaluate h. From Eq. (7.10), $(T_R - T_m)$ and q_0 must then be evaluated. First, we find $(T_R - T_m)$ by performing the integration:

$$T_R - T_m = \frac{\displaystyle\int_0^R v_z (T_R - T) 2\pi r \, dr}{\displaystyle\int_0^R v_z 2\pi r \, dr}. \tag{7.17}$$

Second, we determine q_0 by evaluating the gradient at the wall using Eq. (7.16):

$$q_0 = -k \left(\frac{\partial T}{\partial r}\right)_{r=R}. \tag{7.18}$$

When these operations have been carried out, we can determine the heat-transfer coefficient. The final result is

$$h = 2.18 \frac{k}{R} \tag{7.19}$$

or

$$\frac{hD}{k} = 4.36. \tag{7.20}$$

The dimensionless group resulting from this analysis is the *Nusselt number*. This important dimensionless group for heat flow with *forced* convection will reappear several times as we examine other solutions and correlations. For emphasis, then, the Nusselt number is

$$\mathrm{Nu}_\infty \equiv \frac{hD}{k}. \tag{7.21}$$

Table 7.1 Nusselt numbers for fully developed laminar flow*

Geometry	Velocity distribution†	Condition at wall	$\mathrm{Nu}_\infty = \dfrac{hD_e\ddagger}{k}$
Circular tube	Parabolic	Uniform q_0	4.36
Circular tube	Parabolic	Uniform T_0	3.66
Circular tube	Slug flow	Uniform q_0	8.00
Circular tube	Slug flow	Uniform T_0	5.75
Parallel plates	Parabolic	Uniform q_0	8.23
Parallel plates	Parabolic	Uniform T_0	7.60
Triangular duct	Parabolic	Uniform q_0	3.00
Triangular duct	Parabolic	Uniform T_0	2.35

* From W. M. Rohsenow and H. Y. Choi, *Heat, Mass and Momentum Transfer*, Prentice-Hall, Englewood Cliffs, New Jersey, 1961, page 141.

† Slug flow refers to a flat velocity profile.

‡ D_e is the equivalent diameter, as defined in Chapter 3.

This Nusselt number developed here is for fully developed flow and uniform heat flux with parabolic velocity profile. It is subscripted with ∞ because it represents a limiting case. Many other situations have been analyzed, some of which are given in Table 7.1.

7.2 HEAT TRANSFER WITH LAMINAR FORCED CONVECTION OVER A FLAT PLATE

In Chapter 2, the velocity distribution, within the boundary layer, of a fluid flowing past a plate was determined by two methods—the exact solution, and the approximate integral solution. Here we consider the case of a plate at a higher temperature than the fluid, the plate serving to heat the fluid. Just as a velocity profile continually changes with distance from the leading edge, and results in a *momentum boundary layer* which increases in thickness, there is also a changing temperature profile and development of a *thermal boundary layer* when heat transfer is involved. We depict this situation along with a unit element in Fig. 7.3.

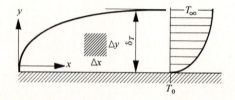

Fig. 7.3 Development of the thermal boundary layer and the temperature distribution over a flat plate.

For the element with a depth of unity perpendicular to the page, we may write the various contributions to the energy balance:

Energy in by conduction across surface at x $\qquad\qquad\qquad$ $q_x|_x \, \Delta y \cdot 1$

Energy out by conduction across surface at $x + \Delta x$ $\qquad$ $q_x|_{x+\Delta x} \, \Delta y \cdot 1$

Energy in by conduction across surface at y $\qquad\qquad\qquad$ $q_y|_y \, \Delta x \cdot 1$

Energy out by conduction across surface at $y + \Delta y$ $\qquad$ $q_y|_{y+\Delta y} \, \Delta x \cdot 1$

Energy in due to fluid flow (sensible heat) across surface at x $\qquad$ $\rho v_x \, \Delta y \cdot 1 \cdot H|_x$

Energy out due to fluid flow (sensible heat) across surface at
$x + \Delta x$ $\qquad\qquad\qquad\qquad\qquad\qquad\qquad\qquad$ $\rho v_x \, \Delta y \cdot 1 \cdot H|_{x+\Delta x}$

Energy in due to fluid flow (sensible heat) across surface at y $\qquad$ $\rho v_y \, \Delta x \cdot 1 \cdot H|_y$

Energy out due to fluid flow (sensible heat) across surface at
$y + \Delta y$ $\qquad\qquad\qquad\qquad\qquad\qquad\qquad\qquad$ $\rho v_y \, \Delta x \cdot 1 \cdot H|_{y+\Delta y}$

Adding all these quantities, dividing through by $\Delta x \, \Delta y$, and taking the limits as $\Delta x \to 0$ and $\Delta y \to 0$, we obtain

$$\frac{\partial q_x}{\partial x} + \frac{\partial q_y}{\partial y} + \frac{\partial (\rho v_x H)}{\partial x} + \frac{\partial (\rho v_y H)}{\partial y} = 0. \tag{7.22}$$

For constant density and conductivity, Eq. (7.22) becomes

$$\rho H \left(\frac{\partial v_x}{\partial x} + \frac{\partial v_y}{\partial y} \right) + \rho \left(v_x \frac{\partial H}{\partial x} + v_y \frac{\partial H}{\partial y} \right) = k \left(\frac{\partial^2 T}{\partial x^2} + \frac{\partial^2 T}{\partial y^2} \right). \tag{7.23}$$

Since continuity requires that $(\partial v_x/\partial x) + (\partial v_y/\partial y) = 0$ and $dH = C_p \, dT$, we finally obtain

$$\rho C_p \left(v_x \frac{\partial T}{\partial x} + v_y \frac{\partial T}{\partial y} \right) = k \left(\frac{\partial^2 T}{\partial x^2} + \frac{\partial^2 T}{\partial y^2} \right). \tag{7.24}$$

The temperature gradient in the y-direction is much steeper than that in the x-direction, therefore the x-directed second derivative term (conduction term) may be neglected. Then Eq. (7.24) simplifies to

$$v_x \frac{\partial T}{\partial x} + v_y \frac{\partial T}{\partial y} = \alpha \frac{\partial^2 T}{\partial y^2}. \tag{7.25}$$

Here

$$\alpha = \frac{k}{\rho C_p},$$

which is called the *thermal diffusivity*. The thermal diffusivity has the units of length squared per time (L^2/t), which are also the units of the kinematic viscosity. In addition, the momentum boundary-layer equation, developed in Chapter 2, is

$$v_x \frac{\partial v_x}{\partial x} + v_y \frac{\partial v_x}{\partial y} = v \frac{\partial^2 v_x}{\partial y^2}. \tag{2.80}$$

Equations (7.25) and (2.80) are analogous. If $v = \alpha$, and if the velocity and thermal boundary conditions are similar, then the temperature and velocity profiles are exactly identical, and the thermal boundary layer δ_T equals the momentum boundary layer δ.

7.2.1 Exact solution

The method of solution for Eq. (7.25) parallels that for the velocity distribution, as given in Section 2.7.1. For the case of a uniform-temperature plate, the following boundary conditions apply

B.C. 1   at $y = 0$, $T = T_0$;

B.C. 2   at $y = \infty$, $T = T_\infty$.

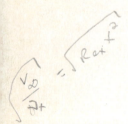

Fig. 7.4 Dimensionless temperature profiles in the laminar boundary layer over a flat plate for various Pr. (From E. Z. Pohlhausen, *Z.f. angew. Math. u. Mech.* **1**, 115 (1921).)

We do not present the details of the exact solution here but Fig. 7.4 gives the final results which are in the form of the temperature T within the fluid being a function of y and x. The temperature is a part of the dimensionless temperature group Θ on the ordinate, which includes the wall temperature T_0 and the bulk fluid temperature T_∞. Space dimensions x and y appear together on the abscissa in exactly the same way, as shown in Fig. 2.7 for describing the velocity profiles. Several curves are shown in Fig. 7.4, each for a different value of the *Prandtl number*, Pr. This number is the ratio of v/α, and for Pr equal to unity, the Θ curve in Fig. 7.4 is exactly the same as v_x/V_∞ in Fig. 2.7. Therefore, the Prandtl number controls how similar the velocity profiles and the temperature profiles are.

Knowing the temperature profile, we can determine the heat-transfer coefficient. From the results given in Fig. 7.4, the *local* heat transfer is

$$h_x = \frac{-k\left(\dfrac{\partial T}{\partial y}\right)_{y=0}}{T_0 - T_\infty} \tag{7.26}$$

$$= 0.332k\, Pr^{0.343}\sqrt{\frac{V_\infty}{vx}}, \qquad Pr \geq 0.6,$$

or, in terms of dimensionless groups,

$$Nu_x = 0.332\, Pr^{0.343}\, Re_x^{0.5}, \tag{7.27}$$

where

$$Nu_x \equiv \frac{h_x x}{k},$$

which is called the *local* Nusselt number. If we wish to know the *average* heat-transfer coefficient, then we can find it by averaging h_x from $x = 0$ to $x = L$:

$$h = \frac{1}{L}\int_0^L h_x\, dx = 0.664k\, Pr^{0.343}\sqrt{\frac{V_\infty}{vL}},$$

or

$$Nu_L = 0.664\, Pr^{0.343}\, Re_L^{0.5}. \tag{7.28}$$

Note the general form of either Eq. (7.27) or (7.28). We shall find in Chapter 8 that the Nusselt number is generally a function of the Reynolds and Prandtl numbers in the problems of *forced convection*.

For the case of $Pr = 1$ ($v = \alpha$), the thermal boundary layer is given by

$$\frac{\delta_T}{x} = \frac{5.0}{\sqrt{V_\infty x/v}}, \tag{7.29}$$

which is the same as the momentum boundary layer.

At this point it is instructive to examine Prandtl numbers for various fluids; approximate figures are given in Table 7.2.

Table 7.2

Substance	Range of Prandtl number (v/α)
Common liquids (water, alcohol, etc.)	2–50
Liquid metals	0.001–0.03
Gases	0.7–1.0

These numbers represent values that include several substances, and cover substantial temperature ranges. The Prandtl numbers of liquids vary significantly with temperature; however, gases show almost no variation in Pr with temperature. From these listed Prandtl numbers, we see that for gases $\delta_T \cong \delta$, for common liquids $\delta_T < \delta$, and for liquid metals—due to their high thermal conductivity— $\delta_T \gg \delta$.

Equations (7.27) and (7.28) are valid only for Pr > 0.5, and thus do not apply to liquid metals. For liquid metals with uniform wall temperature as a boundary condition, the results are approximated by[1]

$$\text{Nu}_x = \sqrt{\text{Re}_x \, \text{Pr}} \left(\frac{0.564}{1 + 0.90\sqrt{\text{Pr}}} \right). \tag{7.30}$$

For the other boundary condition of uniform heat flux at the wall, we present these results:[2]

$$\text{Pr} > 0.5, \qquad \text{Nu}_x = 0.458 \, \text{Pr}^{0.343} \sqrt{\text{Re}_x}; \tag{7.31}$$

$$0.006 \le \text{Pr} \le 0.03, \qquad \text{Nu}_x = \sqrt{\text{Re}_x \, \text{Pr}} \left(\frac{0.880}{1 + 1.317\sqrt{\text{Pr}}} \right). \tag{7.32}$$

7.2.2 Approximate integral method

An alternative solution to this problem is obtained by the approximate integral method. This method, which we introduced in Chapter 2, yields a solution in a relatively straightforward manner.

We write an energy balance over the entire boundary layer (Fig. 7.5), where we assume that $\delta_T < \delta$. We can use the same approach with a slight modification for $\delta_T > \delta$ (liquid metals). Following a similar procedure used for the analogous momentum balance in Chapter 2, the mass of material entering across the surface at x is

$$W_x = \int_0^l \rho v_x \, dy, \tag{2.98}$$

and the mass leaving is

$$W_{x+\Delta x} = W_x + \frac{d}{dx} \left(\int_0^l \rho v_x \, dy \right) \Delta x.$$

Continuity requires that

$$W_l = \frac{d}{dx} \left(\int_0^l \rho v_x \, dy \right) \Delta x. \tag{2.100}$$

[1] E. M. Sparrow and J. L. Gregg, *J. Aero. Sc.* **24**, 852 (1957).

[2] R. J. Nickerson and H. P. Smith, as reported in Rohsenow and Choi, *ibid.*, page 149.

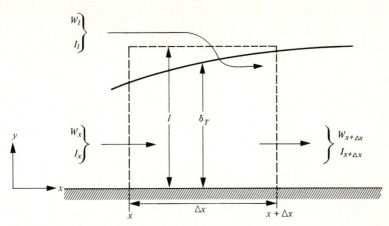

Fig. 7.5 Integrated enthalpy quantities in the boundary layer over a flat plate.

We write the respective total enthalpies I that accompany W_x, $W_{x+\Delta x}$, and W_l, and enter them into the energy balance. We neglect conduction in the x-direction.

$$q_y|_{y=0}\, \Delta x + I_x + I_l = I_{x+\Delta x}, \tag{7.33}$$

or

$$q_y\Big|_{y=0}\, \Delta x + \underbrace{\int_0^l H\rho v_x\, dy}_{I_x} + \underbrace{\frac{d}{dx}\left(H_\infty \int_0^l \rho v_x\, dy\right)\Delta x}_{I_l}$$

$$= \underbrace{\int_0^l H\rho v_x\, dy + \frac{d}{dx}\left(\int_0^l H\rho v_x\, dy\right)\Delta x}_{I_{x+\Delta x}}, \tag{7.34}$$

where H = enthalpy per unit mass.

The integrals in Eq. (7.34) are then split, that is,

$$\int_0^l = \int_0^{\delta_T} + \int_{\delta_T}^l$$

and then simplification yields

$$-k\left(\frac{\partial T}{\partial y}\right)_{y=0} = \frac{d}{dx}\left(\int_0^{\delta_T} H\rho v_x\, dy - \int_0^{\delta_T} H_\infty\, \rho v_x\, dy\right). \tag{7.35}$$

Since $H - H_\infty = C_p(T - T_\infty)$, we obtain, for constant ρ and C_p,

$$\frac{d}{dx}\left(\int_0^{\delta_T} (T_\infty - T)v_x \, dy\right) = \alpha\left(\frac{\partial T}{\partial y}\right)_{y=0}. \tag{7.36}$$

The integral in Eq. (7.36) contains two factors, v_x and $(T_\infty - T)$ which must be known as a function of y within the thermal boundary layer. As before, we arbitrarily select Eq. (2.105) as the velocity distribution, and by similarity also assume a parabolic temperature distribution. Thus

$$\frac{T - T_0}{T_\infty - T_0} = \Theta = \frac{3}{2}\left(\frac{y}{\delta_T}\right) - \frac{1}{2}\left(\frac{y}{\delta_T}\right)^3, \tag{7.37}$$

which satisfies the boundary conditions of interest:

$$\begin{array}{lll} \text{B.C. 1} & \text{at } y = 0, & \Theta = 0; \\ \text{B.C. 2} & \text{at } y = \delta_T, & \Theta = 1. \end{array} \tag{7.38}$$

Substituting the assumed velocity and temperature distributions into Eq. (7.36), and performing the integration yields

$$\frac{d}{dx}\left\{\Theta_\infty V_\infty \delta\left[\frac{3}{20}\left(\frac{\delta_T}{\delta}\right)^2 - \frac{3}{280}\left(\frac{\delta_T}{\delta}\right)^4\right]\right\} = \frac{3}{2}\alpha\frac{\Theta_\infty}{(\delta_T/\delta)\,\delta}. \tag{7.39}$$

Since $(\delta_T/\delta) \le 1$, we may neglect the fourth power term. Then, using Eq. (2.107) to eliminate δ from Eq. (7.39), but at the same time keeping (δ_T/δ), we obtain

$$\left(\frac{\delta_T}{\delta}\right)^3 + \frac{4}{3}x\frac{d(\delta_T/\delta)^3}{dx} = \frac{13}{14}\text{Pr}^{-1}. \tag{7.40}$$

Now, if there exists an unheated length of plate for $0 < x < x_0$, and if for $x \ge x_0$ the plate is maintained at T_0, then $(\delta_T/\delta) = 0$ at $x = x_0$ is a boundary condition used for integrating Eq. (7.40). After separating variables and applying the boundary condition as a lower limit, we obtain

$$\int_0^{\delta_T/\delta} \left[\frac{13}{14}\text{Pr}^{-1} - \left(\frac{\delta_T}{\delta}\right)^3\right]^{-1} d\left(\frac{\delta_T}{\delta}\right)^3 = \frac{3}{4}\int_{x_0}^x \frac{dx}{x}. \tag{7.41}$$

Performing the integration results in

$$\frac{\delta_T}{\delta} = \frac{1}{1.026\sqrt[3]{\text{Pr}}}\sqrt[3]{1 - \left(\frac{x_0}{x}\right)^{3/4}}.$$

For the specific case of no unheated length, that is, the entire plate is maintained at T_0, then $x_0 = 0$, and the ratio of boundary layers is

$$\frac{\delta_T}{\delta} = \frac{1}{1.026}\text{Pr}^{-\frac{1}{3}}. \tag{7.42}$$

From Eq. (7.37), the local heat-transfer coefficient becomes

$$h_x = \frac{-k(\partial T/\partial y)_{y=0}}{(T_0 - T_\infty)} = \frac{3}{2}\frac{k}{\delta_T}. \tag{7.43}$$

Then, with Eqs. (7.42) and (2.107), we obtain

$$h_x = 0.323\frac{k}{x}\sqrt[3]{\text{Pr}}\,\sqrt{V_\infty x/v},$$

or, in terms of dimensionless groups,

$$\text{Nu}_x = 0.323\sqrt[3]{\text{Pr}}\,\sqrt{\text{Re}_x}. \tag{7.44}$$

These solutions are valid when $\delta_T/\delta \le 1$; they are not valid for liquid metals. For liquid metals $\delta_T/\delta > 1$, and

$$\text{Nu}_x = \frac{\sqrt{\text{Re}_x}\,\text{Pr}}{1.55\sqrt{\text{Pr}} + 3.09\sqrt{0.372 - 0.15\,\text{Pr}}}. \tag{7.45}$$

Note the close agreement between the approximate solutions, Eqs. (7.44) and (7.45), with the exact solutions, Eqs. (7.27) and (7.30), respectively.

Example 7.1 Air at 1 atm and 70°F flows parallel to a plate's surface at a rate of 50 ft/sec. The plate, 1 ft long, is initially at a temperature of 200°F. Assume that laminar flow is stable along the entire length.
  a) Calculate the thicknesses of the velocity and thermal boundary layers 6 in. from the leading edge of the plate.
  b) Calculate the initial rate of heat transfer from the entire plate per foot of plate width.
  Here we consider only the initial heat-transfer rate; as time passes, the plate cools, and thus the surface temperature changes. We shall consider the solution to transient problems in Chapter 9.

Solution
  a) We calculate the velocity of the boundary layer using Eq. (2.94),

$$\frac{\delta}{x} = \frac{5.0}{\sqrt{V_\infty x/v}}.$$

For air, evaluating v at an average boundary-layer temperature of $\frac{1}{2}(200 + 70) = 135°F$, the kinematic viscosity is $0.725\ \text{ft}^2/\text{hr}$. Then,

$$\frac{V_\infty x}{v} = \frac{50\ \text{ft}}{\text{sec}}\left|\frac{0.5\ \text{ft}}{}\right|\frac{\text{hr}}{0.725\ \text{ft}^2}\left|\frac{3600\ \text{sec}}{1\ \text{hr}}\right| = 124{,}000,$$

and

$$\delta = (0.5)\left(\frac{5}{\sqrt{124{,}000}}\right) = 7.10 \times 10^{-3}\ \text{ft},$$

or
$$\delta = 0.0852 \text{ in.}$$

For our purposes, Eq. (7.42) is sufficient to give δ_T. Therefore

$$\frac{\delta_T}{\delta} = \frac{1}{1.026} \Pr^{-1/3}.$$

The Prandtl number for air, evaluated at 135°F, is 0.703. Then,

$$\delta_T = \frac{7.10 \times 10^{-3}}{(1.026)\sqrt[3]{0.703}} = 7.75 \times 10^{-3} \text{ ft,}$$

or
$$\delta_T = 0.0930 \text{ in.}$$

b) Eq. (7.28) is used for the average heat-transfer coefficient which applies to the whole plate. This equation is often applied with a slight modification in the exponent of Pr; the value 0.343 is replaced by $\frac{1}{3}$.

$$\mathrm{Nu}_L = 0.664 \, \Pr^{1/3} \mathrm{Re}_L^{1/2},$$

$$\mathrm{Nu}_L = 0.664 \sqrt[3]{0.703} \sqrt{248{,}000} = 295.$$

The thermal conductivity for air at 135°F is 0.017 Btu/hr-ft °F. Then

$$h = \frac{k}{L} \mathrm{Nu}_L = \left(\frac{0.017}{1}\right)(295) = 5.02 \text{ Btu/hr-ft}^2 \text{ °F,}$$

and finally

$$Q = h(T_\infty - T_0)A = (5.02)(200 - 70)(1) = 653 \text{ Btu/hr.}$$

7.3 HEAT TRANSFER WITH NATURAL CONVECTION

In Sections 7.1 and 7.2 we considered two examples of heat transfer with forced convection. In forced convection, the fluid flows as a result of some external forces, resulting in a known velocity distribution which can be entered into the energy equation. The situation is more complex in problems of heat transfer with natural convection. The energy equation still contains terms with velocities, but these velocities are not known *a priori* to solving the energy equation. The momentum balance depends upon the temperature profile which, of course, cannot be solved until the temperature distribution has been established. Hence, the velocity and temperature distributions cannot be treated as separate problems; the temperature distribution, in effect, produces the velocity distribution by causing density differences within the fluid.

Consider specifically the vertical surface in Fig. 7.6; the surface is at T_0, and it heats the neighboring fluid whose bulk temperature is T_∞. In this situation, the velocity component v_y is quite small; the fluid moves almost entirely upward, and

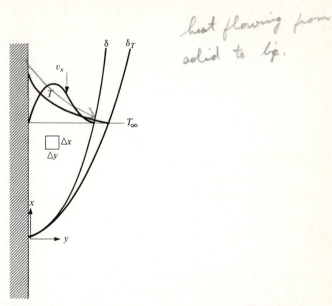

heat flowing from solid to liq.

Fig. 7.6 Thermal and momentum boundary layers for vertical plate natural convection.

therefore we write the momentum equation for the x-component only. In the derivation of the momentum equation for steady forced convection over a flat plate, we ignore the gravity force, and no pressure gradient is involved. We cannot ignore these forces in free convection, and therefore the momentum balance for the present case contains these terms

density = f(T)

$$v_x \frac{\partial v_x}{\partial x} + v_y \frac{\partial v_x}{\partial y} = -\frac{1}{\rho} \frac{\partial P}{\partial x} + v \frac{\partial^2 v_x}{\partial y^2} - g. \qquad (7.46)$$

In the y-direction, $\partial P/\partial y = 0$, or at any x, the pressure is uniform regardless of distance from the wall. The pressure gradient in the x-direction is due to the weight of the fluid itself. Therefore, it can be expressed as $\partial P/\partial x = -\rho_\infty g$, so that in Eq. (7.46) the pressure gradient term becomes

$$-\frac{1}{\rho} \frac{\partial P}{\partial x} - g = \frac{1}{\rho} g(\rho_\infty - \rho) = \beta g(T - T_\infty). \qquad (7.47)$$

The volume expansion coefficient β is defined as

$$\beta = -\frac{1}{\rho} \left(\frac{\partial \rho}{\partial T} \right)_P.$$

The momentum balance in the x-direction then becomes

$$v_x \frac{\partial v_x}{\partial x} + v_y \frac{\partial v_x}{\partial y} = v \frac{\partial^2 v_x}{\partial y^2} + g\beta(T - T_\infty), \qquad (7.48)$$

which is identical with the equation of the forced convection boundary-layer, except for the buoyancy term. Equation (7.48) immediately shows that the momentum balance must be coupled to an appropriate energy equation, in order to treat the buoyancy term.

The energy equation applied to the control volume $\Delta x\, \Delta y$ in this case is identical to that for flow over a flat plate:

$$v_x \frac{\partial T}{\partial x} + v_y \frac{\partial T}{\partial y} = \alpha \frac{\partial^2 T}{\partial y^2}. \tag{7.49}$$

This equation is coupled to Eq. (7.48) by the presence of the velocity terms. The mathematical task at hand is, therefore, to solve coupled Eqs. (7.48) and (7.49), for the boundary conditions:

B.C. 1   at $y = 0$, $v_x = v_y = 0$, $T = T_0$;

B.C. 2   at $y = \infty$, $v_x = v_y = 0$, $T = T_\infty$.

Analytical solutions of such coupled differential equations being beyond the scope of this text, we simply present the results. First, however, let us examine a dimensional analysis approach in order to bring forth pertinent dimensionless groups.

The problem is to determine the conditions for which the velocity profile in a natural convection situation is similar to the velocity profile in another natural convection situation. Both systems have the same boundary conditions, i.e., velocity is zero at the surface and within the bulk fluid removed from the surface. Now employ the dynamic similarity argument introduced in Section 3.1.

First, Eq. (7.48) is written for system 1:

$$v_{x1} \frac{\partial v_{x1}}{\partial x_1} + v_{y1} \frac{\partial v_{x1}}{\partial y_1} = \nu_1 \frac{\partial^2 v_{x1}}{\partial y_1^2} + g_1 \beta_1 (T - T_\infty)_1. \tag{7.50}$$

System 2 is related to system 1 by geometrical and dynamic similarities expressed by the ratios:

$$K_L \equiv \frac{x_1}{x_2} = \frac{y_1}{y_2} = \frac{L_2}{L_2} \qquad K_\nu = \frac{\nu_1}{\nu_2}$$

$$K_v \equiv \frac{v_{x1}}{v_{x2}} = \frac{v_{y1}}{v_{y2}} = \frac{v_1}{v_2} \qquad K_g = \frac{g_1}{g_2} \tag{7.51}$$

$$K_\beta = \frac{\beta_1}{\beta_2} \qquad\qquad K_T = \frac{(T - T_\infty)_1}{(T - T_\infty)_2} = \frac{(T_0 - T_\infty)_1}{(T_0 - T_\infty)_2}.$$

Now, we replace v_{x1}, x_1, v_1, v_{y1}, g_1, etc. in Eq. (7.50) by their equivalents in Eq. (7.51); then we write Eq. (7.48) for system 2:

$$\frac{K_v^2}{K_L} v_{x2} \frac{\partial v_{x2}}{\partial x_2} + \frac{K_v^2}{K_L} v_{y2} \frac{\partial v_{x2}}{\partial y_2} = \frac{K_\nu K_v}{K_L^2} \nu_2 \frac{\partial^2 v_{x2}}{\partial y_2^2} + K_g K_\beta K_T g_2 \beta_2 (T - T_\infty)_2. \tag{7.52}$$

Equation (7.52) if rewritten without all the K's would of course be valid, since it would transform back to Eq. (7.48). Hence,

$$\frac{K_v^2}{K_L} = \frac{K_\nu K_v}{K_L^2} = K_g K_\beta K_T = 1, \tag{7.53}$$

and therefore

$$\frac{v_1^2}{L_1} = \frac{v_2^2}{L_2}, \qquad \frac{v_1 v_1}{L_1^2} = \frac{v_2 v_2}{L_2^2}$$

and

$$g_1 \beta_1 (T_0 - T_\infty)_1 = g_2 \beta_2 (T_0 - T_\infty)_2. \tag{7.54 a, b, c}$$

If we combine Eqs. (7.54 a) and (7.54 b) we get

$$\frac{v_1 L_1}{\nu_1} = \frac{v_2 L_2}{\nu_2}, \tag{7.55}$$

which are Reynolds numbers. The combination of Eqs. (7.54 b) and (7.54 c), yields

$$\frac{g_1 \beta_1 (T_0 - T_\infty)_1 L_1^2}{\nu_1 v_1} = \frac{g_2 \beta_2 (T_0 - T_\infty)_2 L_2^2}{\nu_2 v_2}. \tag{7.56}$$

The group of variables represented in Eq. (7.56) could be considered as an important dimensionless group, but by reflecting on the physical aspects of the problems, we realize that the velocity of the fluid is not an independent quantity, but that it rather depends on the buoyant force. Hence the v's are eliminated from Eq. (7.56), and substituting their equivalents from Eq. (7.55), we obtain

$$\frac{g_1 \beta_1 (T_0 - T_\infty)_1 L_1^3}{\nu_1^2} = \frac{g_2 \beta_2 (T_0 - T_\infty)_2 L_2^3}{\nu_2^2}. \tag{7.57}$$

This dimensionless group of variables which is important in natural convection problems is called the *Grashof number*, Gr. When the buoyancy is the only driving force for convection, the velocity profile is determined entirely by the quantities in the Grashof number, and the Reynolds number is superfluous.

Regarding heat transfer with natural convection, recall that for forced convection, as discussed in Section 7.2, the Nusselt number was correlated in the general form

$$\text{Nu} = f(\text{Pr}, \text{Re}), \qquad \textit{forced convection.}$$

Correspondingly then, for natural convection, the Nusselt number is correlated as

$$\text{Nu} = f(\text{Pr}, \text{Gr}), \qquad \textit{natural convection.}$$

Returning to the complete solution of Eqs. (7.48) and (7.49), we present the results for velocity and temperature distributions (see Fig. 7.7). The curves show

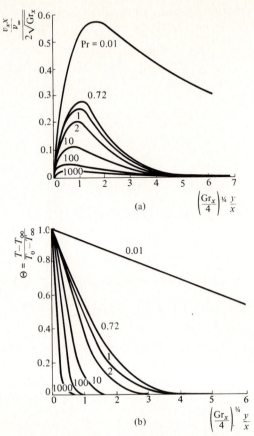

Fig. 7.7 Laminar natural convection for a vertical plate. (a) Dimensionless velocity profiles. (b) Dimensionless temperature profiles. (Calculated by S. Ostrach, *Nat. Advisory Comm. Aeronaut. Tech. Note* 2635, Feb. 1952, as presented in W. M. Rohsenow and H. Y. Choi, *Heat, Mass, and Momentum Transfer*, Prentice–Hall, Englewood Cliffs, New Jersey, 1961, pages 155–159.)

that for $Pr \leq 1$, $\delta_T \cong \delta$, but for $Pr > 1$, $\delta_T < \delta$. For liquid metals, therefore, δ_T is about equal to δ in free convection as contrasted to forced convection in which $\delta_T \gg \delta$. Corresponding to the temperature profile, as shown in Fig. 7.7b, the results for the local Nusselt number correlation are

$$\frac{Nu_x}{\sqrt[4]{Gr_x/4}} = \frac{0.676\, Pr^{1/2}}{(0.861 + Pr)^{1/4}}. \tag{7.58}$$

Equation (7.58) applies for a wide range of Pr numbers ($0.00835 \leq Pr \leq 1000$), and has been shown to apply for laminar flow conditions, $10^4 < Gr_x \cdot Pr < 10^{10}$.

Example 7.2 Calculate the initial heat-transfer rate from a plate at 200°F, 1 ft long × 1 ft wide hung vertically in air at 70°F. Contrast the results with those of Example 7.1.

Solution. Equation (7.58) should be integrated to obtain the average heat-transfer coefficient which can be applied to the whole plate.

In Eq. (7.58), since h_x varies as $x^{-1/4}$, then the average h equals $\frac{4}{3}h_x$. Hence Nu_L defined as hL/k is

$$\frac{\text{Nu}_L}{\sqrt[4]{\text{Gr}_L/4}} = \frac{0.902\,\text{Pr}^{1/2}}{(0.861 + \text{Pr})^{1/4}} \tag{7.59}$$

or

$$\text{Nu}_L = 0.902\,\sqrt[4]{\frac{\text{Gr}_L \cdot \text{Pr}^2}{4(0.861 + \text{Pr})}}. \tag{7.59a}$$

We evaluate the properties at the average boundary temperature of $\frac{1}{2}(200 + 70) =$ 135°F. For air at 135°F,

$$\text{Pr} = 0.703 \quad \text{and} \quad g\beta/\nu^2 = 1.4 \times 10^6 \text{ ft}^{-3}\,°\text{F}^{-1}.$$

The Grashof number is

$$\text{Gr}_L = \frac{g\beta}{\nu^2}(T_0 - T_\infty)L^3 = (1.4 \times 10^6)(200 - 70) = 1.82 \times 10^8.$$

Next, we calculate the product $\text{Gr}_L \cdot \text{Pr}$ to test for laminar flow conditions

$$\text{Gr}_L \cdot \text{Pr} = (1.82 \times 10^8)(0.703) = 1.28 \times 10^8.$$

Since it is between 10^4 and 10^{10}, laminar flow holds, and Eq. (7.59) is valid. When we substitute values of Gr_L and Pr into Eq. (7.59a),

$$\text{Nu}_L = 82.9,$$

from which

$$h = (82.9)\frac{k}{L} = (82.9)\left(\frac{0.017}{1}\right) = 1.41 \text{ Btu/hr-ft}^2\,°\text{F}.$$

Finally, we evaluate the rate of heat transfer Q.

$$Q = h(T_\infty - T_0)A = (1.41)(200 - 70)(1) = 183 \text{ Btu/hr}.$$

For Example 7.1, Q was 653 Btu/hr, that is, the rate of heat transfer for forced convection is considerably higher. This is the usual case.

It is instructive to look at special forms of Eq. (7.59a). First, if $\text{Pr} = 0.702$, then it reduces to

$$\text{Nu}_L = \sqrt[4]{\frac{\text{Gr}_L}{4}}. \tag{7.59b}$$

It so happens that for many gases, including air, O_2, N_2, CO, He (and other inert gases), H_2 and CO_2, Pr is very close to 0.7, and practically constant for temperatures even as high as 3000°F. Thus, we can apply Eq. (7.59b) directly to gases.

Second, if Pr → 0, then Eq. (7.59a) reduces to

$$Nu_L = 0.936 \sqrt[4]{\frac{Gr_L \cdot Pr^2}{4}}.$$

Therefore, a somewhat more approximate, but still adequate, form for most calculations involving natural convection of liquid metals is

$$Nu_L \cong \sqrt[4]{\frac{Gr_L \cdot Pr^2}{4}}. \tag{7.59c}$$

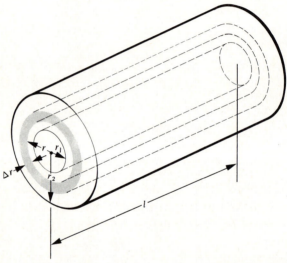

Fig. 7.8 Heat conduction through a solid cylindrical wall. The shaded area depicts the unit volume.

7.4 HEAT CONDUCTION

In this section we consider the problem of heat conduction through the wall of a hollow solid cylinder. Figure 7.8 depicts the situation, and also locates a suitable unit volume with a thickness Δr. From a practical point of view, we may visualize the system as a long cylindrical shaped furnace, and it is desirable to calculate the heat loss through the walls to the surroundings. Suppose the cylinder is long enough so that end effects are negligible; in addition, the system is at steady state, so that both the inside and outside surfaces of the wall are at some fixed temperatures, T_1 and T_2, respectively. For such a system, we develop the energy balance.

Rate of energy in by conduction across surface at r $2\pi r l q_r|_r$

Rate of energy out by conduction across surface at r $2\pi r l q_r|_{r+\Delta r}$

At steady state, these are the only terms that contribute to the energy balance. Thus

$$2\pi l r q_r|_{r+\Delta r} - 2\pi l r q_r|_r = 0.$$

If we divide all terms by $2\pi l\,\Delta r$, and take the limit as Δr approaches zero, we obtain

$$\lim_{\Delta r \to 0} \frac{(rq_r)|_{r+\Delta r} - (rq_r)|_r}{\Delta r} = 0,$$

or

$$\frac{d(rq_r)}{dr} = 0. \tag{7.60}$$

Equation (7.60) requires that

$$rq_r = C_1. \tag{7.61}$$

Note that q_r, the heat flux, is not constant in itself. Since $q_r = -k(dT/dr)$, Eq. (7.61) yields

$$-k\frac{dT}{dr} = \frac{C_1}{r}. \tag{7.62}$$

Integrating once again, we find for constant thermal conductivity

$$T = -\frac{C_1}{k}\ln r - \frac{C_2}{k}. \tag{7.63}$$

By absorbing k in new constants, Eq. (7.63) simplifies even more

$$T = C_3 \ln r + C_4. \tag{7.64}$$

The boundary conditions under consideration are

         B.C. 1   at $r = r_1$, $T = T_1$;

         B.C. 2   at $r = r_2$, $T = T_2$.

Determination of the constants using the boundary conditions yields the temperature distribution

$$\frac{T - T_2}{T_1 - T_2} = \frac{\ln(r/r_2)}{\ln(r_1/r_2)}. \tag{7.65}$$

Determining the heat flux (Btu/hr-ft^2), we note that it is not constant, as we have mentioned.

$$q_r = -k\frac{dT}{dr} = \frac{k}{r}\left[\frac{T_1 - T_2}{\ln(r_1/r_2)}\right]. \tag{7.66}$$

Physically, as the heat flows (Btu/hr) through the wall, it encounters larger areas, so that the flux itself decreases. The heat flow Q (Btu/hr), however, is constant (as it must be for steady state), and is given by

$$Q = q_r(2\pi r L) = \frac{2\pi k L}{\ln(r_1/r_2)}(T_1 - T_2). \tag{7.67}$$

This problem, elementary as it is, demonstrates an interesting engineering characteristic. Suppose we use the cylindrical wall as the insulation of a furnace wall. As increasing thicknesses of insulation are added, the outside layer, due to its greater area, offers less resistance to heat flow than an inner layer of the same thickness. Thus, from a cost point of view, the expense of additional insulation may become greater than the savings associated with reduction in heat losses.

Example 7.3 As part of a proposed continuous annealing process, a rod passes through a cylindrical furnace chamber 4 in. inside diameter and 50 ft long. The inside surface temperature of the furnace wall under operating conditions is predicted to be about 1200°F and the outside surface about 100°F. If it is decided that a heat loss of 250,000 Btu/hr is an acceptable figure, which of the following insulations would you use?

	k, Btu/hr-ft °F	Cost, $ per ft³
Insulation A	0.4	10
Insulation B	0.2	25

Solution. Equation (7.67) can be written

$$\ln(r_2/r_1) = \frac{2\pi k L}{Q}(T_1 - T_2).$$

For A then

$$\ln(r_2/r_1) = \frac{(2\pi)(0.4)(50)(1100)}{(250,000)} = 0.550,$$

so that (for $r_1 = 2$ in.), we have

$$r_2 = 3.48 \text{ in.}$$

Similarly for B, using ratios of conductivities, we get

$$\ln(r_2/r_1) = \left(\frac{0.2}{0.4}\right)(0.550) = 0.275,$$

so that

$$r_2 = 2.64 \text{ in.}$$

We calculate the volume of insulation and the corresponding cost.

$$\text{Cost } A = \frac{\pi(3.48^2 - 2.0^2)\,\text{in}^2}{\left|\begin{array}{c}\end{array}\right|}\frac{1\,\text{ft}^2}{144\,\text{in}^2}\left|\frac{50\,\text{ft}}{}\right|\frac{10\,\$}{\text{ft}^3} = \$88.30. \qquad \tfrac{4}{3}\pi r^3$$

In the same manner,

$$\text{Cost } B = \$32.50.$$

The obvious choice is B.

It should be pointed out that we shall develop a more general presentation of this type of problem in Chapter 9.

7.5 THE GENERAL ENERGY EQUATION

In Sections 7.1–7.4, we determined temperature distributions and heat fluxes for some simple flow systems, by developing pertinent energy balances in differential form. In this section we shall develop more general energy equations which, of course, can be reduced to solve specific problems.

Consider the stationary unit volume $\Delta x\,\Delta y\,\Delta z$ in Figs. 2.4 and 2.5; we apply the law of conservation of energy to the fluid contained within this volume at any given time

$$\begin{pmatrix}\text{rate of accumulation}\\ \text{of internal and}\\ \text{kinetic energy}\end{pmatrix} = \begin{pmatrix}\text{net rate of internal}\\ \text{and kinetic energies}\\ \text{in by convection}\end{pmatrix} + \begin{pmatrix}\text{net rate of}\\ \text{heat in by}\\ \text{conduction}\end{pmatrix} - \begin{pmatrix}\text{net rate of}\\ \text{work done}\\ \text{by fluid}\end{pmatrix} \tag{7.68}$$

This statement of the law of energy conservation is not completely general, because other forms of energy transport, e.g. radiation, and sources such as electrical Joule heating, are not included.

The rate of accumulation of internal and kinetic energy within the unit volume is simply

$$\Delta x\,\Delta y\,\Delta z \frac{\partial}{\partial t}(\rho U + \tfrac{1}{2}\rho v^2), \tag{7.69}$$

where U is the internal energy per unit mass of fluid and v is the *magnitude* of the local fluid velocity.

The *net* rate of internal and kinetic energies in by convection is

$$\Delta y\,\Delta z\{v_x(\rho U + \tfrac{1}{2}\rho v^2)|_x - v_x(\rho U + \tfrac{1}{2}\rho v^2)|_{x+\Delta x}\}$$
$$+ \Delta x\,\Delta z\{v_y(\rho U + \tfrac{1}{2}\rho v^2)|_y - v_y(\rho U + \tfrac{1}{2}\rho v^2)|_{y+\Delta y}\}$$
$$+ \Delta x\,\Delta y\{v_z(\rho U + \tfrac{1}{2}\rho v^2)|_z - v_z(\rho U + \tfrac{1}{2}\rho v^2)|_{z+\Delta z}\}. \tag{7.70}$$

In a similar manner, the net rate of energy in by conduction is

$$\Delta y\,\Delta z\{q_x|_x - q_x|_{x+\Delta x}\} + \Delta x\,\Delta z\{q_y|_y - q_y|_{y+\Delta y}\} + \Delta x\,\Delta y\{q_z|_z - q_z|_{z+\Delta z}\}. \tag{7.71}$$

The work done by the fluid consists of work against gravity, work against pressure, and work against viscous forces. The rate of doing work against the three components of gravity is

$$-\rho \, \Delta x \, \Delta y \, \Delta z (v_x g_x + v_y g_y + v_z g_z). \tag{7.72}$$

The rate of doing work against the pressure at the six faces of the unit volume is

$$\Delta y \Delta z\{(Pv_x)|_{x+\Delta x} - (Pv_x)|_x\} + \Delta x \, \Delta z\{(Pv_y)|_{y+\Delta y} - (Pv_y)|_y\}$$
$$+ \Delta x \Delta y\{(Pv_z)|_{z+\Delta z} - (Pv_z)|_z\}. \tag{7.73}$$

The rate of doing work against the x-directed viscous forces is

$$\Delta y \, \Delta z\{\tau_{xx}v_x|_{x+\Delta x} - \tau_{xx}v_x|_x\} + \Delta x \, \Delta z\{\tau_{yx}v_x|_{y+\Delta y} - \tau_{yx}v_x|_y\}$$
$$+ \Delta x \, \Delta y\{\tau_{zx}v_x|_{z+\Delta z} - \tau_{zx}v_x|_z\}. \tag{7.74}$$

Similar expressions may be written for the work against the y- and z-directed viscous forces

$$\Delta y \, \Delta z\{\tau_{xy}v_y|_{x+\Delta x} - \tau_{xy}v_y|_x\} + \Delta x \, \Delta z\{\tau_{yy}v_y|_{y+\Delta y} - \tau_{yy}v_y|_y\}$$
$$+ \Delta x \, \Delta y\{\tau_{zy}v_y|_{z+\Delta z} - \tau_{zy}v_y|_z\}, \tag{7.75}$$

and

$$\Delta y \, \Delta z\{\tau_{xz}v_z|_{x+\Delta x} - \tau_{xz}v_z|_x\} + \Delta x \, \Delta z\{\tau_{yz}v_z|_{y+\Delta y} - \tau_{yz}v_z|_y\}$$
$$+ \Delta x \, \Delta y\{\tau_{zz}v_z|_{z+\Delta z} - \tau_{zz}v_z|_z\}. \tag{7.76}$$

Substituting all these expressions into Eq. (7.68), dividing by $\Delta x \, \Delta y \, \Delta z$, and taking the limit as Δx, Δy, and Δz approach zero, we obtain one form of the energy equation

$$\frac{\partial}{\partial t}(\rho U + \tfrac{1}{2}\rho v^2) = -\left(\frac{\partial}{\partial x}v_x(\rho U + \tfrac{1}{2}\rho v^2) + \frac{\partial}{\partial y}v_y(\rho U + \tfrac{1}{2}\rho v^2) + \frac{\partial}{\partial z}v_z(\rho U + \tfrac{1}{2}\rho v^2)\right)$$

$$-\left(\frac{\partial q_x}{\partial x} + \frac{\partial q_y}{\partial y} + \frac{\partial q_z}{\partial z}\right) + \rho(v_x g_x + v_y g_y + v_z g_z)$$

$$-\left(\frac{\partial}{\partial x}Pv_x + \frac{\partial}{\partial y}Pv_y + \frac{\partial}{\partial z}Pv_z\right)$$

$$-\left(\frac{\partial}{\partial x}(\tau_{xx}v_x + \tau_{xy}v_y + \tau_{xz}v_z) + \frac{\partial}{\partial y}(\tau_{yx}v_x + \tau_{yy}v_y + \tau_{yz}v_z)\right.$$

$$\left. + \frac{\partial}{\partial z}(\tau_{zx}v_x + \tau_{zy}v_y + \tau_{zz}v_z)\right). \tag{7.77}$$

If all the terms involving $(\rho U + \frac{1}{2}\rho v^2)$ are expanded and combined, we can obtain

$$\rho\left[\frac{\partial}{\partial t}(U + \tfrac{1}{2}v^2) + v_x\frac{\partial}{\partial x}(U + \tfrac{1}{2}v^2) + v_y\frac{\partial}{\partial y}(U + \tfrac{1}{2}v^2) + v_z\frac{\partial}{\partial y}(U + \tfrac{1}{2}v^2)\right]$$

$$+ (U + \tfrac{1}{2}v^2)\left[\frac{\partial \rho}{\partial t} + \frac{\partial}{\partial x}\rho v_x + \frac{\partial}{\partial y}\rho v_y + \frac{\partial}{\partial z}\rho v_z\right]. \qquad (7.78)$$

Continuity (Eq. A, Table 2.1) requires that the second term in the above expression is zero; the remaining term in the expression is the substantial derivative of $(U + \frac{1}{2}v^2)$, so that we may write Eq. (7.77) as

$$\rho\frac{D}{Dt}(U + \tfrac{1}{2}v^2) = \left(\frac{\partial q_x}{\partial x} + \frac{\partial q_y}{\partial y} + \frac{\partial q_z}{\partial z}\right) + \rho(v_x g_x + v_y g_y + v_z g_z)$$

$$- \left(\frac{\partial}{\partial x}Pv_x + \frac{\partial}{\partial y}Pv_y + \frac{\partial}{\partial z}Pv_z\right)$$

$$- \left(\frac{\partial}{\partial x}(\tau_{xx}v_x + \tau_{xy}v_y + \tau_{xz}v_z) + \frac{\partial}{\partial y}(\tau_{yx}v_x + \tau_{yy}v_y + \tau_{yz}v_z)\right.$$

$$\left. + \frac{\partial}{\partial z}(\tau_{zx}v_x + \tau_{zy}v_y + \tau_{zz}v_z)\right). \qquad (7.79)$$

For most engineering problems involving heat flow, it is more convenient to have the equation of energy in terms of heat capacity than in terms of internal energy. The manipulation that leads to the development of the following equation is rather lengthy, and is not given here, but with the aid of the conservation of momentum, we can write Eq. (7.79):

$$\rho\frac{DU}{Dt} = -\left(\frac{\partial}{\partial x}q_x + \frac{\partial}{\partial y}q_y + \frac{\partial}{\partial z}q_z\right) - P\left(\frac{\partial v_x}{\partial x} + \frac{\partial v_y}{\partial y} + \frac{\partial v_z}{\partial z}\right) + \eta\Phi, \qquad (7.80)$$

where the quantity Φ is known as the *dissipation function*,

$$\Phi = 2\left[\left(\frac{\partial v_x}{\partial x}\right)^2 + \left(\frac{\partial v_y}{\partial y}\right)^2 + \left(\frac{\partial v_z}{\partial z}\right)^2\right]$$

$$+ \left[\left(\frac{\partial v_x}{\partial y} + \frac{\partial v_y}{\partial x}\right)^2 + \left(\frac{\partial v_y}{\partial z} + \frac{\partial v_z}{\partial y}\right)^2 + \left(\frac{\partial v_z}{\partial x} + \frac{\partial v_x}{\partial z}\right)^2\right]$$

$$- \frac{2}{3}\left(\frac{\partial v_x}{\partial x} + \frac{\partial v_y}{\partial y} + \frac{\partial v_z}{\partial z}\right)^2. \qquad (7.81)$$

From the definition of enthalpy, $H = U + P/\rho$, we write

$$\frac{DU}{Dt} = \frac{DH}{Dt} - \frac{1}{\rho}\frac{DP}{Dt} + \frac{P}{\rho^2}\frac{D\rho}{Dt}, \tag{7.82}$$

but from continuity,

$$\frac{1}{\rho}\frac{D\rho}{Dt} + \frac{\partial v_x}{\partial x} + \frac{\partial v_y}{\partial y} + \frac{\partial v_z}{\partial z} = 0,$$

so that

$$\rho\frac{DU}{Dt} = \rho\frac{DH}{Dt} - \frac{DP}{Dt} - P\left(\frac{\partial v_x}{\partial x} + \frac{\partial v_y}{\partial y} + \frac{\partial v_z}{\partial z}\right) = 0. \tag{7.83}$$

Substitution of Eq. (7.83) into Eq. (7.80) leads to

$$\rho\frac{DH}{Dt} = -\left(\frac{\partial}{\partial x}q_x + \frac{\partial}{\partial y}q_y + \frac{\partial}{\partial z}q_z\right) + \frac{DP}{Dt} + \eta\Phi. \tag{7.84}$$

From the thermodynamic relations of properties, we can write

$$dH = C_p\, dT + \frac{1}{\rho}(1 - T\beta)\, dP, \tag{7.85}$$

where C_p is the heat capacity, and $\beta \equiv -(1/\rho)(\partial\rho/\partial T)_P$.
    From Eq. (7.85), it follows that

$$\rho\frac{DH}{Dt} = \rho C_p\frac{DT}{Dt} + (1 - T\beta)\frac{DP}{Dt}, \tag{7.86}$$

and from the Fourier rate equations, that

$$q_x = -k\frac{\partial T}{\partial x}, \qquad q_y = -k\frac{\partial T}{\partial y}, \qquad \text{and} \qquad q_z = -k\frac{\partial T}{\partial z}.$$

Substituting Eq. (7.86) and the rate equations into Eq. (7.84) finally gives the energy equation in terms of temperature:

$$\rho C_p\frac{DT}{Dt} = \frac{\partial}{\partial x}\left(k\frac{\partial T}{\partial x}\right) + \frac{\partial}{\partial y}\left(k\frac{\partial T}{\partial y}\right) + \frac{\partial}{\partial z}\left(k\frac{\partial T}{\partial z}\right) + T\beta\frac{DP}{Dt} + \eta\Phi. \tag{7.87}$$

Simplifications of Eq. (7.87) are usually referred to. For example, if k is independent of the space coordinates, we write

$$\rho C_p\frac{DT}{Dt} = k\nabla^2 T + T\beta\frac{DP}{Dt} + \eta\Phi,$$

where

$$\nabla^2 \equiv \frac{\partial^2}{\partial x^2} + \frac{\partial^2}{\partial y^2} + \frac{\partial^2}{\partial z^2}. \tag{7.88}$$

Further, if the system is an *ideal* gas, $\beta = 1/T$:

$$\rho C_p \frac{DT}{Dt} = k\nabla^2 T + \frac{DP}{Dt} + \eta\Phi. \tag{7.89}$$

Many problems involve incompressible fluids where the viscous dissipation term is negligible. In this instance, $\beta = \Phi = 0$, and

$$\rho C_p \frac{DT}{Dt} = k\nabla^2 T. \tag{7.90}$$

In the conduction of heat through *solids*, the velocity is zero; the compressibility term is ignored, so that

$$\rho C_p \frac{\partial T}{\partial t} = k\nabla^2 T. \tag{7.91}$$

7.6 THE ENERGY EQUATION IN CURVILINEAR COORDINATES

In this section, we express some of the relationships for energy balances—already developed in rectangular coordinates (Section 7.5)—in cylindrical and spherical coordinates. Tables 7.3, 7.4, and 7.5 may be used for the problems of heat flow by discarding unnecessary terms, rather than setting up problems by means of shell balances.

Table 7.3   Components of the energy flux, q^*

Rectangular		Cylindrical		Spherical	
$q_x = -k\dfrac{\partial T}{\partial x}$	(A)	$q_r = -k\dfrac{\partial T}{\partial r}$	(D)	$q_r = -k\dfrac{\partial T}{\partial r}$	(G)
$q_y = -k\dfrac{\partial T}{\partial y}$	(B)	$q_\theta = -k\dfrac{1}{r}\dfrac{\partial T}{\partial \theta}$	(E)	$q_\theta = -k\dfrac{1}{r}\dfrac{\partial T}{\partial \theta}$	(H)
$q_z = -k\dfrac{\partial T}{\partial z}$	(C)	$q_z = -k\dfrac{\partial T}{\partial z}$	(F)	$q_\phi = -k\dfrac{1}{r\sin\theta}\dfrac{\partial T}{\partial \phi}$	(I)

* This table and the following two tables are from R. B. Bird, W. E. Stewart, and E. N. Lightfoot, *Transport Phenomena*, Wiley, New York, 1960, pages 317–319.

Table 7.4   The equation of energy in terms of energy and momentum fluxes

Rectangular coordinates

$$\rho C_v \left(\frac{\partial T}{\partial t} + v_x \frac{\partial T}{\partial x} + v_y \frac{\partial T}{\partial y} + v_z \frac{\partial T}{\partial z} \right) = - \left[\frac{\partial q_x}{\partial x} + \frac{\partial q_y}{\partial y} + \frac{\partial q_z}{\partial z} \right]$$

$$- T \left(\frac{\partial P}{\partial T} \right)_\rho \left(\frac{\partial v_x}{\partial x} + \frac{\partial v_y}{\partial y} + \frac{\partial v_z}{\partial z} \right) - \left\{ \tau_{xx} \frac{\partial v_x}{\partial x} + \tau_{yy} \frac{\partial v_y}{\partial y} + \tau_{zz} \frac{\partial v_z}{\partial z} \right\}$$

$$- \left\{ \tau_{xy} \left(\frac{\partial v_x}{\partial y} + \frac{\partial v_y}{\partial x} \right) + \tau_{xz} \left(\frac{\partial v_x}{\partial z} + \frac{\partial v_z}{\partial x} \right) + \tau_{yz} \left(\frac{\partial v_y}{\partial z} + \frac{\partial v_z}{\partial y} \right) \right\} . \qquad \text{(A)}$$

Cylindrical coordinates

$$\rho C_v \left(\frac{\partial T}{\partial t} + v_r \frac{\partial T}{\partial r} + \frac{v_\theta}{r} \frac{\partial T}{\partial \theta} + v_z \frac{\partial T}{\partial z} \right) = - \left[\frac{1}{r} \frac{\partial}{\partial r} (r q_r) + \frac{1}{r} \frac{\partial q_\theta}{\partial \theta} + \frac{\partial q_z}{\partial z} \right]$$

$$- T \left(\frac{\partial P}{\partial T} \right)_\rho \left(\frac{1}{r} \frac{\partial}{\partial r} (r v_r) + \frac{1}{r} \frac{\partial v_\theta}{\partial \theta} + \frac{\partial v_z}{\partial z} \right) - \left\{ \tau_{rr} \frac{\partial v_r}{\partial r} + \tau_{\theta\theta} \frac{1}{r} \left(\frac{\partial v_\theta}{\partial \theta} + v_r \right) \right.$$

$$\left. + \tau_{zz} \frac{\partial v_z}{\partial z} \right\} - \left\{ \tau_{r\theta} \left[r \frac{\partial}{\partial r} \left(\frac{v_\theta}{r} \right) + \frac{1}{r} \frac{\partial v_r}{\partial \theta} \right] + \tau_{rz} \left(\frac{\partial v_z}{\partial r} + \frac{\partial v_r}{\partial z} \right) \right.$$

$$\left. + \tau_{\theta z} \left(\frac{1}{r} \frac{\partial v_z}{\partial \theta} + \frac{\partial v_\theta}{\partial z} \right) \right\} . \qquad \text{(B)}$$

Spherical coordinates

$$\rho C_v \left(\frac{\partial T}{\partial t} + v_r \frac{\partial T}{\partial r} + \frac{v_\theta}{r} \frac{\partial T}{\partial \theta} + \frac{v_\phi}{r \sin \theta} \frac{\partial T}{\partial \phi} \right) = - \left[\frac{1}{r^2} \frac{\partial}{\partial r} (r^2 q_r) \right.$$

$$\left. + \frac{1}{r \sin \theta} \frac{\partial}{\partial \theta} (q_\theta \sin \theta) + \frac{1}{r \sin \theta} \frac{\partial q_\phi}{\partial \phi} \right] - T \left(\frac{\partial P}{\partial T} \right)_\rho \left(\frac{1}{r^2} \frac{\partial}{\partial r} (r^2 v_r) \right.$$

$$\left. + \frac{1}{r \sin \theta} \frac{\partial}{\partial \theta} (v_\theta \sin \theta) + \frac{1}{r \sin \theta} \frac{\partial v_\phi}{\partial \phi} \right) - \left\{ \tau_{rr} \frac{\partial v_r}{\partial r} + \tau_{\theta\theta} \left(\frac{1}{r} \frac{\partial v_\theta}{\partial \theta} + \frac{v_r}{r} \right) \right.$$

$$\left. + \tau_{\phi\phi} \left(\frac{1}{r \sin \theta} \frac{\partial v_\phi}{\partial \phi} + \frac{v_r}{r} + \frac{v_\theta \cot \theta}{r} \right) \right\} - \left\{ \tau_{r\theta} \left(\frac{\partial v_\theta}{\partial r} + \frac{1}{r} \frac{\partial v_r}{\partial \theta} - \frac{v_\theta}{r} \right) \right.$$

$$\left. + \tau_{r\phi} \left(\frac{\partial v_\phi}{\partial r} + \frac{1}{r \sin \theta} \frac{\partial v_r}{\partial \phi} - \frac{v_\phi}{r} \right) + \tau_{\theta\phi} \left(\frac{1}{r} \frac{\partial v_\phi}{\partial \theta} + \frac{1}{r \sin \theta} \frac{\partial v_\theta}{\partial \phi} - \frac{\cot \theta}{r} v_\phi \right) \right\} . \qquad \text{(C)}$$

Note: The terms contained in braces { } are associated with viscous dissipation and may usually be neglected, except for systems with large velocity gradients.

Table 7.5 The equation of energy in terms of the transport properties (*for Newtonian fluids of constant ρ, η, and k; note that the constancy of ρ implies that $C_v = C_p$*)

Rectangular coordinates

$$
\rho C_v \left(\frac{\partial T}{\partial t} + v_x \frac{\partial T}{\partial x} + v_y \frac{\partial T}{\partial y} + v_z \frac{\partial T}{\partial z} \right) = k \left[\frac{\partial^2 T}{\partial x^2} + \frac{\partial^2 T}{\partial y^2} + \frac{\partial^2 T}{\partial z^2} \right]
$$

$$
+ 2\eta \left\{ \left(\frac{\partial v_x}{\partial x} \right)^2 + \left(\frac{\partial v_y}{\partial y} \right)^2 + \left(\frac{\partial v_z}{\partial z} \right)^2 \right\} + \eta \left\{ \left(\frac{\partial v_x}{\partial y} + \frac{\partial v_y}{\partial x} \right)^2 \right.
$$

$$
\left. + \left(\frac{\partial v_x}{\partial z} + \frac{\partial v_z}{\partial x} \right)^2 + \left(\frac{\partial v_y}{\partial z} + \frac{\partial v_z}{\partial y} \right)^2 \right\}.
$$

(A)

Cylindrical coordinates

$$
\rho C_v \left(\frac{\partial T}{\partial t} + v_r \frac{\partial T}{\partial r} + \frac{v_\theta}{r} \frac{\partial T}{\partial \theta} + v_z \frac{\partial T}{\partial z} \right) = k \left[\frac{1}{r} \frac{\partial}{\partial r} \left(r \frac{\partial T}{\partial r} \right) + \frac{1}{r^2} \frac{\partial^2 T}{\partial \theta^2} + \frac{\partial^2 T}{\partial z^2} \right]
$$

$$
+ 2\eta \left\{ \left(\frac{\partial v_r}{\partial r} \right)^2 + \left[\frac{1}{r} \left(\frac{\partial v_\theta}{\partial \theta} + v_r \right) \right]^2 + \left(\frac{\partial v_z}{\partial z} \right)^2 \right\} + \eta \left\{ \left(\frac{\partial v_\theta}{\partial z} + \frac{1}{r} \frac{\partial v_z}{\partial \theta} \right)^2 \right.
$$

$$
\left. + \left(\frac{\partial v_z}{\partial r} + \frac{\partial v_r}{\partial z} \right)^2 + \left[\frac{1}{r} \frac{\partial v_r}{\partial \theta} + r \frac{\partial}{\partial r} \left(\frac{v_\theta}{r} \right) \right]^2 \right\}.
$$

(B)

Spherical coordinates

$$
\rho C_v \left(\frac{\partial T}{\partial t} + v_r \frac{\partial T}{\partial r} + \frac{v_\theta}{r} \frac{\partial T}{\partial \theta} + \frac{v_\phi}{r \sin \theta} \frac{\partial T}{\partial \phi} \right) = k \left[\frac{1}{r^2} \frac{\partial}{\partial r} \left(r^2 \frac{\partial T}{\partial r} \right) \right.
$$

$$
+ \frac{1}{r^2 \sin \theta} \frac{\partial}{\partial \theta} \left(\sin \theta \frac{\partial T}{\partial \theta} \right) + \frac{1}{r^2 \sin^2 \theta} \frac{\partial^2 T}{\partial \phi^2} \right] + 2\eta \left\{ \left(\frac{\partial v_r}{\partial r} \right)^2 \right.
$$

$$
+ \left(\frac{1}{r} \frac{\partial v_\theta}{\partial \theta} + \frac{v_r}{r} \right)^2 + \left(\frac{1}{r \sin \theta} \frac{\partial v_\phi}{\partial \phi} + \frac{v_r}{r} + \frac{v_\theta \cot \theta}{r} \right)^2 \right\}
$$

$$
+ \eta \left\{ \left[r \frac{\partial}{\partial r} \left(\frac{v_\theta}{r} \right) + \frac{1}{r} \frac{\partial v_r}{\partial \theta} \right]^2 + \left[\frac{1}{r \sin \theta} \frac{\partial v_r}{\partial \phi} + r \frac{\partial}{\partial r} \left(\frac{v_\phi}{r} \right) \right]^2 \right.
$$

$$
\left. + \left[\frac{\sin \theta}{r} \frac{\partial}{\partial \theta} \left(\frac{v_\phi}{\sin \theta} \right) + \frac{1}{r \sin \theta} \frac{\partial v_\theta}{\partial \phi} \right]^2 \right\}.
$$

(C)

Note: The terms contained in braces { } are associated with viscous dissipation and may usually be neglected, except for systems with large velocity gradients.

PROBLEMS

7.1 For laminar flow calculate the results given in Table 7.1 for Nu_∞ for slug flow ($V_x =$ uniform) and uniform heat flux in a circular tube.

7.2 Utilize the approximate integral technique and develop an expression for the local Nusselt number for forced convection past a flat plate. Make the following assumptions:
  a) $\delta_T \leq \delta$.
  b) The velocity profile is linear.
  c) The temperature profile is linear.

7.3 A liquid film at T_0 flows down a vertical wall at a higher temperature T_s. Estimate the rate of heat transfer from the wall to the liquid for such contact times that the liquid temperature changes appreciably only in the immediate vicinity of the wall.

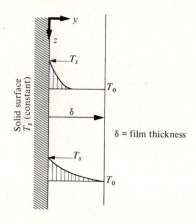

a) Show that the energy equation can be written (state assumptions).

$$\rho C_p v_z \frac{\partial T}{\partial z} = k \frac{\partial^2 T}{\partial y^2}.$$

b) Show, or refer to previous results, that

$$v_z = v_{max}[2y/\delta - (y/\delta)^2],$$

in which

$$v_{max} = \frac{\delta^2 \rho g}{2\eta},$$

and that for near the wall we may write

$$v_z = \frac{\rho g \delta y}{\eta}.$$

c) Combine the above expressions to obtain the differential equation

$$y\frac{\partial T}{\partial z} = \beta\frac{\partial^2 T}{\partial y^2},$$

in which

$$\beta = \frac{\eta k}{\rho^2 C_p g \delta}.$$

d) Write the boundary conditions for such short contact times.

e) Rewrite the differential equation (part c) and the boundary conditions in terms of the following reduced variables

$$\theta \equiv \frac{T - T_0}{T_s - T_0},$$

$$\gamma \equiv y/\sqrt[3]{9\beta z}.$$

Integrate once to obtain

$$\frac{d\theta}{d\gamma} = C^{-\gamma^3}$$

(C is an integration constant). Integrate a second time and use the boundary conditions to obtain

$$\theta = \frac{1}{\Gamma(\frac{4}{3})}\int_\gamma^\infty e^{-\gamma^3}\,d\gamma,$$

in which $\Gamma(\eta)$ is the "Gamma function of n":

$$\Gamma(n) = \int_0^\infty e^{-x}x^{n-1}\,dx \qquad (n > 0).$$

f) Show that the average heat flux to the fluid (for length of solid L) is

$$\bar{q}|_{y=0} = \tfrac{3}{2}k(T_s - T_0)\frac{(9\beta L)^{1/3}}{\Gamma(\frac{4}{3})}.$$

7.4 Calculate

a) the rate of heat flow Q (Btu/hr) across a 2 in. thick slab of stainless steel for the following situations after making reference to the sketch below.

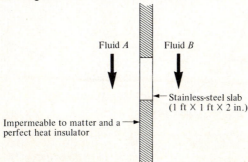

Fluid A   Fluid B

Stainless-steel slab
(1 ft × 1 ft × 2 in.)

Impermeable to matter and a →
perfect heat insulator

Fluid		Bulk velocity, ft/sec	Bulk temperature, °F
1.	A air	50	70
	B air	0	700
2.	A air	0	70
	B sodium	0	700
3.	A air	50	70
	B sodium	0	700
4.	A sodium	50	200
	B sodium	0	700

b) Prepare a table which gives fluid A's bulk temperature, two surface temperatures of the stainless steel (based on average heat-transfer coefficients), and fluid B's bulk temperature for the above four situations.

c) Digest your results (if you feel they are correct) and explain the significance of this problem.

Data: Stainless steel $\rho = 488 \text{ lb}_m/\text{ft}^3$
$C_p = 0.11 \text{ Btu/lb}_m \text{ °F}$
$$k \begin{cases} 68°\text{F} \text{; } 9.4 \text{ Btu/hr-ft °F} \\ 212°\text{F} \text{; } 10 \text{ Btu/hr-ft °F} \\ 1112°\text{F} \text{; } 13 \text{ Btu/hr-ft °F} \end{cases}$$

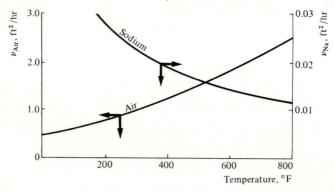

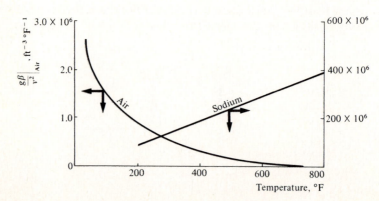

7.5

a) Determine an expression that gives the heat flow Q (Btu/hr) through a solid spherical shell with inside and outside radii of r_1, and r_2, respectively.

b) Examine the results regarding what happens as the shell thickness becomes larger compared with the inside radius.

7.6 A sphere of radius R is in a motionless fluid (no forced or natural convection). The surface temperature of the sphere is maintained at T_R and the bulk fluid temperature is T_∞.

a) Develop an expression for the temperature in the fluid surrounding the sphere.

b) Determine the Nusselt number for this situation. Such a value would be the limiting value for the actual system with convection as the forces causing convection become very small.

7.7 From Fig. 2.1 develop an expression for the temperature distribution in the falling film. Assume fully developed flow, constant properties, and fully developed temperature profile. The free liquid surface is maintained at $T = T_0$ and the solid surface at $T = T_s$, where T_0 and T_s are constants.

a) Ignore viscous heating effects.

b) Include viscous heating effects.

Answer (b)

$$\frac{T - T_0}{T_s - T_0} = \frac{x}{\delta}\left\{1 + \tfrac{3}{4}\,\text{Br}\left[1 - \left(\frac{x}{\delta}\right)^3\right]\right\},$$

where

$$\text{Br} = \frac{\eta \bar{V}^2}{k(T_s - T_0)}.$$

7.8 Develop an expression for the limiting value of Nu (stagnant surroundings) for heat transfer from a cylinder (maintained at a constant temperature) to its surroundings.

7.9 Given that the temperature distribution of a fluid flowing within a tube is

$$\frac{T - T_R}{T_0 - T_R} = \left[1 - \left(\frac{r}{R}\right)^2\right],$$

in which T_R = temperature at the wall ($r = R$), and T_0 = temperature at the center line ($r = 0$); also given that the flow rate may be approximated as the slug flow, that is, $v_z = V$ (constant), write an expression for the heat-transfer coefficient defined as follows using the mixed mean temperature T_m:

$$h \equiv \frac{q_{r=R}}{T_m - T_R},$$

$$T_m = \frac{\int_0^R v_z T 2\pi r\, dr}{\int_0^R v_z 2\pi r\, dr}.$$

[*Hint*: Answer is of the form Nu = (constant).]

7.10 A liquid at a temperature T_0 continuously enters the bottom of a small tank, overflows into a tube, and then flows downward as a film on the inside. At some position down the tube $(z = 0)$ when the flow is fully developed, the pipe heats the fluid with a uniform flux q_R. The heat loss from the liquid's surface is sufficiently small so that it may be neglected.

a) For steady-state laminar flow with constant properties develop by shell balance or show by reducing an equation in Table 7.5 the pertinent differential energy equation that applies to the falling film.

b) Write the boundary conditions for the heat flow.

c) What other information must complement parts (a) and (b) in order to solve the energy equations?

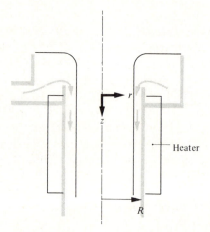

7.11 A liquid metal at the temperatures of interest has a Prandtl number of 0.001. Utilize the approximate integral technique and develop an expression for the local Nusselt number for forced convection parallel to a flat plate at a uniform surface temperature of T_0. Make the following assumptions:

a) The liquid metal approaches the leading edge with uniform temperature T_∞ and uniform velocity V_∞.

b) All flow is laminar.

c) The velocity profile is linear.

d) The temperature profile is linear.

e) The flow of heat parallel to the plate is negligible.

[*Hint*: Before you madly plunge into the problem, think about the fact that the Prandtl number is 0.001. Also, for a linear velocity profile $\delta/x = 3.46\sqrt{v/xV_\infty}$.]

8

CORRELATIONS FOR HEAT TRANSFER WITH CONVECTION

The problems of heat flow with convection, discussed in the preceding chapter, pertain to simple systems with laminar flow. The more complex nature of turbulent flow and its limited accessibility to mathematical treatment requires, for the most part, an empirical and experimental approach to heat transfer. This approach leads to correlations. Please bear in mind, however, that despite the simplicity of laminar flow problems, they should not be underestimated. Many simple solutions have been applied to real systems with approximating assumptions and, besides, the simpler systems provide models for interpretation and presentation of complex systems. On the other hand, the study of turbulent flow is not entirely empirical; it is possible to establish certain theoretical bases for the analyses of turbulent transfer processes. This approach is not presented here, but other texts[1,2] provide an introduction to this subject.

Figure 8.1 shows a physical picture of heat transfer in a bounded fluid. The fluid is artificially subdivided into three regions: the turbulent core, the buffer zone, and the laminar film near the surface. In the turbulent core, thermal energy is transferred rapidly due to the eddy (mixing) action of turbulent flow. Conversely, within the laminar film, energy is transferred by conduction alone—a much slower process than the eddy process. In the buffer region, energy transport by both conduction and by eddies is appreciable. Hence, most of the total temperature drop between the fluid and the surface is within the laminar film; across the buffer zone there is an appreciable temperature drop, but within the turbulent core, the temperature gradients are quite shallow.

In Chapter 3, it was convenient to define a friction factor to deal with momentum transport in fluids in contact with surfaces. Similarly, for energy transport between fluids and surfaces, it is convenient to define a heat-transfer coefficient by

$$h = \frac{q_0}{T_0 - T_f} = \frac{-k(\partial T/\partial y)_0}{T_0 - T_f},$$
(8.1)

---

[1] W. M. Rohsenow and H. Y. Choi, *Heat, Mass, and Momentum Transfer*, Prentice-Hall, Englewood Cliffs, New Jersey, 1961.

[2] R. B. Bird, W. E. Stewart, and E. N. Lightfoot, *Transport Phenomena*, Wiley, New York, 1960.

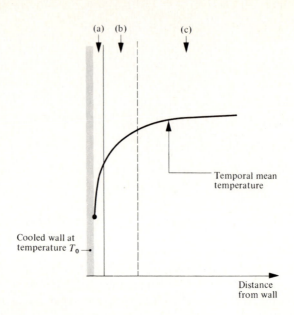

Fig. 8.1 Temperature profile of a flowing fluid bounded by a cooler wall; (a) laminar film, (b) buffer zone, and (c) turbulent core.

where the subscript "0" refers to the respective quantities evaluated at the wall, and T_f is some temperature of the fluid. If the fluid is infinite in extent, we take T_f as the fluid temperature far removed from the surface, and designate it T_∞. If the fluid flows in a confined space, such as inside a tube, T_f is usually the *mixed mean temperature*, denoted by T_m; it is a temperature that would exist if the fluid at a particular cross section were removed and allowed to mix adiabatically.

We define the engineering units of h as Btu/hr-ft^2 °F according to Eq. (8.1); as we shall see, h is a function of the fluid's properties (k, η, ρ, C_p), the system geometry, the flow velocity, and the value of the defining temperature difference. We shall deal in the remainder of this chapter with predicting the dependence of h on these quantities, by presenting some experimental data for various systems in the form of correlations based on dimensional analysis.

8.1 HEAT-TRANSFER COEFFICIENTS FOR FORCED CONVECTION IN TUBES

In Section 3.1, we presented a method of dimensional analysis utilizing a rearrangement of differential equations in order to show that the friction factor was a function of the Reynolds number. We used the same method in Section 7.3 for natural convection. Here we present another method of dimensional analysis in order to bring the reader's attention to a different technique of dimensional analysis, while showing those dimensionless groups which are pertinent to the heat-transfer system of forced convection within a tube.

8.1.1 Dimensional analysis—Buckingham's pi theorem

We may obtain pertinent dimensionless groups without any reference to the basic differential equations at all if we can identify the variables sufficient to specify a given situation. A systematic way of determining the groups is known as *Bucking-ham's pi theorem.*[3] Simply stated, it reads: *The functional relationship among q quantities, whose units may be given in terms of u fundamental units, may be written as a function of q − u dimensionless groups (the π's).*

In the application of the pi theorem, these rules are helpful:

a) Compile a list of the q significant quantities and their respective fundamental units. The fundamental units are L (length), M (mass), Θ (time), and T (temperature).

b) Select what may be called the primary quantities or primary q's. The number of primary q's should equal u, the number of fundamental units.

c) Form each π term by expressing the ratio of the remaining q's (one at a time) to the product of the primary q's, each raised to an unknown power determined as shown below.

We shall now illustrate the method for heat transfer in forced convection. We presume that the heat-transfer coefficient for fully developed forced convection in a tube is a function of the variables

$$h = f(\bar{V}, \rho, k, \eta, C_p, D). \tag{8.2}$$

We can express all these quantities in terms of the four fundamental dimensions M, L, Θ, and T. Specifically,

$$\bar{V}, L\Theta^{-1}, \qquad\qquad \eta, ML^{-1}\Theta^{-1}, \qquad\qquad D, L.$$
$$\rho, ML^{-3}, \qquad\qquad C_p, L^2\Theta^{-2}T^{-1},$$
$$k, ML\Theta^{-3}T^{-1}, \qquad\qquad h, M\Theta^{-3}T^{-1},$$

Note that heat energy is given the units of work (force × distance), namely, $ML^3\Theta^{-1}$.

The number of fundamental units in this case is four. Thus, since $u = 4$, select as the four primary q's: $\bar{V}, D, \rho$, and k, leaving h, η, and C_p. Proceed to form the first π term.

$$\pi_1 = \frac{h}{\bar{V}^a D^b \rho^c k^d} = \frac{M\Theta^{-3}T^{-1}}{(L\Theta^{-1})^a (L)^b (ML^{-3})^c (ML\Theta^{-3}T^{-1})^d}.$$

Equating the exponents of each dimension to zero, so that the π group is dimensionless, we obtain

$$
\begin{aligned}
M: &\quad 1 = c + d, \\
L: &\quad 0 = a + b - 3c + d, \\
\Theta: &\quad -3 = -a - 3d, \\
T: &\quad -1 = -d.
\end{aligned}
$$

---

[3] E. Buckingham, *Trans. ASME* **35**, 262 (1915).

Solving these four equations simultaneously, we obtain

$$a = 0, \qquad b = -1, \qquad c = 0, \qquad \text{and} \qquad d = 1.$$

Thus

$$\pi_1 = \frac{hD}{k} \equiv \text{Nu} \qquad \text{(Nusselt number)}.$$

Similarly, the following π groups can be obtained:

$$\pi_2 = \eta/\rho \bar{V} D \equiv \frac{1}{\text{Re}},$$

$$\pi_3 = C_p \eta/k \equiv \text{Pr} \qquad \text{(Prandtl number)}.$$

Thus, heat-transfer data in forced convection can be correlated in terms of these three dimensional groups:

$$\text{Nu} = \text{Nu (Re, Pr)}. \tag{8.3}$$

Equation (8.3) is not complete enough in some situations. If fully developed flow is not assumed, then the group L/D must be included. In addition, for large temperature differences, the temperature dependence of the fluid viscosity is important, and may be handled approximately by including the group η_m/η_0, where η_m is the viscosity at the mixed mean temperature T_m, and η_0 is the viscosity at the temperature of the solid surface. Hence, a complete correlation is written in the form

$$\text{Nu} = \text{Nu (Re, Pr, } L/D, \eta_m/\eta_0). \tag{8.4}$$

This dimensional analysis is of great use in experimental work involving heat transfer. For example, h depends on eight physical quantities: D, $\bar{V}$, ρ, η_m, η_0, C_p, k, and L. To study all combinations of eight independent variables for ten values of each would require 10^8 tests, whereas, by giving Nu as a function of only four groups (Re, Pr, L/D, η_m/η_0), 10^4 tests would suffice. Thus, a graduate student will have to work on a problem for only five years instead of 50,000 years!

8.1.2   Correlations for forced convection in tubes

For fully developed flow in tubes, the correlation of Seider and Tate[4] for flow in smooth tubes with nearly constant wall temperature is presented in Fig. 8.2. The Reynolds number used here, $\text{Re}_m = D\bar{V}\rho/\eta_m$, is convenient because the laminar-to-turbulent transition is at about 2100 (the same as in Fig. 3.2), even when η_0 differs appreciably from η_m.

For highly turbulent flow ($\text{Re}_m > 10,000$), the equation

$$\text{Nu}_m = 0.026 \, \text{Re}_m^{0.8} \, \text{Pr}_m^{1/3} \left(\frac{\eta_m}{\eta_0} \right)^{0.14} \tag{8.5}$$

---

[4] E. N. Seider and G. E. Tate, *Ind. Eng. Chem.* **28**, 1429 (1936).

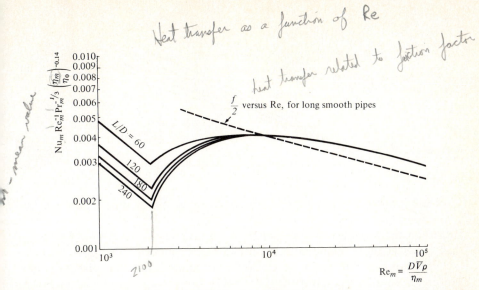

Heat transfer as a function of Re

heat transfer related to friction factor

$\frac{f}{2}$ versus Re, for long smooth pipes

~ mean value

$Re_m = \dfrac{D\bar{V}\rho}{\eta_m}$

Fig. 8.2 Heat-transfer coefficients for fully developed flow in smooth tubes. (From Seider and Tate, *ibid.*)

reproduces experimental data to within about $\pm 20\%$ in the range $10^4 < Re_m < 10^5$, $0.6 < Pr < 100$, and $L/D > 10$. As we shall discuss later, the data need not be restricted to the situations of constant wall temperature.

We superimpose the plot of $f/2$ on Fig. 8.2 for long, hydraulically smooth tubes, which shows that

$$j_H \equiv St_m \left(\frac{C_{p,m}\eta_f}{k_m} \right)^{2/3} = \frac{f}{2}, \tag{8.6}$$

where St_m is the Stanton number, and j_H is often referred to as a *j-factor*. The Stanton number is defined

$$St \equiv \frac{Nu}{Re \cdot Pr} = \frac{h}{C_p \rho \bar{V}}.$$

We refer to Eq. (8.6) as Colburn's analogy between heat transfer and fluid friction. The analogy breaks down for $Re_m < 10{,}000$, and also for rough tubes because f is affected more by roughness than its counterpart j_H. The effect of wall roughness was studied for air by Nunner,[5] and Fig. 8.3 relates the heat-transfer coefficient to the ratio of the friction factor f in the rough pipe to the friction factor f_s in a smooth pipe of the same diameter. The results have not been tested extensively, and should probably be restricted to use with gases.

For liquid metals, where $0.005 < Pr < 0.05$, the following equation represents available experimental data with $Re > 10{,}000$ and for a uniform heat flux:[6]

$$Nu_q = 6.7 + 0.0041(Re \cdot Pr)^{0.793} \exp(41.8\, Pr). \tag{8.7}$$

[5] W. Nunner, *VDI—Forschungsheft* No. 455/1956 (VDI-Verlag GmbH-Dusseldorf).
[6] W. M. Rohsenow and H. Y. Choi, *ibid.*, page 193.

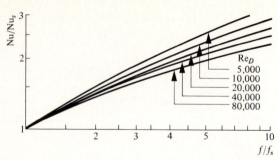

Fig. 8.3 Effect of roughness on heat transfer in turbulent flow. (From Nunner, *ibid.*)

Because of the very high values of h in liquid metals, $(T_0 - T_m)$ is not very large, so that the mixed mean temperature properties can be used without significant error.

Equation (8.7) for liquid metals applies to uniform heat flux along the tube. For uniform wall temperature the difference between Nu_T (uniform wall temperature) and Nu_q (uniform heat flux) is significant for liquid metals (Fig. 8.4); for $\mathrm{Pr} > 0.5$ the difference between Nu_T and Nu_q is negligible, and hence Eq. (8.5) is also valid for constant heat flux boundary conditions.

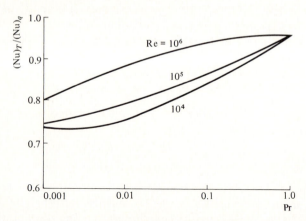

Fig. 8.4 Comparison of uniform wall temperature and uniform heat flux in tubes. (From R. A. Seban and T. T. Shimazaki, *Trans. ASME* **73**, 803 (1951).)

Example 8.1 A 5-ton heat of steel must be decarburized from 0.40 to 0.20% C in 5 min. A convenient method for accomplishing this is to blow oxygen through a submerged lance. A low carbon steel pipe of $\frac{1}{2}$-in. I.D. is used as the lance, despite the fact that the end of the pipe gradually melts. If we estimate the portion of the lance within the furnace to be at 2680°F, calculate the temperature at which the oxygen enters the liquid steel. Neglect the pressure drop through the pipe, i.e., assume that the pressure in the pipe equals 1 atm plus approximately the equivalent of 1 ft of liquid steel, or 1.20 atms.

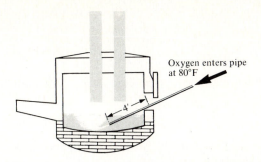

Oxygen enters pipe at 80°F

Solution. The oxygen requirement, for $C + \frac{1}{2}O_2(g) \rightarrow CO(g)$, is

$$\bar{V}\rho = \frac{(10,000)(0.20)\text{lb C}}{(100)(5)\,\text{min}}\left|\frac{1\,\text{lb mol C}}{12\,\text{lb C}}\right|\frac{\frac{1}{2}\,\text{lb mol O}_2}{1\,\text{lb mol C}}\left|\frac{32\,\text{lb O}_2}{1\,\text{lb mol O}_2}\right|\frac{60\,\text{min}}{1\,\text{hr}}\left|\frac{144}{\pi(\frac{1}{4})^2\,\text{ft}^2}\right.$$

$$= 235,000\,\frac{\text{lb O}_2}{\text{hr ft}^2}.$$

The oxygen enters the tube at 80°F, and as it proceeds through the pipe its temperature rises. Consequently, the properties of the oxygen vary with distance down the tube. A commonly employed estimation is to evaluate the properties of the fluid at its average temperature along the heated pipe. The heat-transfer coefficient corresponding to these properties is then calculated, assumed to be constant along the pipe, and used to determine the heat transferred to the gas.

First, arbitrarily assume that the gas leaves the pipe at 1020°F. Then, take the average T_m as $(1020 + 80)°\text{F}/2 = 550°\text{F}$ for the last 4 ft of pipe. At 550°F, the properties of O_2 are

$$\eta = 0.0792\,\text{lb}_m/\text{ft hr},$$
$$k = 0.0270\,\text{Btu/hr-ft °F},$$
$$C_p = 0.235\,\text{Btu/lb}_m\,°\text{F},$$
$$\text{Pr} = 0.690.$$

At 2680°F, $\eta = 0.166\,\text{lb}_m/\text{ft-hr}$ (estimated value). Now calculate Re_m:

$$\text{Re}_m = \frac{D\bar{V}\rho}{\eta_m} = \frac{(1/24)(235,000)}{(0.0792)} = 123,800.$$

Thus Eq. (8.5) applies, so that

$$\text{Nu}_m = (0.026)\,\text{Re}_m^{0.8}\,\text{Pr}_m^{1/3}\left(\frac{\eta_m}{\eta_0}\right)^{0.14}$$

$$= (0.026)(123,800)^{0.8}(0.690)^{1/3}\left(\frac{0.0792}{0.166}\right)^{0.14}$$

$$= 241.$$

Therefore

$$h = \frac{k_m}{D} \mathrm{Nu}_m = \left(\frac{0.0270}{1/24}\right)(241) = 156 \text{ Btu/hr-ft}^2 \text{ °F.}$$

An energy balance applied to the fluid over a differential length of pipe, dx, can be written

$$WC_p \, dT_m = h(T_0 - T_m)(\pi D \, dx),$$

where W = mass flow rate, lb_m/hr, and T_0 = temperature of pipe wall, °F. Integrating for a length of pipe L, with h, C_p, and T_0 taken as constants, yields

$$h \int_0^L dx = -\frac{WC_p}{\pi D} \int_{\Delta T_0}^{\Delta T_L} d \ln (T_m - T_0),$$

or

$$\ln \frac{\Delta T_L}{\Delta T_0} = -\frac{hL\pi D}{WC_p}.$$

Substituting numerical values yields

$$\ln \frac{\Delta T_L}{(2680 - 80)} = -\frac{(156)(4)(\pi)(1/24)}{(320)(0.235)},$$

from which

$$\Delta T_L = 611 \text{°F} \quad \text{or} \quad T_m = 2069 \text{°F} \quad \text{(at } x = L = 4 \text{ ft).}$$

Recalling our initial guess of 1020°F, we see that in terms of estimating properties we have not selected good values.

As a second guess, assume that the gas at the end of the pipe is at 2020°F. Then, we take the average T_m as $(2020 + 80) \text{°F}/2 = 1050 \text{°F}$. At 1050°F, the properties of O_2 are

$$\eta = 0.102 \text{ lb}_m/\text{hr-ft},$$
$$k = 0.0360 \text{ Btu/hr-ft °F},$$
$$C_p = 0.277 \text{ Btu/hr-ft °F},$$
$$\mathrm{Pr} = 0.785.$$

$$\mathrm{Re}_m = \frac{D\bar{V}\rho}{\eta_m} = \frac{(1/24)(235,000)}{0.102} = 96,000.$$

Then, using Eq. (8.5) again, we have

$$\mathrm{Nu}_m = (0.026)(96,000)^{0.8}(0.785)^{1/3}\left(\frac{0.102}{0.166}\right)^{0.14}$$

$$= 243,$$

so that

$$h = \frac{0.0360}{(1/24)}(243) = 210 \text{ Btu/hr-ft}^2 \text{ °F},$$

$$\ln \frac{\Delta T_L}{(2680 - 80)} = -\frac{(210)(4)(\pi)(1/24)}{(320)(0.277)},$$

$$\Delta T_L = 752°F \quad \text{or} \quad T_m = 1928°F \quad (\text{at } x = L = 4 \text{ ft}).$$

We could proceed to repeat the process again using an average T_m of $(1928 + 80)°F/2 = 1004°F$. This differs by only 46°F from 1050°F used for the second approximation. This difference would not alter the properties sufficiently to merit further refinement.

The conclusion that the gas temperature is lower than the metal temperature when it exits from the lance becomes significant when the bubble size is calculated in order to obtain the surface area and residence time for calculations of mass transfer. Since the gas is not fully expanded when it exits from the pipe, the bubble size, calculated on the basis of the pipe size, is smaller than the ultimately expanded bubble size, and the rate of rise is faster than anticipated on the basis of the pipe size.

8.2 HEAT-TRANSFER COEFFICIENTS FOR FORCED CONVECTION PAST SUBMERGED OBJECTS

In the following correlations, the heat-transfer coefficient h is defined for the total surface area of the submerged object, and the defining fluid temperature is that far removed from the surface. We evaluate all the properties, however, at the *film temperature*, $T_f = (T_0 + T_\infty)/2$.

In Fig. 8.5, $j_H = \text{Nu}_f \, \text{Re}_f^{-1} \, \text{Pr}_f^{-1/3}$ is plotted versus Re for a long cylinder perpendicular to the fluid. The figure also shows a plot of $f/2$ versus Re to illustrate that $j_H < f/2$, which is usually the case in flows with curved streamlines. Colburn's analogy breaks down here because of the form drag which has no counterpart in heat transfer.

These results agree closely with McAdams' correlation[7] which is based on experiments with water, oils, and air. Over the range $1 < \text{Re}_f < 10^3$, McAdams specifies

$$\text{Nu}_f \, \text{Pr}_f^{-0.3} = 0.35 + 56 \, \text{Re}_f^{0.52}. \tag{8.8}$$

For higher Reynolds numbers, $10^3 < \text{Re} < 5 \times 10^4$, data collected for air are correlated by

$$\text{Nu}_f \, \text{Pr}_f^{-0.3} = 0.26 \, \text{Re}_f^{0.60}. \tag{8.9}$$

---

[7] W. H. McAdams, *Heat Transmission*, third edition, McGraw-Hill, New York, 1954, page 268.

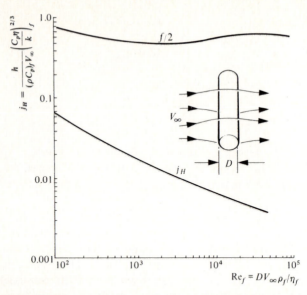

Fig. 8.5 Heat and momentum transfer for a cylinder perpendicularly oriented to flow. (j_H from T. K. Sherwood and R. L. Pigford, *Absorption and Extraction*, second edition, McGraw-Hill, New York, 1952, page 70; f from F. Eisner, see Fig. 3.9.)

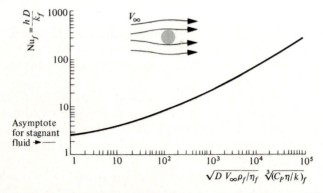

Fig. 8.6 Heat transfer from a sphere to a flowing fluid. (From W. E. Ranz and W. R. Marshall, Jr., *Chem. Eng. Prog.* **48**, 141–146, 173–180 (1952).)

Figure 8.6 gives Nu_f as a function of Re_f and Pr_f for the flow past a sphere. The relationship plotted is

$$hD/k_f = 2.0 + 0.60(DV_\infty \rho_f/\eta_f)^{1/2}(C_p\eta/k)_f^{1/3}. \tag{8.10}$$

This equation predicts that the magnitude of Nu should approach 2 as Re gets smaller and approaches zero; we can calculate this result for pure conduction, from a sphere at a uniform temperature in an infinite stagnant medium.

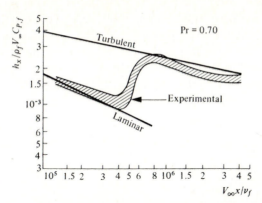

Fig. 8.7 Heat-transfer coefficient for flow over flat plate with turbulent boundary layer and uniform heat flux. Experimental data collected for air, and compared to the analysis for turbulent flow (Eq. 8.12), and laminar flow (Eq. 7.27). (From R. A. Seban and D. L. Doughty, *Trans. ASME* **78**, 217 (1956).)

Figure 8.7 shows results for the flow parallel to an isothermal semi-infinite flat plate. For this flow system, the Colburn analogy holds very well because there is no form drag; therefore, the local heat-transfer coefficient is related to the friction factor by

$$\frac{h_x}{C_p \rho V_\infty} \Pr^{2/3} = \frac{f_x}{2}. \tag{8.11}$$

For turbulent flow (from Eq. 3.28), $f_x = 0.0592 \, (\mathrm{Re}_x)^{-1/5}$; then Eq. (8.11) is

$$\frac{h_x}{C_p \rho V_\infty} \Pr^{2/3} = 0.0296 \, \mathrm{Re}_x^{-1/5}, \tag{8.12}$$

or

$$\frac{h_x x}{k} = \mathrm{Nu}_x = 0.0296 \, \mathrm{Re}^{0.8} \, \Pr^{1/3}. \tag{8.13}$$

The average coefficient h between $x = 0$ and $x = L$ is

$$hL/k = \mathrm{Nu}_L = 0.037 \, \mathrm{Re}_L^{0.8} \, \Pr^{1/3}. \tag{8.14}$$

(Here the initial part of the plate where laminar flow could exist has been ignored.)

Experimental data agree with Eq. (8.13) and with the laminar boundary layer result (Eq. 7.27). Note that Eqs. (8.11)–(8.14) are valid for $\Pr > 0.5$, but not for liquid metals; the equations are valid for either uniform wall temperature or uniform heat flux.

Example 8.2 A hot-wire anemometer is a device which indirectly measures the velocity of a moving fluid. It is usually a platinum wire which is heated electrically and positioned normal to the motion of the fluid. At a steady power input to the wire, its temperature reaches a steady value which can be related to the fluid velocity.

The temperature of the wire is determined by measuring current, voltage drop, hence electrical resistance, and by knowing how the resistance varies with the temperature. Thus, both the heat flux—from the wire to the fluid—and the wire temperature are measured electrically. Given air at 60°F flowing normal to the wire (0.01 ft in diameter) at 140°F, with a heat flux of 2000 Btu/hr-ft², determine the velocity of the air in ft/sec.

Solution. Assume that Eq. (8.9) applies:

$$\mathrm{Nu}_f \, \mathrm{Pr}_f^{-0.3} = 0.26 \, \mathrm{Re}_f^{0.60},$$

$$h = \frac{q_0}{\Delta T} = \frac{2000}{(140 - 60)} = 25 \; \text{Btu/hr-ft}^2 \, °\text{F}.$$

Use the properties of air at $T_f = 100°\text{F}$.

$$\rho = 0.0709 \; \text{lb/ft}^3,$$
$$C_p = 0.241 \; \text{Btu/lb} \, °\text{F},$$
$$k = 0.016 \; \text{Btu/hr-ft} \, °\text{F},$$
$$\eta = 0.046 \; \text{lb/hr-ft},$$

$$\mathrm{Nu}_f = \frac{hD}{k_f} = \frac{(25)(0.01)}{(0.016)} = 15.6,$$

$$\mathrm{Pr}_f = \left(\frac{\eta C_p}{k}\right)_f = \frac{(0.046)(0.241)}{(0.016)} = 0.694,$$

$$\mathrm{Re}_f^{0.60} = \frac{1}{0.26}(15.6)(0.694)^{-0.3} = 67.4$$

$$\mathrm{Re}_f = 1120.$$

Thus Eq. (8.9) applies as assumed, and

$$V_\infty = \left(\frac{\eta_f}{\rho_f}\right)\left(\frac{\mathrm{Re}_f}{D}\right) = \left(\frac{0.046}{0.0709}\right)\left(\frac{1120}{0.01}\right) = 72{,}600 \; \text{ft/hr},$$

or

$$V_\infty = 20.2 \; \text{ft/sec}.$$

8.3 HEAT-TRANSFER COEFFICIENTS FOR NATURAL CONVECTION

In Chapter 7, the discussion of natural convection in laminar flow led to a dimensionless equation of the form

$$\mathrm{Nu} = \mathrm{Nu} \, (\mathrm{Gr}, \mathrm{Pr}).$$

This corresponds to the results of dimensionless analysis in this chapter if the Reynolds number, the important group for forced convection, is replaced by the Grashof number, the important group for natural convection.

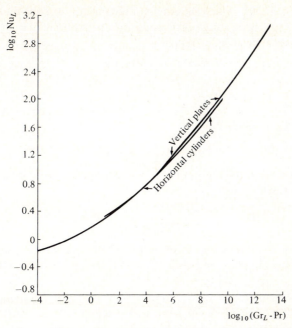

Fig. 8.8 Heat-transfer coefficients for natural convection. (From W. H. McAdams, *ibid.*, pages 173–176.)

Figure 8.8 shows the results of correlating experimental data for free convection from vertical plates and horizontal cylinders to gases and liquids with Prandtl numbers ranging from about 0.5 to 10. The effect of the variation of properties with temperature is included by evaluating properties at $T_f = \frac{1}{2}(T_0 + T_\infty)$. The dimensionless groups are then defined as

$$\mathrm{Nu}_L \equiv \frac{hL}{k_f},$$

$$\mathrm{Pr} \equiv \frac{v_f}{\alpha_f} = \frac{C_p \eta_f}{k_f},$$

$$\mathrm{Gr}_L \equiv \frac{L^3 \rho_f^2 g \beta_f (T_0 - T_\infty)}{\eta_f^2}.$$

The characteristic dimension L is the plate length. For cylinders, the characteristic dimension is one-half the circumference, that is, $\pi D/2$; by defining L as such, note how close the curves for the two different correlations lie. In fact, very often no distinction is made between these two cases, and a third case, namely, that of vertical cylinders with characteristic dimension taken as length, is included in the correlations.

In Chapter 7, Ostrach's analysis of laminar natural convection past vertical plates led to Eq. (7.59). This relationship holds in the laminar flow region $10^4 <$ Gr $\cdot$ Pr $< 10^9$ for a wide range of Prandtl numbers $(0.00835 \leq$ Pr $\leq 1000)$. For the sake of completeness let us repeat it:

$$\frac{\mathrm{Nu}_L}{\sqrt[4]{\mathrm{Gr}_L/4}} = \frac{0.902\,\mathrm{Pr}^{1/2}}{(0.861 + \mathrm{Pr})^{1/4}}. \tag{7.59}$$

In the range of $0.6 <$ Pr < 10, the data can be represented by a much simpler equation (still for laminar flow):*

$$\mathrm{Nu}_L = 0.56\,(\mathrm{Gr}_L \cdot \mathrm{Pr})^{1/4}. \tag{8.15}$$

For liquid metals, the data in the laminar range for horizontal cylinders have been correlated by[8]

$$\mathrm{Nu}_D = 0.53\left(\frac{\mathrm{Pr}}{0.952 + \mathrm{Pr}}\right)^{1/4}(\mathrm{Gr}_D \cdot \mathrm{Pr})^{1/4}. \tag{8.16}$$

The expression is of the same form as Eq. (7.59), and also closely corresponds numerically if the characteristic dimension D (diameter) is replaced by L taken as $\pi D/2$.

In the turbulent range of $10^9 <$ Gr $\cdot$ Pr $< 10^{12}$, the following equation has been proposed as possibly valid for a wide range of Pr for vertical plates[9] and horizontal cylinders with L replaced by $\pi D/2$:

$$\mathrm{Nu}_L = 0.0246\,\mathrm{Gr}_L^{2/5}\,\mathrm{Pr}^{7/15}\,(1 + 0.494\,\mathrm{Pr}^{2/3})^{-2/5}. \tag{8.17}$$

We use a much simpler form of Eq. (8.17) for $0.6 <$ Pr < 10:

$$\mathrm{Nu}_L = 0.13\,(\mathrm{Gr}_L \cdot \mathrm{Pr})^{1/3}. \tag{8.18}$$

For a single *sphere* of diameter D in a body of fluid the relationship

$$\mathrm{Nu}_D = 2 + 0.060\,\mathrm{Gr}_D^{1/4} \cdot \mathrm{Pr}^{1/3} \tag{8.19}$$

agrees with available data for $\mathrm{Gr}_D^{1/4}\,\mathrm{Pr}^{1/3} < 200$. In most instances, if the fluid is a gas, this restriction means that the equation applies to small particles. Note that the relationship yields $\mathrm{Nu}_D = 2$ when the fluid is motionless, as in Problem 7.6 where free convection is neglected.

Example 8.3 Write an expression to relate the growth rate of a single spherical nucleus of solid as it develops in a pure supercooled liquid. Express the result in terms of dR/dt as a function of supercooling ΔT and the properties of the metal. The thermal profile is depicted below; the temperature of the solid may be assumed as the freezing point T_M.

---

* To make a distinction between plates and cylinders in Fig. 8.8, the coefficient 0.56 in Eq. (8.15) should be replaced by 0.59 or 0.53, respectively.

[8] *Reactor Handbook 2*; *Engineering*, AECD-3646 (May, 1955), page 283.

[9] W. M. Rohsenow and H. Y. Choi, *ibid.*, page 204.

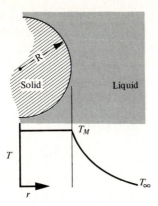

Solution. The rate at which latent heat of fusion evolves is equal to the rate at which heat flows from the interface into the bulk liquid

$$4\pi\rho_s H_f R^2 \frac{dR}{dt} = 4\pi R^2 h\,\Delta T,$$

← growth rate of sphere

where ρ_s = density of the solid, H_f = latent heat of fusion, and $\Delta T = T_M - T_\infty$ = amount of undercooling;

$$\frac{dR}{dt} = \frac{h}{\rho_s H_f}\,\Delta T. \qquad\qquad \text{see } 7.31$$

From Eq. (8.19),

$$\frac{hD}{k} = 2 + 0.060\left(\frac{D^3\rho^2 g\beta\,\Delta T}{\eta^2}\right)^{1/4}\left(\frac{\eta C_p}{k}\right)^{1/3} \cdot \frac{k}{0}$$

When this is substituted into the expression for dR/dt, we obtain

$$\frac{dR}{dt} = \frac{k\,\Delta T}{\rho_s H_f}\left[\frac{1}{R} + 0.0504\left(\frac{\rho^2 g\beta\,\Delta T}{R\eta^2}\right)^{1/4}\left(\frac{\eta C_p}{k}\right)^{1/3}\right].$$

Students of phase transformations are invited to use this as a basis for further discussion and/or refinement.

Fig. 8.9 The convection pattern over a horizontal surface which is warmer than the fluid.

In the previous discussion of this section, heat transfer took place in the presence of convection mostly in the vertical direction. The convection behavior in the vicinity of horizontal surfaces is quite different, as shown in Fig. 8.9, and the correlations of Eqs. (8.15)–(8.19) do not apply. McAdams[10] recommends to use the following equations for some situations of natural convection from *horizontal* surfaces:

1. For a square plate with a surface warmer than the fluid facing upward, or cooler surface facing downward:

$$\mathrm{Nu}_L = 0.14(\mathrm{Gr}_L \cdot \mathrm{Pr})^{1/3}; \tag{8.20}$$

in the turbulent range, $2 \times 10^7 \lesssim \mathrm{Gr}_L \lesssim 3 \times 10^{10}$, and

$$\mathrm{Nu}_L = 0.54(\mathrm{Gr}_L \cdot \mathrm{Pr})^{1/4}; \tag{8.21}$$

in the laminar range, $10^5 \lesssim \mathrm{Gr}_L \lesssim 2 \times 10^7$.

2. For a square plate with a surface warmer than the fluid and facing downward, or cooler surface facing upward:

$$\mathrm{Nu}_L = 0.27(\mathrm{Gr}_L \cdot \mathrm{Pr})^{1/4}. \tag{8.22}$$

In Eqs. (8.20)–(8.22), L is the length of the side of the square. As approximations, we can also apply the equations to horizontal circular disks with L defined as $0.9\,D$ (D being the disk diameter).

8.4 QUENCHING HEAT-TRANSFER COEFFICIENTS

In the next chapter, we present analytical expressions which describe the temperature during heating or cooling as a function of time and position within a solid. The most difficult variable to ascribe to such unsteady-state situations is probably the heat-transfer coefficient governing the energy transport between the solid's surface and the surroundings. The methods of estimating h for some convection systems, which were presented in previous sections of this chapter, can be applied satisfactorily to certain problems of unsteady-state heat flow. For the most part, however, correlations of h-values applicable to quenching operations have not been obtained because of the complexity of the convection systems.

Despite the importance of quenching operations in the heat treatment of metal parts, metallurgists have not considered nearly enough of the quantitative aspects of quenching heat transfer. The most quantitative work related to quenching has actually been developed in the field of boiling heat transfer. Historically, the primary purpose of boiling heat transfer has been simply the conversion of liquid into vapor, and perhaps this is the reason why this problem has not been studied enough by metallurgical engineers. On the other hand, engineers, who are involved with the development of nuclear reactors, rocket nozzles, and spacecraft, have

---

[10] W. H. McAdams, *ibid.*, page 180.

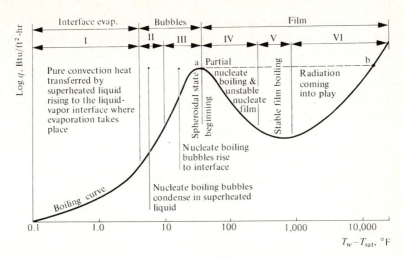

Fig. 8.10 Characteristic boiling curve. (From W. M. Rohsenow, *Developments in Heat Transfer*, MIT Press, Cambridge, Massachusetts, 1964, Chapter 8.)

conducted considerable research into the boiling process, which has resulted in a number of correlations for boiling heat transfer.

Although boiling is a familiar phenomenon, the next section shows that from the point of view of energy transport, it is a complicated process. Figure 8.10 illustrates the complexities of boiling in which several regimes exist. If a metal surface capable of being heated is submerged in a pool of liquid at its saturation temperature, T_{sat}, the following events take place. As the surface temperature T_w is raised above T_{sat}, convection currents circulate the superheated liquid. Heat transfer with convection takes place, and the correlations discussed for natural convection in Section 8.3 apply. However, the liquid cannot be indefinitely superheated so that further increases of $(T_w - T_{sat})$ above approximately 4°F are accompanied by vapor bubbles forming at preferred sites on the surface. This is regime II in which most of the bubbles do not exceed a certain size necessary for their escape. In regime III, $(T_w - T_{sat})$ is large enough so that larger and more stable bubbles grow, and at the same time more bubbles form because more nucleation sites become active on the solid's surface. This mechanism, which is called nucleate boiling, can generate very high heat fluxes. As the bubbles rapidly form and detach themselves from the surface, fresh liquid rushes to the former bubble site, and the very rapid process of bubble formation and detachment repeats itself over and over again. The net result is that all the bubble sites act as micro-pumps and create a large degree of convection at the surface.

If $(T_w - T_{sat})$ is raised to even higher values, for example, 70°F for water, then the nucleation sites on the surface become so numerous that the bubbles actually coalesce, and tend to form a vapor blanketing layer. Beyond this point, sometimes referred to as the *burnout* temperature, the heat flux decreases with increasing

$(T_w - T_{sat})$ because of the vapor's insulating nature. Regime IV encompasses this effect wherein an unstable vapor film forms but collapses and re-forms rapidly. For values of $(T_w - T_{sat})$ greater than about 400°F (for water), a stable film exists (regime V). This is called *film boiling*, with which rather low heat fluxes are associated (contrasted to nucleate boiling); for values of $(T_w - T_{sat})$ greater than 1200°F, radiation across the vapor blanket becomes important, and the rate of heat transfer increases accordingly.

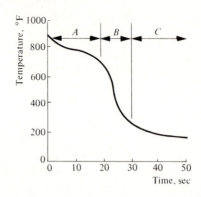

Fig. 8.11 Cooling curve of a metal part quenched in water.

Returning to quenching, we see that Fig. 8.11 shows a typical cooling curve of a metal part plunged into water. Note the three distinct stages during the cooling process. Stage A, called the vapor cooling stage, occurs immediately upon immersing the part in the water. Liquid vaporizes adjacent to the surface and forms a continuous vapor film. Cooling during this stage proceeds by film boiling and, if the part is at a sufficiently high temperature (i.e., steel), also by radiation. The second stage, B, is often denoted by metallurgists as the "vapor transport stage." This term is not really appropriate because during this stage there is a transition from partial nucleate boiling and unstable film boiling to entirely nucleate boiling. During stage C, referred to as the liquid cooling stage, there is no vapor formation, and cooling occurs with natural convection. Thus we see that as the metal cools from its initial elevated temperature to the fluid's temperature, it passes through all the regimes of the boiling curve (Fig. 8.10).

For these and other reasons no comprehensive body of engineering correlations has evolved and been applied to quenching problems. Rather most of the literature gives some average values of h which apply to specific cases.

For example, we can find the values of a parameter H, known as the *severity of quench*, and defined as h/k, which represent quenching practices for steel. Table 8.1 gives some typical values of H.

Data from this table should be applied only to conditions for which the values are derived. For example, h can depend strongly on the temperature of the liquid media alone. The values shown above are for quenchants at room temperature.

Table 8.1 Severity of quench values and heat-transfer coefficients for steel*

Quench media	H-value, ft^{-1}	h, Btu/hr-ft^2 °F
Oil, no agitation	2.4	48
Oil, moderate agitation	4.2	84
Oil, good agitation	6.0	120
Oil, violent agitation	8.4	170
Water, no agitation	12.0	240
Water, strong agitation	18.0	360
Brine, no agitation	24.0	480
Brine, violent agitation	60.0	1200

* Values of H from M. A. Grossman, *Elements of Hardenability*, *ASM*, Cleveland, 1952; values of h calculated using $k = 20$ Btu/hr-ft °F.

In addition, during quenching h varies with the temperature of the solid itself. The values in Table 8.1 are derived specifically to predict the time to cool steel from its austenizing temperature to a temperature midway between the austenizing and original quench temperatures. Therefore all that we can say about the above h-values is that they are average and apply during cooling from 1500°F to about 780°F. It is left to the reader to comprehend the "political" terms—moderate agitation, good agitation, and violent agitation—given in Table 8.1.

For precipitation-hardenable aluminum alloys, an important step in their heat treatment is the cooling rate during quenching from the homogenization temperature. If the cooling rate is too low through a critical range of about 800–500°F, then precipitation occurs, and subsequent ageing cannot be controlled. On the other hand, if the cooling rate during quenching is too great, then untolerable distortion in many parts occurs. In practice, therefore, heat treaters usually quench aluminum alloys in heated water to slow cooling and minimize distortion. In recent years, some heat treaters have substituted water–polyalkylene glycol (PAG) solutions for heated water. These solutions lower the cooling rates, and may be substituted for heated quench water; in addition, more uniform cooling is attained, and distortion can be reduced.

Figure 8.12 gives some values of h for water–PAG solutions based on experimentally measured cooling rates through the range 800–500°F. When solutions of PAG and water are used, h is dependent on composition. In these instances, an immiscible PAG-rich liquid phase forms at about 165°F (the exact temperature depends on the composition). Thus, near the hot solid, the liquid separates into two immiscible liquid phases, and the surface becomes coated with the very viscous PAG-rich phase. Therefore, the rate of heat transfer is not controlled by a water vapor film whose thickness is dependent on bulk water temperature to a large extent, but rather by the layer of the PAG-rich phase whose thickness and properties are presumably more predictable and not so much dependent on the bulk

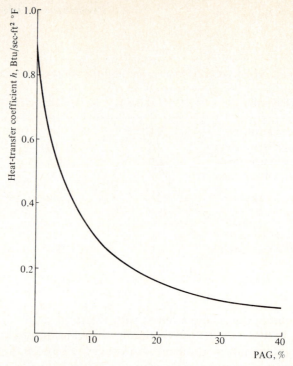

Fig. 8.12 Heat-transfer coefficients for cooling aluminum through the critical range of 800–500°F in water and water–polyalkylene glycol (PAG) solutions. (Reproduced by permission of Progressive Metallurgical Industries, Gardena, California.)

liquid temperature. Hence, more uniform cooling and reduction of distortion are achieved.

Stolz[11] has devised a special numerical technique for obtaining heat-transfer coefficients during quenching from measurements of interior temperatures of a solid sphere. By means of this technique heat-transfer coefficients for quenching oils were evaluated as a function of the surface temperature of the solid. Figure 8.13 shows these data for oils designated as slow, intermediate, and fast. The figure also shows that between 1600°F and 1150°F the heat-transfer coefficients of all three types of oils are similar; over this range all three oils form a continuous vapor film on the solid's surface, and h is only about 100 Btu/hr-ft² °F. From 700° to 130°F, similar heat-transfer coefficients are found for all three oils. The main differences between these oils lie in the regions where the heat-transfer coefficients are greater than 100 Btu/hr-ft² °F, showing that the stable vapor film breaks down first (at a higher surface temperature) for the fast oil, and last for the slow oil.

The results for water and aqueous solutions of 1% NaOH and 5% NaOH are given in Fig. 8.14; all three solutions have a bulk temperature of 110°F. For these aqueous solutions, h is initially between 300 and 900 Btu/hr-ft² °F. In a quench, these initial values of h represent the rapid vaporization of water as the solid plunges into the water. In terms of time, this period is only a fraction of a second,

---

[11] G. Stolz, Jr., *J. Heat Transfer* **82**, 20–26 (1960).

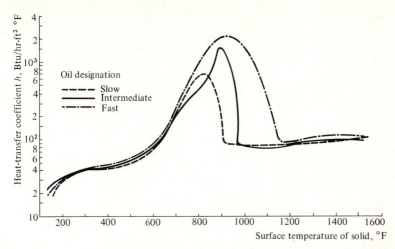

Fig. 8.13 Heat-transfer coefficients in quenching oils. (From G. Stolz, V. Paschkis, C. F. Bonilla, and G. Acevedo, *JISI* **193**, 116–123 (1959).)

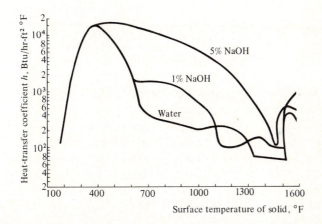

Fig. 8.14 Heat-transfer coefficients in aqueous quenching media. (From V. Paschkis and G. Stolz, Heat and Mass Flow Analyzer Laboratory, Columbia University.)

after which h drops to values indicative of the vapor film stage, but this is short-lived, especially for the 5% NaOH solution. The marked increase in heat transfer in the presence of NaOH over the range 1400–400°F is due to exploding salt crystals that make the vapor film unstable. All three solutions, however, reach the same peak value of h (1.5×10^4 Btu/hr-ft² °F) at approximately 400°F.

In order to quench very long metal shapes, such as the strip from a hot strip mill or a continuously cast slab, we often use a multiplicity of water sprays to achieve cooling. Figure 8.15 shows cooling data with a fan-type spray of water at 70°F which impinges on a hot horizontal surface from above. There is a sharp

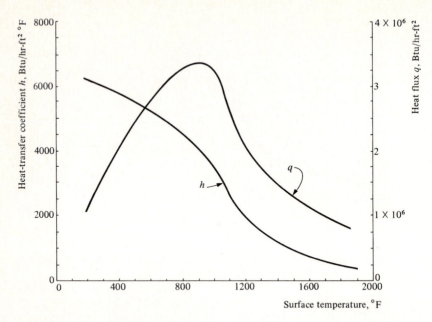

Fig. 8.15 Heat-flux and heat-transfer coefficients for cooling a horizontal surface from above with a fan-type spray of water at 70°F. (Adapted from P. M. Auman, D. K. Griffiths, and D. R. Hill, *Iron and Steel Engineer*, Sept., 1967.)

increase in the heat flux at about 1200–1000°F, which represents a transition from boiling with a vapor film to nucleate boiling.

When the vapor film is present during spray cooling, the rate of heat transfer depends on the degree to which the steam film can be broken down by the impinging droplets. With sprays, the droplets approaching the hot metal surface encounter the vapor film, and their success in penetrating this film depends on their kinetic energy. If the sprays are placed closer to the metal surface or the water velocity is increased, thereby increasing the droplets' kinetic energy, then the probability that the droplets will penetrate the film increases and so does the rate of heat removal.

With this in mind, it has been argued that a continuous stream of water, because of its high kinetic energy associated with its large mass, should penetrate the vapor film and bring about an increase in heat-transfer rates.[12] For this reason, large nozzles in combination with low water pressure have been utilized to produce a falling stream, or jet, of water which does not break into droplets. For such a jet, an increase in the flow rate of water increases the rate of heat transfer up to a point. If the jet-flow rate passes a critical value, then the falling jet breaks up into less effective droplets, and the rate of heat transfer decreases.

In addition to water sprays, oil sprays have been used for quenching. It was

---

[12] E. R. Morgan, T. E. Dancy, and M. Korchynsky, *J. of Metals* **17**, No. 8, 829–831 (1965).

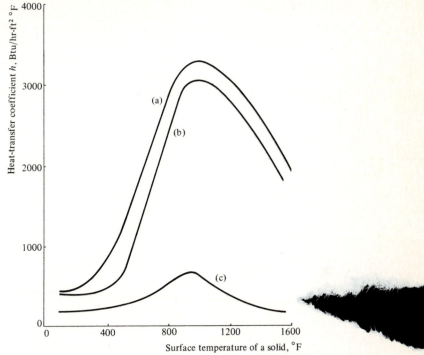

Fig. 8.16 Heat-transfer coefficients for oil-spray quenching and immersion in oil. (a) Very strong spray; flow rate > 17.2 gal/sec-ft^2. (b) Strong spray; flow rate $= 14.7$ gal/sec-ft^2. (c) Still oil. (From Zimin, *ibid.*)

found in a study,[13] that the cooling rate due to properly applied oil sprays was three to five times the cooling rate by immersing in oil in the conventional manner. Figure 8.16 shows some of these data.

8.5 BOILING HEAT TRANSFER

In the past, metallurgists have made hardly any effort to apply boiling heat-transfer theory and correlations to quenching processes. However, they should be aware of this field of technical literature because many of the correlations could be directly applied to quenching, or serve as a basis for a better understanding of the role of properties of fluids used as quenchants.

Because boiling heat transfer is so complex, it cannot be expected that a single equation could possibly correlate data over all the regimes shown in Fig. 8.10. For this reason we present, under separate headings, some available correlations applying to individual boiling regimes.

[13] N. V. Zimin, UDC 621.784.06, pages 854–858, Plenum Press, New York, 1968. Translated from *Metallovedenie i Termicheskaya Obrabotka Metallov*, No. 11, 62–68 (Nov., 1967).

8.5.1 Film boiling

An equation often referred to for calculating the heat-transfer coefficient in film boiling is that of Bromley[14] which applies to horizontal cylinders and vertical plates submerged in liquids at their saturation temperature. The momentum and energy transport involved in the continuous vapor film is a good deal simpler than the transport in the other boiling regimes. For this reason, a simple analytical model has been found as a good basis for analyzing data.

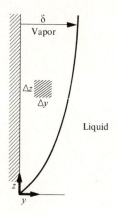

Fig. 8.17 Vapor film model.

Consider the stable film of vapor in laminar flow along the vertical wall in Fig. 8.17. For a steady incompressible vapor flow, a momentum balance applied to the unit element yields

$$\frac{d\tau_{yz}}{dy} = \frac{dP}{dz} + \rho g. \tag{8.23}$$

Since this vapor film is parallel to a liquid column, we can write the pressure term as $dP/dz = -g\rho_l$, where the subscript l here and within the remainder of this argument refers to liquid properties. With this substitution, Eq. (8.23) becomes

$$d\tau_{yz}/dy = g(\rho - \rho_l). \tag{8.24}$$

We solve Eq. (8.24) subject to the boundary condition that the liquid moves freely with the vapor at the vapor–liquid interface, that is,

$$\text{B.C. 1} \qquad \tau_{yz} = 0 \quad \text{at } y = \delta. \tag{8.25}$$

By solving Eq. (8.24) and applying Eq. (8.25), we arrive at an expression for the velocity gradient within the vapor film:

$$\tau_{yz} = -\eta \frac{\partial v_z}{\partial y} = g(\rho_l - \rho)(\delta - y). \tag{8.26}$$

---

[14] L. A. Bromley, *Chem. Eng. Prog.* **46**, No. 5, 221–226 (1950).

We use Eq. (8.26) to determine the velocity profile by subjecting it to the usual case of no slip at the wall, that is,

$$\text{B.C. 2} \qquad v_z = 0 \quad \text{at } y = 0. \tag{8.27}$$

When this is done, we obtain an expression for the velocity profile:

$$v_z = \frac{g(\rho_l - \rho)}{\eta}\left[\delta y - \frac{y^2}{2}\right]. \tag{8.28}$$

From Eq. (8.28), we determine the average velocity of the vapor

$$\bar{V} = \frac{1}{\delta}\int_0^\delta v_z \, dy = \frac{g(\rho_l - \rho)}{3\eta}\delta^2. \tag{8.29}$$

We can write a heat balance over the vapor film thickness δ for a length dz. For illustrative purposes we assume a linear temperature profile across the vapor film. The heat flux across the film is $(k/\delta)(T_w - T_{\text{sat}})$, and this energy goes into the enthalpy change of the liquid in vaporizing to gas and superheating to the average vapor temperature. Thus

$$\frac{k}{\delta}(T_w - T_{\text{sat}}) \, dz = \left[H_v + \frac{1}{\rho\bar{V}\delta}\int_0^\delta \rho v_z C_p(T - T_{\text{sat}}) \, dy\right] dw. \tag{8.30}$$

Here dw represents the mass of vapor formed in the element dz, and H_v is the heat of vaporization. When we substitute Eqs. (8.28) and (8.29) into Eq. (8.30), and carry out the integration, the result is

$$\frac{k}{\delta}(T_w - T_{\text{sat}}) \, dz = H_v\left[1 + \frac{3}{8}\frac{C_p(T_w - T_{\text{sat}})}{H_v}\right] dw. \tag{8.31}$$

The entire coefficient of dw on the right-hand side of Eq. (8.31) is sometimes referred to as the *effective heat of vaporization*, H_v'.

A more detailed analysis of the energy transport gives a somewhat different value of H_v' than that above and should be utilized for calculations. Bromley uses the following definition for the effective heat of vaporization:

$$H_v' = H_v\left[1 + \frac{0.4C_p(T_w - T_{\text{sat}})}{H_v}\right]^2. \tag{8.32}$$

With this in mind, and recognizing that $dw = \rho d(\delta\bar{V})$, the energy balance takes the form

$$\frac{k}{\delta}(T_w - T_{\text{sat}}) \, dz = H_v'\frac{g\rho(\rho_l - \rho)}{3\eta} \, d\delta^3,$$

or

$$\frac{\eta k(T_w - T_{\text{sat}})}{H_v' g\rho(\rho_l - \rho)} \, dz = \delta^3 \, d\delta. \tag{8.33}$$

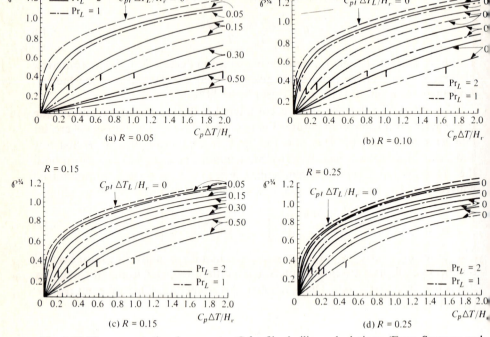

Fig. 8.18 The computational parameter $\mathscr{P}$ for film boiling calculations. (From Sparrow and Cess, *ibid.*)

can be tested at two extremes, specifically for the case of no subcooling and for the case of large subcooling. In the former case, $T_{sat} - T_\infty = 0$ and $\mathscr{P} = C_p(T_w - T_{sat})/H_v$. Substituting this value of $\mathscr{P}$ into Eq. (8.41), we write an expression for film boiling with no subcooling:

$$\mathrm{Nu}_l\left[\frac{16v^2\rho}{g(\rho_l - \rho)L^3}\right]^{1/4} = \frac{4}{3}\left[0.84 + \frac{H_v}{C_p(T_w - T_{sat})}\right]^{1/4}. \tag{8.44}$$

If we now assume that this result for vertical plates of length L is valid for horizontal cylinders with L taken as $\pi D/2$, then we can rearrange Eq. (8.44) (remembering that $\mathrm{Pr} \cong 1$, and hence $k/\eta \cong C_p$) as

$$h_D = 0.60\left[\frac{g\rho(\rho_l - \rho)k^3}{\eta D(T_w - T_{sat})}\right]^{1/4}\left[H_v\left\{1 + 0.84\frac{C_p(T_w - T_{sat})}{H_v}\right\}\right]^{1/4}. \tag{8.45}$$

If we define the effective heat of vaporization as

$$H_v' = H_v\left[1 + 0.84\frac{C_p(T_w - T_{sat})}{H_v}\right], \tag{8.46}$$

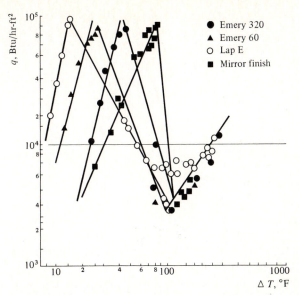

Fig. 8.19 Effect of roughness on boiling heat transfer; test results are for copper–pentane. (From P. Berenson, *J. Heat Transfer, ASME,* Aug. 1961, page 35.)

then, finally, the result obtained by Sparrow and Cess reduces to

$$h_D = 0.60\left[\frac{H'_v g\rho(\rho_l - \rho)k^3}{\eta D(T_w - T_{sat})}\right]^{1/4}.\tag{8.47}$$

This equation agrees almost exactly with Bromley's result—Eq. (8.38)—except that there is a slight difference in the values of H'_v, Eq. (8.32) versus Eq. (8.46). However, in most cases of practical interest $C_p(T_w - T_{sat})/H_v < 0.5$, and the two are almost exactly equal.

Sparrow and Cess examine their result for large subcooling. In Fig. 8.18, vertical markers appear on the curves; these markers indicate *loci*, the left of which actually give results that are exactly equal to the Nusselt–Grashof relationship for purely free convectional heat transfer next to a vertical surface. Thus, for physical conditions lying to the left of the vertical markers, the problem is actually one of natural convection, and Eq. (8.41) need not be used.

8.5.2 Nucleate boiling

In most industrial, but nonquenching, boiling applications, nucleate boiling occurs more frequently than film boiling. Therefore the literature on nucleate boiling is more voluminous than that on film boiling, and there are many correlations available that attempt to predict the rates of nucleate boiling heat transfer. However, the correlations often disagree with one another, probably because nucleate boiling is really an ill-defined process.

Since nucleate boiling involves the nucleation and growth of vapor bubbles that originate at nucleation sites on the metal surface, such as cavities, scratches,

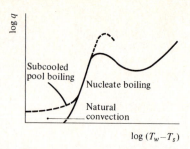

Fig. 8.20 Effect of subcooling on boiling heat transfer.

etc., it is rather complex. Some of the variables which affect the rate of heat transfer are:

1. *Surface condition.* Changing the surface by polishing causes the q versus $(T_w - T_{sat})$ curve to shift toward the right, as shown in Fig. 8.19. Surfaces that have not been freshly machined or polished often require a higher $(T_w - T_{sat})$ for the same q in the nucleate boiling regimes. This may be due to a decrease in cavity sizes resulting from mild oxidation, thus requiring more superheat for equivalent amount of bubble formation.

2. *Gases.* Gases dissolved in the liquid, which come out of solution at the surface well below the normal saturation temperature, agitate the liquid in the same way as vapor bubbles; thus, data show higher heat-transfer rates than those obtained for a degassed liquid.

One correlation for nucleate boiling relies on the supposition that the heat-transfer rate in boiling results from the increased convection of the liquid. With this in mind, Rohsenow[16] proceeds to correlate data by similarity with single-phase forced convection heat-transfer, assuming that

$$\text{Nu}_b = f(\text{Re}_b, \text{Pr}_l). \tag{8.48}$$

The dimensionless groups are defined as

$$\text{Nu}_b \equiv \frac{q_w D_b}{(T_w - T_{sat})k_l} = \frac{hD_b}{k_l}, \tag{8.49}$$

and

$$\text{Re}_b \equiv \frac{G_b D_b}{\eta_l}. \tag{8.50}$$

Here D_b is the diameter of the bubbles as they leave the heating surface, and G_b is the rate of mass of vapor (or bubbles) formed per unit area of heating surface.

---

[16] W. M. Rohsenow, *Developments in Heat Transfer*, Chapter 8, MIT, Cambridge, Massachusetts, 1964.

Table 8.2 Values of C_{sf} for Eq. (8.51)*

Surface–fluid combination	C_{sf}
Water–nickel	0.006
Water–platinum	0.013
Water–copper	0.013
Water–brass	0.006
CCl_4–copper	0.013
Benzene–chromium	0.010
n-pentane–chromium	0.015
Ethyl alcohol–chromium	0.0027
Isopropyl alcohol–copper	0.0025
35% K_2CO_3–copper	0.0054
50% K_2CO_3–copper	0.0027
n-butyl alcohol–copper	0.0030

* From Rohsenow, *ibid.*

We omit the details of Rohsenow's analysis here, but the final equation, based on the above dimensionless groups, which correlates experimental data, is

$$\frac{C_{pl}(T_w - T_{sat})}{H_v} = C_{sf}\left[\frac{q}{\eta_l H_v}\sqrt{\frac{\sigma}{g(\rho_l - \rho)}}\right]^{1/3} Pr_l^{1.7}. \tag{8.51}$$

Here σ is the liquid–vapor surface tension, and C_{sf} is a constant dependent on the nature of the surface s and the fluid f. Values of C_{sf} deduced from the experimental data are given in Table 8.2. This correlation applies when the bulk liquid is at saturation temperature, and is also valid if the liquid is subcooled. The subcooling does, however, shift the characteristic boiling curve in the nucleate boiling regime (Fig. 8.20).

8.5.3 Transition boiling

By using the results of Sections 8.5.1 and 8.5.2, we can describe the heat flux for film boiling and nucleate boiling. The transition between these two, depicted as regime IV in Fig. 8.10, may be described by evaluating the critical heat flux q_{max} at the burnout temperature, and the minimum heat flux q_{min} corresponding to the beginning of stable film boiling. Values of the heat flux in the transition regime lie on the straight line joining q_{max} and q_{min} on a log q versus log ΔT plot (Fig. 8.19).

The magnitude of q_{max} seems to be relatively insensitive to the nature of the surface finish, because the controlling physical phenomena in the boiling now take place at the vapor–liquid interface removed from the solid's surface. One description of what happens at the critical heat flux is attributed to Zuber[17] who used a

---

[17] N. Zuber, *Hydrodynamic Aspects of Boiling Heat Transfer*, U.S. Atomic Energy Commission, AECU-4439, June, 1959.

purely hydrodynamic argument. As the critical heat flux is approached, the amount of vapor entering the liquid as undulating columns increases, and thus reduces the available area for liquid flow toward the surface. Therefore, the relative velocity between the liquid and vapor increases to the point where the vapor–liquid interface becomes unstable and prevents the liquid from contacting the heating surface.

Zuber's hydrodynamic theory leads to the following expression for the maximum heat flux in pool boiling of saturated liquids:

$$\frac{q_{max}}{\rho H_v} = 0.18 \left[\frac{g\sigma(\rho_l - \rho)}{\rho^2} \right]^{1/4} \left[\frac{\rho_l}{\rho_l + \rho} \right]^{1/2}. \tag{8.52}$$

For boiling of subcooled liquids, the problem is to determine the energy transfer from the interface into the bulk liquid. In accordance with the hydro-dynamic aspect of the problem, Zuber reasons that because the vapor is periodically released from the vapor–liquid interface, the temperature distribution within the liquid is also periodically renewed by conduction of energy into the newly arrived subcooled liquid at the interface. The added energy term results in this expression for boiling of subcooled liquids at bulk temperature T_b:

$$q_{max-sub} = q_{max} + 0.696(T_{sat} - T_b)(k_l \rho_l C_{pl})^{1/2} \left[\frac{g(\rho_l - \rho)}{\sigma} \right]^{1/4} \left[\frac{\sigma(\rho_l - \rho)g}{\rho^2} \right]^{1/8}. \tag{8.53}$$

For the minimum heat flux, Zuber's hydrodynamic theory leads to an expression which has wide acceptance:

$$q_{min} = 0.128 \left[\frac{g\sigma(\rho_l - \rho)}{(\rho_l + \rho)^2} \right]^{1/4}. \tag{8.54}$$

The constant 0.128 is recommended by Kesselring et al.[18]

For boiling of subcooled liquids, if the added energy conduction into the liquid is determined for the hydrodynamic conditions corresponding to q_{min}, then one can develop the following equation:

$$q_{min-sub} = q_{min} + 0.775(T_{sat} - T_b)(k_l \rho_l C_{pl})^{1/2} \left[\frac{g(\rho_l - \rho)}{\sigma} \right]^{1/4} \left[\frac{\sigma(\rho_l - \rho)g}{\rho_l^2} \right]^{1/8}. \tag{8.55}$$

Equation (8.55) has not been tested experimentally, but is recommended for use pending verification.

Example 8.4 Estimate the initial heat-transfer coefficient that applies when a $\frac{1}{2}$-in. diam. steel bar at 1500°F is quenched in water at (a) 212°F, and (b) at 70°F. The bar is held in the horizontal position.

---

[18] R. C. Kesselring, P. H. Rosche, and S. G. Bankoff, *AIChE J.* **13**, 669–675 (July, 1967).

Solution

a) For water at 212°F, apply either Eq. (8.38) or (8.47). Using Eq. (8.38), one obtains

$$h_D = 0.62 \left[\frac{H_v' g\rho(\rho_l - \rho)k^3}{(T_w - T_{sat})\eta D} \right]^{1/4},$$

$$H_v' = H_v \left[1 + \frac{0.4 C_p(T_w - T_{sat})}{H_v} \right]^2.$$

("Steam Tables" are a convenient source for many properties of water and steam.)
At 1 atm and 212°F

$$H_v = 970.3 \text{ Btu/lb}, \qquad \rho_l = 59.8 \text{ lb/ft}^3.$$

For water vapor at 1 atm and an average film temperature of $(1500 + 212)°F/2 = 856°F$, the following properties apply

$$C_p = 0.502 \text{ Btu/lb °F},$$
$$\rho = 0.0188 \text{ lb/ft}^3,$$
$$k = 0.034 \text{ Btu/hr-ft °F},$$
$$\eta = 0.062 \text{ lb/hr-ft},$$

$$H_v' = 970.3 \left[1 + \frac{(0.4)(0.502)(1500-212)}{970.3} \right]^2,$$

$$= 1558 \text{ Btu/lb}.$$

Then

$$h_D = 0.62 \left[\frac{(1558)(32.2)(3600)^2(0.0188)(59.8 - 0.02)(0.034)^3}{(1500 - 212)(0.062)(1/24)} \right]^{1/4}$$

$$= 171 \text{ Btu/hr-ft}^2 \text{ °F}.$$

Now determine the effect of radiation with $\varepsilon_w = 0.78$.

$$h_r = \frac{\varepsilon_w \sigma [T_w^4 - T_{sat}^4]}{T_w - T_{sat}}$$

$$= \frac{(0.78)(0.1713)[19.6^4 - 6.72^4]}{(1500 - 212)}$$

$$= 15.3 \text{ Btu/hr-ft}^2 \text{ °F}.$$

Then we find the total heat-transfer coefficient by using Eq. (8.39):

$$h = h_D \left(\frac{h_D}{h} \right) + h_r = 171 \left(\frac{171}{h} \right) + 15.3$$

$$= 179 \text{ Btu/hr-ft}^2 \text{ °F}.$$

b) For water at 70°F, apply Eq. (8.41). First R, as given by Eq. (8.42), must be calculated with vapor properties evaluated at 856°F as before and liquid properties at $(212 + 70)°F/2 = 141°F$. At 141°F, the properties for liquid water are

$$\rho_l = 61.4 \text{ lb}_m/\text{ft}^3,$$
$$\eta_l = 1.21 \text{ lb}_m/\text{hr-ft},$$
$$\beta_l = 0.29 \times 10^{-3} °R^{-1},$$
$$C_{pl} = 1.00 \text{ Btu/lb}_m°F,$$
$$\text{Pr}_l = 2.96,$$
$$k_l = 0.409 \text{ Btu/hr-ft °F}.$$

$$R \equiv \left[\frac{(0.0188)(0.062)}{(61.4)(1.21)}\right]^{1/2}\left[\frac{61.4 - 0.02}{0.0188}\right]^{1/4}\left[\frac{1.00}{(0.29 \times 10^{-3})(970.3)}\right]^{1/4},$$

$$= 0.129.$$

$$\frac{C_{pl}\,\Delta T_l}{H_v} = \frac{(1.00)(142)}{(970.3)} = 0.146,$$

$$\frac{C_p\,\Delta T}{H_v} = \frac{(0.502)(1288)}{970.3} = 0.666.$$

For $R = 0.10$, Fig. 8.18b yields $\mathscr{P}^{1/4} \cong 0.81$; for $R = 0.15$, Fig. 8.18c yields $\mathscr{P}^{1/4} \cong 0.85$. Therefore, at $R = 0.129$, $\mathscr{P}^{1/4} \cong 0.83$, and $\mathscr{P} = 0.954$. Now substitute values in Eq. (8.41) with $L = \pi D/2$.

$$\text{Nu}_l = \tfrac{4}{3}[0.84 + 1.05]^{1/4}\left[\frac{(32.2)(3600)^2(61.4 - 0.02)(\pi/2)(1/24)}{(16)(3.30)^2(0.0188)}\right]^{1/4}$$

$$= 235.$$

$$h_D = \frac{k_l}{L}(235) = \frac{2k_l}{\pi D}(235) = \frac{(2)(0.409)(235)}{(\pi)(1/24)}$$

$$= 1470 \text{ Btu/hr-ft}^2 °F.$$

Then we find the total heat-transfer coefficient by applying Eq. (8.39) again, with $h_r = 15.3 \text{ Btu/hr-ft}^2 °F$:

$$h = 1470\left(\frac{1470}{h}\right) + 15$$

$$= 1477 \text{ Btu/hr-ft}^2 °F.$$

Note the large effect of subcooling; in this particular example, h changes by almost one order of magnitude.

PROBLEMS

8.1 Hot gases flow inside an insulated horizontal tube with dimensions as shown below. Determine the heat-transfer coefficients for both the inside and outside surfaces. The gas at

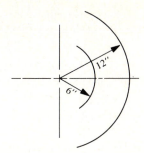

2000°F flows with an average velocity of 50,000 ft/hr. The environment surrounding the tube is air at 100°F, and the outside surface temperature is 150°F.

Data:

Material	T, °F	ρ, lb_m/ft^3	η, $lb_m/hr\text{-}ft$	k, Btu/hr-ft °F	C_p, Btu/lb_m-°F	$g\beta/v^2$, $(ft^3\text{-}°F)^{-1}$
Gas	2000	0.02	0.1*	0.05	0.25	3×10^6
Air	100	0.07	0.049	0.016	0.25	2×10^6
Air	150	0.06	0.053	0.018	0.25	1×10^6

* At these high temperatures, η for the gas does not change much with temperature.

8.2 A vertical surface 5 ft high is at 150°F and the ambient air temperature is 50°F.

a) Calculate the heat-transfer coefficient using information given in the text.

b) Repeat using one of the following simplified equations, which apply reasonably well to air, CO, N_2, and O_2 in the range 100–1500°F. L and D are in ft, ΔT in °F, and h in Btu/hr-ft^2 °F.

Vertical plates of height L:

$$h = 0.29(\Delta T/L)^{1/4}, \quad 10^{-2} < L^3\Delta T < 10^3,$$

$$h = 0.21(\Delta T)^{1/3}, \quad\quad 10^3 < L^3\Delta T < 10^6.$$

Horizontal pipes of diameter D:

$$h = 0.25(\Delta T/D)^{1/4}, \quad 10^{-2} < D^3\Delta T < 10^3,$$

$$h = 0.18(\Delta T)^{1/3}, \quad\quad 10^3 < D^3\Delta T < 10^6.$$

Horizontal square plate with a hot surface facing up or a cold plate facing down:

$$h = 0.27(\Delta T/L)^{1/4}, \quad 10^{-1} < L^3\Delta T < 20,$$

$$h = 0.22(\Delta T)^{1/3}, \quad\quad 20 < L^3\Delta T < 30,000.$$

Horizontal square plate with a hot surface facing down or a cold plate facing up:

$$h = 0.12(\Delta T/L)^{1/4}, \quad 0.3 < L^3\Delta T < 30,000.$$

8.3 Calculate the initial rate of cooling (Btu/hr) of an aluminum plate (4 ft × 4 ft) heated uniformly to 200°F when it is

a) cooled in a horizontal position by a stream of air at 60°F with a velocity of 6 ft/sec, and

b) suspended vertically in stagnant air at the same temperature.

8.4   Air flows at 10 ft/sec through a round tube ($\frac{1}{4}$-in. diameter by 12 ft long) with a uniform tube wall temperature of 200°F. If the air enters at 60°F, at what temperature does it exit from the tube?

8.5   Oil flows in a long horizontal 2-in. I.D. copper tube at an average velocity of 10 ft/sec. If the oil has a bulk temperature of 200°F and the air surrounding the tube is at 70°F, calculate:

    a) the "liquid-side" heat-transfer-coefficient;
    b) the "vapor-side" heat-transfer-coefficient;
    c) the temperature of the copper tube;
    d) the rate of heat transfer to the air.

8.6   Replace the oil with sodium and repeat Problem 8.5 (a)–(d). Compare or contrast the results of the two problems.

8.7   Water flows through a tube 3 ft long by 1-in. I.D. at a velocity of 15 ft/sec. The tube wall temperature is kept constant at 210°F by condensing steam. If the inlet temperature is 60°F, calculate the exit temperature.

8.8   A heat-treating furnace is 20 ft long, 10 ft wide and 6 ft high. If a check with thermocouples indicates that the average wall temperature is 150°F and the top is at 200°F, calculate the heat loss from the furnace in Btu/hr.

8.9   In flow past a flat plate, a laminar boundary layer exists over the forward portion between 0 and L_{tr}, and the turbulent boundary layer exists beyond L_{tr}. With this model, the average h over a plate of length L (with $L > L_{tr}$) can be determined as indicated

$$h = \frac{1}{L}\left(\int_0^{L_{tr}} h_{x(\text{lam})}\, dx + \int_{L_{tr}}^{L} h_{x(\text{turb})}\, dx \right).$$

Take $\text{Re}_{tr} = 3.2 \times 10^5$ and show that

$$hL/k = 0.037\, \text{Pr}^{1/3}(\text{Re}^{0.8} - 15{,}500).$$

[*Hint*: $L_{tr}/L = \text{Re}_{tr}/\text{Re}_L$.]

8.10 Molten aluminum is to be preheated while being transferred from a melting furnace to a casting tundish by pumping it through a heated tube, 2 in. in diameter, at a flow rate of 10,000 lb/hr. The tube wall is kept at a constant temperature of 1400°F.

    a) Calculate the heat-transfer coefficient between the wall and the aluminum.
    b) Using this value of the heat-transfer coefficient, how long would the tube have to be to heat the aluminum from 1250 to 1395°F?

    Data for aluminum at 1300°F:

$$k = 50 \text{ Btu/hr-ft °F},$$
$$\rho = 160 \text{ lb}_m/\text{ft}^3,$$
$$C_p = 0.25 \text{ Btu/lb}_m\text{-°F},$$
$$\eta = 2.9 \text{ lb}_m/\text{hr-ft}.$$

9

CONDUCTION OF HEAT IN SOLIDS

Practicing metallurgists easily recognize that the conduction of heat within solids is fundamental to understanding and controlling many processes. We could cite numerous examples to emphasize the importance of this topic. Some important metallurgical applications that fall in this category include estimating heat losses from process equipment, quenching, or cooling operations where the cooling rate of a part actually controls its metallurgical state and hence its application, and the solidification of ingots or castings.

9.1 THE ENERGY EQUATION FOR CONDUCTION

The general equation for the conduction of heat in a solid free of heat sources or sinks can be written

$$\nabla \cdot k \nabla T = \rho \frac{\partial (C_p T)}{\partial t}. \tag{9.1}$$

The more common form of this equation is written for conductivity independent of position in space and specific heat independent of temperature:

$$\frac{\partial T}{\partial t} = \alpha \nabla^2 T = \alpha \left(\frac{\partial^2 T}{\partial x^2} + \frac{\partial^2 T}{\partial y^2} + \frac{\partial^2 T}{\partial z^2} \right). \tag{9.2}$$

where the thermal diffusivity, as defined in Chapter 7, is

$$\alpha = k/\rho C_p. \tag{9.3}$$

Equation (9.2), although not as rigorous as Eq. (9.1), is used when analytical solutions are sought.

When the temperature in a body is not a function of time but only depends upon position in space, then Eq. (9.2) becomes

$$\nabla^2 T = \frac{\partial^2 T}{\partial x^2} + \frac{\partial^2 T}{\partial y^2} + \frac{\partial^2 T}{\partial z^2} = 0. \tag{9.4}$$

Equation (9.4) therefore applies to *steady-state* conduction in systems free of heat sources and sinks. It is often referred to as the *Laplace equation*.

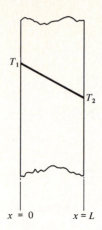

$x = 0$ $x = L$

Fig. 9.1 Steady-state temperature distribution in a plate.

9.2 STEADY-STATE ONE-DIMENSIONAL SYSTEMS

9.2.1 Infinite flat plate

Equation (9.4) for an infinite flat plate, such as that in Fig. 9.1, reduces to

$$d^2 T/dx^2 = 0. \tag{9.5}$$

Boundary conditions are

$$\text{B.C. 1} \quad \text{at } x = 0, \quad T = T_1; \tag{9.6}$$

$$\text{B.C. 2} \quad \text{at } x = L, \quad T = T_2. \tag{9.7}$$

We can solve Eq. (9.5) quite simply, in accordance with the boundary conditions, to yield the temperature profile:

$$\frac{T - T_1}{T_2 - T_1} = \frac{x}{L}. \tag{9.8}$$

In addition the heat flux through the slab may be described as follows:

$$q = -k \frac{dT}{dx} = \frac{k}{L}(T_1 - T_2). \tag{9.9}$$

9.2.2 Series composite wall

Consider a simple series wall made up of two different materials whose thermal conductivities are k_1 and k_2 (Fig. 9.2). There is a flow of heat from the gas at temperature T_i through its boundary layer, the composite wall, and the boundary layer of the gas at T_0.

The unidirectional heat flux through the four parts of the entire circuit is constant because steady-state prevails. Thus

$$Q = Ah_i(T_i - T_1) = \frac{k_1 A}{L_1}(T_1 - T_2) = \frac{k_2 A}{L_2}(T_2 - T_3) = Ah_0(T_3 - T_0). \tag{9.10}$$

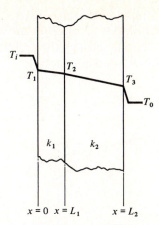

Fig. 9.2 Steady-state temperature distribution in a composite wall.

We can build up a solution based on the four equalities in Eq. (9.10). However, this procedure is rather tedious, and it is much easier to make use of the *resistance concept*.

The flow of heat Q through material, subject to a temperature difference $T_j - T_k$, is analogous to the flow of current I, as a result of a potential difference $E_j - E_k$ through an electrical conductor. From Ohm's law for electricity, the thermal resistance R_t for heat flow is visualized:

$$R_t = \frac{T_j - T_k}{Q}. \tag{9.11}$$

Thus for the composite wall, the four thermal resistances are $1/Ah_i$, L_1/k_1A, L_2/k_2A, and $1/Ah_0$. The total resistance for the whole circuit is simply their sum, so that the heat flow is

$$Q = \frac{T_i - T_0}{1/h_iA + L_1/k_1A + L_2/k_2A + 1/h_0A}. \tag{9.12}$$

With Eq. (9.12), we only need to know the total temperature drop across the system to calculate the heat flux* which we can use to determine the temperature at any position within the composite wall.

Example 9.1 A furnace wall is constructed of 9 in. of fire brick ($k = 0.60$ Btu/hr-ft °F), 6 in. of red brick ($k = 0.40$), 2 in. of glass-wool insulation ($k = 0.04$), and $\frac{1}{8}$ in. steel plate ($k = 26$) on the outside. The heat-transfer coefficients on the inside and

---

* In many texts on heat transfer and in many applications, especially heat-exchanger design, the *overall heat-transfer coefficient*, U, is used, where U is the sum of the thermal conductances. Hence the heat flux through a composite wall could be written as $q = U(T_i - T_0)$ with $U = 1/R_t$.

outside surfaces are 5 and 1 Btu/hr-ft^2 °F, respectively. The gas temperature inside the furnace is 2000°F and the outside air temperature is 90°F.
a) Calculate the heat-transfer rate through the wall (Btu/hr-ft^2).
b) Determine the temperatures at all interfaces.

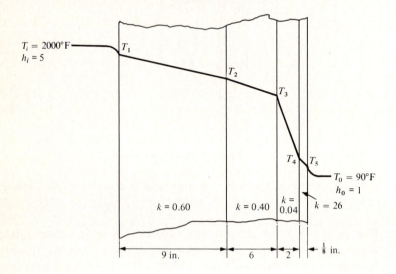

Solution

a)
$$q = \frac{T_i - T_0}{AR_t} = \frac{(2000 - 90)}{\left(\dfrac{1}{5}\right) + \dfrac{(9/12)}{0.60} + \dfrac{(6/12)}{0.40} + \dfrac{(2/12)}{0.04} + \dfrac{(1/96)}{26} + \dfrac{1}{1}}$$

$$= 242 \text{ Btu/hr-ft}^2.$$

b)
$$q = h_i(T_i - T_1).$$

Therefore

$$T_i - T_1 = q/h_i = \frac{242}{5} = 48.4°F,$$

$$T_1 = 1952°F.$$

Similarly,

$$T_1 - T_2 = \frac{q}{(k_1/L_1)} = \frac{(242)(9/12)}{0.60} = 302.6°F,$$

and

$$T_2 = 1649°F.$$

All other temperatures are determined in the same manner:

$$T_1 = 1952°F,$$
$$T_2 = 1649°F,$$
$$T_3 = 1346°F,$$
$$T_4 = 337°F,$$
$$T_5 = 332°F.$$

In addition to the resistances discussed in this section, composite walls often have another type of thermal resistance. When two elements of the composite are in contact, a resistance occurs which depends on the roughness of the two surfaces, the fluid between the surfaces, and the contact pressure. For furnace walls, it is customary to ignore the additional resistance because of the very high thermal resistance of the refractories; for metallic composite walls, however, it is wise to include it. Unfortunately, reliable data are not available because almost each individual situation is different, and so experimental information must be obtained.

An example of a situation where the additional resistance between layers has been determined can be found in the field of steel coil annealing. Olmstead[1] determined that the effective thermal conductivity of the gap between layers of metal in a coil is 0.0446 Btu/ft-hr °F. Even though $k_{steel} = 26.0$ Btu/ft-hr °F, the average thermal conductivity in a radial direction (perpendicular to the sheet) is only 1.79 Btu/ft-hr °F. The result of this study was a redesign of coil heating systems to emphasize heat transfer to the ends of the coil rather than the sides.

9.2.3 Infinite cylinder

For steady radial flow of heat through the wall of a hollow cylinder as depicted in Fig. 9.3, the Laplace equation still applies. However, in this situation, cylindrical coordinates are convenient. The Laplace equation written in cylindrical coordinates can be deduced from Eq. (B), Table 7.5.

$$\frac{1}{r}\frac{\partial}{\partial r}\left(r\frac{\partial T}{\partial r}\right) + \frac{1}{r^2}\frac{\partial^2 T}{\partial \theta^2} + \frac{\partial^2 T}{\partial z^2} = 0. \tag{9.13}$$

For the case at hand, temperature depends only on the radial coordinate; therefore, Eq. (9.13) reduces to:

$$\frac{1}{r}\frac{d}{dr}\left(r\frac{dT}{dr}\right) = 0. \tag{9.14}$$

The boundary conditions under consideration are

$$\text{B.C. 1} \qquad \text{at } r = r_1 \quad T = T_1; \tag{9.15}$$

$$\text{B.C. 2} \qquad \text{at } r = r_2 \quad T = T_2. \tag{9.16}$$

---

[1] C. F. Olmstead. Theory and evolution of coil heating practice in steel mills. *Flat Rolled Products: Rolling and Treatment* **I**, Metallurgical Society Conf., *AIME*, Interscience, New York, 1959.

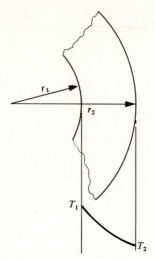

Fig. 9.3 Steady-state temperature distribution in a hollow cylinder.

At this point, the reader should note that this problem and the one discussed in Section 7.4 are identical. By way of illustration, we have used a different starting point here. In any event, the temperature profile is given by

$$\frac{T - T_2}{T_1 - T_2} = \frac{\ln (r/r_2)}{\ln (r_1/r_2)}. \qquad (7.65)$$

It then follows that the heat flux q_r (Btu/hr-ft^2) and the heat flow Q (Btu/hr) are

$$q_r = -\frac{k}{r}\frac{T_1 - T_2}{\ln (r_1/r_2)}, \qquad (7.66)$$

$$Q = \frac{2\pi k L}{\ln (r_2/r_1)}(T_1 - T_2). \qquad (7.67)$$

9.2.4 Composite cylindrical wall

Consider the cylindrical composite wall shown in Fig. 9.4. For steady-state conditions, Q is constant, and for a length L we write

$$Q = h_i(2\pi r_1 L)(T_i - T_1) = \frac{2\pi k_1 L}{\ln (r_2/r_1)}(T_1 - T_2)$$

$$= \frac{2\pi k_2 L}{\ln (r_3/r_2)}(T_2 - T_3) = h_0(2\pi r_3 L)(T_3 - T_0).$$

Again, the series concept is the most convenient method to relate the heat flow to the overall temperature drop:

$$Q = \frac{T_i - T_0}{\dfrac{1}{2\pi L r_1 h_i} + \dfrac{\ln (r_2/r_1)}{2\pi L k_1} + \dfrac{\ln (r_3/r_2)}{2\pi L k_2} + \dfrac{1}{2\pi L r_3 h_0}}. \qquad (9.17)$$

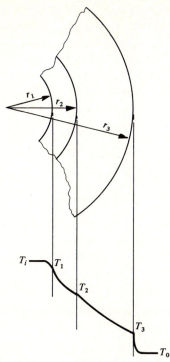

Fig. 9.4 Steady-state temperature distribution in a composite cylindrical wall.

Note that the form of the thermal resistance attributed to the cylindrical wall differs from that of the slab. By comparing Eq. (9.17) with Eq. (9.12), we can contrast the thermal resistances as follows:

$$\text{Infinite slab:} \quad R_t = \frac{L}{kA},$$

$$\text{Infinite cylinder:} \quad R_t = \frac{\ln{(r_2/r_1)}}{2\pi Lk}, \tag{9.18}$$

$$\text{Surface to fluid:} \quad R_t = \frac{1}{Ah}.$$

As an application of Eq. (9.17), consider the design of single-wall tube furnaces to operate at T_i internally while placed in some environment at T_0. The radial heat flow through the furnace wall is then given by Eq. (9.17) written for a tube composed of a single layer with conductivity k;

$$Q = \frac{2\pi L(T_i - T_0)}{\dfrac{1}{h_i r_1} + \dfrac{1}{k}\ln\dfrac{r_2}{r_1} + \dfrac{1}{h_0 r_2}}. \tag{9.19}$$

If we examine Eq. (9.19), we see that as r_2 (the outer radius) increases, there is an increasing resistance to radial heat conduction as embodied in the ln term. Simultaneously, however, as r_2 increases, the outer cooling surface area increases as well. This dual effect suggests that there exists a particular outer radius for which the heat loss (or gain) is a maximum. To examine this proposed effect, fix r_1, and determine that particular value of r_2 for which $dQ/dr_2 = 0$:

$$\frac{dQ}{dr_2} = \frac{-2\pi L(T_i - T_0)(1/kr_2 - 1/h_0 r_2^2)}{\left(\dfrac{1}{h_i r_1} + \dfrac{1}{k}\ln\dfrac{r_2}{r_1} + \dfrac{1}{h_0 r_2}\right)^2} = 0.$$

From this, we obtain the critical outer radius

$$r_{2_c} = k/h_0. \tag{9.20}$$

These concepts are shown in Fig. 9.5 for the situation in which the inside wall temperature T_1 is known or the inner surface thermal resistance $1/h_i$ is zero so that $T_i = T_1$. The curve for $k/h_0 r_1 = 0$ represents the case where both inner and outer surface thermal resistances are zero and the critical radius is infinite. The curve for $k/h_0 r_1 = 1$ is a case for which the critical radius occurs only under the fictitious

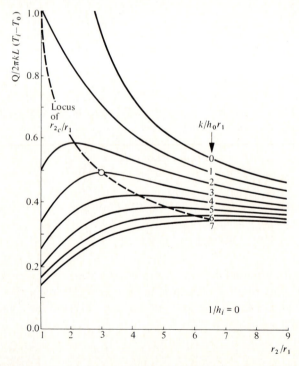

Fig. 9.5 Heat transfer through a single-wall cylindrical tube furnace. (From P. J. Schneider, *Conduction Heat Transfer*, Addison-Wesley, 1955, page 29.)

situation of zero wall thickness. The existence of a critical radius shows that under some realistic conditions, and contrary to common expectations, the heat loss through a tube furnace can actually be decreased by decreasing the insulating wall thickness.

9.3 STEADY-STATE, TWO-DIMENSIONAL HEAT FLOW

9.3.1 Semi-infinite plate

The problem considered in this section introduces the reader to a method of *separation of variables* which is often used to solve the conduction equation (Eq. 9.2), or the Laplace equation (Eq. 9.4).

Figure 9.6 depicts a plate in the xy-plane extending to $y = \infty$ with edges at $x = 0$, $x = L$, and $y = 0$. Such a plate is denoted *semi-infinite* because one of its dimensions, y, is unlimited. The temperature distribution in question is two-dimensional; the plate may be thin enough so that $\partial T/\partial z$ is negligible, or we may consider a cross section of a long bar in which the thermal picture is identical in all planes parallel to the plane under consideration.

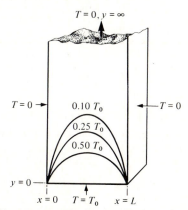

Fig. 9.6 Temperature distribution in a semi-infinite plate.

For such a solid the temperature field is two-dimensional, and under steady-state conditions must satisfy

$$\frac{\partial^2 T}{\partial x^2} + \frac{\partial^2 T}{\partial y^2} = 0. \tag{9.21}$$

For simplicity consider the following boundary conditions:

B.C. 1	at $x = 0$, $T = 0$;	(9.22)
B.C. 2	at $x = L$, $T = 0$;	(9.23)
B.C. 3	at $y = \infty$, $T = 0$;	(9.24)
B.C. 4	at $y = 0$, $T = T_0$ (uniform).	(9.25)

The method of separation of variables consists of seeking *product solutions* in the form

$$T(x, y) = X(x) \cdot Y(y), \tag{9.26}$$

where X is a function of x alone and Y is a function of y alone. Introducing Eq. (9.26) into Eq. (9.21), we have

$$Y\frac{d^2X}{dx^2} + X\frac{d^2Y}{dy^2} = 0 \tag{9.27}$$

or, by *separating the variables*,

$$-\left(\frac{1}{X}\right)\left(\frac{d^2X}{dx^2}\right) = \left(\frac{1}{Y}\right)\left(\frac{d^2Y}{dy^2}\right). \tag{9.28}$$

The right-hand side of Eq. (9.28) is independent of x (by hypothesis); therefore, the left-hand side is also independent of x, and hence must be equal to a constant. Similarly, the left-hand side is independent of y, and demands the right-hand side to be independent of y. Hence, it follows that both sides equal an arbitrary constant; this constant λ^2 is called the *separation constant*.* Therefore

$$\frac{d^2X}{dx^2} + \lambda^2X = 0, \tag{9.29a}$$

$$\frac{d^2Y}{dy^2} - \lambda^2Y = 0. \tag{9.29b}$$

Equations (9.29a) and (9.29b) are homogeneous, linear equations with constant coefficients. This type of equation can be solved by setting $X = e^{ax}$ and $Y = e^{bx}$, and substituting in Eqs. (9.29a) and (9.29b), respectively. For Eq. (9.29a), $a = \pm i\lambda$, and the solution takes the form

$$X = C_1'e^{i\lambda x} + C_2'e^{-i\lambda x}. \tag{9.30}$$

Making use of the identities $e^{\pm i\lambda x} = \cos \lambda x \pm i \sin \lambda x$, the solution takes on a more commonly used form:

$$X = C_1 \cos \lambda x + C_2 \sin \lambda x. \tag{9.31}$$

For Eq. (9.29b), $b = \pm\lambda$, and the solution is

$$Y = C_3e^{\lambda y} + C_4e^{-\lambda y}. \tag{9.32}$$

The general solution of the Laplace equation is thus assumed to be the product of Eqs. (9.31) and (9.32).

In applying boundary conditions, experience shows that it is somewhat easier to apply the "toughest" boundary condition last. In this situation, B.C. 4 is the "toughest"; this will become apparent as we develop the solution.

---

* The constant λ^2 is given this form now because we know this yields a useful result.

First, we examine the boundary conditions for x. For Eq. (9.31) to satisfy Eq. (9.22), X must vanish at $x = 0$; therefore, $C_1 = 0$. Similarly, X must vanish at $x = L$ to satisfy Eq. (9.23). Therefore

$$\sin \lambda L = 0. \tag{9.33}$$

Equation (9.33) requires $\lambda L = 0$, π, 2π, 3π, etc., or, in general, $\lambda_n = n\pi/L$, where $n = 0, 1, 2, 3$, etc. The equation $\sin \lambda L = 0$ is called the *eigenfunction* and the values of λ_n, the *eigenvalues*. We have now considered both boundary conditions for x. At this point we may write

$$X = C_2 \sin \frac{n\pi x}{L}. \tag{9.34}$$

Equation (9.34) obviously satisfies (9.29a) for any *eigenvalue*; it is also true that the sum of all the *eigenfunctions* satisfies Eq. (9.29a). Therefore, we write

$$X = \sum_{n=0}^{\infty} C_n \sin \frac{n\pi x}{L}. \tag{9.35}$$

When the boundary conditions for y are applied, we see that Eq. (9.24) requires $C_3 = 0$ in Eq. (9.32). Thus,

$$Y = C_4 e^{-\lambda y} = C_4 e^{-(n\pi/L)y}. \tag{9.36}$$

The product solution now takes the form

$$T = X \cdot Y = \sum_{n=0}^{\infty} A_n e^{-(n\pi/L)y} \sin \frac{n\pi x}{L}, \tag{9.37}$$

where the A_n's have absorbed the constants involved.

For the final boundary condition (Eq. 9.25), Eq. (9.37) becomes

$$T_0 = \sum_{n=0}^{\infty} A_n \sin \frac{n\pi x}{L}. \tag{9.38}$$

To determine all the various values of A_n, we multiply both sides of this equation by $\sin m\pi x/L$, where m is one particular integral value of n; we then integrate between $x = 0$ and $x = L$:

$$T_0 \int_{x/L=0}^{1} \sin m\pi \left(\frac{x}{L}\right) d\left(\frac{x}{L}\right) = \int_{x/L=0}^{1} \sum_{n=0}^{\infty} A_n \sin n\pi \left(\frac{x}{L}\right) \sin m\pi \left(\frac{x}{L}\right) d\left(\frac{x}{L}\right). \tag{9.39}$$

This procedure might not at all be obvious to the reader who has not met with these types of problems before. Actually, this procedure parallels the development of the *Fourier theorem*.

A table of definite integrals indicates that all integrals on the right-hand side of Eq. (9.39) are zero for all values of n, except for $n = m$, when it equals $A_n/2$. The integral on the left is $2/n\pi$, where $n = 1, 3, 5, \ldots$, odd. Therefore,

$$A_n = \frac{4T_0}{n\pi}, \qquad n \text{ odd.} \tag{9.40}$$

The final solution is

$$\frac{T}{T_0} = \sum_{\substack{n=1 \\ n,\,\text{odd}}}^{\infty} \frac{4}{n\pi} e^{-(n\pi/L)y} \sin \frac{n\pi x}{L}.$$
(9.41)

Figure 9.6 shows some isotherms corresponding to Eq. (9.41).

Now consider the temperature distribution in the same plate with different boundary conditions.

B.C. 1	at $x = 0$,	$T = T_1$;	(9.42)
B.C. 2	at $x = L$,	$T = T_1$;	(9.43)
B.C. 3	at $y = \infty$,	$T = T_1$;	(9.44)
B.C. 4	at $y = 0$,	$T = f(x)$.	(9.45)

The development of Eq. (9.41) is possible due to the *homogeneous boundary* conditions, namely, Eqs. (9.22) and (9.23), that is, the method depends on the temperature being zero at the two extremes of x. For the case at hand, we simply transform the temperature variable T into θ, where $\theta = T - T_1$. With θ, then, the Laplace equation becomes

$$\frac{\partial^2 \theta}{\partial x^2} + \frac{\partial^2 \theta}{\partial y^2} = 0,$$
(9.46)

and the boundary conditions are then

$$\theta(0, y) = \theta(L, y) = \theta(x, \infty) = 0 \quad \text{and} \quad \theta(x, 0) = f(x) - T_1 = F(x).$$

Again, assuming product solutions, $\theta = X \cdot Y$, and proceeding in exactly the same manner as before, we can obtain by referring to Eq. (9.37):

$$\theta = \sum_{n=0}^{\infty} A_n e^{-(n\pi/L)y} \sin \frac{n\pi x}{L}.$$
(9.47)

By applying the "toughest" boundary condition, we get

$$F(x) = \sum_{n=0}^{\infty} A_n \sin \frac{n\pi x}{L},$$
(9.48)

and proceeding as before, we have

$$\int_{x/L=0}^{1} F(x) \left[\sin m\pi \frac{x}{L} \right] d\left(\frac{x}{L}\right) = \frac{A_n}{2}.$$
(9.49)

Therefore, we evaluate the arbitrary constants:

$$A_n = \frac{2}{L} \int_{x=0}^{x=L} F(x) \sin \frac{n\pi x}{L} \, dx.$$
(9.50)

Noting that $A_0 = 0$, we obtain the final solution

$$T = T_1 + \sum_{n=1}^{\infty} A_n e^{-(n\pi/L)y} \sin \frac{n\pi x}{L}, \tag{9.51}$$

where

$$A_n = \frac{2}{L} \int_{x=0}^{x=L} [f(x) - T_1] \sin \frac{n\pi x}{L} dx. \tag{9.52}$$

9.3.2 Rectangular plate

In this section we go on to consider the finite plate with the three edges $x = 0$, $x = L$, and $y = 0$ maintained at zero temperature,

$$T(0, y) = T(L, y) = T(x, 0) = 0, \tag{9.53a, b, c}$$

and the fourth edge $y = H$ maintained at a temperature distribution $f(x)$,

$$T(x, H) = f(x). \tag{9.54}$$

Again Eq. (9.21) must be satisfied, and we employ the method of separation of variables. Equations (9.53a) and (9.53b) will be satisfied if

$$X(0) = X(L) = 0, \tag{9.55a, b}$$

whereas Eq. (9.53c) implies the condition

$$Y(0) = 0. \tag{9.56}$$

As before, when Eqs. (9.55a) and (9.55b) are satisfied,

$$X = \sum_{n=0}^{\infty} C_n \sin \frac{n\pi x}{L}. \tag{9.35}$$

This time we choose to write the general solution of Eq. (9.29b) as

$$Y = C_3 \sinh \lambda y + C_4 \cosh \lambda y. \tag{9.57}$$

(This is another form of Eq. (9.32) if the definitions of the hyperbolic sine and hyperbolic cosine are substituted.) The boundary condition for Y, Eq. (9.56), eliminates C_4, and as before, $\lambda_n = n\pi/L$. Thus it follows that the product solution takes the form

$$T = X \cdot Y = \sum_{n=1}^{\infty} A_n \sinh \frac{n\pi y}{L} \sin \frac{n\pi x}{L}. \tag{9.58}$$

We must now determine the coefficients A_n in such a way that the remaining condition, Eq. (9.54), is satisfied. For this boundary condition, the solution becomes

$$f(x) = \sum_{n=1}^{\infty} \left(A_n \sinh \frac{n\pi H}{L} \right) \sin \frac{n\pi x}{L}. \tag{9.59}$$

Following the same procedure as before to determine coefficients, we obtain

$$A_n \sinh \frac{n\pi H}{L} = \frac{2}{L} \int_0^L f(x) \sin \frac{n\pi x}{L} \, dx. \tag{9.60}$$

and writing $B_n = A_n \sinh (n\pi H/L)$, we find that the solution (Eq. 9.59) takes the form

$$T = \sum_{n=1}^{\infty} B_n \frac{\sinh \dfrac{n\pi y}{L}}{\sinh \dfrac{n\pi H}{L}} \sin \frac{n\pi x}{L}, \tag{9.61}$$

where

$$B_n = \frac{2}{L} \int_0^L f(x) \sin \frac{n\pi x}{L} \, dx. \tag{9.62}$$

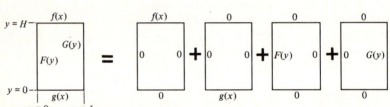

Fig. 9.7 Method of solving the temperature distribution in a rectangular plate with general boundary conditions, by adding the solutions of four simple cases.

We can obtain the solution of the more general problem of dealing with the temperature prescribed along all four edges by superimposing four solutions, such as the one obtained here, each corresponding to a problem in which zero temperatures are prescribed along three of the four edges. Figure 9.7 schematically depicts this method of solution.

9.4 TRANSIENT SYSTEMS, FINITE DIMENSIONS

In this section, we consider conduction heat transfer in solids, in which the temperature varies not only with position in space, but also undergoes a continuous change with time at any position. In particular, this topic is of vital importance in quenching or cooling operations.

9.4.1 Newtonian heating or cooling

When a solid such as a flat plate, initially at a uniform temperature T_i, is cooled by a fluid at temperature T_f, the temperature distribution varies with time, as shown

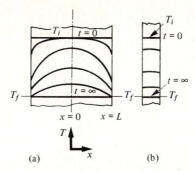

Fig. 9.8 Transient temperature distributions during cooling of (a) a thick plate, and (b) a thin plate.

in Fig. 9.8(a). If the plate is thin (Fig. 9.8b), and/or its thermal conductivity high, then the temperature gradients within the plate are negligible, and we may consider the temperature only as a function of time. In the next section, we shall examine when this criterion is met; we find that if $hL/k \lesssim 0.1$, then the analysis that follows is valid.

For this situation, the heat balance takes a rather simple form. We equate the rate of heat lost by the plate to the rate of heat transfer to the fluid:

$$- V\rho C_p \frac{dT}{dt} = hA(T - T_f), \qquad (9.63)$$

where V = volume of the plate, and A = area of the plate exposed to the fluid. Rearranging for integration, we have

$$\int_{T_i}^{T} \frac{dT}{T - T_f} = -\frac{hA}{V\rho C_p} \int_0^t dt, \qquad (9.64)$$

$$\frac{T - T_f}{T_i - T_f} = \exp\left(\frac{-hAt}{\rho C_p V}\right). \qquad (9.65)$$

The analysis applies only to the cases in which internal gradients are negligible. If this requirement is met, we speak of Newtonian cooling. Note that no geometric restrictions are imposed, since Eq. (9.65) just specifies A and V; also note that the solution does not contain the conductivity of the metal.

The temperature history of such objects during cooling is illustrated in Fig. 9.9, which shows that for the same *surface radius*, R, or semithickness, L, the time it takes an infinite cylinder (or infinite square rod) to cool within 1% of thermal equilibrium with its surrounding fluid is almost 50% longer than for a sphere (or cube) of the same material. On the other hand, an infinite plate of the same material and semithickness equal to the cylinder radius requires 100% more time.

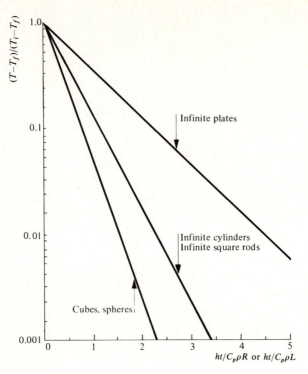

Fig. 9.9 Cooling (or heating) of bodies with negligible temperature gradients. R is radius, and L is semithickness.

Example 9.2 A plate of aluminum with dimensions $\frac{1}{4}$ in. $\times$ 12 in. is solutionized at $1000°$F, and then quenched in water at $200°$F. Over the temperature range 1000–$500°$F, the heat-transfer coefficient for quenching in the water may be assumed constant as 90 Btu/hr-ft^2 °F. In the range 500–$200°$F, the heat-transfer coefficient may be approximated as 450 Btu/hr-ft^2 °F.

a) Calculate the cooling rate during quenching at that instant when the plate is at $600°$F.

b) Repeat, when the plate is at $300°$F.

c) Determine the time it takes the plate to cool to $250°$F. The properties of the aluminum are:

$$k = 45 \text{ Btu/hr-ft °F},$$
$$C_p = 0.24 \text{ Btu/lb}_m,°\text{F},$$
$$\rho = 180 \text{ lb}_m/\text{ft}^3.$$

Solution. We can write an expression for the cooling rate by simply rearranging Eq. (9.63):

$$\frac{dT}{dt} = \frac{h}{\rho C_p} \frac{A}{V} (T - T_f).$$

For a plate or slab whose thickness is much less than its width and length, A/V may be assumed equal to A/V for an infinite plate. Therefore

$$\frac{A}{V} = \frac{1}{L} = \frac{1}{\frac{1}{8}\text{ in.}}\left|\frac{12\text{ in.}}{1\text{ ft}}\right. = 96\text{ ft}^{-1}.$$

a) $\dfrac{dT}{dt} = \dfrac{(90)(96)}{(180)(0.24)}(600 - 200) = -80,000°\text{F/hr}$

or

$$\frac{dT}{dt} = -22.2°\text{F/sec} \quad \text{at} \quad T = 600°\text{F}.$$

b) $\dfrac{dT}{dt} = \dfrac{(450)(96)}{(180)(0.24)}(300 - 200) = -100,000°\text{F/hr}$

or

$$\frac{dT}{dt} = -27.8°\text{F/sec} \quad \text{at} \quad T = 300°\text{F}.$$

c) First calculate time t_1 to reach 500°F. Use Eq. (9.65):

$$\frac{T - T_f}{T_i - T_f} = \exp\left[-\frac{h}{\rho C_p}\frac{A}{V}t_1\right],$$

$$\frac{500 - 200}{1000 - 200} = \exp\left[-\frac{(90)(96)}{(180)(0.24)}t_1\right].$$

Solving for t_1, we have

$$t_1 = 0.00491\text{ hr.}$$

Calculate time t_2 to cool from 500 to 250°F. Now T_i is interpreted as 500°F:

$$\frac{250 - 200}{500 - 200} = \exp\left[-\frac{(450)(96)}{(180)(0.24)}t_2\right].$$

Solving for t_2, we get

$$t_2 = 0.00184\text{ hr.}$$

Therefore total time to cool to 250°F is given by

$$t = t_1 + t_2 = (0.00491 + 0.00184)\text{ hr}$$

or

$$t = 24.2\text{ sec.}$$

We should point out that the critical assumption embodied in Newtonian cooling is that the internal temperature gradients are non-existent. The reader is assured that the assumption, as applied to this particular example, is valid.

9.4.2 Bodies with internal temperature gradients

For the flat plate of Fig. 9.8(a), the applicable differential equation by reference to Eq. (9.2) is

$$\frac{\partial T}{\partial t} = \alpha \frac{\partial^2 T}{\partial x^2}. \tag{9.66}$$

We find the solution $T(x, t)$ of this equation for the boundary conditions

$$T(x, 0) = T_i \quad \text{(uniform)}, \tag{9.67}$$

$$\frac{\partial T(0, t)}{\partial x} = 0, \tag{9.68}$$

$$\frac{\partial T(L, t)}{\partial x} + \frac{h}{k}[T(L, t) - T_f] = 0, \tag{9.69}$$

by again employing the method of separation of variables.

The boundary condition in time (Eq. 9.67), is usually referred to as the initial condition. For this case, the initial condition is some uniform temperature; we could, however, consider some arbitrary temperature distribution, $f(x)$, existing at time zero.

Equation (9.68) results from the symmetry of the problem, and we develop the final boundary condition (Eq. 9.69) by equating the heat flux at the surface to the rate of heat transfer to the fluid. Before proceeding, we give some attention to these two boundary conditions.

While applying the method of separation of variables to the two-dimensional heat flow situation in Section 9.3.1, we noted that a suitable boundary condition for mathematical purposes was homogeneous, that is, $T = 0$. Similarly, for mathematical convenience, Eqs. (9.68) and (9.69) should be homogeneous. In summary, we state the three basic types of boundary conditions that can be treated analytically:

$$T(\text{boundary}) = 0, \tag{9.70a}$$

$$\frac{\partial T}{\partial x}(\text{boundary}) = 0, \tag{9.70b}$$

or

$$\frac{\partial T}{\partial x}(\text{boundary}) \pm \frac{h}{k} T(\text{boundary}) = 0. \tag{9.70c}$$

Comparing Eqs. (9.68) and (9.69) with Eqs. (9.70b) and (9.70c), respectively, we see that we must make a modification because T_f is not necessarily zero. However, the modification is easily recognized; specifically, let $\theta = T - T_f$, so that the differential equation and boundary conditions become

$$\frac{\partial \theta}{\partial t} = \alpha \frac{\partial^2 \theta}{\partial x^2}, \tag{9.71}$$

$$\theta(x, 0) = T_i - T_f = \theta_i, \tag{9.72}$$

$$\frac{\partial \theta}{\partial x}(0, t) = 0, \tag{9.73}$$

$$\frac{\partial \theta(L, t)}{\partial x} + \frac{h}{k} \theta(L, t) = 0. \tag{9.74}$$

We seek a product solution of the form $\theta(x, t) = X(x) \cdot G(t)$. Then Eq. (9.71) becomes

$$\frac{1}{X} \frac{d^2 X}{dx^2} = \frac{1}{\alpha G} \frac{dG}{dt} = -\lambda^2, \tag{9.75}$$

from which

$$\frac{d^2 X}{dx^2} + \lambda^2 X = 0, \tag{9.76}$$

and

$$\frac{dG}{dt} + \alpha \lambda^2 G = 0. \tag{9.77}$$

We write the solution for Eq. (9.76)

$$X = c_1 \cos \lambda x + c_2 \sin \lambda x, \tag{9.78}$$

and for Eq. (9.77), the solution is

$$G = \exp(-\lambda^2 \alpha t). \tag{9.79}$$

Boundary condition (9.73) requires that $c_2 = 0$, and Eq. (9.74) can be shown to require that

$$\cot \lambda_n L = \frac{1}{(h/k)L} (\lambda_n L). \tag{9.80}$$

Equation (9.80) is analogous to Eq. (9.33) in that λ_n takes on an infinite number of *eigenvalues*. Figure 9.10 indicates this for the first three *eigenvalues*.

Taking the product of Eqs. (9.78) and (9.79) with $c_2 = 0$, and realizing that all values of λ_n satisfying Eq. (9.80) are suitable, we have

$$\theta = \sum_{n=1}^{\infty} A_n \exp(-\lambda_n^2 \alpha t) \cos \lambda_n x, \tag{9.81}$$

where the A_n's have absorbed the constants involved. The initial condition (9.72)

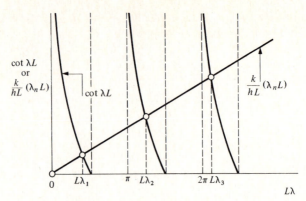

Fig. 9.10 Solutions of λ_n roots of Eq. (9.80).

still remains to be satisfied. When Eq. (9.72) is substituted into Eq. (9.81), we get

$$\theta_i = \sum_{n=1}^{\infty} A_n \cos \lambda_n x. \tag{9.82}$$

Multiplying both sides of this equation by $\cos \lambda_m x \, dx$, and integrating from $x = 0$ to $x = L$, we obtain

$$\theta_i \int_0^L \cos \lambda_m x \, dx = \int_0^L \sum_{n=1}^{\infty} A_n \cos \lambda_n x \cos \lambda_m x \, dx. \tag{9.83}$$

When $m \neq n$, all integrals on the right-hand side of Eq. (9.83) are zero, and when $m = n$ the integral has the nonzero value

$$A_n \left[\frac{L}{2} + \frac{1}{2\lambda_n} \sin \lambda_n L \cos \lambda_n L \right].$$

The integral on the left-hand side of Eq. (9.83) is $(1/\lambda_n) \sin \lambda_n L$. Therefore

$$A_n = \frac{2\theta_i \sin \lambda_n L}{\lambda_n L + \sin \lambda_n L \cos \lambda_n L}. \tag{9.84}$$

The final solution is then

$$\frac{\theta}{\theta_i} = \frac{T - T_f}{T_i - T_f} = 2 \sum_{n=1}^{\infty} \frac{\sin \lambda_n L}{\lambda_n L + \sin (\lambda_n L) \cos (\lambda_n L)} \exp(-\lambda_n^2 \alpha t) \cos(\lambda_n x), \tag{9.85}$$

where the λ_n's are the roots of Eq. (9.80), given in Table 9.1.

Graphical evaluations of Eq. (9.85) and the analogous solutions for infinitely long cylinders and spheres have been presented in many graphical forms for practical use. All these graphical presentations are given in terms of a *relative temperature* as a function of the *Fourier number*, *Biot number*, and *relative position*.

Table 9.1 The first four roots of Eq. (9.80)*

Bi	k/hL	$\lambda_1 L$	$\lambda_2 L$	$\lambda_3 L$	$\lambda_4 L$
100	0.01	1.5552	4.6658	7.7764	10.8871
10	0.10	1.4289	4.3058	7.2281	10.2003
1	1.0	0.8603	3.4256	6.4373	9.5293
0.1	10.0	0.3111	3.1731	6.2991	9.4354

* A more comprehensive table can be found in H. S. Carslaw and J. C. Jaeger, *Conduction of Heat in Solids*, second edition, Oxford University Press, 1959, page 491.

These four dimensionless groups of variables are defined as follows:

$$1.\ \text{relative temperature} \equiv \frac{T - T_f}{T_i - T_f},$$

$$2.\ \text{Fourier number (Fo)} \equiv \frac{\alpha t}{L^2},$$

$$3.\ \text{Biot number (Bi)} \equiv \frac{hL}{k},$$

$$4.\ \text{relative position} \equiv \frac{x}{L}.$$

Here for plates, L is the semithickness, and x is the distance out from the center. For cylinders and spheres, the radius R replaces L, and r replaces x in the above definitions.

Among the earliest graphs prepared were the so-called Gurney–Lurie charts.[2] These charts are still referred to in metallurgical engineering literature. However, the charts are limited to a small range of Fo and Bi values. Other diagrams commonly used are the Heisler charts[3] for $0.01 \lesssim \text{Bi} < \infty$ and large values of Fo. For more convenience, Figs. 9.11, 9.12, and 9.13 have been constructed for the temperature response in infinite plates, infinite cylinders, and spheres, respectively.

Example 9.3 As an example of the use of Figs. 9.11, 9.12, and 9.13, consider a very long cylindrical stainless steel bar, 5 in. in diameter, which is heated to 400°F uniformly across its diameter. The bar is then cooled in a blast of fan forced air at 80°F with $h = 25$ Btu/hr-ft^2 °F.

a) Find the time it takes for the center to reach 100°F.

b) When the center reaches 100°F, what is the surface temperature?

c) What would be the minimum possible time for the center of the bar to reach 100°F if it were cooled in an ideal quenchant ($H = \infty$) at 80°F?

[2] H. P. Gurney and J. Lurie, *Ind. Eng. Chem.* **15**, 1170–1172 (1923).
[3] M. P. Heisler, *Trans ASME* **69**, 227–236 (1947).

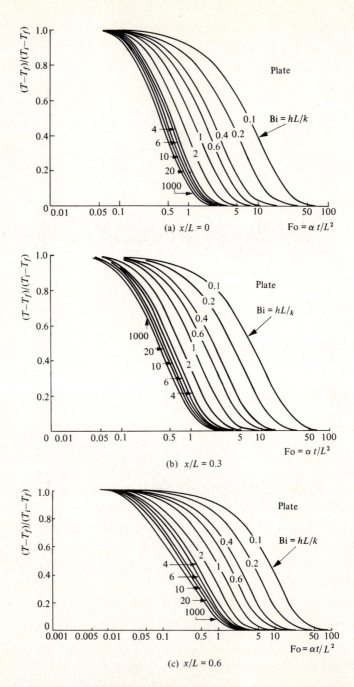

Fig. 9.11 Temperature response of an infinite plate initially at a uniform temperature T_i, and then subjected to a convective environment at T_f.

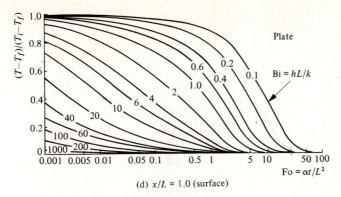

(d) $x/L = 1.0$ (surface)

Fig. 9.11   (*continued*)

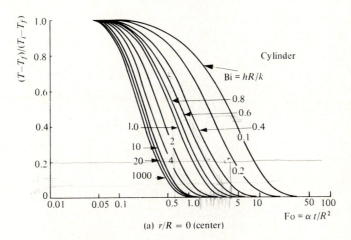

(a) $r/R = 0$ (center)

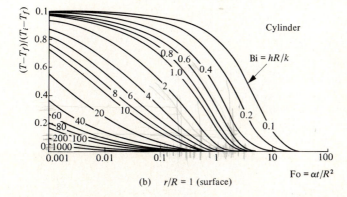

(b) $r/R = 1$ (surface)

Fig. 9.12   Temperature response of an infinite cylinder initially at a uniform temperature T_i, and then subjected to a convective environment at T_f.

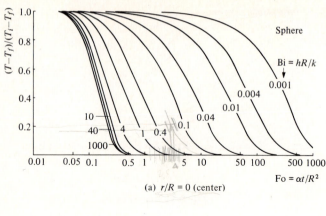

(a) $r/R = 0$ (center)

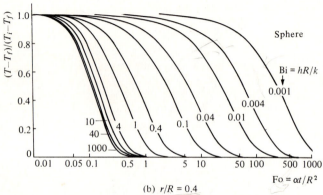

(b) $r/R = 0.4$

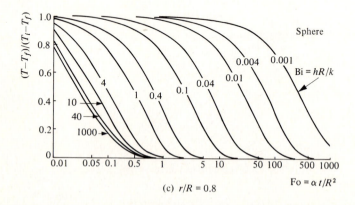

(c) $r/R = 0.8$

Fig. 9.13 Temperature response of a sphere initially at a uniform temperature T_i, and then subjected to a convective environment at T_f.

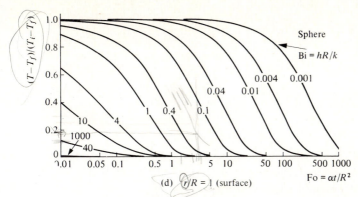

(d) $r/R = 1$ (surface)

Fig. 9.13 (*continued*)

The thermal properties of the stainless steel are $k = 9$ Btu/hr-ft °F and $\alpha = 0.158$ ft²/hr.

Solution

a) When $T = 100°$F at the center, we have

$$\frac{T - T_f}{T_i - T_f} = \frac{100 - 80}{400 - 80} = 0.0625.$$

Also

$$\text{Bi} = \frac{hR}{k} = \frac{(25)(5/24)}{9} = 0.579.$$

From Fig. 9.12(a), $\alpha t/R^2 \cong 3.1$. Then

$$t = \frac{(3.1)(5/24)^2}{(0.158)} = 0.84 \text{ hr}$$

or

$$t = 50.5 \text{ min.}$$

b) For the surface temperature, we refer to Fig. 9.12(b). When Fo $= 3.1$ and Bi $= 0.579$, we find that

$$\frac{T - T_f}{T_i - T_f} \cong 0.04.$$

Therefore

$$T = 0.04(400 - 80) + 80 = 93°\text{F.}$$

c) The minimum possible time would correspond to the ideal quench, or the situation in which the surface temperature is equal to the temperature of the cooling medium, that is, Bi → ∞. It is sufficient to use Fig. 9.12(a) with Bi $= 1000$.

Thus

$$\alpha t / R^2 \cong 0.57,$$

and

$$t = \frac{(0.57)(5/24)^2}{0.158} = 0.179 \text{ hr},$$

or

$$t = 10.7 \text{ min}.$$

Example 9.4 The process of austempering requires that a steel be quenched to just above its M_s temperature, and then isothermally transformed to a lower bainite structure that resembles tempered martensite. For a steel with the TTT diagram below, estimate the maximum thickness of plate that can be completely austempered by quenching into molten salt at 400°F from a 1600°F austenitizing temperature with $h = 50 \text{ Btu/hr-ft}^2 \text{ °F}$. For the steel, assume $\alpha = 0.46 \text{ ft}^2/\text{hr}$ and $k = 20 \text{ Btu/hr-ft °F}$.

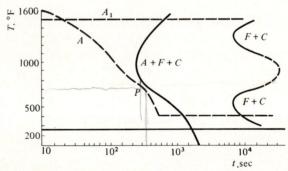

Since we must cool at such a rate as to bring the center of the plate past the "nose" of the curve without undergoing any transformation, we can start by assuming that point P is the critical point. This can be tested later. Since point P is at 700°F, then

$$\frac{T - T_f}{T_i - T_f} = \frac{700 - 400}{1600 - 400} = 0.25.$$

Solution. Figure 9.11(a) gives us the required result. What is needed, is the value of the semithickness L that will result in agreement between Bi, Fo, and reduced temperature. Since

$$t \cong 420 \text{ sec}, \quad \text{or} \quad 0.116 \text{ hr},$$

then

$$\text{Fo} = (0.46)(0.116)/L^2 = 0.0536/L^2$$

and

$$\text{Bi} = \frac{50}{20} L = 2.5L.$$

Try $L = 0.1$ ft. In this case, Bi $= 0.25$ and, from Fig. 9.11(a), Fo should be 6.6. However, with $L = 0.1$ ft, Fo $= 5.36$, so that this value of L does not satisfy Fo and Bi simultaneously.

Try $L = 0.05$ ft. Now Bi $= 0.125$, and Fo should be 12. With $L = 0.05$ ft, Fo is calculated as 20.

We have bracketed the actual value, and after more trial and error, the value of $L = 0.092$ ft leads to agreement between Bi $= 0.23$ and Fo $= 6.35$.

As a check, the times to reach various temperatures with $L = 0.092$ ft should be calculated to see that the corresponding time–temperature curve does not intersect the nose of the TTT curve. The result of this calculation is shown in the figure.

Thus we can say that a plate 2.2 in. thick can be fully austempered. Strictly speaking, since the TTT diagram is developed using small samples isothermally transformed, the "nose" of the curve should be moved *slightly* to the right for continuous cooling applications, but as a first approximation, the isothermal TTT diagram may be used.

In conclusion, it is important to indicate when limiting cases are valid, such as Newtonian cooling. This, of course, simplifies our work, as Fig. 9.14 demonstrates.

Figure 9.14 presents the solution for the infinite plate such that temperature distributions within plates are shown for different times during cooling in media of various Biot numbers. Examination of Fig. 9.14 shows that the temperature gradients within the slab decrease as Bi decreases. A low value of Bi, physically interpreted, reflects low resistance to heat flow within the body, L/k, relative to the resistance of the cooling media h^{-1}. In practice, when Bi $\lesssim 0.1$, the temperature is nearly uniform within the plate. For such cases, we approximate the cooling or heating processes as being controlled solely by surface resistance, and we can apply the Newtonian relationship, Eq. (9.65), as a very close approximation. We may also apply the same approximation to the heating or cooling of cylinders and spheres, with the criterion Bi $\lesssim 0.1$ still being in effect.

At the other extreme, when the cooling or heating process can be considered to be completely controlled by the internal resistance of the body, then the surface temperature, in effect, immediately changes to T_f, the temperature of the fluid, and remains constant at this temperature. This situation can be considered as a special case with Bi $\to \infty$. Alternatively, a new solution to Eq. (9.66) could be developed for the same boundary conditions as those of Eqs. (9.67) and (9.68), with $T(L, t) = T_f$ replacing Eq. (9.69). The solution to this problem is given by

$$\frac{T - T_f}{T_i - T_f} = \frac{4}{\pi} \sum_{n=0}^{\infty} \frac{(-1)^n}{2n + 1} \exp\left[\frac{-(2n + 1)^2 \pi^2}{4} \frac{\alpha t}{L^2}\right] \cos \frac{(2n + 1)\pi}{2} \frac{x}{L}. \quad (9.86)$$

Examination of Fig. 9.14 shows that the criterion, $T(L, t) = T_f$, is closely approximated when Bi $= 100$. Thus when Bi $\gtrsim 100$, Eq. (9.85) can be very closely approximated by Eq. (9.86).

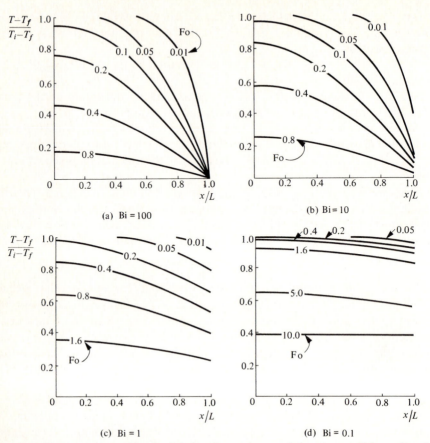

Fig. 9.14 Temperature profiles for infinite plates when cooled under conditions of various Biot numbers.

9.4.3 Long-time and short-time solutions

The solutions to many problems of unsteady-state heat conduction are in the form of infinite series, such as Eqs. (9.85) and (9.86). This type of series converges rapidly for large Fourier numbers. For short times (small $\alpha t/L^2$), however, the convergence is very slow, requiring an expansion of a great number of terms in the series to obtain sufficiently accurate answers. If we require a solution for short times, alternative series that evolve by the method of Laplace transforms when solving Eq. (9.66) are more convenient. These alternative series have the advantage of converging rapidly for small $\alpha t/L^2$, but the disadvantage of converging very slowly for large $\alpha t/L^2$. Thus the two kinds of solutions complement each other, depending on what value of $\alpha t/L^2$ is of interest. These considerations are important when we wish to use the actual equations, rather than the graphical solutions, as might be the case in computer programming, for example.

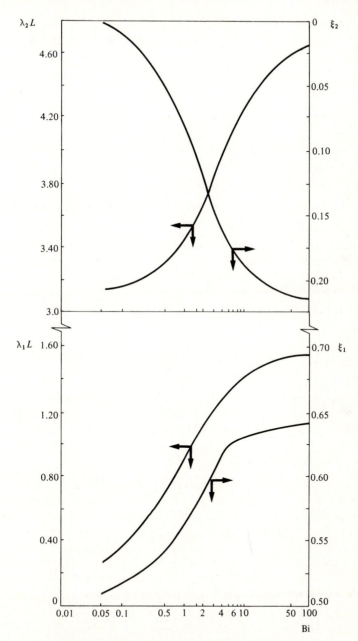

Fig. 9.15 Values of constants for Eq. (9.87). Note that values of ξ_2 are negative.

If we desire a long-time solution, rather than the complete series of Eq. (9.85), it is convenient to use only the first two terms of the series. For this case, we write

$$\frac{T - T_f}{T_i - T_f} = 2 \sum_{n=1}^{2} \xi_n \exp\left[-(\lambda_n L)^2 \text{Fo}\right] \cos\left[(\lambda_n L)\left(\frac{x}{L}\right)\right]. \qquad (9.87)$$

Here

$$\xi_n = \frac{\sin(\lambda_n L)}{\lambda_n L + \sin(\lambda_n L)\cos(\lambda_n L)}.$$

The value of $\lambda_n L$ depends on the value of the index n and the Biot number; hence ξ_n is also a function of the same. Values of $(\lambda_n L)$ and ξ_n for $n = 1$ and $n = 2$ are plotted in Fig. 9.15 for $0.05 \le \text{Bi} \le 100$.

Example 9.5 A slab of aluminum 4 in. thick at 975°F is quenched in a bath of water at 75°F. The heat-transfer coefficient is estimated to be 2000 Btu/hr-ft² °F.

a) Calculate the temperature at the center of the slab 30 sec after being plunged into the water. Use Eq. (9.87).

b) Repeat, using Fig. 9.11.

The properties of the aluminum are the same as those listed in Example 9.2.

Solution

$$\text{Bi} = \frac{hL}{k} = \frac{(2000)(2/12)}{45} = 7.41.$$

From Fig. 9.15

$$\lambda_1 L = 1.39, \qquad \xi_1 = 0.628$$

and

$$\lambda_2 L = 4.22, \qquad \xi_2 = -0.187.$$

Also

$$\text{Fo} = \frac{\alpha t}{L^2} = \frac{(45)(30/3600)}{(180)(0.24)(2/12)^2} = 0.313.$$

Substituting values into Eq. (9.87), we write

$$\frac{T - T_f}{T_i - T_f} = 2\{0.628 \exp\left[-(1.39^2)(0.313)\right] - 0.187 \exp\left[-(4.22^2)(0.313)\right]\}.$$

The first term within the brackets is much larger than the second; this indicates that Eq. (9.87) is converging rapidly, and therefore a good answer results by use of this equation. Solving, we have

a)
$$\frac{T - T_f}{T_i - T_f} = 0.686,$$

and

$$T = (0.686)(975 - 75) + 75 = 693°F.$$

b)
$$\frac{T - T_f}{T_i - T_f} = 0.695$$

and

$$T = (0.695)(975 - 75) + 75 = 701°F.$$

A difference of only 8°F lies between the two methods. This difference would have been smaller if we had considered longer times, but would have been larger for shorter times. Usually, Eq. (9.87) suffices for problems of this type if Fo > 0.25; it is wise, however, to compare results with those obtained by using Figs. 9.11–9.13.

When the long-time solution is not appropriate, we may utilize the alternative series mentioned above. Here we only present the first several terms of the series without going through the analysis of solving the differential equation. We may deduce the following expression from an equation given by Carslaw and Jaeger[4]

$$\frac{T - T_f}{T_i - T_f} = 1 - \left[\text{erfc} \frac{(1 - x/L)}{2\sqrt{\text{Fo}}} + \text{erfc} \frac{(1 + x/L)}{2\sqrt{\text{Fo}}} \right.$$

$$+ \exp[\text{Bi}(1 - x/L) + \text{Bi}^2 \text{Fo}] \cdot \text{erfc} \left[\text{Bi}\sqrt{\text{Fo}} + \frac{1 - x/L}{2\sqrt{\text{Fo}}} \right]$$

$$+ \exp\left[\text{Bi}(1 + x/L) + \text{Bi}^2 \text{Fo}\right] \cdot \text{erfc} \left[\text{Bi}\sqrt{\text{Fo}} + \frac{1 + x/L}{2\sqrt{\text{Fo}}} \right]. \quad (9.88)$$

Equation (9.88) contains the *complementary error function*, erfc. The erfc N is related to the erf N (the error function) simply by

$$\text{erfc } N = 1 - \text{erf } N,$$

and the erf N is defined as the value of a definite integral. The definite integral is

$$\text{erf } N = \frac{2}{\sqrt{\pi}} \int_0^N e^{-\beta^2} \, d\beta.$$

---

[4] H. S. Carslaw and J. C. Jaeger, *Conduction of Heat in Solids*, second edition, Oxford University Press, 1959, page 310.

The error function is commonly encountered, and hence it has been tabulated. An abbreviated compilation is given in Table 9.2.

Table 9.2 Tabulation of the error function

N	erf N	N	erf N	N	erf N
0.00	0.00000	0.50	0.5205	1.0	0.8427
0.05	0.05637	0.55	0.5633	1.1	0.8802
0.10	0.1125	0.60	0.6039	1.2	0.9103
0.15	0.1680	0.65	0.6420	1.3	0.9340
0.20	0.2227	0.70	0.6778	1.4	0.9523
0.25	0.2763	0.75	0.7112	1.5	0.9661
0.30	0.3286	0.80	0.7421	1.6	0.9763
0.35	0.3794	0.85	0.7707	1.7	0.9838
0.40	0.4284	0.90	0.7969	1.8	0.9891
0.45	0.4755	0.95	0.8209	1.9	0.9928
		1.00	0.8427	2.0	0.9953

Notes

a) $\text{erf } N = \dfrac{2}{\sqrt{\pi}} \displaystyle\int_0^N e^{-\beta^2}\, d\beta,$

b) $\text{erf } 0 = 0; \text{ erf } \infty = 1,$

c) $\text{erfc } N$ (complementary error function) $= 1 - \text{erf } N,$

d) $N < 0.2, \text{ erf } N \cong \dfrac{2N}{\sqrt{\pi}},$

e) $N > 2.0, \text{ erfc } N \cong \dfrac{e^{-N^2}}{\sqrt{\pi N}},$

f) $\text{erf}(-N) = -\text{erf}(N).$

Example 9.6 Repeat Example 9.5, but calculate the temperature at the center of the slab 12 sec after being plunged into the water.

Solution

$$\text{Fo} = (0.313)\left(\frac{12}{30}\right) = 0.125.$$

Then

$$\text{erfc }\frac{1 - x/L}{2\sqrt{\text{Fo}}} = \text{erfc }\frac{1}{2\sqrt{0.125}} = \text{erfc } 1.416 = 0.0452.$$

$$\exp\left[\mathrm{Bi}(1 - x/L) + \mathrm{Bi}^2\,\mathrm{Fo}\right] = \exp\left[7.41 + 7.41^2 \times 0.125\right]$$

$$= 1.55 \times 10^6.$$

$$\mathrm{erfc}\left[\mathrm{Bi}\sqrt{\mathrm{Fo}} + \frac{1 - x/L}{2\sqrt{\mathrm{Fo}}}\right] = \mathrm{erfc}\left[7.41\sqrt{0.125} + \frac{1}{2\sqrt{0.125}}\right]$$

$$= \mathrm{erfc}\,[4.04] = 0.230 \times 10^{-7}.$$

Therefore

$$\frac{T - T_f}{T_i - T_f} = 1 - (2 \times 0.0452) + (2 \times 1.55 \times 10^6 \times 0.230 \times 10^{-7}) = 0.981,$$

$$T = 0.981(975 - 75) + 75 = 924°F.$$

As the reader probably suspects, long-time and short-time solutions are also available for cylinders and spheres. These solutions can be deduced by reference to Carslaw and Jaeger.[5]

9.4.4 Cooling and heating rates

In metallurgy, the rates of heating and/or cooling are often more important than the determination of the temperature itself. For cases where Newtonian cooling applies, the determination of cooling rates is not too difficult; we discussed this in Section 9.4.1, and illustrated it in Example 9.2. However, when Newtonian cooling does not apply, it is necessary to resort to other means. If analytical expressions are desired, then solutions such as those discussed in Section 9.4.3 may be used to determine $\partial T/\partial t$. If approximate and graphical solutions are needed, then the charts in this section may be consulted. Figures 9.16, 9.17, and 9.18 include these charts for infinite plates, infinitely long cylinders, and spheres, respectively. In order to save space, we do not present the surface rates for cylinders and spheres. They are almost the same as those for infinite plates; for all values of **Bi**, the surface cooling rates for spheres are only 3–4% higher than for plates. However, the cooling or heating rates of plates, cylinders, and spheres do differ greatly in the internal positions.

Example 9.7 Determine the cooling rate at the center of the cylindrical bar, described in Example 9.3, when the center reaches 300°F.

Solution. From Example 9.3, the Biot number is 0.579. For the center, we have

$$\frac{T - T_f}{T_i - T_f} = \frac{300 - 80}{400 - 80} = 0.688.$$

From Fig. 9.12(a), we deduce that

$$\mathrm{Fo} \cong 0.50.$$

---

[5] H. S. Carslaw and J. C. Jaeger, *ibid.*

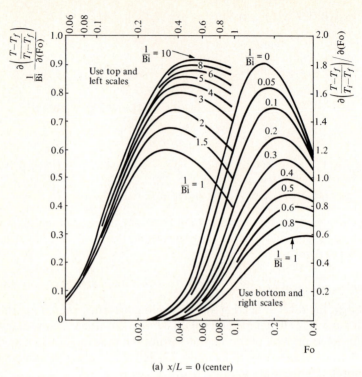

(a) $x/L = 0$ (center)

Fig. 9.16 Rate of temperature increase of an infinite plate initially at a uniform temperature, and then exposed to a uniform-temperature convective environment. (From V. Paschkis, *Welding Research Supplement*, Sept., 1946, pages 497–502.)

Then by using Fig. 9.17(a), we arrive at

$$\frac{1}{\text{Bi}} \frac{\partial\left(\dfrac{T - T_f}{T_i - T_f}\right)}{\partial(\text{Fo})} \cong 1.17.$$

Therefore

$$\frac{\partial T}{\partial t} = (1.17)(\text{Bi})(T_i - T_f)\left(\frac{\alpha}{R^2}\right)$$

$$= (1.17)(0.579)(400 - 80)\left(\frac{0.158}{25/576}\right)$$

$$= 791\ °\text{F/hr} \quad \text{or} \quad 0.22\,°\text{F/sec}.$$

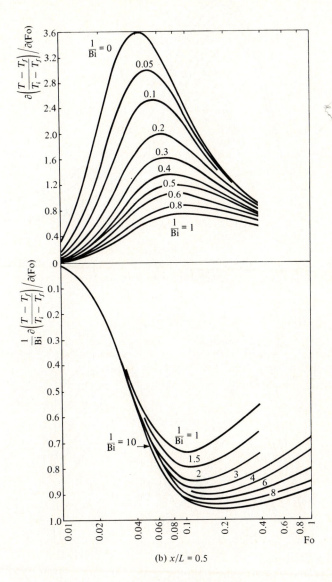

(b) $x/L = 0.5$

Fig. 9.16 (*continued*)

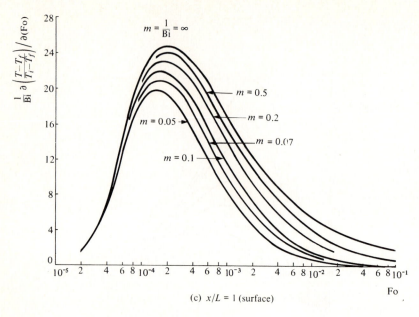

(c) $x/L = 1$ (surface)

Fig. 9.16 (*continued*)

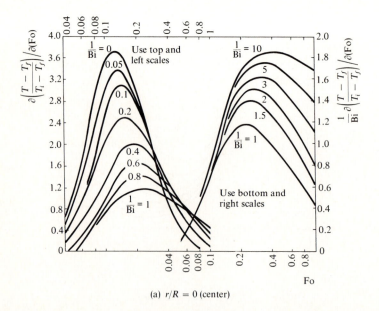

(a) $r/R = 0$ (center)

Fig. 9.17 Rate of temperature increase of infinitely long cylinders initially at a uniform temperature, and then exposed to a convective environment at constant temperature. (From V. Paschkis, *Welding Research Supplement*, Sept., 1946, pages 497–502.) For the surface refer to Fig. 9.16(c).

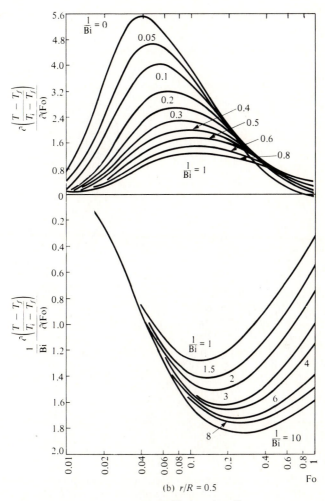

(b) $r/R = 0.5$

Fig. 9.17 (*continued*)

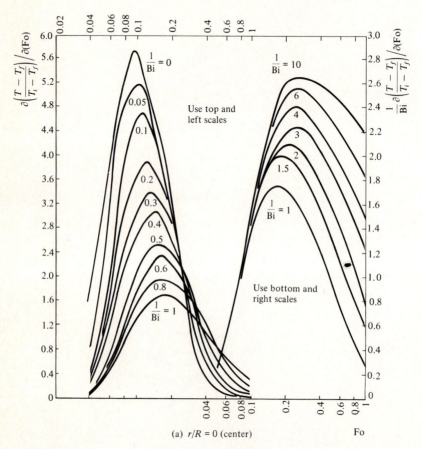

Fig. 9.18 Rate of temperature increase of spheres initially at a uniform temperature, and then exposed to a convective environment at constant temperature. For the surface refer to Fig. 9.16(c).

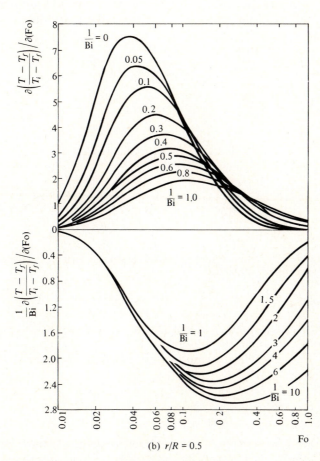

(b) $r/R = 0.5$

Fig. 9.18   *(continued)*

9.5 TRANSIENT CONDITIONS, INFINITE AND SEMI-INFINITE SOLIDS

In Section 9.4, all the solids we considered had at least one dimension of finite extent. In this section, we consider the so-called infinite and semi-infinite solids. The solutions we shall deal with can be applied when the time involved in transient situations is very short, or during the time of interest the depth of material affected by the boundary condition is less than the thickness of material itself. There will be many other applications, such as heat transfer in certain solidification problems, and also in diffusion problems which are discussed in Part III, Mass Transport.

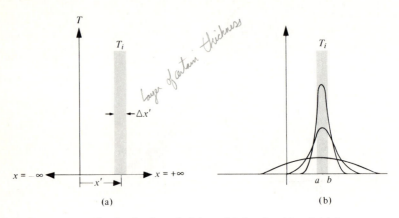

Fig. 9.19 Temperature distribution in an infinite solid showing (a) an initial temperature peak, and (b) the decay of the peak with time.

9.5.1 Infinite solid

Consider the infinite solid in Fig. 9.19(a) where at time zero a thin slice of material $\Delta x'$ thick at $x = x'$ is at some temperature T_i. This temperature peak decays with time, as shown in Fig. 9.19(b). Again for transient heat conduction, Eq. (9.66) applies, and the solution is

$$T(x, t) = \frac{T_i}{2\sqrt{\pi\alpha t}} \exp\left[-\frac{(x - x')^2}{4\alpha t}\right]\Delta x'. \tag{9.89}$$

If another slice of material of the same thickness exists at time zero with a different temperature T_i, then the temperature at x and t is the sum of contributions from both peaks. For this case then, Eq. (9.89) applied to the peaks at x'_1 and x'_2 is given by

$$T(x, t) = \sum_{n=1}^{2} \frac{T_{in}}{2\sqrt{\pi\alpha t}} \exp\left[-\frac{(x - x'_n)^2}{4\alpha t}\right]\Delta x'. \tag{9.90}$$

Now think of many such slices side by side occupying the whole space. If the thicknesses of the individual slices are allowed to approach zero, then all the various T_{in}'s can be considered to form a function of x', $f(x')$. In this case, examine the series in the limits as the slices all approach zero thickness

$$T(x, t) = \lim_{\Delta x' \to 0} \sum_{n=1}^{\infty} \frac{f(x')}{2\sqrt{\pi \alpha t}} \exp\left[\frac{-(x - x')^2}{4\alpha t}\right] \Delta x'. \tag{9.91}$$

As $\Delta x' \to 0$, the infinite summation becomes an integral

$$T(x, t) = \int_{x' = -\infty}^{\infty} \frac{f(x')}{2\sqrt{\pi \alpha t}} \exp\left[\frac{-(x - x')^2}{4\alpha t}\right] dx'. \tag{9.92}$$

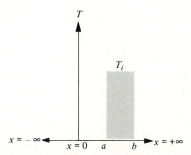

Fig. 9.20 The initial distribution of uniform temperature T_i between $x = a$ and $x = b$.

As an example consider an initial distribution (Fig. 9.20). For this case, we develop the solution by recognizing that $f(x') = 0$ for all x except $a < x < b$ where $f(x') = T_i$. Substituting this information into Eq. (9.92) yields

$$T = \int_{-\infty}^{a} \emptyset \, dx + \int_{a}^{b} \frac{T_i}{2\sqrt{\pi \alpha t}} \exp\left[\frac{-(x - x')^2}{4\alpha t}\right] dx' + \int_{b}^{+\infty} 0 \, dx. \tag{9.93}$$

Let $\beta \equiv (x' - x)/2\sqrt{\alpha t}$; this transforms Eq. (9.93) into

$$T = \frac{T_i}{\sqrt{\pi}} \int_{\beta = (a-x)/2\sqrt{\alpha t}}^{(b-x)/2\sqrt{\alpha t}} e^{-\beta^2} \, d\beta = \frac{T_i}{2}\left(\frac{2}{\sqrt{\pi}} \int_{0}^{(b-x)/2\sqrt{\alpha t}} e^{-\beta^2} \, d\beta - \frac{2}{\sqrt{\pi}} \int_{0}^{(a-x)/2\sqrt{\alpha t}} e^{-\beta^2} \, d\beta\right). \tag{9.94}$$

Therefore we write Eq. (9.94) as

$$T = \frac{T_i}{2}\left[\text{erf}\left(\frac{b - x}{2\sqrt{\alpha t}}\right) - \text{erf}\left(\frac{a - x}{2\sqrt{\alpha t}}\right)\right]. \tag{9.95}$$

9.5.2 Semi-infinite solid

A semi-infinite solid has an extent of $0 \leq x < \infty$, that is, a very thick solid with a bounding surface at $x = 0$. At this surface the transient is put into effect. For example, suppose we wish to solve the problem with the following boundary conditions for the region $0 \leq x < \infty$:

$$T(x, 0) = f(x);$$ (9.96)

$$T(0, t) = 0.$$ (9.97)

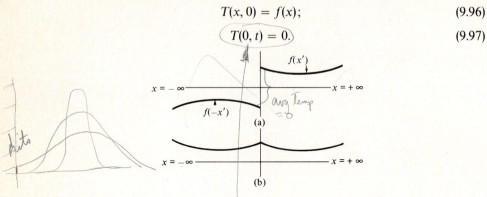

Fig. 9.21 Temperature distributions in infinite solids that satisfy boundary conditions for semi-infinite solids. (a) Odd function, $T(0, t) = 0$. (b) Even function, $\partial T(0, t)/\partial x = 0$.

We develop the solution to this problem by referring to an infinite solid with the initial condition depicted in Fig. 9.21(a). For the *odd function* where $f(-x') = -f(x')$, the boundary condition (9.97) is automatically satisfied; therefore we use Eq. (9.92) as follows to satisfy this boundary condition:

$$T = \int_{-\infty}^{0} \frac{-f(-x')}{2\sqrt{\pi\alpha t}} \exp\left[\frac{-(x-x')^2}{4\alpha t}\right] dx' + \int_{0}^{\infty} \frac{f(x')}{2\sqrt{\pi\alpha t}} \exp\left[\frac{-(x-x')^2}{4\alpha t}\right] dx'. \quad (9.98)$$

The problem with the flux equal to zero at the surface is set up in a similar manner. For this case, the boundary conditions for the region $0 \leq x < \infty$ become

$$T(x, 0) = f(x);$$ (9.99)

$$\frac{\partial T}{\partial x}(0, t) = 0.$$ (9.100)

We indicate the method of setting up this situation in Fig. 9.21(b), where an *even function*, $f(x') = f(-x')$, gives symmetry to the temperature field about $x = 0$, so that condition (9.100) is automatically satisfied. Thus we use Eq. (9.92) in the following form

$$T = \int_{-\infty}^{0} \frac{f(-x')}{2\sqrt{\pi\alpha t}} \exp\left[\frac{-(x-x')^2}{4\alpha t}\right] dx' + \int_{0}^{+\infty} \frac{f(x)}{2\sqrt{\pi\alpha t}} \exp\left[\frac{-(x-x')^2}{4\alpha t}\right] dx'. \quad (9.101)$$

Now as an application of Eq. (9.98), consider the important and often encountered problem of the semi-infinite solid with the boundary conditions

$$T(x, 0) = T_i;$$ (9.102)

$$T(0, t) = T_s.$$ (9.103)

If we define $\theta = T - T_s$, the boundary conditions become

$$\theta(x, 0) = T_i - T_s = \theta_i; \tag{9.104}$$

$$\theta(0, t) = 0. \tag{9.105}$$

Thus Eq. (9.98) applies for θ, where $f(x') = \theta_i$ (uniform initial distribution). First, we put Eq. (9.98) in a more convenient form:

$$\theta = \frac{1}{2\sqrt{\pi\alpha t}} \int_0^\infty f(x') \left\{ \exp\left[\frac{-(x - x')^2}{4\alpha t}\right] - \exp\left[\frac{-(x + x')^2}{4\alpha t}\right] \right\} dx'. \tag{9.106}$$

Next we change variables, $\beta = (x' - x)/2\sqrt{\alpha t}$ and $\beta' = (x' + x)/2\sqrt{\alpha t}$, and substitute $f(x') = \theta_i$:

$$\frac{\theta}{\theta_i} = \frac{1}{\sqrt{\pi}} \int_{\beta = -x/2\sqrt{\alpha t}}^\infty e^{-\beta^2} d\beta - \frac{1}{\sqrt{\pi}} \int_{\beta' = +x/2\sqrt{\alpha t}}^\infty e^{-\beta'^2} d\beta', \tag{9.107}$$

or noting that primes are no longer necessary, we have

$$\frac{\theta}{\theta_i} = \frac{1}{\sqrt{\pi}} \left(\int_{\beta = -x/2\sqrt{\alpha t}}^\infty e^{-\beta^2} d\beta + \int_{\beta = \infty}^{+x/2\sqrt{\alpha t}} e^{-\beta^2} d\beta \right). \tag{9.108}$$

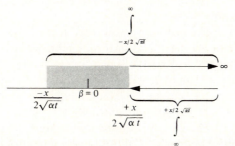

Fig. 9.22   Schematic representation of the integral in Eq. (9.108).

Figure 9.22 schematically indicates these integrals, and shows that their sum results in

$$\frac{\theta}{\theta_i} = \frac{1}{\sqrt{\pi}} \int_{-x/2\sqrt{\alpha t}}^{+x/2\sqrt{\alpha t}} e^{-\beta^2} d\beta = \frac{2}{\sqrt{\pi}} \int_0^{x/2\sqrt{\alpha t}} e^{-\beta^2} d\beta. \tag{9.109}$$

The solution in its final form is

$$\frac{T - T_s}{T_i - T_s} = \text{erf}\,\frac{x}{2\sqrt{\alpha t}}. \tag{9.110}$$

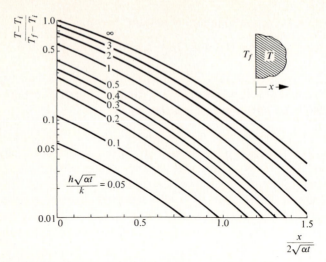

Fig. 9.23 Temperature history in a semi-infinite solid with surface resistance. (From P. J. Schneider, *Conduction Heat Transfer*, Addison-Wesley, 1955, page 266.)

To summarize, Eq. (9.110) describes the temperature as a function of position and time in a semi-infinite solid initially at a uniform temperature T_i, with its surface temperature suddenly raised or lowered to T_s at $t = 0$ and maintained for $t > 0$. In practice, we consider a solid to be semi-infinite until such time as all of the material differs appreciably from T_i.

Equation (9.110) is applied to many problems in the engineering and metallurgical literature. We apply it to solidification problems in Chapter 10 and to numerous diffusion problems in Part III.

Another important problem of heat flow for a semi-infinite solid remains to be discussed. For the cooling (or heating) of slabs, Eq. (9.85) was developed; however, as mentioned previously, this equation is unwieldy for "short times." For short times we found that Eq. (9.88) was more appropriate. Here, we present a solution that is a slightly simplified version of Eq. (9.88), which applies for the very short times when the center of the slab has not yet felt the effects of the changing temperature field. For such a situation, we may consider the solid to be semi-infinite, and the following initial and boundary conditions must be satisfied:

$$T(x, 0) = T_i; \tag{9.111}$$

$$k\frac{\partial T(0, t)}{\partial x} = h[T(0, t) - T_f]. \tag{9.112}$$

We present only the final solution:

$$\frac{T - T_i}{T_f - T_i} = \text{erfc}\left(\frac{x}{2\sqrt{\alpha t}}\right) - e^{\gamma}\,\text{erfc}\left[\frac{x}{2\sqrt{\alpha t}} + \frac{h}{k}\sqrt{\alpha t}\right], \tag{9.113}$$

where

$$\gamma = \frac{h}{k}\sqrt{\alpha t}\left[\frac{x}{\sqrt{\alpha t}} + \frac{h}{k}\sqrt{\alpha t}\right]. \tag{9.114}$$

The solution given by Eq. (9.113) is also shown in Fig. 9.23.

9.6 SIMPLE MULTIDIMENSIONAL PROBLEMS

In this section we present methods of obtaining solutions for solid shapes such as cubes, rectangular bars, and short cylinders. These solutions can be obtained directly by simply combining solutions for the semi-infinite solid, the infinitely long cylinder, and the infinite slab, which are given in Sections 9.4 and 9.5.

As an example, consider an infinitely long bar with a rectangular cross section $2L$ by $2l$, using coordinates as illustrated by Fig. 9.24. The bar is initially at the uniform temperature T_i, and then it is suddenly exposed to a convective environment at T_f. In such a bar, $T(x, y, t)$ must satisfy the partial differential equation

$$\frac{\partial T}{\partial t} = \alpha\left(\frac{\partial^2 T}{\partial x^2} + \frac{\partial^2 T}{\partial y^2}\right). \tag{9.115}$$

We can prove that the solution $T(x, y, t)$ is the simple product

$$T(x, y, t) = X(x, t) \cdot Y(y, t). \tag{9.116}$$

Here, $X(x, t)$ is the solution for the temperature response in the infinite slab of thickness $2L$, and $Y(y, t)$ is the solution for the infinite slab of thickness $2l$. The reader is invited to pursue Problem 9.14 at the end of this chapter to satisfy himself of the validity of Eq. (9.116).

We may also make use of the solution for the semi-infinite solid in combination with solutions to solids such as the infinite plate or semi-infinite cylinder. Specifically, consider the semi-infinite cylinder for which we seek the solution of $T(r, y, t)$. If $S(y, t)$ represents the solution to the semi-infinite solid in the regime $0 \le y < \infty$, and $C(r, t)$ is the solution for the infinitely long cylinder, then the solution we seek is simply their product, that is,

$$T(r, y, t) = S(y, t) \cdot C(r, t). \tag{9.117}$$

By means of this method, many so-called *product solutions* can be developed for a large number of solid shapes, some of which are depicted in Fig. 9.25, with their respective solutions indicated.

Example 9.8 A short cylindrical bar of stainless steel, 5 in. in diameter and 6 in. long, is heated to 400°F. The bar is then cooled in a blast of fan-forced air at 80°F with $h = 25$ Btu/hr-ft² °F. Use the same thermal properties as in Example 9.3.

a) After the bar has been cooled for 10 min, what is the temperature at its geometrical center?

b) What is the surface temperature midway between the ends after 10 min of cooling?

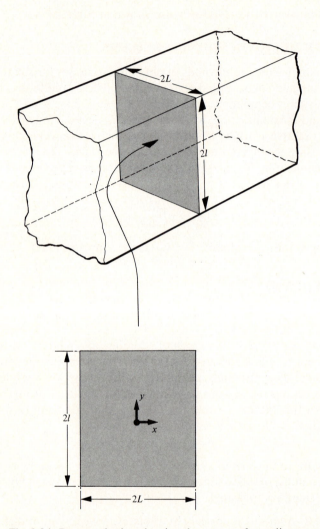

Fig. 9.24 Rectangular bar showing the system of coordinates.

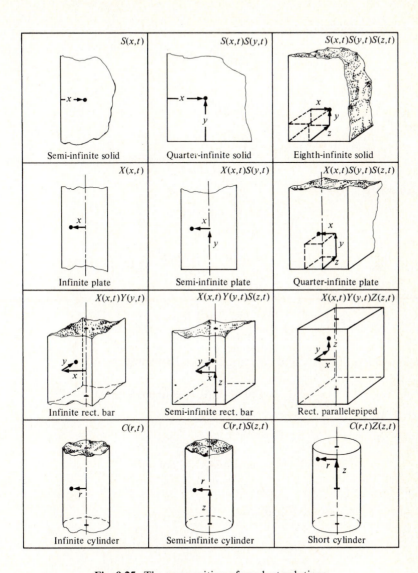

Fig. 9.25 The composition of product solutions.

Solution

a) From Example 9.3,

$$Bi_R = \frac{hR}{k} = 0.579,$$

$$Fo_R = \frac{\alpha t}{R^2} = 3.1,$$

and

$$C(0, t) = \left(\frac{T_0 - T_f}{T_i - T_f}\right)_{\text{inf.cyl.}} = 0.0625.$$

Additionally,

$$Bi_L = \frac{hL}{k} = \frac{(25)(3/12)}{9} = 0.695,$$

$$Fo_L = \frac{\alpha t}{L^2} = \frac{(0.158)(0.84)}{(3/12)^2} = 2.13,$$

and from Fig. 9.11(a),

$$X(0, t) = \left(\frac{T_0 - T_f}{T_i - T_f}\right)_{\text{inf.plate}} = 0.31.$$

The answer we seek is found by using the product solution.

$$\frac{T - T_f}{T_i - T_f} = C(0, t) \cdot X(0, t) = (0.0625)(0.31)$$

$$= 0.0194.$$

Then

$$T = (0.0194)(400 - 80) + 80$$

$$= 86.2°F.$$

b) For this case we use Fig. 9.12(b):

$$C(R, t) = \left(\frac{T_R - T_f}{T_i - T_f}\right)_{\text{inf.cyl.}} = 0.040.$$

$$\frac{T - T_f}{T_i - T_f} = C(R, t) \cdot X(0, t) = (0.040)(0.31)$$

$$= 0.0124.$$

Then

$$T = (0.0124)(400 - 80) + 80$$

$$= 84.0°F.$$

PROBLEMS

9.1 A furnace wall is constructed of 7 in. of fire brick ($k = 0.60$ Btu/hr-ft °F), 4 in. of red brick ($k = 0.40$), 1 in. of glass-wool insulation ($k = 0.04$), and $\frac{1}{8}$ in. steel plate ($k = 26$) on the outside. The heat-transfer coefficients on the inside and outside surfaces are 9 and 3 Btu/hr-ft² °F, respectively. The gas temperature inside the furnace is 2500°F and the outside air temperature is 90°F.

a) Calculate the heat-transfer rate through the wall (Btu/hr-ft²).
b) Determine the temperatures at all interfaces.

9.2 Consider the flow of heat through a spherical shell. For steady state conditions, the inside surface ($r = R_1$) is at temperature T_1, and the outside surface ($r = R_2$) is at T_2.

a) Write the pertinent differential energy equation that applies.
b) Write the boundary conditions and develop an expression for the temperature distribution in the shell.
c) Develop an expression for the heat flow (Q, Btu/hr) through the shell.

9.3 A semi-infinite plate with edges at $x = 0$, $x = L$, $y = 0$ and $y = \infty$ is subjected to the following boundary conditions at steady state.

$$x = 0, \quad T = 0$$
$$x = L, \quad T = 0$$
$$y = \infty, \quad T = 0$$
$$y = 0, \quad T = T_A \sin\frac{\pi x}{L} \quad (T_A = \text{constant}).$$

a) Write an expression for $T(x, y)$.
b) Write an expression for the heat flux along the edge $y = 0$.

9.4 The temperature T is maintained at 0°F along the three sides of a long bar with a square cross section (1 ft × 1 ft). The fourth side is maintained at 100°F.

a) At steady-state, find an expression for the temperature T at any point (x, y) in the bar.
b) Calculate the rate at which heat is transferred through the hot face ($k = 50$ Btu/hr-ft °F). Express your answer in Btu/hr.
c) What is the temperature at the center of the bar?

9.5 A thin wire is extruded at a fixed velocity through dies, and the wire temperature at the die is a fixed value T_0. The wire then passes through the air for some distance before it is rolled onto large spools where the temperature has been reduced to T_L. It is desired to investigate the relationship between wire velocity and the distance between the extrusion nozzle and roll for the specific values of T_0 and T_L.

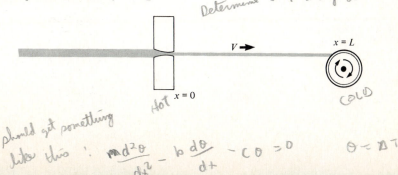

a) Derive the differential equation for determining wire temperature as a function of distance from the nozzle. [*Hint*: Since temperature gradients across the wire are certainly negligible, a slice between x and $x + \Delta x$ may be chosen that includes the wire surface. The heat balance then includes heat lost to the surroundings at T_a].

b) State boundary conditions and solve for the temperature in the extended wire.

Answer

$$\frac{T - T_a}{T_0 - T_a} = \exp\left[-\left(\beta - \sqrt{\beta^2 + \frac{2h}{Rk}}\right)x\right]; \qquad \beta \equiv \frac{V}{2\alpha}.$$

9.6 Steel ball bearings ($\frac{1}{2}$ in. diameter) are austenitized at 1600°F and then quenched in a large tank of oil at 100°F. Calculate:

a) The time to cool the center of the bearings to 400°F.

b) The surface temperature when the center is at 400°F.

c) The space-mean temperature when the center is at 400°F.

d) If 10,000 balls are quenched per hour, calculate the rate of heat removal from the oil that must be accomplished in order to maintain its temperature at 100°F.

Data:

$$h = 300 \text{ Btu/hr-ft}^2 \text{ °F},$$

$$\rho = 450 \text{ lb}_m/\text{ft}^3,$$

$$C_p = 0.15 \text{ Btu/lb}_m \text{ °F},$$

$$k = 25 \text{ Btu/hr-ft °F}.$$

(handwritten annotations: $\frac{m}{v} = \rho$; $\Delta H = C_p \Delta T$; $0.15(1600 - 374)$; $\rho = \frac{m}{v}$)

9.7 Copper shot is made by dropping molten droplets into water at 100°F. The droplets may be approximated as spheres with a diameter of 0.2 in. Calculate the total time for the droplets to cool to 200°F if they enter the water at 2200°F.

Data for Cu, in units of Btu, lb_m, ft, °F, etc.:

$$\text{Freezing point} = 1985\text{°F}$$

$$C_p \text{ (solid)} = 0.09$$

$$C_p \text{ (liquid)} = 0.12$$

$$\text{Heat of fusion} = 89$$

$$\rho \text{ (solid)} = 560$$

$$\rho \text{ (liquid)} = 530$$

$$k \text{ (solid)} = 200$$

$$k \text{ (liquid)} = 150$$

Quench data for water:

Temperature range	h, Btu/hr-ft^2 °F
2200–1200°F	80
1200–200°F	400

9.8 Steel ball bearings (0.2 ft in diameter) are austenitized at 1500°F and then quenched in fluid X at 100°F. It is known by utilizing a thermocouple that a continuous vapor film surrounds the bearings for 72 sec until the surface temperature drops to 500°F and at the same time the center temperature is 700°F. Knowing these results, determine the time it takes for the center of smaller bearings (0.02 ft diameter) of the same steel to reach 1200°F when quenched from 1500°F into fluid X at 100°F.

9.9 An open-ended cylindrical section of a steel pressure vessel 10 ft in diameter with 8-in. thick walls is being heat-treated. The wall temperature is brought to a uniform value of 1750°F. Then the vessel is quenched in slow oil at 70°F.

a) How long does it take for the surface to reach 1000°F?

b) What is the temperature at the center of the wall at that time?

9.10 A cylindrical piece of steel 2 in. in diameter and initially at 1600°F is quenched into 70°F water ($H = 18.0$). Calculate the temperature at the surface of the piece after 1 min, 2 min, and 5 min. Compare your results with the temperature at the same location if the piece had been quenched in oil ($H = 6.0$).

TABLE 8.1
PG 259

9.11 Compute the temperature, as a function of time, across a slab of steel 4 in. thick, cooled from 1600°F by water sprays from both sides.

9.12 Consider a short cylinder 6 in. high and with a diameter of 6 in. The cylinder is initially at a uniform temperature of 500°F and cools in ambient air at 80°F. Assume steel.

a) Write the partial differential equation that describes the temperature within the cylinder.

b) Calculate the temperature at the geometric center after 1 hr of cooling.

c) Calculate the temperature on the cylindrical surface midway between the end faces after 1 hr of cooling.

d) In answering parts (b) and (c), show why your calculation procedure was justified, that is, demonstrate that the differential equation in part (a) is satisfied.

9.13 A steel blank, 1 ft in diameter and 2 ft long, is heated in a preheating furnace maintained at 2080°F.

a) Calculate the temperature in the center of the blank after the blank has been heated for 2 hr from an initial temperature of 80°F.

b) Calculate the time required to heat a smaller blank, $\frac{1}{2}$ ft in diameter and 1 ft long, to the same center temperature as the larger blank in Part (a).

Data:

$$h = 20 \text{ Btu/hr-ft}^2 \text{ °F},$$

$$k = 20 \text{ Btu/hr-ft °F},$$

$$\rho = 480 \text{ lb}_m/\text{ft}^3,$$

$$C_p = 0.10 \text{ Btu/lb}_m \text{°F}.$$

9.14 The temperature field $T(x, y, t)$ in an infinitely long rectangular ($2L \times 2l$) bar must satisfy the partial-differential equation

$$\frac{\partial^2 T}{\partial x^2} + \frac{\partial^2 T}{\partial y^2} = \frac{1}{\alpha} \frac{\partial T}{\partial t}.$$

Prove that $T(x, y, t)$ can be found by the product

$$\cdot T(x, y, t) = T_l(x, t) \cdot T_L(y, t),$$

where $T_l(x, t)$ is the solution for the temperature history in the semi-infinite plate bounded by $-l < x < +l$, and $T_L(y, t)$ is the solution for the temperature history in the semi-infinite plate bounded by $-L < y < L$.

9.15 In the flame hardening of surfaces of thick (semi-infinite) steel parts a very hot flame is played on the surface for a short time and then a water quench follows directly. If the surface of a steel part is brought essentially instantaneously to 2400°F, and the flame 1 in. wide moves at a rate of 1 in. per min with a water quench following directly, to what depth can a steel with the continuous cooling transformation diagram given below be hardened to a 100% martensite structure? This will have to be an estimate of a limiting case.

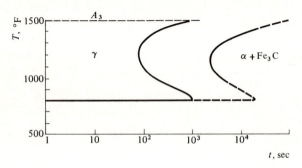

use solution curve with Bi # around 1000

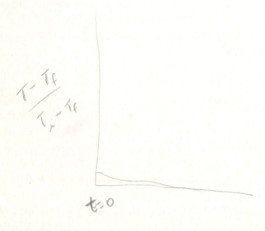

$$\frac{T - T_f}{T_i - T_f}$$

$t = 0$

SOLIDIFICATION HEAT TRANSFER

The production of most metal parts, except of articles produced by powder-metallurgy techniques, involves solidification. Castings obviously entail solidification; forgings and wrought products are also castings that have been hot worked, and their behavior in many cases can be traced back to the method of solidification. In particular, the solidification rate of alloys is an extremely important processing variable. The solidification rate relates directly to the coarseness—or fineness—of dendritic structures and hence controls the spacing and distribution of micro-segregates, such as coring, second phases, and inclusions. Thermal gradients during freezing are also of great significance, being related to the formation of microporosity in alloys. For these metallurgical reasons and from a process engineering viewpoint, solidification heat transfer should be recognized as an important topic.

The analysis of heat transfer during solidification is more complex than that of conduction heat transfer presented in Chapter 9. This is one of the reasons for the paucity of the literature devoted to this topic compared to that available for conduction heat flow in solids. However, sufficient theory has evolved to treat many practical problems, and the metallurgist should be aware of some of the analyses.

10.1 SOLIDIFICATION IN SAND MOLDS

The largest quantity of metal is cast in sand molds, excepting the tonnage of steel cast in ingot molds. The following analysis applies when the metal solidifies in sand molds, or more generally, when the predominant resistance to heat flow is within the mold itself, e.g., the mold is made of plaster, granulated zircon, mullite, or various other materials that are poor conductors of heat.

Consider pure liquid metal with no superheat poured against a flat mold wall of a poor conductor. Figure 10.1 shows the temperature distribution in the metal and the mold at some time during solidification. Because all the resistance to heat flow is almost entirely within the mold, the surface temperature T_s nearly equals the melting temperature of the metal T_M. This means that during freezing the temperature drop through the solidified metal is small, and at the metal–mold interface a constant temperature of $T_s \cong T_M$ is maintained. Under these conditions,

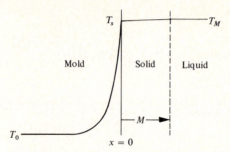

Fig. 10.1 The temperature distribution during solidification of a metal in a sand mold.

the temperature history in the mold is given by Eq. (9.110) (i.e., the solution for a semi-infinite body):

$$\frac{T - T_M}{T_0 - T_M} = \text{erf} \frac{x}{2\sqrt{\alpha t}}, \tag{10.1}$$

where x is the distance into the mold, α is its thermal diffusivity, and T_0 is the initial uniform temperature (usually T_0 is *room temperature*). The use of this equation certainly implies that the mold is sufficiently thick to satisfy the boundary condition $T(\infty, t) = T_0$. In practice, this requirement is often met because the heat-affected zone in the mold is confined to a layer of sand only about one-quarter of the casting thickness.

Of primary interest is not the mold's temperature history, but rather the rate at which heat is extracted from the solidifying metal, which ultimately leads to a determination of the total solidification time. Equation (10.1) is used to obtain the amount of heat which flows into the mold, and this quantity of heat must equal the latent heat evolved during solidification.

The heat flux into the mold follows from Eq. (10.1):

$$q|_{x=0} = -k\left(\frac{\partial T}{\partial x}\right)_{x=0} = \frac{k(T_M - T_0)}{\sqrt{\pi \alpha t}}. \tag{10.2}$$

Remembering that $\alpha = k/\rho C_p$, we rewrite Eq. (10.2) as

$$q|_{x=0} = \frac{\sqrt{k\rho C_p}}{\sqrt{\pi t}}(T_M - T_0). \tag{10.3}$$

The product $k\rho C_p$ represents the ability of the mold to absorb heat at a certain rate and is called the *heat diffusivity*.

The rate at which latent heat is evolved per unit area can be written

$$\rho' H_f \frac{dM}{dt}, \tag{10.4}$$

where ρ' = density of solidifying metal, lb_m/ft^3, H_f = latent heat of fusion of the metal, Btu/lb_m, and M = thickness of metal solidified, ft.

Equating Eq. (10.3) to Eq. (10.4) yields the rate at which the interface advances into the liquid:

$$\frac{dM}{dt} = \frac{(T_M - T_0)\sqrt{k\rho C_p}}{\rho'H_f\sqrt{\pi t}}. \qquad (10.5)$$

Integration follows with the limits

$$M = 0 \qquad \text{at } t = 0 \qquad (10.6a)$$

and

$$M = M \qquad \text{at } t = t, \qquad (10.6b)$$

$$M = \frac{2}{\sqrt{\pi}}\left(\frac{T_M - T_0}{\rho'H_f}\right)\sqrt{k\rho C_p}\sqrt{t}. \qquad (10.7)$$

SAND MOLD
thickness eqn.

Thus, we see that the amount of solidification depends on certain metal characteristics, $(T_M - T_0)/\rho'H_f$, and the mold's heat diffusivity, $k\rho C_p$.

10.1.1 Effect of contour on solidification time

Freezing from a planar mold wall, as discussed above, is not the usual problem engineers encounter in practice. It is often important to evaluate the freezing times of complex shapes, in which the contour of the mold wall has some influence on solidification time. For example, contrast heat flow into the convex and concave walls to the plane mold wall situation shown in Fig. 10.2. Heat flow into the convex surface is divergent and, therefore, slightly more rapid than into the plane mold. In contrast, heat flow into the concave surface is convergent and less rapid than into the plane mold wall.

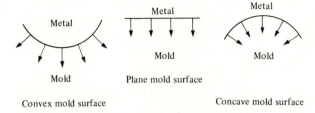

Convex mold surface   Concave mold surface

Fig. 10.2 Effect of contour on the heat flux into molds.

As a first approximation, such effects are sometimes neglected because the heated zone in the mold is shallow, and the difference in heat flow between a plane mold wall and a contoured wall is small. As such, we visualize that a given mold surface area has the ability to absorb a certain amount of heat in a given time regardless of its contour. Thus, we generalize Eq. (10.3) for all contours, and for a given surface area A, the mold absorbs an amount of heat Q in time t:

$$Q = \int_0^t Aq|_{x=0}\, dt = \frac{Ak(T_M - T_0)}{\sqrt{\pi\alpha}} \int_0^t \frac{dt}{\sqrt{t}}$$

$$= \frac{2Ak(T_M - T_0)}{\sqrt{\pi\alpha}}\sqrt{t}. \tag{10.8}$$

For a casting of volume V to completely solidify, all its latent heat must be removed; hence, the total latent heat Q evolved is

$$Q = \rho' V H_f. \tag{10.9}$$

Equations (10.8) and (10.9) are then combined to yield the solidification time of a casting in terms of its volume-to-surface area ratio:

$$t = C\left(\frac{V}{A}\right)^2, \tag{10.10}$$

where

$$C \equiv \frac{\pi}{4}\left(\frac{\rho' H_f}{T_M - T_0}\right)^2 \left(\frac{1}{k\rho C_p}\right).$$

Equation (10.10) is often referred to as Chvorinov's rule, and C, as Chvorinov's constant. It permits comparison of freezing times of castings with different shapes and sizes. The relationship works best for casting geometries in which none of the mold material becomes saturated with heat, such as in internal corners or internal cores. The success of this relationship hinges on the mold material absorbing the same amount of heat per unit area exposed to the metal. This is strictly true only for castings which have similar shapes but different sizes.

In some applications when more precision is required, it is necessary to account for the effect of mold contour on solidification. To quantify some contour effects, let us examine differences between castings of three basic shapes, namely, the infinite plate, the infinitely long cylinder, and the sphere. First, we define two dimensionless parameters, β and γ:

$$\beta \equiv \frac{V/A}{\sqrt{\alpha t}},$$

and

$$\gamma \equiv \left(\frac{T_M - T_0}{\rho' H_f}\right)\rho C_p.$$

With these parameters, the freezing times for the three basic shapes may be compared.

For the infinite plate,

$$\beta \equiv \gamma\left(\frac{2}{\sqrt{\pi}}\right).$$

(10.11)

For the cylinder,

$$\beta \equiv \gamma\left(\frac{2}{\sqrt{\pi}} + \frac{1}{4\beta}\right).$$

(10.12)

For the sphere,

$$\beta = \gamma\left(\frac{2}{\sqrt{\pi}} + \frac{1}{3\beta}\right).$$

(10.13)

We can deduce Eq. (10.11) from Eq. (10.10) by a rearrangement. Equations (10.12) and (10.13) have resulted by rearranging expressions presented by Adams and Taylor;[1] their expression for the cylinder is approximate while that for the sphere is exact.

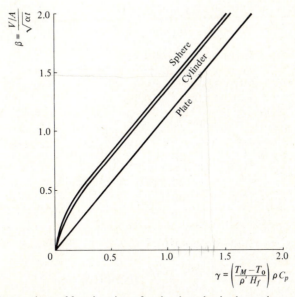

Fig. 10.3 Comparison of freezing times for the three basic shapes in sand molds.

These expressions show the error of using Chvorinov's rule without regard to the mold contours. For example, let us refer to Fig. 10.3 which relates freezing times for the three basic shapes according to Eqs. (10.11)–(10.13). By calculating a

---

[1] C. M. Adams and H. F. Taylor, *Trans. AFS* **65**, 170–176 (1957).

value of γ from properties of the metal and mold, we read a value of β corresponding to the different shapes from the curves of Fig. 10.3. For metal–sand combinations, γ is approximately unity, so that

$$\beta \text{ (plate)} = 1.13,$$

$$\beta \text{ (cylinder)} = 1.32,$$

$$\beta \text{ (sphere)} = 1.38.$$

We see that neglecting the contour can lead to an error of as much as 40–50% in freezing time.

The expression for solidification of a sphere, Eq. (10.13), may be used for other chunky shapes such as cubes with improvement in accuracy over the simple relation, Eq. (10.11). Similarly, the expression for solidification time of a cylinder may be used to approximate the freezing time of bars of square cross section.

Example 10.1 Determine the solidification time of the following iron castings, both poured with no superheat into sand molds:

a) a slab-shaped casting 4 in. thick;

b) a spherically shaped casting 4 in. in diameter.

Iron data:

> Freezing temperature = 2802°F,
> Heat of fusion = 117 Btu/lb$_m$,
> Solid density = 490 lb$_m$/ft^3,
> Liquid density = 460 lb$_m$/ft^3,
> Heat capacity of liquid = 0.18 Btu/lb$_m$ °F.

Sand data:

> Heat capacity = 0.28 Btu/lb$_m$ °F,
> Thermal conductivity = 0.50 Btu/ft-hr °F,
> Density = 100 lb$_m$/ft^3.

Solution

(Assume the mold is initially at 82°F.)

$$\gamma \equiv \left[\frac{2802 - 82}{(490)(117)} \right] (100)(0.28) = 1.33,$$

Fig 10.3

$$\alpha = \frac{0.50}{(100)(0.28)} = 0.0178 \text{ ft}^2/\text{hr}.$$

a) For plate castings, from Fig. 10.3,

$$\beta = \frac{V/A}{\sqrt{\alpha t}} = 1.51,$$

and for an infinite plate, $V/A = L$ where L is the semithickness. Therefore

$$t = \frac{L^2}{1.51^2\alpha} = \frac{(2/12)^2}{(1.51)^2(0.0178)} = 0.685 \text{ hr.}$$

b) For a spherical casting, from Fig. 10.3,

$$\beta = \frac{V/A}{\sqrt{\alpha t}} = 1.75,$$

and for a sphere, $V/A = R/3$, in which R is the radius. Therefore

$$t = \frac{R^2}{9(1.75^2)\alpha} = \frac{(2/12)^2}{(9)(1.75^2)(0.0178)} = 0.0564 \text{ hr.}$$

The sphere solidifies in less than one-tenth the time required for the slab to solidify.

10.1.2 Effect of superheat on solidification time

We can assess the effect of superheat on the solidification time by realizing that, in addition to absorbing latent heat, the sand must also absorb the superheat. Again, we assume that the temperature gradients within the casting are negligible, and at the time solidification is complete, the entire casting is close to its freezing point. In this case, the total quantity of heat to be removed from the casting is

$$Q = \rho' V H_f + \rho'_l V C_{p,l} \Delta T_s. \tag{10.13a}$$

The subscript l denotes liquid phase properties, and ΔT_s is the amount of superheat in degrees.

We now consider infinite plate castings, and in order to make the analysis simple, yet sufficiently accurate, we assume that Eq. (10.8) is valid even though the interface temperature of the mold is not constant while the liquid phase loses its superheat. In view of this approximation, it is certainly acceptable not to distinguish differences in the density of the liquid and solid phases. Thus $\rho'_l \cong \rho'$, and when Eq. (10.13a) is set equal to Eq. (10.8), we obtain

$$t = \frac{\pi}{4} \frac{1}{k\rho C_p}\left(\frac{\rho' H'_f}{T_M - T_0}\right)^2 \left(\frac{V}{A}\right)^2. \tag{10.14}$$

In this expression, H'_f is the *effective heat of fusion*, and represents the sum of the latent heat of fusion and the liquid's superheat, that is,

$$H'_f = H_f + C_{p,l} \Delta T_s. \tag{10.15}$$

Note that the solidification time is still proportional to $(V/A)^2$.

10.2 SOLIDIFICATION IN METAL MOLDS

When poured into metal molds, castings freeze rapidly, and temperatures change drastically in both the mold and the casting. An understanding of the variables affecting solidification in metal molds is important because most ingots, all permanent mold castings, and all die castings are made in metal molds. Also many sand castings are made in molds that incorporate metal inserts at strategic positions to increase the rate of solidification.

The analysis of heat transfer when metal is poured against a chill wall is more complicated than that when metal is poured into a sand mold; this is due to the fact that metal molds are much better heat conductors than sand molds. The added complexities are illustrated by the casting-mold situation shown in Fig. 10.4. At the solidified metal–mold interface, a temperature drop exists, due to thermal contact resistance. The condition of no contact resistance would exist only if the mold–casting contact were so intimate that *wetting* would occur, that is, the casting would become soldered to the mold.

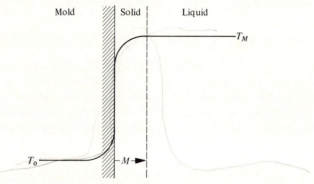

Fig. 10.4 The temperature distribution during the solidification of a metal from a chill wall.

In addition to the contact resistance, there are other differences between solidification in sand and in chill molds:

a) The thermal conductivity of the metal being cast forms an important portion of the overall resistance to heat flow. This results in the surface temperature being well below the melting point, while appreciable thermal gradients exist within the solidifying metal.

b) More total heat is removed during solidification because of the solidified metal which is subcooled. Thus, the heat capacity of the solidifying metal is important.

In the following sections, we shall discuss selected problems of solidification, each of them being of some practical importance.

10.2.1 Constant casting surface temperature

Consider a mass of pure liquid metal, initially at its freezing temperature, which has its surface suddenly cooled to a constant temperature T_s. After some solidification has occurred, the temperature profile in the solidifying metal will appear as the solid profile (Fig. 10.5a). The temperature profile is identical to the temperature profile in the semi-infinite solid depicted in Fig. 10.5(b), except that the temperature

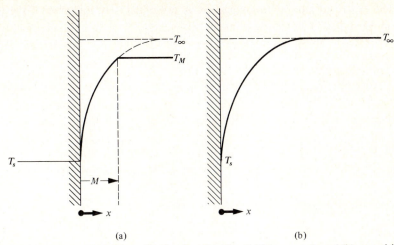

(a) (b)

Fig. 10.5 Analogous temperature distributions in (a) solidifying metal, and (b) a semi-infinite solid.

field in the solidifying metal is between T_M and T_s rather than actually extending to T_∞. However, the temperature "reaches" for T_∞ just as in the semi-infinite solid, and the temperature distribution within the solidified metal takes the form

$$\frac{T - T_s}{T_\infty - T_s} = \mathrm{erf}\, \frac{x}{2\sqrt{\alpha' t}}. \tag{10.16}$$

In Fig. 10.5(a), T_∞ is not known *a priori*. It is an imaginary temperature which makes the temperature distribution analogous to the case of the semi-infinite solid, or it may be thought of as an integration constant.

We now develop an expression for the rate of solidification by applying the boundary conditions

$$T(M, t) = T_M, \tag{10.17}$$

i.e., the temperature at the solid–liquid interface is the freezing point. In addition, we recognize that the rate of evolution of latent heat of fusion equals the heat flux into the solid at the interface, that is,

$$k' \frac{\partial T}{\partial x}(M, t) = H_f \rho' \frac{dM}{dt}. \tag{10.18}$$

When applied to the temperature distribution, Eq. (10.17) yields

$$\frac{T_M - T_s}{T_\infty - T_s} = \mathrm{erf}\, \frac{M}{2\sqrt{\alpha' t}}. \tag{10.19}$$

Since the left-hand side of this equation is constant, the argument of the error function must also be constant. Hence

$$M = 2\beta\sqrt{\alpha' t}. \qquad (10.20)$$

Once again the thickness solidified is proportional to $t^{\frac{1}{2}}$.

To evaluate the constant β, we evaluate the heat flux at the solid–liquid interface which we obtain from Eq. (10.16):

$$k'\frac{\partial T}{\partial x}(M, t) = \frac{(T_\infty - T_s)\sqrt{k'\rho' C_p'}}{\sqrt{\pi}\sqrt{t}} \exp\left[-\frac{M^2}{4\alpha' t}\right]$$

$$= \frac{(T_M - T_s)\sqrt{k'\rho' C_p'}}{\sqrt{\pi}\sqrt{t}\,\operatorname{erf}\beta} \exp(-\beta^2). \qquad (10.21)$$

The latent heat evolved at the interface is written

$$H_f\rho'\frac{dM}{dt} = H_f\rho'\beta\frac{\sqrt{\alpha'}}{\sqrt{t}}. \qquad (10.22)$$

Substituting Eqs. (10.21) and (10.22) into Eq. (10.18), and simplifying, we have

$$\beta e^{\beta^2} \operatorname{erf}\beta = (T_M - T_s)\frac{C_p'}{H_f\sqrt{\pi}}. \qquad (10.23)$$

We now have an expression to calculate β. We may use Fig. 10.6 to evaluate β, rather than using Eq. (10.23) which entails trial and error.

To summarize, β can be determined from Fig. 10.6. Thus the rate of solidification is known (Eq. 10.20), T_∞ can be determined (Eq. 10.19), and the temperature distribution can be computed (Eq. 10.16), if so desired.

This analysis is, of course, valid for unidirectional heat flow; its results can be applied to slab-shaped castings. If the solidification time is sought, then evaluate β and use Eq. (10.20) with $M = L$, the semithickness of the slab.

The above method of solution is not systematic and cannot be extended to the solidification of other shapes. However, Adams[2] has presented a method utilizing a power series that can be extended to the freezing of spheres and cylinders. His results for solidification times of spheres and cylinders freezing with a constant surface temperature, T_s, are given in Figs. 10.7 and 10.8.

While application of the foregoing is limited because it is difficult to imagine practical situations in which a constant surface temperature is maintained, an example of a case of practical importance is the determination of the solidification rate in a large steel ingot poured against a copper, water-cooled mold wall, except for the initial stage of solidification. The solutions are also useful for indicating the maximum freezing rate that can possibly be obtained by convective cooling, since the boundary condition of constant surface temperature corresponds to a case of $h \to \infty$ at the surface.

[2] C. M. Adams, "Thermal Considerations in Freezing," *Liquid Metals and Solidification*, ASM, Cleveland, Ohio, 1958.

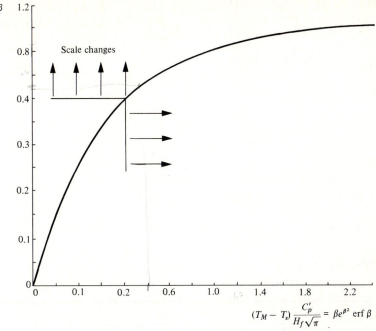

$$(T_M - T_s) \frac{C_p'}{H_f \sqrt{\pi}} = \beta e^{\beta^2} \, \text{erf} \, \beta$$

Fig. 10.6   Evaluation of β for Eq. (10.23).

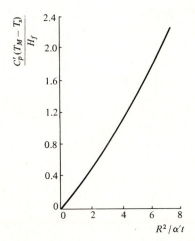

Fig. 10.7   Solidification times for spheres with constant surface temperature. (From C. M. Adams, "Thermal Considerations in Freezing," *Liquid Metals and Solidification*, *ASM*, Cleveland, Ohio, 1958.)

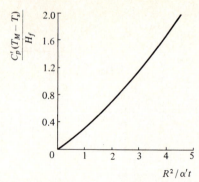

Fig. 10.8 Solidification times for cylinders with constant surface temperatures.   (From C. M. Adams, *ibid.*)

10.2.2 Gradients within mold and casting, no interface resistance

We depict this case in Fig. 10.9. Both the mold and metal form barriers to heat flow. The mold is initially at room temperature, and the liquid metal at its melting point. The mold is thick enough so that no temperature rise occurs on its exterior surface, and we can consider it to be semi-infinite. This case is applicable, for example, in determining the solidification rate of a large ingot against a heavy metal mold; it applies after sufficient material has frozen, so that interface resistance is no longer important. This analysis is also useful in deciding if a particular metal–sand mold combination is such that T_s does, or does not, approximate T_M.

In the previous problem, T_s was fixed as the boundary condition of the situation. In the case at hand, T_s is established at a particular level, depending upon the thermal properties of both the mold and the solidifying metal.

We now proceed to develop the solution which satisfies the requirements

$$\lim_{\xi \to 0} \left[k \left(\frac{\partial T}{\partial x} \right)_{x=0-\xi} - k' \left(\frac{\partial T}{\partial x} \right)_{x=0+\xi} \right] = 0; \qquad (10.24)$$

that is, the heat flux into the casting–mold interface from the solidifying metal must equal the flux away from the interface into the mold. As before, two additional

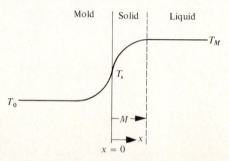

Fig. 10.9 Temperature distribution during solidification with no interface resistance.

boundary conditions must be satisfied at the freezing interface:

$$T(M, t) = T_M; \tag{10.17}$$

$$k' \frac{\partial T}{\partial x} (M, t) = \rho' H_f \frac{dM}{dt}. \tag{10.18}$$

The mold is obviously semi-infinite in the negative x-domain with some unknown surface temperature T_s. Thus

$$\frac{T - T_s}{T_0 - T_s} = \text{erf} \frac{-x}{2\sqrt{\alpha t}}, \tag{10.25}$$

where T_0 is the initial uniform temperature of the mold. For the solidifying metal, Eq. (10.16) applies again; however, note at this point that T_∞ and T_s are both unknown,

$$\frac{T - T_s}{T_\infty - T_s} = \text{erf} \frac{x}{2\sqrt{\alpha' t}}. \tag{10.16}$$

When we apply Eq. (10.17) to Eq. (10.16), we realize that the argument of the error function is constant, and again defined as β, so that Eq. (10.20) applies.

On differentiating Eqs. (10.25) and (10.16), and applying Eqs. (10.24), (10.17), and (10.18), we obtain

$$\frac{(T_M - T_s)C'_p}{H_f \sqrt{\pi}} = \beta e^{\beta^2} \text{ erf } \beta, \tag{10.23}$$

$$\frac{(T_M - T_0)C'_p}{H_f \sqrt{\pi}} = \beta e^{\beta^2} \left(\sqrt{\frac{k' \rho' C'_p}{k \rho C_p}} + \text{erf } \beta \right), \tag{10.26}$$

$$\frac{(T_\infty - T_s)C'_p}{H_f \sqrt{\pi}} = \beta e^{\beta^2}, \tag{10.27}$$

$$\frac{T_s - T_0}{T_\infty - T_s} = \sqrt{\frac{k' \rho' C'_p}{k \rho C_p}}. \tag{10.28}$$

To summarize the results of this section, we calculate the temperature profiles in both the mold and the solidifying metal; we also determine the solidification rate.

We calculate the temperature profile in the mold by completing the following steps:

1) Calculate the mold–casting interface temperature T_s from the known thermal properties

$$(T_M - T_0)\left(\frac{C'_p}{H_f}\right) \quad \text{and} \quad \sqrt{\frac{k' \rho' C'_p}{k \rho C_p}},$$

and Fig. 10.10, which was derived by calculating β on a trial and error solution of Eq. (10.26), and then determining T_s from Eq. (10.23).

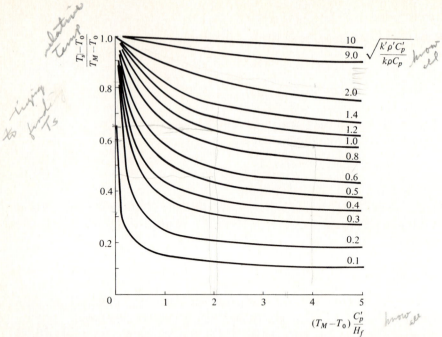

Fig. 10.10 Relative mold–casting interface temperatures for unidirectional freezing with no interface resistance. (From C. C. Reynolds, *Trans. AFS* **72**, 343 (1964).)

2) The value of T_s thus obtained can be used in Eq. (10.25) for the temperature profile in the mold.

We calculate the temperature profile in the solidifying metal by performing these steps:

a) Calculate T_s from Fig. 10.10.
b) Determine T_∞ using either Eq. (10.27) or Eq. (10.28).
c) The values of T_∞ and T_s thus obtained can be used in Eq. (10.16) for the temperature profile in the metal.

If we wish to calculate the thickness of the solidified metal, then we determine T_s from Fig. 10.10 and use Fig. 10.6 for a value of β. With this value of β, Eq. (10.20) can be applied.

Example 10.2 Determine the freezing time of an iron slab-shaped casting which is 4 in. thick. Assume no interface resistance, and consider the cases of pure iron being poured at its freezing point into (a) a sand mold, (b) a water-cooled copper mold, and (c) a very thick copper mold.

Mold data:

Material	Heat capacity, Btu/lb$_m$ °F	Density, lb$_m$/ft^3	Thermal conductivity, Btu/ft-hr °F
Sand	0.28	100	0.50
Copper	0.09	560	230

Iron data:

$$\text{Freezing temperature} = 2802°\text{F},$$
$$\text{Heat of fusion} = 117 \, \text{Btu/lb}_m,$$
$$\text{Solid density} = 490 \, \text{lb}_m/\text{ft}^3,$$
$$\text{Liquid density} = 460 \, \text{lb}_m/\text{ft}^3,$$
$$\text{Solid heat capacity} = 0.16 \, \text{Btu/lb}_m \, °\text{F},$$
$$\text{Liquid heat capacity} = 0.18 \, \text{Btu/lb}_m \, °\text{F},$$
$$\text{Thermal conductivity} = 48 \, \text{Btu/hr-ft} \, °\text{F}.$$

Solution

a) $(T_M - T_0)\dfrac{C'_p}{H_f} = (2802 - 80)\dfrac{0.16}{117} = 3.72;$

$$\sqrt{\frac{k'\rho'C'_p}{k\rho C_p}} \equiv \sqrt{\frac{(48)(490)(0.16)}{(0.50)(100)(0.28)}} = 16.4.$$

From Fig. 10.10, we obtain a value of T_s:

$$\frac{T_s - T_0}{T_M - T_0} \cong 1.0.$$

Therefore $T_s \cong T_M$, and the analysis used in Example 10.1 is valid. Therefore

$$t(\text{sand}) = 0.685 \, \text{hr}.$$

b) In this case, T_s equals the temperature of the water-cooled mold which we take to be 80°F.

$$(T_M - T_s)\frac{C'_p}{H_f\sqrt{\pi}} = (2802 - 80)\frac{0.16}{117\sqrt{\pi}} = 2.10.$$

From Fig. 10.6, we see that $\beta = 0.98$, which is applied to Eq. (10.20). Hence

$$t = \left(\frac{M}{2\beta}\right)^2\frac{1}{\alpha'} = \frac{M^2}{4\beta^2}\cdot\frac{\rho'C'_p}{k'}$$

$$= \frac{(2/12)^2}{(4)(0.98)^2}\cdot\frac{(490)(0.16)}{(48)} = 0.0118 \, \text{hr}.$$

c) $(T_M - T_0)\dfrac{C'_p}{H_f} = 3.72,$ as in part (a);

$$\sqrt{\frac{k'\rho'C'_p}{k\rho C_p}} = \sqrt{\frac{(48)(490)(0.16)}{(230)(560)(0.09)}} = 0.570.$$

From Fig. 10.10, we get

$$\frac{T_s - T_0}{T_M - T_0} \cong 0.42,$$

$$T_s = (0.42)(2802 - 80) + 80 = 1223°\text{F}.$$

Then we resort to Fig. 10.6 to obtain a value of β:

$$(T_M - T_s)\frac{C_p'}{H_f\sqrt{\pi}} = (2802 - 1223)\left(\frac{0.16}{117\sqrt{\pi}}\right) = 1.22$$

and

$$\beta = 0.87.$$

Using Eq. (10.20) as in part (b) yields

$$t = 0.0150 \text{ hr.}$$

10.2.3 Interface resistance

In the previous sections, we depicted situations in which the surface temperature T_s remained constant. When we examine the more likely situation of some interface resistance, then the surface temperature of the casting varies with time, and the analysis becomes more complex. Before analyzing the temperature history in the mold and casting in a general case (Fig. 10.4), let us first consider a simpler case (Fig. 10.11).

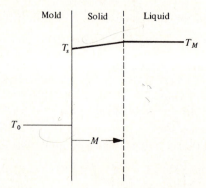

Fig. 10.11 Temperature distribution during solidification with a predominating interface resistance.

In Fig. 10.11, the interface resistance predominates over the resistances offered by both the solidifying metal and the metal. Practical importance is attached to this case when solidification time is short; the analysis is useful for estimating solidification times of small thin-section parts cast in heavy metal molds such as die and permanent mold castings.

In this case, the temperature gradients within the mold and the casting are negligible, and the heat escapes the casting as if a heat-transfer coefficient applies at the surface. Thus the total quantity of heat Q that crosses the mold–casting interface in time t is

$$Q = hA(T_s - T_0)t. \tag{10.29}$$

First case

As with sand castings, if the temperature gradients are negligible, then $T_s \cong T_M$, and only latent heat is removed from the casting during solidification. Therefore by combining Eqs. (10.9) and (10.29), we can easily show that

$$\frac{V}{A} = \frac{h(T_M - T_0)}{\rho' H_f} t. \tag{10.30}$$

Note that shape has no effect on the applicability of Eq. (10.30). Shape was also not specified in the case of the cooling or heating of a solid body with negligible internal temperature gradients. If we wish to apply Eq. (10.30) to unidirectional solidification, we see that

$$M = \frac{h(T_M - T_0)}{\rho' H_f} t. \tag{10.31}$$

The thickness solidified, M, is proportional to time rather than the square root of time.   *Second Case*

   Now consider the case in which $T_s \neq T_M$, and the heat leaves the casting via h at the surface to a water-cooled mold maintained at T_0. Here we simplify the analysis by approximating the temperature profile within the solidifying metal as a linear function. Then we write the heat flux at the casting–mold interface as

$$q|_{x=0} = k' \frac{T_M - T_s}{M}. \tag{10.32}$$

Also

$$q|_{x=0} = h(T_s - T_0). \tag{10.33}$$

We then eliminate the surface temperature T_s, which varies, by combining Eqs. (10.32) and (10.33); because of the linear temperature profile, we express the flux at the solid–liquid interface simply as

$$q|_{x=0} = q|_{x=M} = \frac{T_M - T_0}{1/h + M/k'}. \tag{10.34}$$

In addition, at $x = M$, the latent heat is evolved so that

$$q|_{x=M} = \rho' H_f \frac{dM}{dt}. \tag{10.35}$$

growth rate

By combining Eqs. (10.34) and (10.35) and integrating with $M = 0$ at $t = 0$, and $M = M$ at $t = t$, we obtain

$$M = \frac{h(T_M - T_0)}{\rho' H_f} t - \frac{h}{2k'} M^2. \tag{10.36}$$

Adams[3] has solved the problem in a more rigorous manner in which the temperature profile is not assumed to be linear as above. The more refined analysis is similar to Eq. (10.36) with an additional factor a:

$$M = \frac{h(T_M - T_0)}{\rho' H_f a} t - \frac{h}{2k'} M^2. \tag{10.37}$$

---

[3] C. M. Adams, *ibid.*

In this expression

$$a \equiv \frac{1}{2} + \sqrt{\frac{1}{4} + \frac{C_p'(T_M - T_0)}{3H_f}}.$$

Equation (10.37) is almost exact for $hM/k \gtrless 1/2$. For $hM/k < 1/2$, the thickness solidified is overestimated by approximately 10–15%.

10.2.4 Gradients within mold and casting with interface resistance

In tackling this situation, we extend the concept of the mold–metal interface temperature introduced in Section 10.2.2. However, in this case there is no constant temperature at the casting–mold interface; rather, there are two surface temperatures, neither of which is constant. To handle this problem, consider an imaginary reference plane between the mold and casting which is at T_s, where T_s is constant and determined by the method of Section 10.2.2 using Eq. (10.28). The contact resistance is then apportioned on both sides of the imaginary plane in accordance with the equations

$$h_M = \left(1 + \sqrt{\frac{k\rho C_p}{k'\rho'C_p'}}\right)h, \tag{10.38}$$

$$h_C = \left(1 + \sqrt{\frac{k'\rho'C_p'}{k\rho C_p}}\right)h. \tag{10.39}$$

In these expressions, h is the total heat-transfer coefficient across the interface, h_M is the coefficient on the mold side of the plane, and h_C is the coefficient on the casting side of the plane.

Figure 10.12 depicts the entire situation, showing the surface temperatures, T_{sC} and T_{sM}, of the casting and the mold, respectively. We handle the problem as follows:

1) We first solve for T_s as if there were no interface resistance.
2) Using T_s obtained in (1) and applying Eq. (10.39), we isolate the casting half and study it by using the analysis given in Section 10.2.3. For the casting side, heat is transferred from T_{sC} to T_s via h_C.
3) If we wish to study the mold half, we resort to the results of Section 9.5.2 and use Eq. (9.113). For the mold side, heat is supplied to the mold from a source at T_s to the surface at T_{sM} via h_M.

We present Fig. 10.13 to illustrate the effect of the contact thermal resistance. First, in order to solidify the same amount of metal, it takes more time with some resistance. Note, also, that the two curves become parallel after the early stages of solidification, and M varies linearly with $\sqrt{t}$ for both cases. For the early stage of solidification, M does not vary linearly with $\sqrt{t}$ when there is some contact resistance.

Example 10.3   a) Determine the freezing time of the 4 in. thick slab of iron discussed

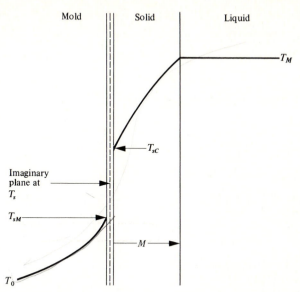

Fig. 10.12 Temperature distribution in the casting and mold when the interface resistance does not predominate.

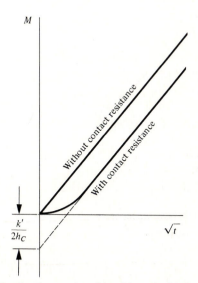

Fig. 10.13 Rate of solidification in metal molds with and without contact resistance.

in Example 10.2 when it is cast in a very thick copper mold. The total heat transfer-coefficient across the casting–mold interface may be taken as 250 Btu/hr-ft² °F.
b) When the casting has completely solidified, what is the surface temperature of the mold?

Solution

a) First solve the problem as if there were no interface resistance, and obtain T_s. This was done in Example 10.2 with the result

$$T_s = 1790°F.$$

Using Eq. (10.39), determine the coefficient for the casting side of the interface:

$$h_C = \left(1 + \sqrt{\frac{(48)(490)(0.16)}{(230)(560)(0.09)}}\right)(250)$$

$$= 392 \text{ Btu/hr-ft}^2 \text{ °F}.$$

Now use Eq. (10.37), but in a modified form for the case at hand; it is

$$M = \frac{h_C(T_M - T_s)}{\rho' H_f a}t - \frac{h_C}{2k'}M^2.$$

Substituting properties and $M = \frac{1}{6}$ ft, we have

$$\frac{1}{6} = \frac{(392)(2802 - 1790)}{(490)(117)a}t - \frac{(392)}{(2)(48)}\left(\frac{1}{6}\right)^2,$$

in which

$$a = \frac{1}{2} + \sqrt{\frac{1}{4} + \frac{(0.16)(2802 - 1790)}{(3)(117)}} = 1.34.$$

Solving for t yields

$$t = 0.0543 \text{ hr}.$$

b) To solve for the surface temperature of the mold we may use Eq. (9.113) or Fig. 9.23) with h replaced by h_M, T_f replaced by T_s, and T_i replaced by T_0.

For the surface,

$$x/(2\sqrt{\alpha t}) = 0.$$

We obtain h_M by using Eq. (10.38),

$$h_M = \left(1 + \sqrt{\frac{(230)(560)(0.09)}{(48)(490)(0.16)}}\right)250$$

$$= 688 \text{ Btu/hr-ft}^2 \text{ °F}.$$

Then

$$\frac{h_M}{k}\sqrt{\alpha t} = \frac{688}{230}\sqrt{\frac{(230)(0.0543)}{(560)(0.09)}} = 1.49.$$

From Fig. 9.23, we have

$$\frac{T - T_0}{T_s - T_0} \cong 0.69,$$

so that

$$T = (0.69)(1223 - 80) + 80$$

$$= 869°F.$$

Up to this point, we have recognized that a thermal resistance exists at the mold–casting interface in all cases unless soldering occurs. Since no physical bonding takes place at the interface, then the casting and the mold are free to move due to thermal–physical effects. In fact, it is well known that when a metal is cast against a metal mold, a gap forms at the interface. This gap is formed as a result of the mold expanding due to its absorption of heat, and to the solid skin of metal shrinking due to its lowering temperature. Usually, the prediction of the rate of heat transfer across the gap is not reliable, and so we rely upon empirical measurements of surface temperatures and the heat absorbed by the mold to deduce appropriate heat-transfer coefficients. Some heat-transfer coefficients for various casting situations are given in Table 10.1. Since such values are strongly dependent on specific processes and geometrical situations, the data in Table 10.1 should serve only as general guidelines.

Table 10.1 Heat-transfer coefficients across casting–mold interfaces

Casting situation	Heat-transfer coefficient, Btu/hr-ft^2 °F	Reference listed below
Steel in continuous casting mold	50–400*	(a)
Steel in continuous casting mold, 4 × 4 in.		(b)
withdrawal rate of 20 in./min	85	
withdrawal rate of 100 in./min	140	
withdrawal rate of 175 in./min	190	
Ductile iron in gray iron mold (coated with amorphous carbon)	300	(c)
Steel in static cast iron mold	180	(c)
Copper in centrifugal steel mold	40–60	(d)
Aluminum alloy in small permanent copper molds	300–450	(e)

* The authors state that h depends on section size, withdrawal speed, and the shrinkage characteristics of the metal.

a) A. W. D. Hills and M. R. Moore, *Heat and Mass Transfer in Process Metallurgy*, Inst. Min. and Met., London, 1967.

b) E. Y. Kung and J. C. Pollock, *Simulation* 29–36 (Jan. 1968).

c) C. C. Reynolds, *ibid.*

d) R. W. Ruddle, *The Solidification of Castings*, second edition, The Institute of Metals, London, 1957.

e) B. Bardes and M. C. Flemings, *Trans. AFS* **74**, 406 (1966).

10.3 INTEGRAL SOLUTION FOR SOLIDIFICATION

In Chapter 7 we used an integral method to examine a problem involving heat transfer with convection, in which all the thermal gradients existed within a thermal boundary layer. A *thermally affected zone* in conduction problems can be examined in a similar manner. The temperature transient within a body is caused by some boundary condition at the surface ($x = 0$). The thermally affected zone is defined between $x = 0$ and x', an interior position at which the body has just begun to feel the transient introduced by the boundary condition. As long as the system is in an unsteady state, x' is a function of time.

Before proceeding to solve a specific solidification problem, we first develop the *integral equation of conduction heat transfer*. Consider the differential equation that applies to one-dimensional conduction heat transfer with constant thermal diffusivity, namely,

$$\alpha \frac{\partial^2 T}{\partial^2 x} = \frac{\partial T}{\partial t}. \tag{10.40}$$

We multiply both sides of Eq. (10.40) by dx, and integrate from $x = 0$ to x':

$$\int_0^{x'} \alpha \frac{\partial^2 T}{\partial x^2}\, dx = \int_0^{x'} \frac{\partial T}{\partial t}\, dx. \tag{10.41}$$

We readily visualize the left-hand side of Eq. (10.41):

$$\int_0^{x'} \alpha \frac{\partial^2 T}{\partial x^2}\, dx = \alpha \left[\frac{\partial T}{\partial x}(x', t) - \frac{\partial T}{\partial x}(0, t) \right]. \tag{10.42}$$

We rely on a mathematical identity to represent the right-hand side of Eq. (10.41):

$$\int_0^{x'} \frac{\partial T}{\partial t}\, dx = \frac{d}{dt} \int_0^{x'} T\, dx - T(x', t) \frac{dx'}{dt}. \tag{10.43}$$

Substituting Eqs. (10.42) and (10.43) into Eq. (10.41) yields the conduction heat transfer integral equation

$$\alpha \left[\frac{\partial T}{\partial x}(x', t) - \frac{\partial T}{\partial x}(0, t) \right] = \frac{d}{dt} \int_0^{x'} T\, dx - T(x', t) \frac{dx'}{dt}. \tag{10.44}$$

Several problems involving phase changes have been solved by the integral method which we can use to obtain approximate solutions. To apply this method, consider solidification when the surface of an ingot is maintained at some constant temperature below the freezing point.[4] This is the same problem as we considered in Section 10.2.1.

We apply Eq. (10.44) to the freezing problem at hand by taking x' as the thickness of material solidified, $M(t)$. The conditions to be met are

$$M(0) = 0, \tag{10.45}$$

$$T(0, t) = T_s, \tag{10.46a}$$

$$T(M, t) = T_M, \tag{10.46b}$$

and at the liquid–solid interface,

$$k' \frac{\partial T}{\partial x}(M, t) = \rho' H_f \frac{dM}{dt}. \tag{10.47}$$

Let $\theta = T - T_M$; then the boundary conditions transform into

$$\theta(0, t) = T_s - T_M = \theta_s, \tag{10.48a}$$

$$\theta(M, t) = 0, \tag{10.48b}$$

and

$$k' \frac{\partial \theta}{\partial x}(M, t) = \rho' H_f \frac{dM}{dt}. \tag{10.49}$$

Expressing the integral equation in θ, and applying Eqs. (10.48a), (10.48b), and (10.49) yields

$$\alpha' \left[\frac{\rho' H_f}{k'} \frac{dM}{dt} - \frac{\partial \theta}{\partial x}(0, t) \right] = \frac{d}{dt} \int_0^M \theta \, dx. \tag{10.50}$$

As usual, when employing an integral technique, we assume the temperature distribution. In this case, let us assume that the temperature distribution is a second-degree polynomial in x. Then

$$\theta = a(x - M) + b(x - M)^2, \tag{10.51}$$

which automatically satisfies Eq. (10.48b). Two further boundary conditions must be satisfied: Eqs. (10.48a) and (10.49). However, Eq. (10.49) in its present form is not suitable because the coefficients in the polynomial would involve dM/dt. In turn, the left-hand side of Eq. (10.50) would involve dM/dt, and the integral equation would then yield a second-order differential equation for $M(t)$, whereas there is only one initial condition for $M(t)$, namely, $M(0) = 0$. To overcome this difficulty, we transform Eq. (10.49) into a different form before applying it to the assumed polynomial.

---

[4] The remainder of this section is from T. R. Goodman, "Integral Methods for Nonlinear Heat Transfer," *Advances in Heat Transfer* **1**, Academic Press, New York, 1964.

From Eq. (10.48), we write

$$\frac{d\theta}{dt}(M, t) = \left(\frac{\partial\theta}{\partial x}\right)_M \frac{dM}{dt} + \left(\frac{\partial\theta}{\partial t}\right)_M = 0. \tag{10.52}$$

From Eq. (10.49), we substitute (dM/dt) into Eq. (10.52), and obtain

$$\frac{k'}{\rho'H_f}\left(\frac{\partial\theta}{\partial x}\right)_M^2 + \left(\frac{\partial\theta}{\partial t}\right)_M = 0. \tag{10.53}$$

Then, since $\partial\theta/\partial t = \alpha'\,\partial^2\theta/\partial x^2$, we finally obtain

$$\frac{k'}{\rho'H_f}\left(\frac{\partial\theta}{\partial x}\right)_M^2 + \alpha'\left(\frac{\partial^2\theta}{\partial x^2}\right)_M = 0. \tag{10.54}$$

We use Eq. (10.54) rather than Eq. (10.49) to evaluate the arbitrary constants.

Substituting the assumed temperature distribution into Eqs. (10.54) and (10.48a) yields

$$a = \frac{\rho'\alpha'H_f}{k'M}(1 - \sqrt{1 + \mu}) \tag{10.55}$$

and

$$b = \frac{aM + \theta_s}{M^2}, \tag{10.56}$$

where

$$\mu \equiv -\frac{2\theta_s k'}{\alpha'\rho'H_f}.$$

Substituting the completely described profile into Eq. (10.50), we obtain the differential equation for M:

$$M\frac{dM}{dt} = \frac{6\alpha'[1 - (1 + \mu)^{1/2} + \mu]}{5 + \mu + (1 + \mu)^{1/2}}. \tag{10.57}$$

Integrating Eq. (10.50), with the initial condition $M(0) = 0$, leads to

$$M = 2\beta\sqrt{\alpha't}. \tag{10.58}$$

Here

$$\beta = \sqrt{3}\left[\frac{1 - (1 + \mu)^{1/2} + \mu}{5 + (1 + \mu)^{1/2} + \mu}\right]^{1/2}. \tag{10.59}$$

We can write the parameter μ in a simpler form:

$$\mu \equiv \frac{2(T_M - T_s)C_p'}{H_f}.$$

Thus β is a function of the same group of variables as found for the exact solution (Eq. 10.23). Agreement between this approximate value of β and the exact value of β is about 7%.

10.4 CONTINUOUS CASTING

The basic features of a continuous casting machine are depicted in Fig. 10.14. The metal passes through a water-cooled metal mold forming a thin solidified skin ($\frac{1}{4}$ to $\frac{3}{4}$ in. thick), and is then subjected to a water spray for the remainder of solidification. Figure 10.15 schematically shows the temperature distribution within the partially frozen casting as it moves downward with a velocity u in the y-direction. While in the mold, the conduction of heat in the thin shell of solid is much greater in the x-direction than in the y-direction. Therefore the conduction of heat in the withdrawal direction may be ignored. Under these circumstances, the analysis used for static casting as presented in Section 10.2.3 may be applied with only minor modifications.

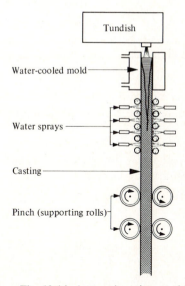

Fig. 10.14 A strand casting machine.

As the metal passes through the mold, we assume that a constant heat-transfer coefficient applies to account for the mold–casting interface resistance as the heat is removed by the water-cooled mold maintained at T_0. Equation (10.37) applies if the time t is recast as the distance down the mold wall y, divided by the casting velocity u. We account for the effect of liquid superheat by realizing that on passing down from the tundish, enough turbulence prevails in the liquid core, at least between the mold walls, so that we may assume that the liquid temperature is uniform with y at T_P. Then, at the solid–liquid interface, latent heat plus the

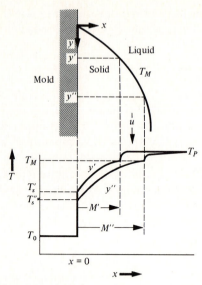

Fig. 10.15 The partially frozen portion of a continuous casting and the temperature profile as a function of the distance down the mold.

liquid superheat is conducted toward the mold wall through the solid skin, and in the formulation the effective latent heat of fusion H'_f applies in which

$$H'_f = H_f + C_{p,l}(T_P - T_M).$$

The results are convenient when presented in dimensionless groups. From Eq. (10.37), we deduce the groups

1) $$\frac{h^2 y}{u k' \rho' C'_p},$$

gives n with respect to y

2) $$\frac{C'_p(T_M - T_0)}{H'_f},$$

$\frac{dM}{dt}$

3) $$\frac{hM}{k'}.$$

T_s

These groups and others are used in Figs. 10.16–10.18 to present the results. We can determine the thickness of solid metal at various positions down the mold from Fig. 10.16. Figure 10.17 shows how the surface temperature T_s of the cast metal varies with the mold position, and we use Fig. 10.18 to calculate the total rate of heat removed entering the mold of length L. Skin thickness and surface temperature are important parameters, since the solid skin on exit from the mold must be of some minimum strength. The rate of heat removal is useful to compute cooling water requirements.

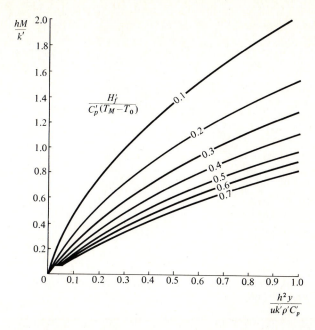

Fig. 10.16 Thickness solidified, M, versus distance down the mold. (Figures 10.16–10.18 are from A. W. D. Hills and M. R. Moore, *ibid.*)

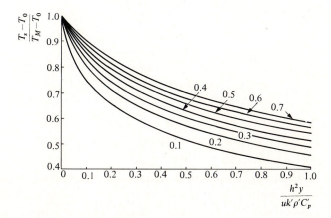

Fig. 10.17 Surface temperature versus distance down the mold. Numbers on the curves are the same as in Fig. 10.16.

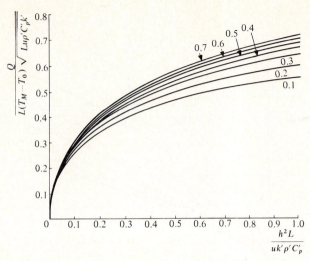

Fig. 10.18 Rate of heat removal by mold cooling water Q versus mold length L. Numbers on the curves are the same as in Fig. 10.16.

Example 10.4 Determine (a) the mold length, and (b) the cooling water requirements to produce steel slab (2 ft wide by 3 in. thick) at a withdrawal rate of 120 in./min. The solid skin on exit from the mold should be 0.50 in. thick, and a heat-transfer coefficient of 250 Btu/hr-ft^2 °F may be assumed to apply. To minimize thermal pollution of the water source, only a 10°F temperature rise of the water is allowed.

Data for steel:

$$\text{Freezing temperature} = 2730°F,$$
$$\text{Pouring temperature (from tundish)} = 2820°F,$$
$$\text{Latent heat of fusion} = 115 \text{ Btu/lb}_m,$$
$$\text{Solid density} = 480 \text{ lb}_m/\text{ft}^3,$$
$$\text{Solid heat capacity} = 0.16 \text{ Btu/lb}_m \text{ °F},$$
$$\text{Liquid heat capacity} = 0.18 \text{ Btu/lb}_m \text{ °F},$$
$$\text{Solid thermal conductivity} = 44 \text{ Btu/lb}_m\text{-ft °F}.$$

Solution

a)
$$\frac{hM}{k'} = \frac{(250)(0.50/12)}{44} = 0.238.$$

$$H'_f = H_f + C_{p,l}(T_P - T_M)$$
$$= 115 + 0.18(2820 - 2730) = 131 \text{ Btu/lb}_m,$$

and

$$\frac{H'_f}{C'_p(T_M - T_0)} = \frac{131}{0.16(2730 - 70)} = 0.308.$$

From Fig. 10.16, we get

$$\frac{h^2 y}{uk' \rho' C_p} \cong 0.11.$$

Therefore

$$y = L = (0.11) \frac{(120/12 \times 60)(44)(480)(0.16)}{(250)^2},$$

$$L = 3.6 \text{ ft}.$$

b) From Fig. 10.18,

$$\frac{Q}{L(T_M - T_0)\sqrt{Lu\rho' C_p' k'}} \cong 0.28.$$

$$Q = (0.28)(3.6)(2730 - 70)\sqrt{(3.6)(600)(480)(0.16)(44)}$$

$$= 7.25 \times 10^6 \text{ Btu/hr}.$$

Then we determine the flow rate of water required (heat capacity of water is 1 Btu/°F lb_m).

$$\text{Flow rate of water} = \frac{7.25 \times 10^6 \text{ Btu}}{\text{hr}} \left| \frac{lb_m \, °F}{1 \text{ Btu}} \right| \frac{1}{10°F} \left| \frac{1 \text{ gal}}{8.33 \, lb_m} \right.$$

$$= 8.70 \times 10^4 \text{ gal/hr}$$

$$= 1450 \text{ gpm}.$$

PROBLEMS

Data for Problems 10.1–10.4:

Mold material	k, Btu/ft-hr °F	ρ, lb_m/ft^3	C_p, Btu/lb_m °F
Green sand	0.40	100	0.28
Mullite	0.22	100	0.18
Copper	230.0	560	0.10

Casting material	T_M, °F	H_f, Btu/lb_m	ρ', lb_m/ft^3	C_p', Btu/lb_m °F	k, Btu/hr-ft °F
Iron	2802	117	460	0.15	25
Aluminum	1220	170	161	0.25	150

10.1 Plot distance solidified versus the square root of time for the following metals (in each case the pure metal is poured at its melting point against a flat mold wall):

a) Iron in a green sand mold.
b) Aluminum in a green sand mold.
c) Iron in a mullite mold heated to 1800°F.

10.2 How long does it take to freeze a 4-in. diameter sphere of pure iron in a green sand mold assuming

    a) no superheat and neglecting the heat flow divergence?

    b) no superheat and realizing that a sphere is being cast?

    c) 200°F superheat and realizing that a sphere is being cast?

10.3 Plot distance solidified versus time for iron poured at its melting point into a heavy copper mold, assuming that

    a) there is a large flat mold wall and no resistance exists to heat flow at the mold–metal interface;

    b) interface resistance to heat flow predominates ($h = 100$ Btu/hr-ft^2 °F);

    c) interface resistance predominates in the early stages of solidification and then becomes negligible.

10.4 Show whether or not iron can be cast against a very thick aluminum mold wall without causing the aluminum to melt.

10.5 Slab-shaped steel castings are prone to center-line porosity, which—for our purposes—is simply an alignment of defects along the plane of last solidification. The sketch below shows the solidification of a slab cast in sand and the location of the defects (centerline porosity).

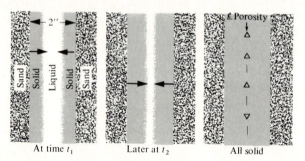

At time t_1 Later at t_2 All solid

The solidification time for the 2-in. slab cast in sand is known to be 6 min; when cast in an insulating mullite mold, the time is 60 min. If a casting is made in the composite mold depicted below, determine the thickness T the casting should have to yield $1\frac{7}{8}$ in. of sound metal after machining.

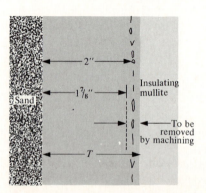

10.6 A 2-in. thick slab of aluminum is cast in a mold made of sand (forming one face) and a proprietary material (forming the other face). The aluminum is poured with no superheat, and the as-cast structure of the slab is examined after solidification and cooling. The examination shows that the plane of last solidification (i.e., the plane where the two solidification fronts meet) is located $1\frac{1}{2}$ in. from the sand side. Knowing this and using the following data, calculate the *heat diffusivity* (not the thermal diffusivity) of the proprietary material.

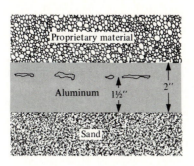

Data:

Freezing point of Al = 1220°F

	Al	Sand
k, Btu/hr-ft °F	45.00	0.40
ρ, lb$_m$/ft^3	180.00	100.00
C_p, Btu/lb$_m$	0.25	0.25
H_f, Btu/lb$_m$	170.00	—

10.7 A continuous casting machine forms molten steel into a slab up to 76 in. wide and 9 in. thick at a production rate of 1.5 million tons per year. Determine the vertical length of the mold if the solid shell must be 0.5 in. thick at the mold exit, and calculate the cooling water requirement (lb$_m$/hr) if its temperature rise is from 80°F to 180°F.

Data for low carbon steel:

$$k = 20 \text{ Btu/hr-ft °F,}$$

$$C_p = 0.16 \text{ Btu/lb}_m \text{ °F (liquid or solid),}$$

$$H_f = 120 \text{ Btu/lb}_m,$$

$$T_M = 2770°F,$$

$$\rho = 480 \text{ lb}_m/\text{ft}^3.$$

Heat-transfer coefficient: $h = 200$ Btu/ft^2-hr °F.

Pouring temperature: 2850°F.

10.8 The dwell time in the mold of a continuous casting machine is defined as the interval which the metal spends in the mold during solidification, that is, $t = L/u$, in which t = dwell time, L = length of mold over which solidification is occurring, and u = velocity of metal through the mold. Since the skin solidified in the mold is not thick, a simple analysis might be expected to apply.

a) Assume that the temperature distribution within the solidified metal is linear at any position down the mold and show that

$$\rho' H_f \frac{dM}{dt} = \frac{1}{1/h + M/k'}(T_M - T_0).$$

b) From part (a) obtain an expression for thickness solidified versus time.
c) Compare the results for dwell time in Problem 10.7 and the dwell time calculated from part (b) for the same conditions.

10.9 A junction-shaped casting as depicted below is made in a sand mold. The junction may be considered infinitely long in the z-direction. For the upper right quadrant of sand:

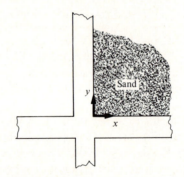

a) Write a differential equation for temperature.
b) Write the boundary conditions (for time and space) that apply.
c) Write the solution yielding temperature as a function of position in the sand and time.

11

RADIATION HEAT TRANSFER

In Chapter 6, we alluded to energy transfer by radiation as a mechanism for heat flow, but due to the entirely different natures of conduction and radiation, only a limited discussion of radiation appeared prior to this chapter. Energy transport by conduction depends on the existence of a conducting material. On the other hand, radiation is electromagnetic energy in transport; therefore, energy travels through empty space via radiation. The rate equations for conduction and radiation reflect their very different characters, and show no similarity. The energy flux by conduction is proportional to the thermal gradient, that is, $q_x = -k(\partial T/\partial x)$. In radiant heat transfer, the energy flux—the so-called emissive power—is proportional to the fourth power of the absolute temperature, that is, $e \propto T^4$.

Solids, liquids, and some gases emit radiant energy composed of many wavelengths. Figure 11.1 depicts the spectrum of electromagnetic radiation and the atomic mechanisms responsible for the various forms of radiation. The excitation process causing radiation may vary and can be classified as fluorescence, x-radiation, radio waves, etc. If the excitation comes from molecular motion which characterizes temperature, the radiation is called thermal. The thermal radiation is of primary interest here and its range of wavelengths is taken to be $0.1–100\,\mu$.

11.1 BASIC CHARACTERISTICS

Thermal radiation is defined as the energy transferred by electromagnetic waves that originate from a body because of its temperature. The rate at which energy is emitted depends on the substance itself, its surface condition, and its surface temperature.

Figure 11.2 depicts a surface emitting and receiving thermal radiation. The total emissive power e is the flux of thermal radiation energy *emitted* into the entire volume above the surface. In terms of electromagnetic waves, the total emissive power is the sum of all the energy carried by all wavelengths emitted from the surface per unit area per unit time. The quantity e is also called *emissive power*, *emittance*, *total hemispherical radiation intensity*, or *radiant-flux density*.

We define the *total irradiation, G*, of a body as the flux of thermal radiant energy incoming to the surface, and the *total radiosity, J*, as the total flux of radiant

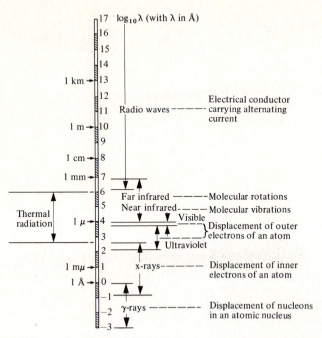

Fig. 11.1 Electromagnetic radiation spectrum (1 μ is 10^{-3} mm; 1 mμ is 10^{-3} μ).

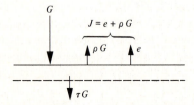

Fig. 11.2 The distribution of thermal radiation at a surface.

energy leaving the surface of a body. The radiosity includes both the energy emitted and the energy reflected by the surface. As fluxes, the units of G and J are in Btu/hr-ft^2.

The incoming energy (total irradiation) can be absorbed, reflected, or transmitted, so that

$$G = \alpha G + \rho G + \tau G,$$

or

$$\alpha + \rho + \tau = 1, \tag{11.1}$$

where α = absorptivity (fraction of G absorbed), ρ = reflectivity (fraction of G

reflected), and τ = transmissivity (fraction of G transmitted). Most liquids and solids are opaque to thermal radiation, so that $\tau = 0$, and

$$\alpha + \rho = 1. \tag{11.2}$$

11.2 THE BLACK RADIATOR AND EMISSIVITY

In the study of real surfaces, it is convenient to define a hypothetical ideal surface called a black radiator. A black radiator is a surface which absorbs all incoming radiation; thus $\alpha = 1$, and $\rho = 0$. Another important characteristic of the black radiator is that it is a perfect emitter, since it emits radiation of all wavelengths, and its total emissive power is theoretically the highest that can be achieved at a given temperature. It follows that any real body emitting thermal radiation has a total emissive power of some fraction of the emissive power of a black radiator, which is often called a black body. This fraction is defined as the emissivity ε. Therefore

$$e = \varepsilon e_b. \tag{11.3}$$

Imagine a region in a space completely filled with black body radiation. A real body 1 emitting radiation at a rate of e_1 is placed in this region; the net rate of energy transferred from the body is

$$q_{1,\text{net}} = \underset{\substack{\text{energy} \\ \text{emitted}}}{e_1} - \underset{\substack{\text{energy} \\ \text{absorbed}}}{\alpha_1 e_b}. \tag{11.4}$$

By utilizing Eq. (11.3), we can also write

$$q_{1,\text{net}} = \varepsilon_1 e_b - \alpha_1 e_b. \tag{11.5}$$

If the body is in thermal equilibrium with the black body radiation, then $q_{\text{net}} = 0$, and $\alpha_1 = \varepsilon_1$. The same result can be obtained for any other body placed in this space, and thus the absorptivity and emissivity of any body *at thermal equilibrium* are equal.

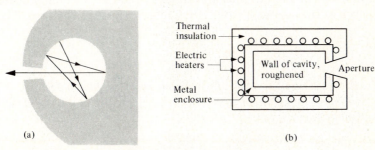

(a)

Thermal insulation

Electric heaters

Metal enclosure

Wall of cavity, roughened

Aperture

(b)

Fig. 11.3 Laboratory black bodies.

It is important to note that although black body radiation is hypothetical, radiation very closely approximating that of a black radiator can be realized. Figure 11.3(a) depicts a small hole emanating from a hollow space having nonblack walls at a uniform temperature. Since the hole is small, only a very small fraction of the radiation entering the hole and diffusely reflected off the nonblack walls will escape; hence the escaping energy is entirely representative of the radiation within the cavity.

The energy emitted from any point on the walls of the enclosure is εe_b; when this radiation strikes other parts of the enclosure, a certain fraction is reflected, namely, $\rho\varepsilon e_b$. Similarly, a second reflection equals $\rho(\rho\varepsilon e_b)$, and the third one is $\rho^3\varepsilon e_b$. Since the hole is small, probability favors radiation that ultimately passes through the hole to be made up of an infinite number of such reflections. Therefore

$$e\,(\text{hole}) = \varepsilon e_b(1 + \rho + \rho^2 + \rho^3 + \cdots), \tag{11.6}$$

or

$$e\,(\text{hole}) = \varepsilon e_b \frac{1}{1 - \rho}, \tag{11.7}$$

and since $\alpha = 1 - \rho$, and thermal equilibrium exists within the cavity, then $\varepsilon = 1 - \rho$ for the isothermal cavity. Therefore Eq. (11.7) reduces to

$$e\,(\text{hole}) = e_b. \tag{11.8}$$

Since radiation from a cavity is very nearly black, the *hohlraum* (heated cavity), shown in Fig. 11.3(b), is normally used as a radiation standard. Such an enclosure may be constructed with heavy metal internal walls that are roughened and oxidized to provide a surface of high emissive power. The walls are heated electrically, and the exterior is well insulated.

11.3 THE ENERGY DISTRIBUTION AND THE EMISSIVE POWER

The thermal radiation from a solid body is composed of a continuous spectrum of wavelengths forming an energy distribution. Figure 11.4 depicts the spectrum of a black body at various temperatures. For most engineering calculations, the total emissive power of a black body at a particular temperature is an important quantity, which is given by the area under the curve applied at that temperature. Therefore

$$e_b = \int_0^\infty e_{b,\lambda}\, d\lambda, \tag{11.9}$$

where λ is the wavelength, and $e_{b,\lambda}$ is the *monochromatic emissive power. Planck's equation* or *Planck's distribution law* relates $e_{b,\lambda}$ erg/cm^3 sec to the wavelength (cm) and the absolute temperature:

$$e_{b,\lambda} = \frac{2\pi h c^2 \lambda^{-5}}{\exp\!\left(\dfrac{ch}{\kappa_B \lambda T}\right) - 1}, \tag{11.10}$$

where h = Planck's constant, 6.6238×10^{-27} erg-sec, c = velocity of light, 2.9979×10^{10} cm/sec, and κ_B = Boltzmann's constant, 1.3803×10^{-16} erg/°K.

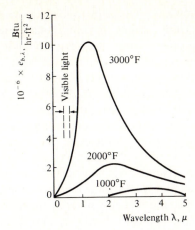

Fig. 11.4 Black body spectral energy distribution.

Substituting Eq. (11.10) into Eq. (11.9) and integrating result in the *Stefan-Boltzmann* equation for black body radiation, which is often written as:

$$e_b = \sigma T^4, \tag{11.11}$$

where the constant σ (in engineering units) is

$$\sigma = \frac{2\pi^5 \kappa_B^4}{15 c^2 h^3} = 0.1713 \times 10^{-8} \text{ Btu/hr-ft}^2 \text{ }^\circ R^4. \tag{11.12}$$

In view of the Stefan-Boltzmann equation, the total emissive power e of a real body is

$$e = \varepsilon \sigma T^4. \tag{11.13}$$

According to Fig. 11.5, the ratio e/e_b of a typical real body varies with wavelength; thus the quantity ε used in Eq. (11.13) is in fact the mean value of ε_λ, the monochromatic emissivity. Qualitatively, the monochromatic emissivity for metals decreases with increasing wavelength. In contrast, the monochromatic emissivity

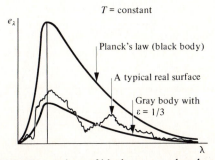

Fig. 11.5 Comparison of black, gray, and real surfaces.

of electric nonconductors has a tendency to increase with increasing wavelength, but the variation of e_λ with λ can be quite irregular. Figure 11.6 depicts experimental results showing these trends for aluminum having three different surface finishes. For the bare metal surface (commercial finish), the emissivity decreases almost steadily with increasing wavelength. The anodized aluminum behaves as a nonmetal, since the anodizing process produces a thick* oxide coating on the surface. Furthermore, by comparing the commercial finish surface with the polished surface, we see that roughness increases emissivity. But since the descriptions such as polished, rough, or commercial finish are rather vague, emissivity data should be used judiciously with regard to the surface description.

Not only does emissivity vary with wavelength, but in the case of well-polished surfaces, it is also a function of the view angle from the normal to the surface (Fig. 11.7). Therefore, a directional emissivity ε_ϕ exists, but for our purposes the ε used for calculations is the average value for a substance emitting radiation in all directions as well as all wavelengths. In this sense, ε used for most calculations is properly called the *total hemispherical emissivity*; usually, however, it is simply called emissivity.

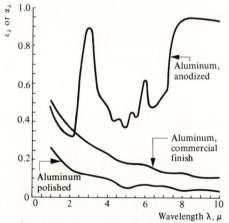

Fig. 11.6 Typical wavelength dependence of ε_λ and α_λ for metals.

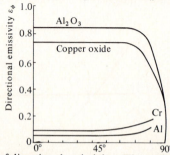

Fig. 11.7 Variation of directional emissivity. The surfaces are well polished.

* Relative to the depth of participating material which is only a fraction of a micron for metals.

Figures 11.8 and 11.9 give the emissivity of several substances including polished metals. Additional data are given in Table 11.1; several oxides and some refractory materials are listed. More extensive listings of emissivity data are available in texts on radiation heat transfer.[1,2,3] In metals, emissivity increases with electrical resistance, and, correspondingly, with temperature; ε is nearly proportional to $T^{1/2}$ for many pure metals. For electric nonconductors, the emissivities are much higher than for *clean* metal surfaces, and they often, but not always, decrease with temperature.

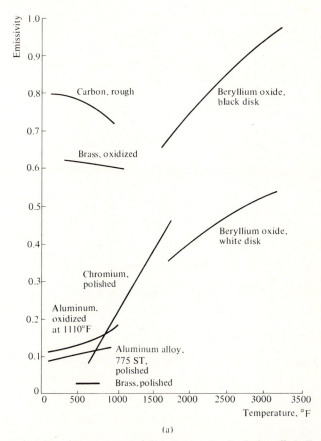

(a)

Fig. 11.8 The emissivity of several materials. (From G. G. Gubareff, J. E. Janssen, and R. H. Torborg, *Thermal Radiation Properties Survey*, Honeywell Research Center, Minneapolis, 1960.)

[1] H. C. Hottel and A. F. Sarofim, *Radiative Transfer*, McGraw-Hill, New York, 1967.

[2] E. M. Sparrow and R. D. Cess, *Radiation Heat Transfer*, Brooks/Cole, Belmont, California, 1966.

[3] T. J. Love, *Radiative Heat Transfer*, Merrill, Columbus, Ohio, 1968.

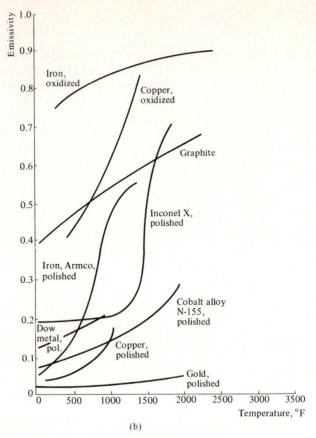

(b)

Fig. 11.8 (*continued*)

Table 11.1 The emissivity of some ceramic materials. (From H. C. Hottel and A. F. Sarofim, *ibid*.)

Material	Temperature, °F	Emissivity
Cuprous oxide	1470–2010	0.66–0.54
Magnesium oxide	530–1520	0.55–0.20
	1650–3100	0.20
Nickel oxide	1200–2290	0.59–0.86
Thorium oxide	530– 930	0.58–0.36
	930–1520	0.36–0.21
Alumina–silica–iron oxide		
58–80% Al_2O_3, 16–38% SiO_2, 0.4% Fe_2O_3	1850–2850	0.61–0.43
26–36% Al_2O_3, 50–60% SiO_2, 1.7% Fe_2O_3	1850–2850	0.73–0.62
61% Al_2O_3, 35% SiO_2, 3% Fe_2O_3	1850–2850	0.78–0.68
Fireclay brick	1832	0.75
Magnesite refractory brick	1832	0.38
Quartz (opaque)	570–1540	0.92–0.80
Zirconium silicate	460– 930	0.92–0.80
	930–1530	0.80–0.52

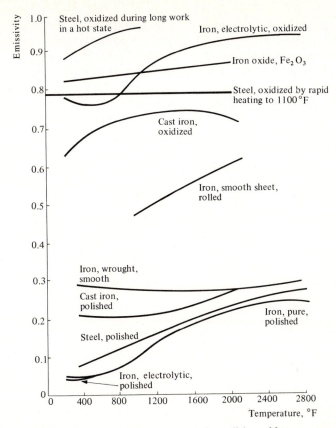

Fig. 11.9 Variation of emissivity with temperature and condition of ferrous materials. (From G. G. Gubareff *et al.*, *ibid.*)

11.4 GRAY BODIES AND ABSORPTIVITY

The absorptivity of a real surface is even more difficult to evaluate with precision than its emissivity. In addition to the factors which affect emissivity, the absorptivity depends on the character of the incoming radiation; therefore, in general, we cannot regard the absorptivity as dependent on the surface alone. Hence, we assign two subscripts to α, the first to represent the temperature of the absorbing surface, and the second—the source temperature of the incoming radiation. We have already seen in Section 11.2 that the emissivity ε_1 of a surface at T_1 is equal to the absorptivity α_1 which the surface exhibits for black radiation from a source at the same temperature T_1; for this reason, it is customary in a wide variety of engineering problems to assume that $\alpha = \varepsilon$. We shall now discuss the conditions for which it is permissible to deduce α-values from ε-values.

If α_λ is independent of λ, the surface is said to be *gray*, and its total absorptivity is independent of the incoming spectral energy distribution; then for a surface at

T_1 receiving radiation from an emitter at T_2, $\alpha_{12} = \alpha_{11}$. Since $\alpha_{11} = \varepsilon_1$, we see that the emissivity may be substituted for α, even though the temperatures of the source radiation and the receiver are not the same. There are very few substances with α_λ independent of λ, but many real surfaces may be considered gray. For example, we show the emissivity of Inconel-X in Fig. 11.10. The emissivity is almost constant in the range 150–1000°F, and if this surface were in an enclosure with temperatures between 150 and 1000°F, the gray assumption would be good. However, if either the enclosure temperature or the Inconel surface temperature were between 1000°F and 1600°F, where the emissivity of the Inconel varies rapidly with temperature, the material would be very much nongray. Thus, when available, a convenient criterion for grayness versus nongrayness is the emissivity versus temperature curve.

Under conditions of radiation from a source at T_1 incident on a *metallic* surface at a lower temperature T_2, we prefer to evaluate the absorptivity of the metallic surface as

$$\alpha_{21} = \varepsilon_2(T^*), \tag{11.14}$$

in which $T^* = \sqrt{T_1 T_2}$. We illustrate the use of these approximations in the following section.

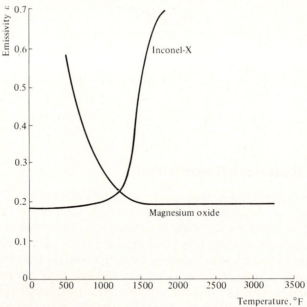

Fig. 11.10 Two materials that exhibit a range of constant ε and a range of rapidly varying ε.

11.5 EXCHANGE BETWEEN INFINITE PARALLEL PLATES

When a system is of simple geometry such as infinite parallel plates, concentric cylinders, or concentric spheres, the evaluation of radiant heat transfer is simplified

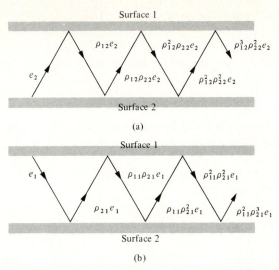

Fig. 11.11 Radiation exchange between parallel surfaces. (a) Irradiation from surface (2). (b) Irradiation from surface (1).

because the whole amount of the radiation leaving one of the surfaces is completely intercepted by the other surface. We shall treat more complex geometrical aspects in the next section, but here we consider a case of simple geometry to show the utilization of the system properties. Specifically, consider the two parallel surfaces of infinite extent in Fig. 11.11. In this schematic representation, the energy is viewed as two parts: the radiation originating from surface 1 is shown in Fig. 11.11(b); radiation originating from surface 2 is shown in Fig. 11.11(a).

The irradiation of surface 1 from surface 2 consists of all the upward pointing arrows in Fig. 11.11(a), or

$$e_2 + \rho_{12}\rho_{22}e_2 + \rho_{12}^2\rho_{22}^2e_2 + \cdots \rho_{12}^n\rho_{22}^n e_2. \tag{11.15}$$

We write this infinite series as

$$\frac{e_2}{1 - \rho_{12}\rho_{22}}. \tag{11.16}$$

In a similar manner, the summation of the upward pointing arrows in Fig. 11.11(b) is

$$\frac{\rho_{21}e_1}{1 - \rho_{11}\rho_{21}}. \tag{11.17}$$

The sum of these two terms is the total irradiation of surface 1; thus

$$G_1 = \frac{e_2}{1 - \rho_{12}\rho_{22}} + \frac{\rho_{21}e_1}{1 - \rho_{11}\rho_{21}}. \tag{11.18}$$

The net heat flux from surface 1 is

$$q_{1,net} = J_1 - G_1, \tag{11.19}$$

but the radiosity J_1 is composed of the radiation emitted at surface 1 plus that reflected,

$$J_1 = e_1 + \rho_{1x}G_1. \tag{11.20}$$

(The second subscript of ρ is not denoted at this point.) By combining Eqs. (11.18)–(11.20), we find that the net heat flux from surface 1 is

$$q_{1,net} = e_1 - (1 - \rho_{1x})G_1 = e_1 - \alpha_{1x}G_1$$

or

$$q_{1,net} = e_1 - \left(\frac{\alpha_{12}e_2}{1 - \rho_{12}\rho_{22}} + \frac{\alpha_{11}\rho_{21}e_1}{1 - \rho_{11}\rho_{21}} \right). \tag{11.21}$$

Here α_{1x} is denoted α_{12} and α_{11} for the first and second terms on the right side of Eq. (11.21) because it is associated with the radiations originating at surfaces 2 and 1, respectively.

The basic practical difficulty in the analysis of systems with nonblack surfaces is what absorptivity value should be used. In the light of this problem, which was discussed in Section 11.4, let us analyze the various forms that Eq. (11.21) takes depending on the manner in which we evaluate α_{11}. The gray assumption which would apply for Inconel-X plates, if both were between 150 and 1000°F, and for refractory surfaces at high temperatures such as magnesia in Fig. 11.10 for temperatures greater than 1500°F, makes $\alpha_{12} = \alpha_{11} = \varepsilon_1$, $\alpha_{21} = \alpha_{22} = \varepsilon_2$, and $\rho_{12} = \rho_{11} = 1 - \varepsilon_1$, $\rho_{21} = \rho_{22} = 1 - \varepsilon_2$. In this case, Eq. (11.21) simplifies to

$$q_{1,net} = (e_{b1} - e_{b2}) \frac{1}{1/\varepsilon_1 + 1/\varepsilon_2 - 1}. \tag{11.22}$$

For metal plates, we can improve Eq. (11.22) by the approximation of Eq. (11.14); we evaluate the emissivity of the cooler surface at $T^* = \sqrt{T_1 T_2}$. For example, if surface 2 is hotter than 1, we write Eq. (11.22) as

$$q_{1,net} = (e_{b1} - e_{b2}) \frac{1}{1/\varepsilon_1(T^*) + 1/\varepsilon_2 - 1}. \tag{11.23}$$

We accomplish the evaluation of radiant heat transfer between nongray surfaces with the most precision only by considering the best approximation for their absorptivities, and, also the variation in properties with direction. Fortunately, many industrial materials, especially refractories used in furnaces, behave as gray diffuse radiators so that these details may be omitted from analysis. We shall consider nonblack surfaces as gray in this chapter. More details regarding nongray bodies and directional properties can be found in more advanced treatments of radiation heat transfer.

Example 11.1 We can use the above results to illustrate the principle of *radiation shields.* Multiple shields are very effective when used as high-temperature insulation against thermal radiation. To illustrate the principle, develop an expression for the reduction in radiant heat transfer between two infinite and parallel refractory walls, when a thin plate of aluminum is placed between them. Assume that all surfaces are gray; the emissivity of the walls is 0.8 and the emissivity of the aluminum is 0.2.

The system with appropriate subscripts is shown below.

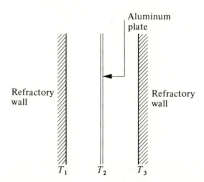

Solution. First, we examine the case in the absence of the aluminum sheet, that is, direct radiant transfer from surface 1 to surface 3. Equation (11.22) applies with appropriate change of subscripts. Therefore

$$q_{1,\text{net}}(\text{no shield}) = (e_{b1} - e_{b3})\frac{1}{1/\varepsilon_1 + 1/\varepsilon_3 - 1};$$

Now in the presence of the shield, we apply Eq. (11.22) to the net radiant flux from surface 1 to surface 2. Hence

$$q_{1,\text{net}} = (e_{b1} - e_{b2})\frac{1}{1/\varepsilon_1 + 1/\varepsilon_2 - 1}.$$

Similarly, we write the radiant flux from surface 2 to surface 3 as

$$q_{2,\text{net}} = (e_{b2} - e_{b3})\frac{1}{1/\varepsilon_3 + 1/\varepsilon_2 - 1}.$$

Since $q_{2,\text{net}} = q_{1,\text{net}}$ (steady state), we eliminate e_{b2} from these equations:

$$q_{1,\text{net}} = \frac{(e_{b1} - e_{b3})}{(1/\varepsilon_1 + 1/\varepsilon_2 - 1) + (1/\varepsilon_2 + 1/\varepsilon_3 - 1)}.$$

The ratio of radiant flux with the shield to that without the shield is

$$\frac{q_{1,\text{net}}\,(\text{with shield})}{q_{1,\text{net}}\,(\text{no shield})} = \frac{1/\varepsilon_1 + 1/\varepsilon_3 - 1}{(1/\varepsilon_1 + 1/\varepsilon_2 - 1) + (1/\varepsilon_2 + 1/\varepsilon_3 - 1)}.$$

With $\varepsilon_1 = \varepsilon_3 = 0.8$, and $\varepsilon_2 = 0.2$, this ratio equals 0.143. Insertion of more shields would lower the ratio even more considerably.

11.6 VIEW FACTORS

Having discussed the complexities of property evaluation, we shall now consider the problem of geometric arrangement in problems involving transfer of thermal radiation. In this connection, we introduce the *view factor*. In the engineering literature different terms are used for this factor, such as *configuration factor*, *direct-exchange factor*, *angle factor*, or simply *factor*, and perhaps some others which are unknown to us.

The simplest case for calculating the net loss of energy from a body due to radiation is to visualize a body at temperature T_1 which is completely surrounded by an environment at T_2. Since the environment completely surrounds the body, the radiation impinging on surface 1 is black (as in a cavity). The thermal flux (or emissive power) from the body is $e_1 = \varepsilon_1 \sigma T_1^4$, and that absorbed by the body from the large isothermal cavity is $\alpha_{12} e_{b2} = \alpha_{12} \sigma T_2^4$. Then

$$q_{1,\text{net}} = \sigma[\varepsilon_1 T_1^4 - \alpha_{12} T_2^4], \tag{11.24}$$

and in the instance where we assume that the body is gray, $\alpha_{12} = \varepsilon_1$, and we write Eq. (11.24) as

$$q_{1,\text{net}} = \sigma \varepsilon_1 (T_1^4 - T_2^4). \tag{11.25}$$

The more general case of radiation exchange is a system composed of several surfaces at different temperatures and each of different emissivities. Such a situation involves the use of the *view factor*, F_{ij}. We define it as the fraction of the radiation which leaves surface i in all directions and is intercepted by surface j.

To evaluate F_{12}, imagine the two black surfaces, of areas A_1 and A_2, shown in Fig. 11.12. A small surface element dA_1 emits radiation in all directions (from one

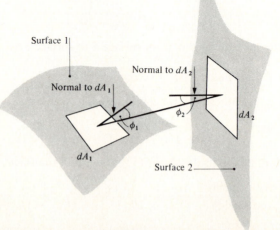

Fig. 11.12 Exchange of radiation between elemental areas.

side), and a small surface element dA_2 on the other surface intercepts some of this radiation. Let r be the distance between the surface elements, and ϕ_1 and ϕ_2 the angles that r makes with the two normals, as depicted in the figure. The intensity of the radiation transferred from dA_1 to dA_2, called E_{12}, is proportional to:

1. The apparent area of dA_1 viewed from dA_2, $\cos \phi_1 \, dA_1$.
2. The apparent area of dA_2 viewed from dA_1, $\cos \phi_2 \, dA_2$.
3. The reciprocal of the distance squared, $1/r^2$, between the two surface elements, as in all radiation problems.

Thus

$$dE_{12} = I \frac{\cos \phi_1 \, dA_1 \cos \phi_2 \, dA_2}{r^2}, \text{ Btu/hr,} \tag{11.26}$$

where I is a proportionality constant, Btu/hr-ft^2. By inspecting Eq. (11.26), I must be the intensity leaving dA_1 in a particular direction, that is, if $\phi_1 = \phi_2 = 0$, then it is the same problem as projecting light onto a screen.

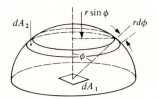

Fig. 11.13 Hemispherical radiation intensity.

At this point it may appear that I and e_b (the emissive power of a black body) are identical. This is *not* the case, and here we digress from the major argument. Consider the radiation received from the entire field of view to dA_1, that is, the hemisphere in Fig. 11.13. Visualize dA_2 as the unit surface described by the ring on the hemisphere. Since the normals to a spherical surface are coincident to the radii, ϕ_2 in Eq. (11.26) must be 0, so that $\cos \phi_2 = 1$. The elemental area dA_2 is given by the product of $r \, d\phi_1$ and its length, $2\pi \sin \phi_1 r$. If we integrate over the entire hemisphere, Eq. (11.26) yields (for isotropic I) per unit area dA_1:

$$dE_{12} \text{ (hemisphere)} = 2\pi I \int_0^{\pi/2} \frac{\cos \phi_1 \sin \phi_1 r^2 \, d\phi_1}{r^2}, \text{ Btu/hr-ft}^2,$$

or

$$dE_{12} \text{ (hemisphere)} = \pi I. \tag{11.27}$$

Note that

$$dE_{12} \text{ (hemisphere)} = \frac{dE_{12}}{dA_1} = e_1. \tag{11.28}$$

Since we have been discussing a black body, the result states that the emissive power of a black body e_b is the product of π and the intensity of the emitted radiation:

$$e_b = \pi I. \tag{11.29}$$

Thus, by combining Eqs. (11.26) and (11.29), we return to the major argument, and express the general equation for the transfer of radiant energy from one black body to another as

$$E_{12} = \int_{A_1} dE_{12} = \frac{e_b}{\pi} \int_{A_1} \int_{A_2} \frac{\cos \phi_1 \cos \phi_2}{r^2} dA_2 \, dA_1. \tag{11.30}$$

For convenience, the view factor F_{12}, which is dimensionless, is introduced here; it is defined by the equation

$$A_1 F_{12} = \frac{1}{\pi} \int_{A_1} \int_{A_2} \frac{\cos \phi_1 \cos \phi_2}{r^2} dA_2 \, dA_1. \tag{11.31}$$

This integral contains nothing but geometrical aspects of radiant exchange, and has fortunately been evaluated for many geometries, and one does not have to perform the integration oneself. Now we can write a simple equation to express the energy emitted by a black surface 1 and intercepted by surface 2, using the definition of F_{12},

$$E_{12} = e_{b1} A_1 F_{12}. \tag{11.32}$$

In exactly the same manner, we could have developed an analogous expression for the energy originating at 2 and arriving at 1.

$$E_{21} = e_{b2} A_2 F_{21}. \tag{11.33}$$

Then, the net exchange of energy from 1 to 2 is $A_1 q_{1,\text{net}}$ or $Q_{1,\text{net}}$:

$$Q_{1,\text{net}} = E_{12} - E_{21}. \tag{11.34}$$

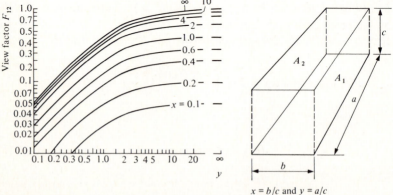

Fig. 11.14 View factors for identical, parallel, directly-opposed rectangles. (From T. J. Love, *Radiative Heat Transfer*, ibid.)

At thermal equilibrium, $Q_{1,net} = 0$, and from Eqs. (11.32) and (11.33) and the fact that the emissive power of a black body depends only upon temperature, we see that $A_2 F_{21} = A_1 F_{12}$. In general, we can say that

$$A_i F_{ij} = A_j F_{ji}. \tag{11.35}$$

Equation (11.35), though simple, is very useful, and the reader should keep it in mind.

Values of F_{12} have been calculated for various geometric arrangements of two surfaces. Figures 11.14–11.16 provide such view factor charts. More extensive listings can be found in texts on radiation heat transfer. Although we have restricted the discussion leading up to the view factor charts to black surfaces, it is apparent that for nonblack surfaces, the emissivities of which are independent of emission

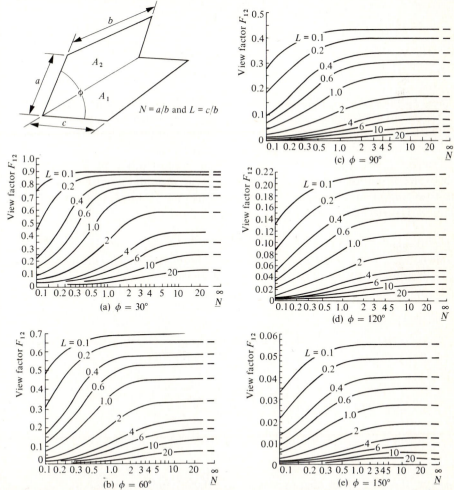

Fig. 11.15 View factors for two rectangles with a common edge. (From T. J. Love, *Radiative Heat Transfer, ibid.*)

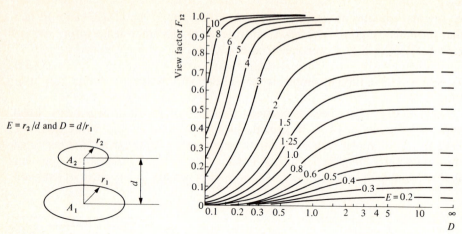

$E = r_2/d$ and $D = d/r_1$

Fig. 11.16 View factors for parallel, directly-opposed disks. (From T. J. Love, *Radiative Heat Transfer, ibid.*)

angles, F_{12} calculated by the method discussed above will again represent the fractional radiation from A_1 intercepted by A_2. Thus such a calculation is exact for black surfaces, and very useful for most nonmetallic, oxidized or rough metal surfaces which exhibit nearly isotropic surface properties.

Since the view factor is the fraction of radiation emitted by a surface and intercepted by another surface, it is obvious that all the radiation from any surface must be ultimately intercepted when there are many surfaces involved. For example, if surface 1 is *seen* by surfaces 2, 3, 4 ... n, then

$$F_{12} + F_{13} + F_{14} + \cdots F_{1n} = 1. \tag{11.36}$$

Example 11.2 Calculate the net heat flow (Btu/hr) by radiation to the furnace wall at 500°F from the furnace floor at 1000°F. Both surfaces can be considered to be black radiators.

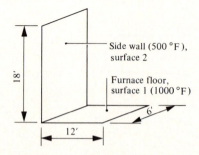

Solution. The energy emitted by the black surface 1 and intercepted by 2 is

$$E_{12} = e_{b1} A_1 F_{12}.$$

Similarly, the energy emitted by surface 2 and intercepted by surface 1 is

$$E_{21} = e_{b2}A_2F_{21}.$$

The net exchange of energy from surface 1 to surface 2 is

$$Q_{1,net} = E_{12} - E_{21},$$

and, since $A_2F_{21} = A_1F_{12}$, we can finally write the net exchange of energy from surface 1 to surface 2 as

$$Q_{1,net} = A_1F_{12}(e_{b1} - e_{b2}) = A_1F_{12}\sigma[T_1^4 - T_2^4].$$

Figure 11.15(c) is used to evaluate F_{12} with

$$N = \frac{18}{6} = 3, \qquad L = \frac{12}{6} = 2, \qquad \text{and} \qquad \phi = 90°.$$

The view factor F_{12} is 0.165. Recalling that $\sigma = 0.171 \times 10^{-8}$ Btu/hr-ft^2 °R^4, we determine that

$$Q_{1,net} = (72)(0.165)(0.171)\left[\left(\frac{1460}{100}\right)^4 - \left(\frac{960}{100}\right)^4\right]$$

$$= 76{,}000 \text{ Btu/hr.}$$

11.7 ELECTRIC CIRCUIT ANALOGY FOR RADIATION PROBLEMS

Consider an all-black surface enclosure (Fig. 11.17a) in which

$$F_{12} + F_{13} + F_{14} = 1. \tag{11.37}$$

The net heat flow (Btu/hr) from surface 1 is

$$Q_{1,net} = (E_{12} - E_{21}) + (E_{13} - E_{31}) + (E_{14} - E_{41})$$

$$= A_1F_{12}(e_{b1} - e_{b2}) + A_1F_{13}(e_{b1} - e_{b3}) + A_1F_{14}(e_{b1} - e_{b4}). \tag{11.38}$$

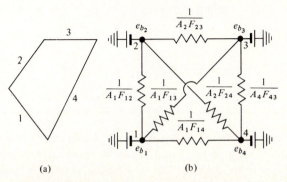

(a)   (b)

Fig. 11.17 (a) A black enclosure. (b) Its electrical analog.

We can also develop Eq. (11.38) by analyzing the situation with an electric analog. We think of each surface as a node in the electric analog (Fig. 11.17b) and represent the potential at each node by e_{bi}, the current by the heat flow, and the resistance between two potential nodes i and j by $1/A_iF_{ij}$. We draw the circuit, and connect each node by resistors to all the other nodes. We can then easily recognize the net heat flow from any surface. For example, we obtain the net heat flow from surface 4 by simply summing the analog currents through the three resistors connected to node 4:

$$Q_{4,net} = A_1F_{14}(e_{b4} - e_{b1}) + A_2F_{24}(e_{b4} - e_{b2}) + A_4F_{43}(e_{b4} - e_{b3}).$$

It may be desirable to express the net heat flow only in terms of A_4. In this case, we utilize Eq. (11.35) to yield

$$Q_{4,net} = A_4F_{41}(e_{b4} - e_{b1}) + A_4F_{42}(e_{b4} - e_{b2}) + A_4F_{43}(e_{b4} - e_{b3}). \quad (11.39)$$

The electric analog for a gray surface is also very useful in solving many problems. The total energy flux radiating from a surface is its radiosity, or

$$J = \rho G + \varepsilon e_b, \quad (11.40)$$

where G is the irradiation. The net heat flux transferred from the surface is

$$q_{net} = J - G, \quad (11.41)$$

and by eliminating G from these last two equations, we get

$$q_{net} = \frac{\varepsilon}{1 - \varepsilon}(e_b - J).$$

Therefore, for the net heat flow from a surface of area A, we can write

$$Q_{net} = \frac{\varepsilon}{1 - \varepsilon} A(e_b - J). \quad (11.42)$$

Thus, we can represent a gray surface with a potential of a black surface reduced by a resistance $(1 - \varepsilon)/A\varepsilon$, as in Fig. 11.18. The use of an electric analog is particularly useful when dealing with gray surfaces. We illustrate this in Example 11.3.

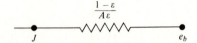

Fig. 11.18   Electric analog of a gray surface.

Example 11.3   Consider the exchange of energy between two parallel and gray plates. For infinitely long plates, 1 and 2, use the method of electric analogs and develop an expression for the heat flux if the plates are at T_1 and T_2, and have emissivities of ε_1 and ε_2, respectively.

Solution. The system and the appropriate analog are shown below.

By examining the analog circuit, the reader can see that each surface is represented by the nodes shown in Fig. 11.18. The analog circuit is completed by connecting the radiosity nodes with the resistance between the surfaces. Therefore, when dealing with gray surfaces, connect the radiosities; when dealing with black surfaces, connect the black body emissive powers, as shown in Fig. 11.17(b).

To solve the problem, simply determine the total resistance between e_{b1} and e_{b2}. The total resistance is

$$\frac{1 - \varepsilon_1}{A_1 \varepsilon_1} + \frac{1}{A_1 F_{12}} + \frac{1 - \varepsilon_2}{A_2 \varepsilon_2}.$$

Then by using Ohm's law, we can immediately write an expression for the net heat flow from 1 to 2:

$$Q_{1,net} = (e_{b1} - e_{b2}) \frac{1}{\dfrac{1 - \varepsilon_1}{A_1 \varepsilon_1} + \dfrac{1 - \varepsilon_2}{A_2 \varepsilon_2} + \dfrac{1}{A_1 F_{12}}}.$$

In this case, $A_1 = A_2$, and $F_{12} = 1$. Therefore

$$q_{1,net} = \frac{Q_{1,net}}{A_1}$$

$$= (e_{b1} - e_{b2}) \frac{1}{1/\varepsilon_1 + 1/\varepsilon_2 - 1}.$$

This expression is the same as Eq. (11.22). After having treated a few situations by using electric analogs, the reader will become quite apt at developing expressions involving radiation exchange between gray and/or black surfaces.

11.8 FURNACE ENCLOSURES

As an application of radiant heat transfer, consider furnace design. We often encounter a situation in which energy is transferred from a *heat source* to a *heat sink* with intermediate refractory walls. For example, the heat source might be a row of

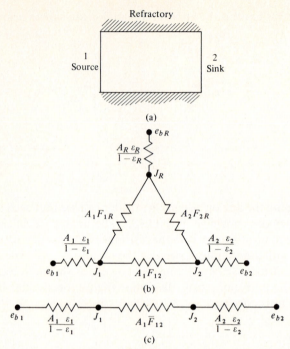

Fig. 11.19 (a) Furnace enclosure. (b) and (c) Its electric analogs. (In this case, conductances rather than resistances are indicated.)

electric resistors to heat a metal part placed in a furnace. In normal furnaces, the radiation incident on the refractory walls is so large compared to the heat conduction through the walls that we may approximate the conduction as zero when calculating radiant fluxes. Such walls are examples of *no-net-flux surfaces*. This assumption greatly simplifies the problem of transferring radiant heat from surfaces to sinks.

As an example of this problem, consider the transfer from face 1 to the opposite face 2 with intermediate refractory walls (Fig. 11.19a). All four walls are assumed to be gray, and the electric analog is given in Fig. 11.19(b). First, simplify the circuit by combining all the resistances between J_1 and J_2, remembering that in parallel circuits conductances are directly additive, and in series circuits the same applies to resistances. The equivalent conductance between J_1 and J_2 is

$$A_1 \bar{F}_{12} = A_1 F_{12} + \frac{1}{1/A_1 F_{1R} + 1/A_2 F_{2R}}, \qquad (11.43)$$

and the circuit reduces to Fig. 11.19(c). The symbol $\bar{F}_{12}$ is often used when the several view factors in a system are all combined, as illustrated here. Finally, the equivalent resistance for the entire circuit is

$$\frac{1}{A_1 \mathscr{F}_{12}} = \frac{1}{A_1} \frac{1 - \varepsilon_1}{\varepsilon_1} + \frac{1}{A_1 \bar{F}_{12}} + \frac{1}{A_2} \frac{1 - \varepsilon_2}{\varepsilon_2}. \qquad (11.44)$$

The symbol $\mathscr{F}_{12}$ is used when the entire circuit has been reduced; thus, $\mathscr{F}_{12}$ depends not only on the individual view factors of the elements in the system but also on the individual emissivities. Since we are approximating the refractory walls as *no-net-flux* surfaces, then the current from J_R to e_{bR} is zero, and the current that goes from e_{b1} to e_{b2} in the equivalent circuit of Fig. 11.19(c) represents the net flow of energy from the source to the sink. In this sense, we refer to $\overline{F}_{12}$ as the *total exchange factor* because it takes into account reradiating surfaces between the two surfaces (source and sink) of primary interest.

In this case, then

$$Q_{1,\text{net}} = A_1\mathscr{F}_{12}(e_{b1} - e_{b2}) = A_1\mathscr{F}_{12}\sigma(T_1^4 - T_2^4),\qquad(11.45)$$

in which $A_1\mathscr{F}_{12}$ is given by Eq. (11.44).

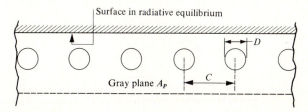

Fig. 11.20 A refractory-backed row of cylindrical elements.

In furnaces, electric resistors very often provide the source of heat, and can be mounted on a back-insulated refractory wall in a parallel array, as shown in Fig. 11.20. It is convenient to replace such a system by an equivalent continuous gray plane, having effective emissivity ε_P and temperature equal to that of the resistors. The gray plane is really imaginary and we show it by the dotted line in Fig. 11.20. To calculate ε_P, we make use of Fig. 11.21 to determine $\overline{F}_{PE}$, this is equivalent to $\overline{F}_{12}$ in Eq. (11.43). Then, by rearranging Eq. (11.44) with $\varepsilon_P = 1$, we can determine $\mathscr{F}_{PE}$ from the expression

$$\mathscr{F}_{PE} = \cfrac{1}{\cfrac{1}{\overline{F}_{PE}} + \cfrac{C}{\pi D}\left(\cfrac{1}{\varepsilon_E} - 1\right)},\qquad(11.46)$$

where C and D are defined in Fig. 11.21. If no reradiating surface (no back-insulated refractory wall) exists, then $\mathscr{F}_{PE} = \overline{F}_{PE}$ in this expression. Finally, we can consider the elements equivalent to the plane A_P with an emissivity $\varepsilon_P = \mathscr{F}_{PE}$ at the operating temperature of the elements.

Example 11.4 A furnace has a floor 3 ft × 6 ft, and located at the top, 5 ft away, are heating elements, 2 in. in diameter, 3 ft long on 4-in. centers backed by a well-insulated refractory roof. The heating elements are Nichrome IV ($\varepsilon = 0.74$), which operate at a surface temperature of 2000°F. The side walls are insulated refractory

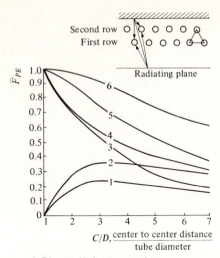

Fig. 11.21 Total exchange factors $\bar{F}_{PE}$ between a black plane (P) and one or two rows of cylindrical elements (E) backed by a no-net-flux refractory wall. Direct view factors F_{PE} are also shown for comparison. Note that the elements form an array of equilateral triangles. (From H. C. Hottel and A. F. Sarofim, *ibid.*)

(no-net-flux) surfaces. Find the rate at which heat can be transferred to a cold sheet of aluminum (3 ft × 6 ft, $\varepsilon = 0.15$) placed on the floor of the preheated furnace.

Solution. To demonstrate only the aspects of radiant heating, we shall ignore convective heating on the upper surface of the aluminum sheet and any heating that might occur by conduction across minute contact points between the sheet and the furnace floor.

To calculate the rate at which heat is transferred by radiation to the upper surface of the sheet, we divide the system into four zones: (1) an equivalent gray plane to represent the elements, (2) the side walls of dimensions 5 ft × 6 ft, (3) the side walls of dimensions 5 ft × 3 ft, and (4) the upper surface of the aluminum sheet. The analog circuit (showing conductances) is

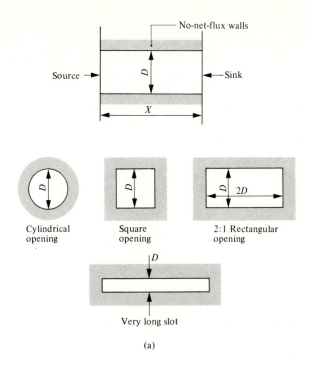

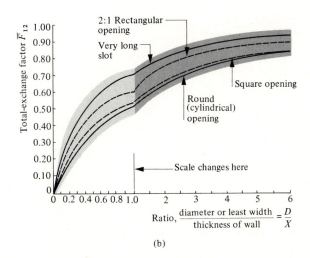

Fig. 11.22 Total exchange factors between parallel source and sink connected by no-net-flux walls. (a) Various geometries included and (b) values of the total exchange factor. (From H. C. Hottel and A. F. Sarofim, *ibid*.)

In this particular problem, to solve for $A_1 \bar{F}_{14}$, the equivalent conductance for the entire circuit between J_1 (source) and J_4 (sink), would be a messy task if done by algebraic means. A 4×4 matrix could be set up and solved on a computer rather easily, or one could set up an actual analog circuit. For our purposes, we refer the reader to Fig. 11.22, from which $\bar{F}_{14}$ can be found directly. In Fig. 11.22, the *total exchange factor* is presented for several geometries of source–sink combinations connected by perfect reradiating walls. (Naturally, we have selected a geometry represented on the figure!) By using Fig. 11.22, we can represent the circuit very simply as

Using Fig. 11.22, we find, for the geometry of a 2:1 rectangular opening with $D = 3$ ft and $X = 5$ ft, that

$$\bar{F}_{14} = 0.48.$$

Next we proceed to determine ε_1. From Fig. 11.21 (curve 5) with $C/D = 2$, we obtain $\bar{F}_{PE} = 0.88$. Then, from Eq. (11.46),

$$\varepsilon_1(\varepsilon_P) = \mathscr{F}_{PE} = \cfrac{1}{\cfrac{1}{0.88} + \cfrac{2}{\pi}\left(\cfrac{1}{0.74} - 1\right)} = 0.741.$$

Now using Eq. (11.44) with $A_1 = A_4$, we get

$$\frac{1}{\mathscr{F}_{14}} = \frac{1 - \varepsilon_1}{\varepsilon_1} + \frac{1}{\bar{F}_{14}} + \frac{1 - \varepsilon_4}{\varepsilon_4}$$

$$= \frac{1 - 0.741}{0.741} + \frac{1}{0.48} + \frac{1 - 0.15}{0.15}$$

$$= 0.124.$$

Therefore

$$Q_{1,\text{net}} = A_1 \mathscr{F}_{14} \sigma (T_1^4 - T_4^4)$$

$$= (3 \times 6)(0.124)(0.171)\left[\left(\frac{2460}{100}\right)^4 - \left(\frac{540}{100}\right)^4\right]$$

$$= 141,000 \text{ Btu/hr.}$$

To this, we could add the heat transferred to the bottom of the plate; assume that the furnace floor had been preheated to 2000°F prior to the insertion of the aluminum plate, and that all the heat transferred across the furnace floor–aluminum sheet interface is via radiation across a gap of two infinitely long parallel plates. Using Eq. (11.22), and assuming that the emissivity of the refractory floor is 0.8, we obtain

$$Q_{R,net} = \frac{(6 \times 3)(0.171)\left[\left(\dfrac{2460}{100}\right)^4 - \left(\dfrac{540}{100}\right)^4\right]}{1/0.8 + 1/0.15 - 1}$$

$$= 165,000 \text{ Btu/hr.}$$

Therefore

$$Q_{total} = 141,000 + 165,000 = 306,000 \text{ Btu/hr.}$$

The heat flow rates calculated above are the initial values, and apply only when the cold sheet has been inserted in the oven. We deal with the more general problem of describing the transient heating-up of the sheet in Section 11.12.

11.9 RADIATION COMBINED WITH CONVECTION

In the preceding sections of this chapter we considered energy transfer by radiation as an isolated phenomenon. Indeed, in many high-temperature applications, problems should be tackled in this manner because, at high temperatures, radiation can completely dominate since radiant heat flow depends on the fourth power of the absolute temperature. In many practical situations, however, we cannot neglect convective heat transfer, and it is necessary to consider both modes of energy transport.

When we include radiation in a calculation of a situation which also involves convection, we realize that the total heat flow is the sum of the convective heat flow and the radiant heat flow. Thus, we conveniently use the total heat-transfer coefficient h_t, which is composed of the heat-transfer coefficient h in the usual sense, and the radiant heat-transfer coefficient h_r:

$$h_t = h + h_r. \tag{11.47}$$

We define the radiant heat-transfer coefficient as

$$h_r = \frac{q_{1,net}}{(T_1 - T_2)} = \mathscr{F}_{12}\left[\frac{\sigma(T_1^4 - T_2^4)}{T_1 - T_2}\right], \tag{11.48}$$

in which $T_1 - T_2$ is a temperature difference in which T_2 may be chosen as some convenient temperature in the system.

We often encounter the situation in which surface 1 is completely surrounded by or exposed only to some fluid whose bulk temperature is at T_f. In this case, it is convenient to select T_2 as T_f, and, since $\mathscr{F}_{12} = \varepsilon_1$, we write

$$h_r = \varepsilon_1\left[\frac{\sigma(T_1^4 - T_2^4)}{T_1 - T_2}\right]. \tag{11.49}$$

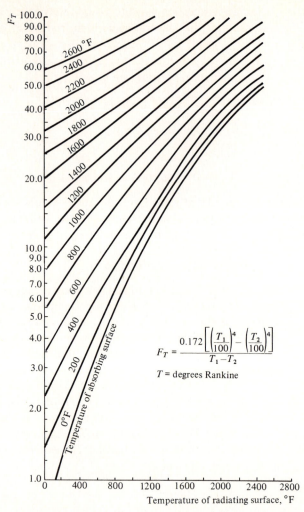

Fig. 11.23 The temperature factor, F_T, for use in Eqs. (11.48) or (11.49). (From F. Kreith, *Principles of Heat Transfer*, second edition, International Textbook Co., 1965, page 231.)

The square-bracketed part of Eqs. (11.48) and (11.49) is called the temperature factor, F_T, and its values for various surface temperatures, T_1, and reference temperatures, T_2, are given in Fig. 11.23.

Example 11.5 A thermocouple with an emissivity of 0.7 measures the temperature of a gas flowing in a long duct whose internal wall surfaces are at 500°F. The temperature indicated by the thermocouple is 1000°F, and the convective heat-transfer coefficient between the gas and the surface of the thermocouple is 20 Btu/hr-ft² °F. Determine the true temperature of the gas.

Solution. The temperature of the thermocouple is less than that of the gas because the thermocouple radiates to the wall. We can write an energy balance for steady-state conditions in which the radiant heat flow from the thermocouple to the wall equals the convective heat flow from the gas to the couple. Assuming that the thermocouple can be approximated as a gray surface, we write

$$hA_1(T_f - T_1) = A_1 F_{12} \varepsilon_1 \sigma(T_1^4 - T_2^4),$$

where A_1 is the thermocouple surface area; T_f, T_1, and T_2 are the temperatures of the fluid, thermocouple, and duct walls, respectively. Since the thermocouple is completely surrounded by the duct walls, $F_{12} = 1$. The gas temperature is then found by

$$T_f - T_1 = \frac{\varepsilon_1 \sigma}{h}(T_1^4 - T_2^4)$$

$$= \frac{(0.7)(0.171)}{(20)}\left[\left(\frac{1460}{100}\right)^4 - \left(\frac{960}{100}\right)^4\right]$$

$$= 222°R \text{ or } °F,$$

and

$$T_f = 1222°F.$$

Errors such as those indicated by the above problem have induced engineers to pursue thermocouple designs which incorporate radiation shields and/or means of increasing h between the fluid and the thermocouple. Some devices are illustrated in Fig. 11.24.

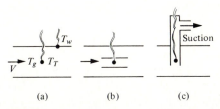

(a)   (b)   (c)

Fig. 11.24 Measurement of flowing gas temperatures. (a) Simplest device most liable to errors. (b) Incorporation of a radiation shield. (c) Incorporation of a radiation shield and use of the increased fluid velocity past the thermocouple's surface.

11.10 RADIATION FROM GASES

The methods given in the preceding sections are applicable only to systems involving gases which are transparent to radiation. Gases with simple symmetrical molecules, such as H_2, He, O_2, and N_2, are essentially transparent to radiation, but heteropolar gases (CO, CO_2, H_2O, SO_2, NH_3, HCl), which interact with radiation sufficiently, must be made part of the system for calculations.

Solids emit radiation at all wavelengths, whereas gases emit and absorb radiation only between the narrow regions of wavelengths called bands. Analysis of radiation participation should be made for each band, and the total effect can be obtained by summation. However, it is more convenient to solve the problems in terms of the total radiation and emission characteristics of the gases.

As a first approximation, the total number of radiating gas molecules present in a space determines the emissivity and absorptivity of the gas. This approximation is best at lower pressures with corresponding large molecular mean free paths permitting each molecule to radiate as though it is alone, with a minimal shielding effect by the other molecules present in space. Gas radiation differs from the radiation of solids whose emission and absorption are surface phenomena, because when calculating emission and absorption for a gas layer at a particular temperature, its thickness and pressure must be accounted for. In fact, the emissivity and absorptivity of a gas is a function of the active gas species, temperature, thickness and shape, total pressure, and the partial pressure of the active gas. To handle all these complexities, a series of charts has been developed based on experimental data for evaluating the emissivities and absorptivities of various gases.

Consider a hemisphere of radius L, containing a heteropolar gas of a given partial pressure; the problem is the radiant heat exchange between the gas at temperature T_g and a unit element at the center of the base of the hemisphere. The emission of the gas to the surface is $\varepsilon_g \sigma T_g^4$ per unit surface area, where ε_g denotes gas emissivity. Specifically, in Figs. 11.25–11.27 the *reduced* emissivities ε_g' for CO_2, CO, and H_2O, are presented as a function of the absolute temperature and the product of L and partial pressure. These values are called reduced emissivities because they are plotted for a reduced pressure; that is, all data are at near-zero partial pressure and at a total pressure of 1 atm. The graphs apply strictly to the hemispherical body of gas defined above; we shall discuss briefly how to treat other shapes.

The effect of total pressure on ε_g for CO_2 and H_2O is given in Figs. 11.28 and 11.29, where the ratio of the actual emissivity, at the state being considered, to the reduced emissivity is presented as C_c and C_w for CO_2 and H_2O, respectively. When both H_2O and CO_2 are present in the gas, we must make another correction for the mixture composition using Fig. 11.30. We subtract the value of $\Delta\varepsilon$ in this chart from the sum of emissivities for each component. This correction is due to the fact that both gases emit radiation in overlapping wavelength bands.

For shapes other than hemispheres, we take L as the *effective beam length*, and calculate it for use with Figs. 11.25–11.30. Table 11.2 lists L for shapes other than hemispheres. For approximate calculations of shapes which are not listed, we can take L as $3.4 \times$ volume/surface area. For a presentation and discussion of radiation data for many other gas species, the reader should consult Chapter 6 of the text by Hottel and Sarofim.[4]

---

[4] H. C. Hottel and A. F. Sarofim, *ibid.*

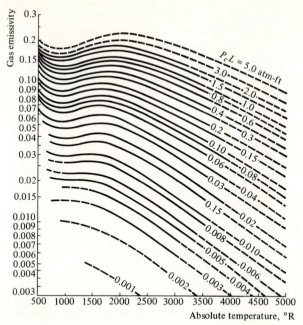

Fig. 11.25 Reduced emissivity of carbon dioxide. (Figures 11.25–11.27, 11.29, and 11.30 are adapted from H. C. Hottel and A. F. Sarofim, Chapter 6, *ibid.*)

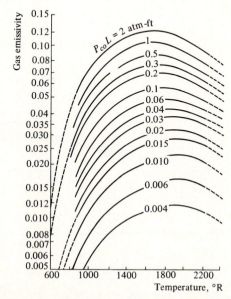

Fig. 11.26 Reduced emissivity of carbon monoxide.

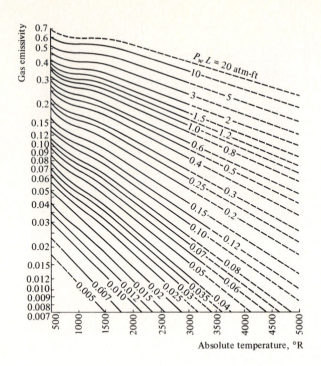

Fig. 11.27 Reduced emissivity of water vapor.

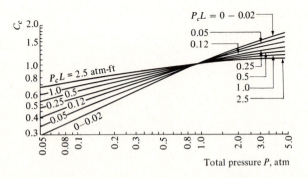

Fig. 11.28 Correction factor for CO_2 emissivity. (From F. Kreith, *Principles of Heat Transfer*, *ibid.*)

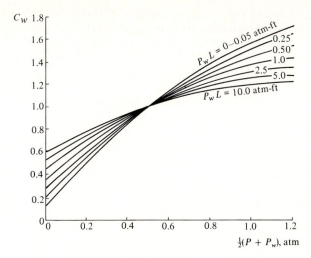

Fig. 11.29 Correction factor for H_2O emissivity.

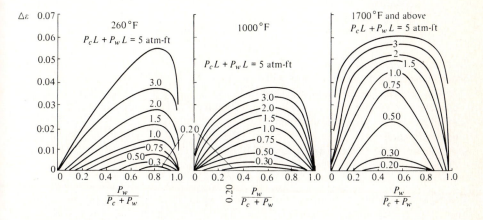

Fig. 11.30 Correction for band overlap when CO_2 and H_2O coexist.

Table 11.2 Beam lengths for gas radiation. (From H. C. Hottel, "Radiation," Chapter IV of *Heat Transmission*, by W. H. McAdams, McGraw-Hill, 1954)

Shape	Characterizing dimension, X	When $p_g L = 0$	For average values of $p_g L$
		Factor by which X is multiplied to obtain mean beam length, L	
Sphere	Diameter	2/3	0.60
Infinite cylinder	Diameter	1	0.90
Semi-infinite cylinder, radiating to center of base	Diameter	—	0.90
Right-circular cylinder, height = diameter, radiating to center of base	Diameter	—	0.77
Same, radiating to whole surface	Diameter	2/3	0.60
Infinite cylinder of half-circular cross section. Radiating to spot on middle of flat side	Radius	—	1.26
Rectangular parallelepipeds			
1:1:1 (cube)	Edge	2/3	—
1:1:4, radiating to 1 × 4 face		0.90	—
radiating to 1 × 1 face	Shortest edge	0.86	—
radiating to all faces		0.89	—
1:2:6, radiating to 2 × 6 face		1.18	—
radiating to 1 × 6 face	Shortest edge	1.24	—
radiating to 1 × 2 face		1.18	—
radiating to all faces		1.20	—
1:∞:∞ (infinite parallel planes)	Distance between planes	2	—
Space outside			
Space outside infinite bank of tubes with centers on equilateral triangles; tube diameter = clearance	Clearance	3.4	2.8
Same as preceding, except tube diameter = one-half clearance	Clearance	4.45	3.8
Same, except tube centers on squares; diameter = clearance	Clearance	4.1	3.5

Example 11.6 Determine the emissivity of a gas containing 50% N_2, 40% CO_2, and 10% H_2O at a total pressure of 2 atm with $T_g = 2000°R$. The gas is contained in a very long cylindrical space having a 2-ft diameter.

Solution. From Table 11.2:

$$L = (0.90)(2) = 1.8 \text{ ft.}$$

Then

$$P_cL = (0.40)(2)(1.8) = 1.44 \text{ atm ft},$$

$$P_wL = (0.10)(2)(1.8) = 0.36 \text{ atm ft},$$

$$\tfrac{1}{2}(P + P_w) = \tfrac{1}{2}(2 + 0.20) = 1.10 \text{ atm},$$

and

$$\frac{P_w}{P_c + P_w} = \frac{0.20}{0.80 + 0.20} = 0.20.$$

From Figs. 11.25 and 11.27–11.29, $\varepsilon_c' = 0.17$, $\varepsilon_w' = 0.12$, $C_c = 1.10$, $C_w = 1.55$, and $\Delta\varepsilon = 0.032$.

Therefore

$$\varepsilon_g = C_w\varepsilon_w' + C_c\varepsilon_c' - \Delta\varepsilon$$

$$= (1.55)(0.12) + (1.10)(0.17) - 0.032 = 0.34.$$

11.11 ENCLOSURES FILLED WITH RADIATING GASES

By evaluating the emissivity and absorptivity of a gas by the method outlined in Section 11.10, we can evaluate the interchange of radiant energy between a gas and the surface of its enclosure. First, consider a black bounding surface. The gas emits energy which is entirely absorbed by the black surface, but, on the other hand, the gas absorbs only a fraction of the radiation from the surface. Thus we write the flux of energy from the gas g to the surface 1 as

$$q_{g \to 1} = e_g - \alpha_{g1}e_{b1}$$

$$= \varepsilon_g e_{bg} - \alpha_{g1}e_{b1}$$

$$= \sigma(\varepsilon_g T_g^4 - \alpha_{g1} T_1^4). \tag{11.50}$$

When the bounding surface is gray, it reflects part of the energy emitted by the gas which is partially absorbed in successive passes back through the gas. In this case, we give the flux of energy by the right side of Eq. (11.50) multiplied by a factor between ε_1 and 1. For surfaces with ε_1 of 0.7 or greater, it is sufficient to use the following equation:

$$q_{g \to 1} = \sigma\left(\frac{\varepsilon_1 + 1}{2}\right)(\varepsilon_g T_g^4 - \alpha_{g1} T_1^4), \tag{11.51}$$

where the factor is simply $(\varepsilon_1 + 1)/2$.

In order to incorporate participating gases into a system containing surfaces, we consider the gas to be a floating node of potential e_{bg} and of conductance $A_i\varepsilon_{gi}$ between the gas node and the surface i. Due to the presence of the gas which absorbs some of the radiation between surface nodes, the conductances between the surfaces themselves are multiplied by the gas transmissivity; for example, the conductance between surfaces i and j is $A_i F_{ij} \tau_{igj}$. The following example demonstrates the use of such an electric analog for a system with a participating gas.

Example 11.7 Two large parallel plates (with emissivity of 0.6) at temperatures of 1200°F and 800°F transfer radiation across a 3-in. gap filled with CO_2 at 1-atm pressure. Calculate the rate of heat transfer from the hot plate to the cold plate.

Solution. We show the system and the equivalent network below.

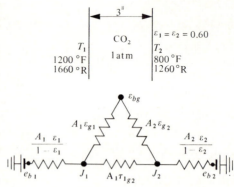

Figure —Example 11.7: (a) Radiation between parallel plates with a participating gas, and (b) the electric analog.

In order to evaluate the emissivities and transmissivities of the gas, we must know its temperature. Assume the gas temperature is 1000°F (this can be checked later). From Table 11.1, $L = (2)(3/12) = 0.5$ ft. Then, $P_c L = (1)(0.5) = 0.5$ ft atm, and from Fig. 11.25, $\varepsilon_g = 0.12$; the correction for nongrayness is small in this particular problem, so that $\varepsilon_{g1} = \varepsilon_{g2} = \alpha_{1g2} = 0.12$, and $\tau_{1g2} = 0.88$. The network can be reduced to a single equivalent resistance for current (heat flow) between 1 and 2. Here $A_1 = A_2 = A$ and $F_{12} = 1$, so that

$$
\frac{1}{\mathscr{F}_{12}} = \frac{1 - \varepsilon_1}{\varepsilon_1} + \frac{1}{F_{12}\tau_{1g2} + \dfrac{1}{1/\varepsilon_{g1} + 1/\varepsilon_{g2}}} + \frac{1 - \varepsilon_2}{\varepsilon_2}
$$

$$
= \frac{0.4}{0.6} + \frac{1}{(1)(0.88) + \dfrac{1}{1/0.12 + 1/0.12}} + \frac{0.4}{0.6}
$$

$$
= 2.39.
$$

Then,

$$
q_{1 \to 2} = \mathscr{F}_{12}\sigma(T_1^4 - T_2^4) = \frac{(0.1713)(16.6^4 - 12.2^4)}{2.39} = 3620 \text{ Btu/hr ft}^2.
$$

Now check the assumption that $T_g = 1000°F$.

$e_{b1} = \sigma T_1^4 = (0.1713)(16.6)^4 = 13,000;$ similarly, $e_{b2} = 4300$ Btu/hr/ft^2.

Then

$$J_1 = e_{b1} - \frac{q_{1 \to 2}}{\varepsilon_1/(1 - \varepsilon_1)} = 10,590 \text{ Btu/hr ft}^2,$$

$$J_2 = e_{b2} + \frac{q_{1 \to 2}}{\varepsilon_2/(1 - \varepsilon_2)} = 6710 \text{ Btu/hr ft}^2.$$

Since $\varepsilon_{g1} = \varepsilon_{g2}$, the potential e_{bg} must be the average of J_1 and J_2, so that

$$e_{bg} = \frac{10,590 + 6710}{2} = 8650 \text{ Btu/hr ft}^2,$$

and

$$T_g = 100 \sqrt[4]{\frac{e_{bg}}{0.1713}} = 1490°R = 1030°F.$$

The problem could be solved again with this as the gas temperature; in this case, however, 1030°F is sufficiently close to the assumed 1000°F, so that no significant improvement in the result would be gained.

It is instructive to extend the above example, by comparing it to the case of no participating gas (see Example 11.3). In this case

$$\frac{1}{\mathscr{F}_{12}} = \frac{1}{\varepsilon_1} + \frac{1}{\varepsilon_2} - 1 = \frac{1}{0.6} + \frac{1}{0.6} - 1$$

$$= 2.33.$$

This is hardly different from the factor (2.39) determined in Example 11.7. Therefore, even though we have selected an example involving an atmosphere of 100% CO_2, we could still consider the atmosphere to be transparent for many calculations. Only at partial pressures greater than about 4 atm of the participating species and/or with very large enclosure dimensions, would the effect of the gas be more evident.

The likely application of these principles to metallurgical processes, which ordinarily involve combustion gases at 1 atm total pressure, requires only small corrections to heat fluxes. For example, a typical combustion gas at 3000°R could contain 12% CO_2 and 20% H_2O. From Figs. 11.25 and 11.27, the reduced emissivities for a rather large L (e.g., 3 ft) are only 0.08 and 0.11, respectively.

11.12 TRANSIENT CONDUCTION WITH RADIATION AT THE SURFACE

In Chapter 9, we discussed some solutions to transient problems involving solids with heat applied to, or withdrawn from, the surface via convective h. In several operations, notably heat treating processes, stock is heated via radiation. The condition of heat input at the receiver (stock) surface from a heat source at constant temperature T_s is

$$\mathcal{F}_{s1}\sigma[T_s^4 - T^4(0, t)] = -k\frac{\partial T(0, t)}{\partial x}. \tag{11.52}$$

Here all temperatures are absolute, and we determine $\mathcal{F}_{s1}$ according to the methods outlined above, depending on the spatial relationship between the source and stock and their respective emissivities. We can write Eq. (11.53) in the dimensionless form

$$M[\theta_s^4 - \theta^4(0, \mathrm{Fo})] = -\frac{\partial\theta(0, \mathrm{Fo})}{\partial(x/l)}, \tag{11.53}$$

where

$$M = \frac{\mathcal{F}_{s1}\sigma T_0^3 l}{k}, \qquad \mathrm{Fo} = \frac{\alpha t}{l^2},$$

$$\theta = \frac{T}{T_0}, \qquad \theta_s = \frac{T_s}{T_0}.$$

T_0 is the initial temperature of the stock, and l is its characteristic dimension.

Consider specifically the temperature response of a sheet, $0 \le x \le l$, thin enough for Newtonian conditions (no internal gradients) to apply, which is heated at $x = 0$ by a radiation heat source at constant temperature. The back face ($x = l$) loses a negligible amount of heat so that $[\partial T(l, t)/\partial x] = 0$. We give the solution to this problem in Fig. 11.31(a) and (b) for heating and cooling, respectively. For other closely related situations we refer the reader to Schneider.[5,6]

Example 11.8 (from Schneider[5]) We heat treat a continuous sheet of steel (0.041 in. thick) by passing it through a radiant furnace designed to provide a heat-up and cool-down cycle. The furnace consists of a 10 ft long ceiling heat-source section ($\mathcal{F}_{s1} = 0.72$) at 2200°F followed by a ceiling heat-sink section ($\mathcal{F}_{s1} = 0.86$) at -200°F. If the sheet enters the furnace at 70°F, must be heated to 2000°F, and then cooled to 500°F before leaving the oven, calculate (a) the required feed velocity and (b) the length of cool-down section. Data: density of steel = 486 lb/ft³; heat capacity = 0.11 Btu/lb-°F.

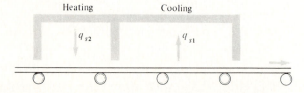

[5] P. J. Schneider, *Temperature Response Charts*, Wiley, New York, 1963.

[6] P. J. Schneider, *J. of the Aero/Space Sciences* **27**, No. 7, 546–548 (1960).

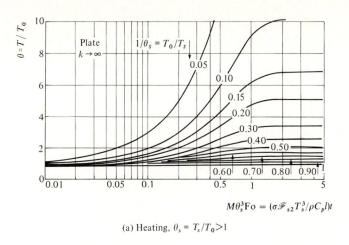

(a) Heating, $\theta_s = T_s/T_0 > 1$

(b) Cooling, $\theta_s = T_s/T_0 < 1$

Fig. 11.31 Temperature response of a plate, $0 \le x \le l$, with no internal thermal gradients and an insulated back face, at $x = l$, after sudden exposure to (a) a radiation heat source (heating) or (b) a radiation heat sink (cooling). (From P. J. Schneider, *Temperature Response Charts*, Wiley, New York, 1963.)

Solution

a) For the heating section, we apply Fig. 11.31(a) with

$$\frac{1}{\theta_s} = \frac{T_0}{T_s} = \frac{(70 + 460)°\text{R}}{(2200 + 460)°\text{R}} = 0.199,$$

$$\theta = \frac{T \text{ (leaving)}}{T_0} = \frac{(2000 + 460)°\text{R}}{(70 + 460)°\text{R}} = 4.64.$$

From Fig. 11.31(a)

$$M\theta_s^3 \, \text{Fo} = \frac{\mathscr{F}_{s1}\sigma T_0^3 l}{k} \cdot \left(\frac{T_s}{T_0}\right)^3 \cdot \frac{k}{\rho C_p} \frac{t}{l^2}$$

$$= \frac{\mathscr{F}_{s1}\sigma T_s^3 t}{\rho C_p l} = 0.980.$$

Then the heat-up exposure time t must be equal to

$$\frac{0.980\rho C_p l}{\mathscr{F}_{s1}\sigma T_s^3} = \frac{(0.980)(486)(0.11)(0.041/12)}{(0.72)(0.1713 \times 10^{-8})(2660^3)}$$

$$= 0.00764 \text{ hr} = 27.5 \text{ sec.}$$

Thus the feed velocity is

$$\frac{10 \text{ ft}}{27.5 \text{ sec}} = 0.364 \text{ ft/sec.}$$

b) We use Fig. 11.31(b) for the cooling section with

$$\frac{1}{\theta_s} = \frac{T_0}{T_s} = \frac{(2000 + 460)^\circ \text{R}}{(-200 + 460)^\circ \text{R}} = 9.46$$

$$\theta = \frac{T(\text{leaving})}{T_0} = \frac{(500 + 460)^\circ \text{R}}{(2000 + 460)^\circ \text{R}} = 0.39.$$

Then from Fig. 11.31(b)

$$M\theta_s^3 \, \text{Fo} = 0.0064,$$

from which we determine the cool-down exposure time

$$t = \frac{(0.0064)(486)(0.11)(0.041/12)}{(0.86)(0.1713 \times 10^{-8})(260^3)} = 0.0448 \text{ hr} = 161 \text{ sec.}$$

Thus the required length of the cool-down section is

$$(0.364 \text{ ft sec}^{-1})(161 \text{ sec}) = 59 \text{ ft.}$$

PROBLEMS

11.1 A radiation heat-transfer coefficient is often defined as $q = h_r(T_1 - T_2)$.

a) For a gray solid at T_1, completely surrounded by an environment at T_2, show that h_r is given by

$$h_r = \sigma\varepsilon_1(T_1^2 + T_2^2)(T_1 + T_2).$$

b) Calculate the rate of heat transfer (by radiation and convection) from a vertical surface 5 ft high and the percentage by radiation for the following cases:

Surface ($\varepsilon = 0.8$) at:	Air at:
500°F	80°F
1000°F	80°F
1500°F	80°F

11.2 A metal sphere at 1800°F is suddenly placed in a vacuum space whose walls are at 180°F. The sphere is 2 in. in diameter and we may assume that its surface is perfectly black.

a) Calculate the time it takes to cool the sphere to 300°F. The specific gravity of the sphere is 7.0, and its heat capacity is 0.3 Btu/lb$_m$-°F. Assume that uniform temperature exists in the sphere at each instant.

b) Discuss the validity of the assumption in part (a) with quantitative reasoning using a conductivity of 30 Btu/hr-ft °F.

11.3 Steel sheet is preheated in vacuum for subsequent vapor deposition. The steel sheet is placed between two sets of cylindrical heating elements (1 in. in diameter). The heating efficiency is kept high by utilizing radiation reflectors of polished brass. The steel sheet and radiation reflectors may be considered to be infinite parallel plates.

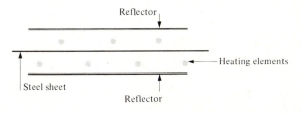

a) Calculate the rate of heat transfer from the heating elements (maintained at 2800°F) to the steel when the steel is at 80, 250, 500, 750, and 1000°F. The reflectors may be assumed to be at 100°F.

b) Plot the results of part (a) and determine by graphical integration the average value of the heat transfer rate as the steel sheet is heated from 80 to 1000°F.

c) From part (b) determine the time it takes to heat a steel sheet $\frac{1}{4}$ in. thick from 80 to 1000°F.

Data:

ε (heating elements)	$= 0.9$
ε (steel sheet)	$= 0.3$
ε (of reflectors)	$= 0.03$
ρ of steel	$= 480$ lb$_m$/ft^3
C_p of steel	$= 0.15$ Btu/lb$_m$
Area of steel to area of elements	$= 10/1$

11.4 A large furnace cavity has an inside surface temperature of 1500°F. The walls of the furnace are 1 ft thick. A hole 6 in. × 6 in. is open through the furnace wall to the room at 70°F.

a) Calculate the heat loss by radiation (Btu/hr) through the open hole.

b) Calculate the heat loss if a sheet of nickel 6 in. × 6 in. is fitted across the hole on the outside surface.

Data: emissivity of furnace refractories = 0.9, and emissivity of nickel sheet = 0.4.

[*Note.* The furnace refractories should be approximated as perfect heat insulators.]

11.5 Cast iron is continually tapped from the bottom of a cupola into an open refractory channel. The metal enters at 2800°F and runs down the channel at a rate of 4000 lb/hr. The dimensions of the channel are shown below.

Neglect the heat loss by conduction through the refractory and estimate the metal discharge temperature.

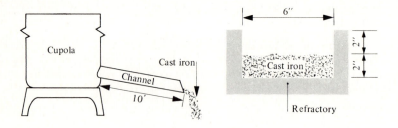

Data for molten cast iron:

$$C_p = 0.20 \text{ Btu/lb-°F}$$
$$\rho = 430 \text{ lb/ft}^3$$
$$\varepsilon = 0.30$$

11.6 A method of melting metal so as to avoid crucible contamination is levitation melting. A metal sample is placed in an electromagnetic field from a coil wound as a cone. The field

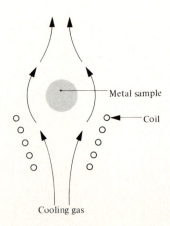

not only supplies power to melt the metal but also levitates it. But with this set-up, the strength of field necessary to keep the sample levitated is sometimes such that the metal gets overheated. A means of preventing this is to pass a cooling gas past the sample.

Develop an expression that relates the steady-state temperature of the metal to the heating power supplied by the field (Q_p, Btu/hr), the gas temperature T_0, the gas velocity V, and any other parameters which you think are essential. Metal temperatures of interest are 2000–3000°F, and the convection heat-transfer coefficient can be expressed as

$$h_c = KV^{0.7},$$

in which K is a known constant.

11.7 A steel sheet $\frac{1}{2}$ in. thick and having the shape of a square 5 ft × 5 ft comes out of a heat-treating furnace at 1500°F. During heat treatment its surface was oxidized so that its emissivity is 0.8.

a) Calculate its initial cooling rate if it is suspended freely by a wire in a room at 80°F. Neglect all heat transfer with convection, i.e., deal only with radiation heat transfer.

b) Calculate its initial cooling rate if it is supported vertically on a horizontal surface (also at 80°F) which has an emissivity of 0.2. Again deal only with radiation heat-transfer.

Data (all units in Btu, hr, ft, °F, etc.):

	Supporting surface	Steel sheet
Thermal conductivity	100	10
Density	500	540
Heat capacity	1.0	0.20
Emissivity	0.2	0.8

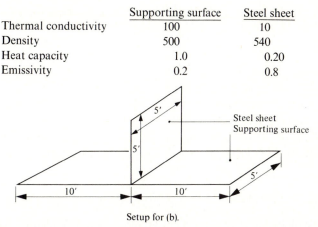

Setup for (b).

11.8 Brass sheet, $\frac{1}{16}$ in. thick, is to be tempered in a continuous tempering furnace. The first portion of the furnace is filled with gases at 1200°F, moving against the strip direction at a speed of 60.0 ft/min. The strip itself moves at a speed of 10.0 ft/min. The walls are of SiO_2 brick each 1 ft from the brass sheet as it moves vertically, and the gases are 40% H_2O, 10% CO_2, and 50% N_2. How many ft of strip would have to be in the first portion of the furnace so that the strip could exit into the second section at a temperature of 800°F?

Data:

$C_{p\,\text{brass}} = 0.10 \text{ Btu/lb-°F}$ $C_{p\,\text{gas}} = 0.26$ $\eta_{\text{gas}} = 0.1 \text{ lb/ft-hr}$

$\rho_{\text{brass}} = 574 \text{ lb/ft}^3$ $\rho_{\text{gas}} = 0.03$

$k_{\text{brass}} = 100 \text{ Btu/hr-ft °F}$ $k_{\text{gas}} = 0.05$

12

THERMAL BEHAVIOR OF METALLURGICAL PACKED-BED REACTORS

This chapter deals with the transfer of heat under turbulent flow conditions between a packed column of solid particles and a gas flowing through the packing. Counter-current flow or static-bed situations are the only cases we shall study, since they represent virtually all the metallurgical applications of this type of heat transfer. Metallurgical processes utilizing the counter-current flow of solids and gases include, for example, iron and lead blast furnaces, shaft pelletizing and/or calcining furnaces, iron cupolas, and copper blast furnaces. The equations developed for these cases can be easily modified to describe the thermal behavior of crossflow processes, such as iron-ore sintering and pelletizing and zinc-ore roasting and sintering. In most of these cases, the complexity of the equations makes analytical solutions almost impossible, and most studies which have attempted to do so have ultimately relied on numerical integration methods to provide final results. However, we can make some simplifications which lead to analytical expressions that are at least indicative of the behavior of real systems, and show the effects that may occur due to changes in operating conditions. In the following sections, we shall present some of the analytical expressions and discuss their applications.

12.1 INITIAL DEFINITIONS AND ASSUMPTIONS

Figure 12.1 depicts the general physical situation as a bed of slowly moving solids contained within a vertical shaft with a stream of gas flowing counter-current to the solid. Depending on whether the solids are being heated or cooled, they can enter the bed as a cold stream or a hot stream, respectively. In the following text, the temperatures of the hot stream, whether solid or fluid, are subscripted h and the temperatures of the colder stream are denoted by subscript c. The entering condition is superscripted 0 and the exiting or leaving condition l. The depth of the solid bed is denoted by L.

In order to handle the differential equations and boundary conditions assumed, we make several initial assumptions which are applicable to the entire chapter:

1. *Plug flow in the axial direction.* This assumes that there is no back-mixing of gases.

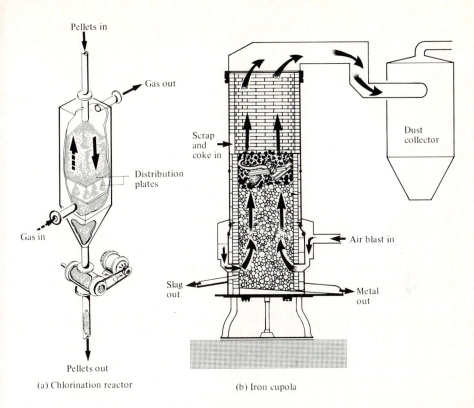

Pellets in

Gas out

Distribution plates

Gas in

Pellets out

(a) Chlorination reactor

Scrap and coke in

Air blast in

Slag out

Metal out

Dust collector

(b) Iron cupola

Fig. 12.1 Examples of metallurgical counter-current packed-bed reactors.

2. *No radial heat transfer.* This is a reasonable assumption in beds of a large cross section, such as in sinter plants, blast furnaces and large cupolas, but may not be a good assumption for small-diameter cupolas, or for cases with significant channeling.

3. *Adiabatic system.* This means no heat losses to the containing walls are allowed. As in assumption (2), this is reasonable except in the case of small-diameter cupolas or other shaft furnaces.

4. *No viscous heating effects.* Since gas velocities are well below those where frictional heating becomes important, this is reasonable to assume.

5. *No radiation effects within the gas phase.* Following the approach in Chapter 6, we assume that the thermal conductivity of the bed includes the radiant heat-transfer mechanism between particles.

6. *No thermal gradients within particles.* This amounts to Newtonian heating or cooling conditions, which, for all practical purposes, arise when the Biot number of the particles is less than 0.10 approximately, although for many packed bed designs, it has been accepted to be as high as 0.25.

In general, we may write energy balances across a unit thickness of bed, dz, for both the solid and gas phases, keeping in mind that either one may be the hotter or colder phase.

Energy balance on gases

$$V_{0_g} \rho_g c_g \left(\frac{\partial T_g}{\partial z} \right) + \rho_g c_g \omega \left(\frac{\partial T_g}{\partial t} \right) + hS(T_g - T_s) - Q_R = 0. \tag{12.1}$$

Energy balance on solids

$$V_s \rho_s c_s (1 - \omega) \left(\frac{\partial T_s}{\partial z} \right) + \rho_s c_s (1 - \omega) \left(\frac{\partial T_s}{\partial t} \right) + hS(T_g - T_s)$$

$$- \frac{\partial}{\partial z} \left[k_{\text{eff}} \left(\frac{\partial T_s}{\partial z} \right) \right] - Q_R = 0, \tag{12.2}$$

where S = total surface area of particles per unit volume of bed, ft^2/ft^3,

h = heat-transfer coefficient, between the gas and solids, Btu/min-ft^2 °F,

Q_R = heat released by reactions, Btu/min-ft^3,

ω = void fraction,

k_{eff} = effective thermal conductivity of bed,

ρ_s, ρ_g = densities of solid and gas phases, respectively, lb/ft^3,

c_s, c_g = heat capacity of solid and gas, respectively, Btu/lb °F,

V_{0_g} = *superficial* velocity of the gas stream, ft/min,

V_s = *actual* velocity of solids, ft/min.

For convenience in the following sections, we define two new terms.

a) Thermal capacities per unit cross-sectional area of bed:

$$G_s = \rho_s c_s (1 - \omega), \text{ Btu/ft}^3 \text{ °F},$$

$$G_g = \rho_g c_g \omega, \text{ Btu/ft}^3 \text{ °F}.$$

b) Thermal flows per unit cross-sectional area of bed:

$$W_s = V_s G_s, \text{ Btu/min-ft}^2 \text{ °F},$$

$$W_g = V_{0_g} G_g / \omega, \text{ Btu/min-ft}^2 \text{ °F}.$$

12.2 STEADY-STATE, COUNTER-CURRENT FLOW

In this section, we examine the resulting temperature profiles of gases and solids when both V_g and V_s have steady, non-zero values. In order to consider the simplest possible case, mathematically and physically, we make the additional assumptions:

1. The volumetric heat-transfer coefficient hS is constant.
2. Conduction within the bed is negligible.
3. No reactions occur ($Q_R = 0$).

This might be the situation, for example, in a shaft pelletizing furnace in which pellets of agglomerated finely-ground hematite are fed into the top of a shaft, and descend against a rising stream of hot gases. The gases heat the solid pellets, thus providing the necessary temperature and time for the sintering of particles and strengthening of pellets. Both the temperature profile and the solid's velocity in the shaft are important, because the sintering rate of the particles is a function not only of the temperature reached, but also of the length of time they remain at that temperature.

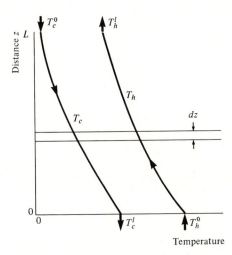

Fig. 12.2 Notation and definitions for the analysis of counter-flow heat transfer.

Figure 12.2 illustrates the conditions and provides a reference framework for the following development.

First, let us develop a simple expression for the length of bed needed to effect a specified exchange of heat. In Figure 12.2, consider a slice of bed of thickness dz. Within a unit area of that slice, the hot stream (the gases in a solid-heating operation) loses an amount of heat:

$$dq = -W_h \, dT_h, \text{ Btu/min ft}^2, \qquad (12.3)$$

while the cold stream gains the same amount of heat:

$$dq = W_c \, dT_c. \qquad (12.4)$$

(The negative sign indicates that the temperature gradient is negative with respect to the z-direction).

We may also say that the heat transferred is

$$dq = hS(T_h - T_c) \, dz. \qquad (12.5)$$

From Eqs. (12.3) and (12.4) we see that

$$d(T_h - T_c) = -dq\left(\frac{1}{W_h} - \frac{1}{W_c}\right).\tag{12.6}$$

Substituting dq from Eq. (12.5), we get

$$d(T_h - T_c) = -hS(T_h - T_c)\left(\frac{1}{W_h} - \frac{1}{W_c}\right)dz,$$

or

$$\frac{d(T_h - T_c)}{(T_h - T_c)} = -hS\left(\frac{1}{W_h} - \frac{1}{W_c}\right)dz.\tag{12.7}$$

Integrating Eq. (12.7) with

$$(T_h - T_c) = (T_h^0 - T_c^l)\qquad\text{at } z = 0$$

finally yields

$$\ln\frac{(T_h - T_c)}{(T_h^0 - T_c^l)} = -hSz\left(\frac{1}{W_h} - \frac{1}{W_c}\right).\tag{12.8}$$

If a specified entering temperature for the cold stream, T_c^0, is available, along with T_h^0 and T_c^l, then we can calculate T_h^l for a given bed length ($z = L$), heat-transfer coefficient, and thermal flow condition. On the other hand, if we predetermine all the temperatures of the cold and hot streams at the ends of the bed, then we find the necessary bed length, since

$$\ln\frac{(T_h^l - T_c^0)}{(T_h^0 - T_c^l)} = -hSL\left(\frac{1}{W_h} - \frac{1}{W_c}\right),$$

and

$$L = \ln\frac{(T_h^l - T_c^0)}{(T_h^0 - T_c^l)}\bigg/ hS\left(\frac{1}{W_c} - \frac{1}{W_h}\right).\tag{12.9}$$

This is a useful expression for design purposes, particularly where the only specifications are on the final and initial temperatures, and the only design parameter is the length of shaft desired.

On the other hand, there are situations in which we need to know the temperature profiles of the gas and the solid. This is often the case when it is necessary to know how long the solids are above some specified temperature. In order to obtain the profiles, we use Eq. (12.8), and write it as

$$(T_h - T_c)\bigg|_z = (T_h^0 - T_c^l)\exp\left[-hSz\left(\frac{1}{W_h} - \frac{1}{W_c}\right)\right],$$

which gives the local temperature *difference* at any level z. We know, however, from a heat balance between the entrance and any point in the bed, that

$$W_h[T_h^0 - T_h(z)] = W_c[T_c^l - T_c(z)].$$

Applying these equations, and the relationship

$$T_h = T'_h \qquad \text{at } z = L,$$

we develop an equation for the temperature of the hotter stream, at any distance z from its entrance:

$$T_h(z) = T_h^0 - (T_h^0 - T_c^0)\left\{\frac{1 - \exp\left[-\frac{hS}{W_h}\left(1 - \frac{W_h}{W_c}\right)z\right]}{1 - \frac{W_h}{W_c}\exp\left[-\frac{hS}{W_h}\left(1 - \frac{W_h}{W_c}\right)L\right]}\right\}. \qquad (12.10)$$

Similarly, the local temperature of the colder medium is

$$T_c(z) = T_h^0 - (T_h^0 - T_c^0)\left\{\frac{1 - \frac{W_h}{W_c}\exp\left[-\frac{hS}{W_h}\left(1 - \frac{W_h}{W_c}\right)z\right]}{1 - \frac{W_h}{W_c}\exp\left[-\frac{hS}{W_h}\left(1 - \frac{W_h}{W_c}\right)L\right]}\right\}. \qquad (12.11)$$

The equations given below are useful for
a) the outlet temperature of the hot stream:

$$T'_h = T_h^0 - (T_h^0 - T_c^0)\frac{W_c}{W_h}\left\{1 - \frac{1 - \frac{W_h}{W_c}}{1 - \frac{W_h}{W_c}\exp\left[-\frac{hS}{W_h}\left(1 - \frac{W_h}{W_c}\right)L\right]}\right\}; \qquad (12.12)$$

b) the outlet temperature of the cold stream:

$$T'_c = T_h^0 - (T_h^0 - T_c^0)\left\{\frac{1 - \frac{W_h}{W_c}}{1 - \frac{W_h}{W_c}\exp\left[-\frac{hS}{W_h}\left(1 - \frac{W_h}{W_c}\right)L\right]}\right\}. \qquad (12.13)$$

Equations (12.10)–(12.13) appear to be indefinite when $W_h = W_c$, which is a condition often met in practice. We overcome this difficulty by applying L'Hopitals rule, differentiating both numerator and denominator as $W_h/W_c \to 1$. The results, for the special case where $W_h/W_c = 1$, are then

$$T_h(z) = T_h^0 - (T_h^0 - T_c^0)\left|\frac{z}{L + \frac{W_h}{hS}}\right|, \qquad (12.10a)$$

$$T_c(z) = T_h^0 - (T_h^0 - T_c^0)\left|\frac{1 + \frac{hSz}{W_h}}{1 + \frac{hSL}{W_h}}\right|, \qquad (12.11a)$$

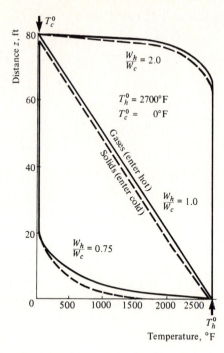

Fig. 12.3 Effect of change in thermal flow ratio W_h/W_c on the region of heat transfer in a counter-current packed bed. Volumetric heat-transfer coefficient hS is constant (2.5 Btu/min-ft^3 °F) in all cases. Note that the region of active heat transfer is pushed from one end of the shaft to the other as the ratio changes.

$$T_h^l = T_h^0 - (T_h^0 - T_c^0)\left(\frac{1}{1 + \dfrac{W_h}{hSL}}\right), \qquad (12.12a)$$

$$T_c^l = T_h^0 - (T_h^0 - T_c^0)\left(\frac{1}{1 + \dfrac{hSL}{W_h}}\right). \qquad (12.13a)$$

All of these equations may be used when initial conditions of both streams, and their respective flow rates are available. Using some reasonable values for W_h, W_c, hS, T_h^0, and T_c^0, we illustrate the effects of changes in these parameters on thermal profiles in a shaft furnace in Figs. 12.3 and 12.4.

Example 12.1 One way to increase the productivity of an electric-arc furnace for melting steel is to preheat the scrap steel charge. It is usual to aim for a scrap temperature of 1200°F. If scrap at room temperature is fed into the top of a shaft furnace and removed at the bottom, and nitrogen, preheated to 1500°F, is introduced at the bottom, how long will the shaft be?

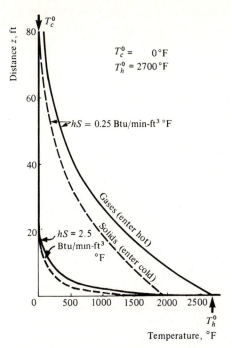

Fig. 12.4 Effect of change in the volumetric heat-transfer coefficient hS on the region of heat transfer in a counter-current packed bed. Thermal flow ratio is constant at $W_h/W_c = 0.75$ for both cases. Note that in the case of the low coefficient, T_h never reaches T_c^0.

Data:

$$S = 20 \text{ ft}^2/\text{ft}^3,$$
$$h = 0.3 \text{ Btu/min-ft}^2 \text{ °F},$$
$$W_c \text{ (solids)} = 10.0 \text{ Btu/min-°F ft}^2 = W_s,$$
$$W_h \text{ (gases)} = 12.0 \text{ Btu/min-°F ft}^2 = W_g.$$

Solution. According to Eq. (12.9) we see that all of the necessary information is available, except for the outlet gas temperature. Since $W_s(T_s^l - T_s^0)$ is the heat absorbed by the charge per min per sq ft of shaft area, and $W_g(T_g^0 - T_g^l)$ is equal to this value, we can write

$$W_s(T_s^l - T_s^0) = 10.0(1200 - 70) = 11,300 \text{ Btu/min-ft}^2,$$
$$W_g(T_g^0 - T_g^l) = 11,300 \text{ Btu/min-ft}^2.$$

Therefore

$$T_g^l = T_g^0 - \frac{11,300}{W_g} = 1500 - \frac{11,300}{12.0}$$

$$= 558°\text{F}.$$

Now using Eq. (12.9), we obtain

$$L = 2.303 \log \frac{(558 - 70)}{(1500 - 1200)} \bigg/ (0.3)(20)\left(\frac{1}{10.0} - \frac{1}{12.0}\right)$$

$$= 4.75 \text{ ft.}$$

12.3 HEAT-TRANSFER COEFFICIENTS IN PACKED BEDS

Up to this point, we have presented the heat-transfer coefficient h, but we have not mentioned how to obtain this value for a flow through packed beds. As might be expected, we do not obtain h in a packed bed from any of the correlations presented in Chapter 8; therefore new correlations are needed. Since

$$\text{Nu} = f(\text{Re, Pr}),$$

is still true, we could look for an expression of the form

$$\text{Nu} = A\, \text{Re}^n\, \text{Pr}^m.$$

Here we are dealing exclusively with gases as the fluid phase, and since the Prandtl numbers of gases are all about equal over a wide range of temperatures, there is not any strong dependence of Nu on Pr for gas–solid heat transfer. So we are left with finding the relationship between Nu and Re. However, this is much easier said than done, as most of the experimental work on this subject has been confined to much lower temperatures than those of most metallurgical processes.

The highest temperatures—1100°C—were employed by Furnas[1] in his classic research in 1932. Recently, Kitaev and coworkers[2] have summarized the results of the pertinent studies at elevated temperatures, and also some of the lower temperature results. Their review, which is probably the best ever made on the subject, includes a complete reanalysis of Furnas' data. Since the effects of void volume and temperature are so difficult to obtain, and as yet have not been firmly established, the volumetric heat-transfer coefficient, hS or h_v, is presented by them in the form

$$h_v = \frac{A V_{0g}^{0.9} T^{0.3} f(\omega)}{D_p^{0.75}}, \tag{12.14}$$

where $h_v = hS$ = volumetric heat-transfer coefficient, k cal/m^3-hr °C,
  A = a coefficient dependent on bed material,
  T = temperature, °C,
  D_p = particle diameter, mm,
  V_{0g} = superficial gas velocity, m/sec,
  $f(\omega)$ = function of void fraction.

For natural, lump, materials, $A = 160$ and $f(\omega) = 0.5$, so that

$$h_v = \frac{80 V_{0g}^{0.9}\, T^{0.3}}{D_p^{0.75}}.$$

[1] C. C. Furnas, *U.S. Bureau of Mines Bull.*, 261 (1932).
[2] B. I. Kitaev, Y. G. Yaroshenko, and V. D. Suchkov, *Heat Exchange in Shaft Furnaces*, Pergamon Press, New York, 1968.

A nomogram has been developed, Fig. 12.5, which facilitates the determination of h_v.

For perfect spheres, the equation

$$h_v = \frac{12 V_{0_g} T^{0.3}}{D_p^{1.35}}$$ (12.15)

agrees with Furnas' results and those of Saunders and Ford[3], when D_p is in meters.

Example 12.2 Estimate the volumetric heat-transfer coefficient in a shaft furnace in which hot gases (1200°C) are used to heat harden iron oxide pellets, 1 in. in diameter. The gas velocity is 1.0 m/sec.

Solution. Entering the nomograph at 1.0 m/sec, going up to 1200°C, across to 25 mm, and down, we find that $h_v = 12,500$ kcal/m³-hr °C. This becomes

$$h_v = 1.25 \times 10^4 \times 1.45 \times 10^{-3}$$

$$= 18.5 \text{ Btu/ft}^3\text{-min °F.}$$

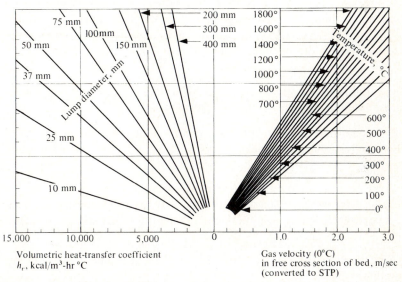

Fig. 12.5 Nomograph for the volumetric heat-transfer coefficient hS or h_v, based on knowledge of the superficial gas velocity V_{0g}, the gas temperature, and the particle diameter D_p. (Adapted from B. I. Kitaev *et al.*, *ibid.*)

Multiply	by	to obtain
ft/sec	0.3048	m/sec
in.	25.4	mm
kcal/m³-hr °C	0.001045	Btu/ft³-min °F

[3] O. H. Saunders and H. Ford, *J. Iron & Steel Inst.* **I**, 292 (1940).

12.4 STATIONARY BED, INFINITE HEAT TRANSFER

In many cross-flow situations, such as in grate-type pelletizing and sintering plants, the packed bed moves very slowly in the x-direction while the gas flow is in the z-direction, perpendicular to the bed surface. If we pick out a unit area of bed as it moves into the gas contacting zone, we may analyze its unsteady-state behavior as if it were stationary, at least until it passes out of the zone at the other end. In this situation (Fig. 12.6), V_s is zero.

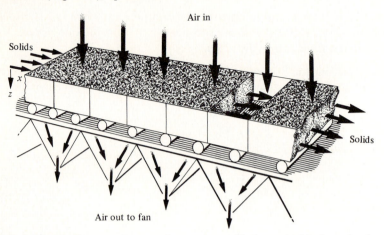

Fig. 12.6 Schematic diagram of a cross-flow system, with the packed bed moving from left to right, and the gases being drawn through it from top to bottom.

Let us consider first the situation where the assumed conditions, in addition to those in Section 12.1, include:

1. No reactions occur in the bed.
2. No conduction occurs within the bed.
3. The heat-transfer coefficient is so large that the solid and gas each have the same temperature at any point in the bed at any time. Mathematically, this means that $T_s = T_g$, and $(\partial T_s/\partial t) = (\partial T_g/\partial t)$.

Returning to Eqs. (12.1) and (12.2), we see that under the present set of assumptions

$$W_g \frac{\partial T_g}{\partial z} + G_g \frac{\partial T_g}{\partial t} = 0,$$

or

$$W_g\left(\frac{\partial T_s}{\partial z}\right) + G_g\left(\frac{\partial T_s}{\partial t}\right) = 0, \tag{12.16}$$

and

$$G_s\left(\frac{\partial T_s}{\partial t}\right) = 0. \tag{12.17}$$

Combining Eqs. (12.16) and (12.17), we arrive at

$$\left(\frac{\partial T_s}{\partial t}\right) + \frac{W_g}{G_g + G_s}\left(\frac{\partial T_s}{\partial z}\right) = 0, \tag{12.18}$$

which is recognized as a quasi-linear partial differential equation of the first order, with the form

$$P\left(\frac{\partial T_s}{\partial t}\right) + Q\left(\frac{\partial T_s}{\partial z}\right) = R. \tag{12.19}$$

The general solution is of the form

$$U_2 = f(U_1),$$

where $U_1(T_S, t, z) = C_1$, and $U_2(T_S, t, z) = C_2$ are the solutions of any two of the following relationships:

$$\frac{dt}{P} = \frac{dz}{Q} = \frac{dT_S}{R}.$$

In this case, two convenient relationships are

$$dz = \left(\frac{W_g}{G_g + G_s}\right) dt, \tag{12.20}$$

and

$$dT_s = 0 \cdot dt = 0. \tag{12.21}$$

Integration of Eqs. (12.20) and (12.21) results in

$$z - \left(\frac{W_g}{G_g + G_s}\right) t = U_1,$$

and $T_s = U_2$. The general solution is then

$$T_s = f\left[z - \left(\frac{W_g}{G_g + G_s}\right) t\right]. \tag{12.22}$$

We find the particular solution by specifying the initial condition of the temperature profile of the bed as a function of distance z along the bed at time zero; at a later time this function describes the same profile with the distance increased by the value of $[W_g/(G_g + G_s)]t$. This means that the original temperature profile is propagated, unchanged, through the bed at a rate of $[W_g/(G_g + G_s)]$ if the inlet gas temperature remains constant.

Example 12.3 A bed of magnetite (Fe_3O_4) particles, 1 ft high which has a cross-sectional area of 1 ft^2 is heated from a uniform temperature of 500°F to 1800°F with dry, heated nitrogen. The nitrogen enters at the rate of 1 lb/min for every 3 lb of Fe_3O_4 to be heated. The bed density is 150 lb/ft^3, and the void fraction 0.52.

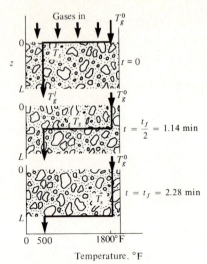

Fig. 12.7 Propagation of a thermal front through a bed of solids. T_g^0 is constant at all times.

a) Illustrate the thermal profile of the solids in the bed at time zero, at the time when all the Fe_3O_4 is heated to 1800°F, and when 50% of the charge is heated, assuming an infinite heat-transfer coefficient.

b) Calculate the time to heat all of the Fe_3O_4 to 1800°F.

Solution. In Fig. 12.7 we see the answer to part (a). A vertical temperature *front* passes through the bed at a velocity $[W_g/(G_g + G_s)]$. From the definition of the example, we can calculate this velocity.

$$W_g = 0.29\left(\frac{Btu}{lb\text{-}N_2\ °F}\right)\left(\frac{1\ lb\ N_2}{3\ lb\ Fe_3O_4}\bigg|\frac{}{min}\bigg|\frac{150\ lb\ Fe_3O_4}{}\bigg|\frac{}{ft^2}\right)$$

$$= 14.5\ Btu/min\text{-}ft^2\ °F.$$

$$G_g = \rho_g c_g \omega = \left(\frac{0.017\ lb}{ft^3}\bigg|\frac{0.29\ Btu}{lb\ °F}\bigg|\frac{0.52}{}\right) = 0.00256\ Btu/ft^3\ °F,$$

$$G_s = \rho_s c_s(1 - \omega) = \frac{0.22\ Btu}{lb\ °F}\bigg|\frac{150\ lb\ Fe_3O_4}{ft^3\ bed}$$

$$= 33.0\ Btu/ft^3\ °F.$$

Therefore, the velocity at which the front moves is

$$\frac{14.5}{33.0 + \sim 0} = 0.44\ ft/min,$$

and the time, t_f, to heat the entire bed to 1800°F is

$$\frac{1\ ft}{0.44\ ft/min} = 2.28\ min.$$

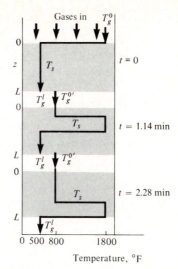

Fig. 12.8 Propagation of a thermal wave through a bed of solids. T_g^0 decreases to $T_g^{0'}$ after 20 sec.

Suppose that the temperature of the incoming gases, instead of being kept at 1800°F, has dropped to 800°F after 20 sec. The profile at the same times would be as shown in Fig. 12.8. Here the inlet gas temperature changes; after the change is effected, the resultant profile propagates, unchanged, through the bed.

12.5 STATIONARY BED, INFINITE HEAT-TRANSFER COEFFICIENT, AND HEAT OF REACTION

Let us now assume the same conditions as in Section 12.4, except that a reaction liberating Q_R Btu/min ft^3 occurs in the gas phase within the bed. The energy balance equations, in terms of T_s, are thus

$$W_g\left(\frac{\partial T_s}{\partial z}\right) + G_g\left(\frac{\partial T_s}{\partial t}\right) - Q_R = 0,$$

and

$$G_s\left(\frac{\partial T_s}{\partial t}\right) = 0.$$

Since we may combine the above two equations to yield

$$\left(\frac{\partial T_s}{\partial t}\right) + \left(\frac{W_s}{G_g + G_s}\right)\left(\frac{\partial T_s}{\partial z}\right) = \frac{Q_R}{G_g + G_s}, \tag{12.23}$$

we can obtain a solution in the same manner as shown in Section 12.4. In this case,

$$U_2(T, t, z) = T_s - \left(\frac{Q_R}{G_g + G_s}\right)t,$$

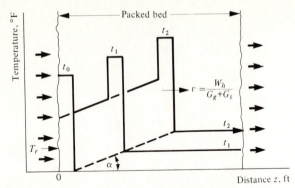

Fig. 12.9 Temperature profiles in a stationary bed with gas at T_r entering a bed with an initial hot zone, and heat generation rate Q_R. Note that $\tan \alpha = Q_R/(G_g + G_s)$.

and

$$U_1(T, t, z) = z - \left(\frac{W_g}{G_g + G_s}\right)t,$$

so that the general solution is now

$$T_s = \left(\frac{Q_R}{G_g + G_s}\right)t + f\left[z - \left(\frac{W_g}{G_s + G_g}\right)t\right]. \qquad (12.24)$$

Thus, any initial or impressed temperature profile is propagated through the bed with a velocity $W_g/(G_g + G_s)$, but, in addition, there is a continual increase in temperature. The rise is at a rate of $Q_R/(G_s + G_g)$, deg/min, or, if we plot the temperature as a function of distance down the bed, we give the thermal gradient in the heat-affected zone of the bed by

$$\tan \alpha = Q_R/(G_g + G_s).$$

For example, in Fig. 12.9, a bed of solids is heated for a short time, $0 < t < t_0$, with a hot, inert gas, and then, at time t_0, the incoming gas changes to a reactive gas at a lower temperature T_r. The profile is propagated as in Fig. 12.8, except that now, as time progresses, the temperatures at all points increase proportionally to the time, according to Eq. (12.24).

Example 12.4 Consider the same bed as described in Example 12.3, but now replace the nitrogen with the same amount by weight of dry air at the same temperature. In this case the magnetite is oxidized to hematite (Fe_2O_3), and heat is liberated. Under these conditions, a magnetite particle is completely oxidized in 5 min.

a) Graph the temperature distribution in the bed at times $t = 1.0$ and 2.0 min, using the same assumptions as before.

b) Calculate the maximum temperature reached in the bed when all of the magnetite has been oxidized.

Solution. To start with, let us calculate the velocity of the front. This in turn requires that we recalculate W_g, using the heat capacity of air, 0.28 Btu/lb °F, rather than that of nitrogen. The difference is slight, and

$$W_g = 14.0 \text{ Btu/min-ft}^2 \text{ °F.}$$

As before, $G_s = 33.0$ Btu/ft³ °F, and we will neglect G_g.
Therefore, the velocity at which the front moves is

$$\frac{14.0}{33} = 0.424 \text{ ft/min,}$$

and

$$t_f = \frac{1}{0.424} = 2.36 \text{ min.}$$

Now, in order to obtain "thermal front" temperatures, we need to evaluate Q_R. The heat of oxidation of Fe_3O_4 to Fe_2O_3 is 210 Btu/lb Fe_3O_4.

$$Q_R = \frac{150 \text{ lb } Fe_3O_4}{ft^3} \left| \frac{210 \text{ Btu}}{\text{lb } Fe_3O_4} \right| \frac{}{5 \text{ min}}$$

$$= 6300 \text{ Btu/ft}^3\text{-min.}$$

Therefore

$$\frac{Q_R}{G_g + G_s} = \frac{6300 \text{ Btu/ft}^3 \text{ min}}{33 \text{ Btu/ft}^3 \text{ °F}}$$

$$= 191 \text{°F/min.}$$

Figure 12.10 illustrates the results. The maximum temperature reached is

$$T_g^0 + \left(\frac{Q_R}{G_g + G_s}\right) t_f = 1800 + (191)(2.36)$$

$$= 2239 \text{°F.}$$

This temperature rise can have disastrous consequences if the maximum temperature exceeds the melting point of the material being processed (Fe_2O_3 in this case), or is so high as to damage the equipment.[*]

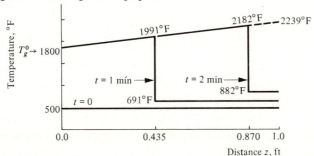

Fig. 12.10 Temperature profiles as a function of time for the conditions described in Example 12.4.

---

[*] For further examples of the use of these equations, we refer the reader to J. Humbert and J. F. Elliott, *Proc. Blast Furnace, Coke Oven, and Raw Materials Conf., AIME* **20**, 130 (1961).

12.6 STATIONARY BED, EFFECT OF THERMAL CONDUCTIVITY WITHIN THE BED

A real phenomenon, thus far neglected, is the effective thermal conduction within the bed itself. Since radiation heat-transfer increases this value significantly at elevated temperatures (see Fig. 6.13), it is quite important in the region of a *flame front*. Assuming that the heat-transfer coefficient is still infinite, we can combine the basic equations (12.1) and (12.2), as in the previous case, with $Q_R = 0$ and constant k_{eff}, to yield

$$(G_s + G_g)\left(\frac{\partial T_s}{\partial t}\right) + W_g\left(\frac{\partial T_s}{\partial z}\right) - k_{eff}\left(\frac{\partial^2 T_0}{\partial z^2}\right) = 0. \qquad (12.25)$$

This is a linear, second order, partial differential equation with constant coefficients. The solution to this equation, for the conditions

$$\text{I.C.} \qquad T_s = T_g = T_s^0, \qquad 0 < z < \infty, \qquad t < 0,$$

$$\text{B.C.1} \qquad T_g = T_g^0, \qquad z = 0, \qquad t > 0,$$

$$\text{B.C.2} \qquad T_s = T_s^0, \qquad z = \infty, \qquad t > 0,$$

is

$$\theta = \theta_g = \theta_s = \frac{T - T_s^0}{T_g^0 - T_s^0} = \frac{1}{2}\left[1 - \text{erf }\phi + \exp\left(\frac{W_g z}{k_{eff}}\right)\text{erfc }\phi\right], \qquad (12.26)$$

where

$$\phi = \sqrt{\frac{W_g G_g t}{4 G_s k_{eff}}} - z\sqrt{\frac{G_s}{4 k_{eff} t}}.$$

Elliott[4] has computed a useful graph for application of this equation to modify the shape of a thermal wave moving within the bed, and we present it in Fig. 12.11. The application of this graph is as follows. Suppose that a cold bed, $\theta_s = 0$, is subjected to a hot gas, $\theta_g = 1$. Since we are assuming an infinite heat-transfer coefficient, the profile would be propagated down the bed as in Section 12.4. If the conditions shown in Fig. 12.12 are used, after 9 min the pulse looks as shown by the solid line in Fig. 12.12(b). To see how conduction would affect the profile, we now refer to Fig. 12.11, and compute the distance term for various distances around the leading edge (e.g., $2 \lesssim z \lesssim 4.5$ ft, $t = 9$ min, $W_g/G_s = 0.4$ ft/min, and an assumed $k_{eff} = 0.01$ Btu/min-ft °F). Using the curve for the leading edge of the pulse, we find the reduced temperature in the bed which is shown in Fig. 12.12. Figure 12.13 shows a similar result for an initial pulse of hot gas followed by cold gas.

The value of k_{eff} used in Fig. 12.12 is reasonable for many oxide materials. We could hardly expect a higher value, although it could be lower. Thus it becomes

[4] J. F. Elliott, *Trans. AIME* **227**, 806 (1963).

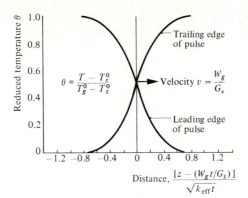

Fig. 12.11 Graph for correction of thermal profile for conduction within bed. (From J. F. Elliott, *ibid.*)

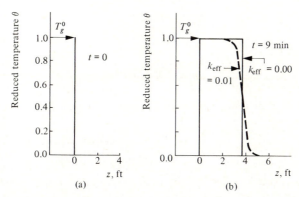

Fig. 12.12 Reduced temperature graph for the shape of thermal wave when W_g/G_s is 0.4 ft/min and k_{eff} is 0.01 Btu/min-ft °F.

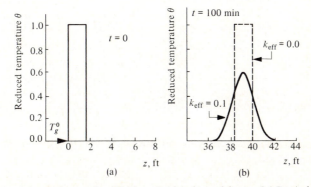

Fig. 12.13 Graph of thermal wave for $W_g/G_s = 0.4$ ft/min and $k_{eff} = 0.1$ Btu/min-ft °F. The incoming gas is reduced to the original low temperature after a short initial period of heating. (--- indicates profile if $k_{eff} = 0.0$ instead of 0.1.)

obvious that the effect of thermal conductivity in spreading the pulse is usually small, but its effect on the maximum temperature reached by a pulse may be sizeable, and should be considered when looking at the temperature history of material in a packed bed.

12.7 THE EFFECT OF A FINITE HEAT-TRANSFER COEFFICIENT— STATIONARY BED

If we remove the condition that hS is infinite, so that $T_s \neq T_g$ at all z and t, and if we continue to assume that the thermal conductivity of the solid is infinite, and again consider the case of no reaction taking place, then we must deal with the solution of the entire basic energy equations as written with $Q_R = 0$. We can obtain an analytical solution,[5] but it is in the form of Bessel functions and is not easy to use, except when carrying out computations on a computer.

However, if we define new variables

$$Y = \frac{hSz}{W_g}$$

and

$$Z = \frac{hS}{G_s}\left(t - \frac{\omega z}{V_{0g}}\right),$$

a graphical solution to the energy equations is available, Fig. 12.14. We describe the use of this graph in the following example.

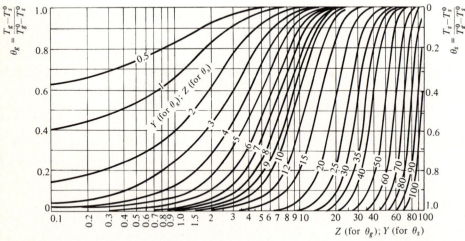

Fig. 12.14 Graph for determination of gas and solid temperatures as a function of time, distance, and heat-transfer coefficient in a stationary packed bed. (Adapted from B. I. Kitaev *et al., ibid.*)

[5] See J. F. Elliott, *ibid.*, or B. I. Kitaev *et al., ibid.*, or H. S. Mickley, T. K. Sherwood, and C. E. Reed, *Applied Mathematics in Chemical Engineering*, McGraw-Hill, New York, 1957.

Example 12.5 A bed of granular material, 1 ft deep, must be heated from 70 to 1800°F.

a) Calculate the time required to accomplish this under the assumed conditions:

$$W_g = 10.0 \text{ Btu/min-ft}^2 \text{ °F},$$
$$G_s = 20.0 \text{ Btu/ft}^3 \text{ °F},$$
$$hS = 50.0 \text{ Btu/min-ft}^3 \text{ °F},$$
$$V_{0g} = 500 \text{ ft/min},$$
$$T_s^0 = 70°F,$$
$$T_g^0 = 1900°F.$$

b) Prepare a graph of the temperature profile of the gas and solid after 3 min.

Solution
a) Calculate Y, with $z = 1$ ft.

$$Y = \frac{hSz}{W_g} = \frac{50 \times 1}{10.0} = 5.0.$$

The reduced temperature required is

$$\theta_s = \frac{T_s - T_s^0}{T_g^0 - T_s^0} = \frac{1800 - 70}{1900 - 70} = 0.945.$$

According to Fig. 12.14, we find that for $Y = 5.0$, and $\theta_s = 0.945$, Z must equal 13. Since

$$Z = \frac{hS}{G_s}\left(t - \frac{\omega z}{V_{0g}}\right)$$

and because

$$\omega z/V_{0g} < < t,$$

then

$$t = \frac{G_s Z}{hS} = \frac{(20)(13)}{(50)}$$
$$= 5.2 \text{ min}.$$

b) Let $z = 0.25, 0.5, 0.75,$ and 1.0 ft. Then

z	Y	Z	θ_s	T_s	θ_g	T_g
0.25 ft	1.25	7.5	0.97	1845°F	0.99	1880°F
0.50	2.50	7.5	0.93	1770	0.95	1810
0.75	3.75	7.5	0.84	1605	0.91	1735
1.00	5.0	7.5	0.71	1370	0.81	1550

The results are plotted below.

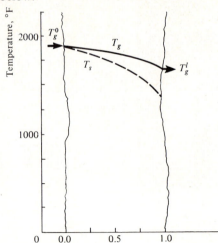

Figure—Example 12.5: Solid and gas temperatures are given as functions of distance after 3 min.

Figure 12.15 indicates experimental temperature profiles[6] from sintering research studies, and the photographs show graphically how sharp the flame front can be. In this case, the gases are drawn down through the bed by suction. A small amount of coke, 6%, is mixed with the ore and burns, providing the necessary continuation of the flame front; in fact, the temperature increases slightly. As the front progresses, it spreads out, primarily because of the heat-transfer coefficient being less than infinite.

Finally, consider another stationary-bed heating situation, where scrap pre-heating buckets utilize hot waste-gases to preheat scrap for electric arc and basic oxygen steelmaking furnaces. Thomas[7] has investigated this process, and developed Fig. 12.16 which relates the heat transferred, Q_t, to the heat required to bring the entire charge to the entering temperature of the heating gases, Q_{max}, and to the physical parameters in the system:

V_{0g} = superficial gas velocity (STP),
t = heating time,
L = bed height,
$\bar{c}_g$ = mean specific heat of gas per unit volume of gas,
$\bar{c}_s$ = mean specific heat of solids per unit volume of solids,
ω = void fraction in bed.

For further details, we refer the reader to the original paper.

[6] R. A. Limons and H. M. Kraner, paper in *Agglomeration*, Interscience–Wiley, 1962.

[7] C. G. Thomas, article in *Energy Management in Iron and Steelworks*, Special Publication No. 105, Iron and Steel Inst., London, 1968, page 87.

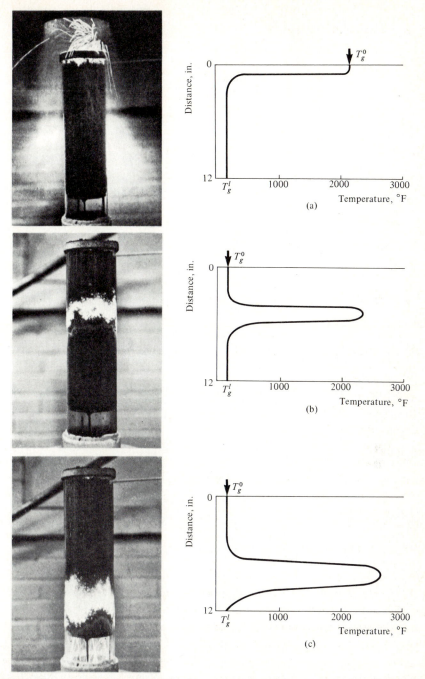

Fig. 12.15 Photographs of iron-ore sintering process in a laboratory study.   Unsintered mixture of ore and coke is placed in glass tube, ignited (top photo), and then after 30 sec, the cold air is drawn down through the bed until the flame front reaches the bottom. (From R. A. Limons and H. M. Kraner, *ibid*.)

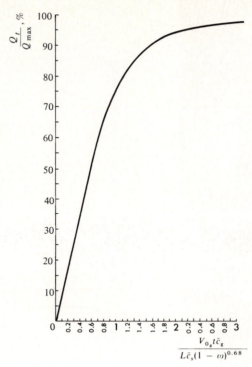

Fig. 12.16 Thermal-energy transfer calculation chart. (From C. G. Thomas, *ibid.*)

PROBLEMS

12.1 Consider the problem outlined in Example 12.1.

a) Develop the temperature profile for the metal and gas as a function of distance.

b) How long would the shaft have to be if the gas thermal flow W_g were reduced to 11.0 Btu/min-°F ft^2?

c) What happens if $W_g = 10.0$ Btu/min-°F ft^2?

12.2

a) Estimate hS for the conditions in Example 12.1 from Fig. 12.5. Assume a particle size of 3 in.

b) Compute L using the new hS ($= h_v$) value.

12.3 A sinter bed contains 6% by weight of coke (assume 100% C). The particle size is such that combustion of particles to CO_2 is complete in 5 min. Calculate the rate of airflow required in order to propagate an initial 2200°F-flame front 1 in. in depth through a 14-in. bed in ½ hr. If the total area of the sinter bed is 1000 sq ft, what capacity must the suction fan have? Data: $\rho_s = 250$ lb/ft^3; $C_s = 0.25$ Btu/lb-°F; $\omega = 0.4$.

12.4 Calculate the time to cool a bed 18 in. deep, which has uniformly sized $\frac{1}{2}$-in. diameter spheres ($\omega = 0.4$), from a uniform initial temperature of 2400°F to a maximum temperature of 200°F. The airflow rate is 100 ft^3/min per ft^2 of bed surface, and the entering gas temperature is 70°F: $\rho_s = 250$ lb/ft^3.

12.5 A shaft furnace to heat pellets of ore to 800°C is needed. If the process parameters are such that $W_h = 2.0$ and $W_c = 2.5$, $T_c^\ell = 1000°$ and $T_h^\ell = 1000°$, and $hS = 1.0$, find ℓ, T_h^0 and T_c^0 such that T_c ($z = \ell/2$) = 800°.

PART THREE

MASS TRANSPORT

In Chapter 1, we introduced Newton's law of viscosity describing momentum transport due to a velocity gradient, and in Chapter 6 Fourier's law of heat conduction was described as energy transfer due to a temperature gradient. In Chapter 13, we present Fick's law of diffusion which describes the transport of a chemical species through a phase due to the concentration gradient of the species.

The rate at which many metallurgical processes occur is determined by diffusion, and it is most desirable to treat these processes in a quantitative manner when possible. The carburizing of steel, some instances of metal oxidation, metal–cladding, and slag–metal reactions are just a few examples of the various diffusion-controlled processes encountered in metallurgy.

On the other hand, there are many metallurgical processes whose rate is determined rather by the rate of reaction at an interface between two phases than by diffusional transport. Kinetics describes the rates of reactions in contrast to Fick's law of diffusion. As you may suspect, many processes are of mixed control, that is, neither the reaction rate nor the diffusional transport process alone controls the overall transfer of material.

This is illustrated by an example where an iron sheet is to be carburized in a furnace at an elevated temperature (above 932°C) utilizing the reaction

$$CH_4(g) = \underline{C} + 2H_2(g).^*$$

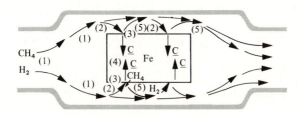

___

* By underlining carbon we have specified its state (just as (g) means gaseous state). In this instance, $\underline{C}$ means that carbon is in solution in the iron.

The rate at which the iron carburizes is related to the system variables, such as the gas flow rate, gas composition, pressure, temperature, sheet thickness, etc. The following consecutive steps are necessarily involved in the process:

1. Transport of methane (carbon source) in the direction of the gas flow.
2. Transport of methane normal to the surface of the iron.
3. The phase boundary reaction, possibly occurring in several steps.
4. Diffusion of carbon from the surface into the interior of the iron sheet.
5. Counter-diffusion of hydrogen gas back into the gas stream.

Step (1) involves supplying the necessary methane to the sample by bulk gas flowing into the tube and past the sample. When the gas stream flows past the iron sheet, a velocity profile is established, and if the reaction proceeds, the methane is depleted at and near the solid–gas boundary; thus step (2) involves diffusion with convection. Then the methane having reached the iron surface decomposes there, in some manner, with carbon atoms entering the lattice—step (3)—and finally— step 4—carbon diffuses into the bulk of the iron. In step (5), the gaseous reaction product, H_2, must diffuse back out into the bulk gas stream, in order that the reaction may proceed.

In general, most metallurgical processes involve such a variety of different steps occurring in gaseous, solid, or liquid phases, and at different phase boundaries. When one step is much slower than the others, a limiting condition is reached and this *rate determining step* determines the overall reaction rate; if none of the steps is much slower than the others, a more general condition—*mixed control*—is achieved. This parallels heat flow problems in that the overall diffusion of heat into or out of a body may be controlled entirely by its thermal diffusivity or by the heat-transfer coefficient at its surface, depending on their relative magnitudes. By way of review, a good example of this is the cooling of solids which, as limiting conditions, either obey Newton's law of cooling, or, if the heat-transfer coefficient approaches infinity, cool at a rate entirely controlled by the solid's ability to diffuse heat.

In Chapter 13 we present Fick's law of diffusion and deal with various types of diffusion coefficients, relating them to each other. Then we discuss their prediction in various phases and conditions of applicability. Chapter 14 contains a variety of solutions to Fick's first and second laws under diffusion-controlled conditions which are often encountered in metallurgical situations in the solid state. Chapters 15 and 16 provide an introduction to mass transfer in fluid systems and the combination of chemical kinetics with mass transfer.

13

FICK'S LAW AND DIFFUSIVITY OF MATERIALS

Transport of mass by diffusion is complex. When considering diffusion processes in multi-component situations, we must realize that the rates of movement of the various components can be different from each other, and that the rates to a large extent depend on the nature of other components present and on their concentrations. With this in mind, let us consider first the rate equations for diffusion, and then discuss the values of the diffusivity and the prediction of these values in various materials.

13.1 DEFINITION OF FLUXES—FICK'S FIRST LAW

Diffusion is the movement of a species from a region of high concentration to a region of low concentration; in general, the rate of diffusion is proportional to the concentration gradient. Consider Fig. 13.1, which depicts a thin plate of pure iron.

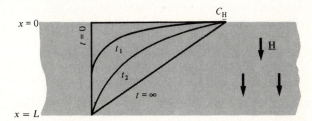

Fig. 13.1 Establishment of steady-state concentration gradient; Fick's first law.

Both sides of the iron plate are exposed to the same pressure of hydrogen which means that after equilibration, the concentration of hydrogen dissolved in the iron is fixed and uniform across the plate. At some instant, $t = 0$, the upper surface is subjected to a much higher pressure of the gas, which establishes a new hydrogen concentration at that surface. The material beneath this surface is gradually enriched as the hydrogen diffuses from the high concentration at the upper surface into the low concentration region. A steady-state concentration profile is eventually reached when a constant rate of hydrogen mass is required from the gas phase at

$x = 0$, in order to maintain the concentration difference across the plate. This example is, of course, analogous to Figs. 1.4 and 6.1, and may also be described by a rate equation.

If the concentration of component A is given in mass units, then the rate equation for diffusion may be written

$$W_{Ax} = -D_A\left(\frac{\partial \rho_A}{\partial x}\right),$$ (13.1)

where W_{Ax} = mass flux of A in the x-direction, g(of A)-cm^{-2} sec^{-1}, ρ_A = mass concentration of A, g(of A)-cm^{-3} (of total material), and D_A = diffusion coefficient, or diffusivity of A, cm^2-sec^{-1}.

In metallurgical applications the rate equation is usually written in terms of molar concentrations:

$$j_{Ax} = -D_A\left(\frac{\partial C_A}{\partial x}\right),$$ (13.2)

where j_{Ax} = molar flux of A in the x-direction, mol (of A)-cm$^{-2}\cdot$ sec^{-1}, and C_A = molar concentration of A, mol (of A)-cm^{-3} (of total material).

Equations (13.1), (13.2), and other alternative statements of the rate equation are all referred to as *Fick's first law of diffusion* which states that species A diffuses in the direction of decreasing concentration of A, similarly as heat flows by conduction in the direction of decreasing temperature, and momentum is transferred in viscous flow in the direction of decreasing velocity.

Equations (13.1) and (13.2) are convenient forms of the rate equation when the density of the total solution is uniform, that is, in solid or liquid solutions, or in a dilute gaseous mixture. When the density is not uniform, other forms of Fick's first law may be preferable. For example, the mass flux may be given by

$$W_{Ax} = -\rho D_A\left(\frac{\partial \rho_A^*}{\partial x}\right),$$ (13.3)

where ρ is the density of the entire solution, g-cm^{-3}, and ρ_A^* is the mass fraction of A. Similarly, the molar flux may take the form

$$j_{Ax} = -C D_A\left(\frac{\partial X_A}{\partial x}\right),$$ (13.4)

in which C is the local molar concentration in the solution at the point where the gradient is measured, mol (of all components)-cm^{-3} (of total solution), and X_A is the mole fraction of A (C_A/C) in the solution.

The diffusivities D_A, appearing in Eqs. (13.1)–(13.4), are of a particular type, known as the *intrinsic diffusion coefficients*, since they are defined in the presence of a concentration gradient of A.

13.2 DIFFUSION IN SOLIDS

Over the past decades, a large emphasis has been placed on determining diffusion mechanisms in solids, in the hope that eventually diffusion coefficients can be predicted for a given set of conditions without the necessity for carrying out the long experiments involved in their determination. But we have to admit that that goal has not been reached. However, it behooves metallurgists to understand the proposed mechanisms, and especially the implications of the terms self-diffusion, intrinsic diffusion, interdiffusion, interstitial diffusion, vacancy diffusion, etc., so that they can make intelligent estimates of diffusivities, and/or the effects of alloying, for practical uses. In the following sections, we shall consider the various types of diffusion coefficients and some of the proposed mechanisms for diffusion.

13.2.1   Self-diffusion

If we could stand inside the lattice of a solid, we would see a continual motion of the atoms, each vibrating about its normal lattice point. Furthermore, we would see occasional unoccupied sites, that is, vacancies. If we focus on a vacancy and its immediate neighboring atoms, we will eventually see the site suddenly become occupied and one of its neighboring sites become vacant. In this way, a particular atom can slowly progress through the lattice. Another way to think of it is to consider that the vacancy wanders randomly through the lattice. At any rate, the net effect is a random motion of the atoms themselves.

    The rate at which an atom meanders through the lattice of a pure metal is the *self-diffusion rate*. We can measure it using radioactive atoms (tracers), as illustrated in Fig. 13.2, which depicts self-diffusion between a central region initially with a uniform concentration C_A^T of radioactive atoms and two adjoining regions initially containing only normal atoms. It is assumed that the diffusion behavior of normal and of radioactive atoms is virtually the same, and by using a scheme as in Fig. 13.2

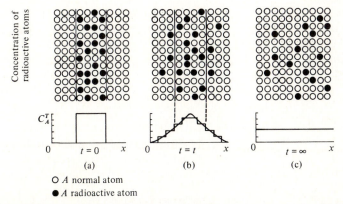

O *A* normal atom
● *A* radioactive atom

Fig. 13.2 Self-diffusion in materials. (a) Condition before diffusion. (b) Condition after some diffusion. (c) Uniform condition after prolonged diffusion.

we can measure the rate at which the atoms diffuse among themselves, that is, self-diffusion.

We can also speak of self-diffusion rates in homogeneous alloys in the same manner as we do for pure metals. In an alloy of gold and nickel, for example, we can determine the self-diffusion coefficient of gold atoms D_{Au}^*, if the central layer of the homogeneous alloy initially contains radioactive gold atoms in the same proportion to Ni atoms as in the outer layers (Fig. 13.3). Similarly, if radioactive nickel atoms are initially present in the central layer, then we can determine its self-diffusion coefficient, D_{Ni}^*. Figure 13.4 gives self-diffusion data for the entire range of compositions in the gold–nickel system. It should be emphasized that *self-diffusion data apply to homogeneous alloys in which there is no gradient in chemical composition.* Also, D_i^* values in an alloy depend on the composition of the alloy, and the temperature.

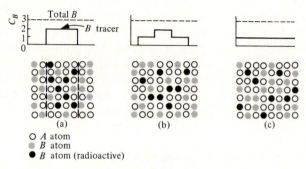

O A atom
◐ B atom
● B atom (radioactive)

Fig. 13.3 Self-diffusion in an alloy.

You may recall from Chapter 1 that Einstein proposed the equation

$$D^* = B^* \kappa_B T \qquad (1.17)$$

to describe self-diffusion, where B^* is the self-diffusion "mobility", a measure of the ability of an atom to migrate within the structure without any external field or chemical free energy gradient providing a driving force.

The probabilistic approach has shown that the self-diffusion coefficient in a simple cubic lattice may be expressed as[1]

$$D^* = \tfrac{1}{6}\delta^2 v, \qquad (13.5)$$

where δ is the interatomic spacing, and v is the jump frequency. From measurements of D^* and δ, the jump frequency appears to be approximately 10^8–10^{10} sec^{-1}, which is 1 in every 10^4 or 10^5 vibrations per atom. Most studies of self-diffusion are concerned with this sort of information, that is, with the structure and motion

---

[1] See, for example, P. G. Shewmon, *Diffusion in Solids*, McGraw-Hill, New York, 1963, or R. E. Reed-Hill, *Physical Metallurgy Principles*, Van Nostrand, Princeton, New Jersey, 1964, pages 255–256.

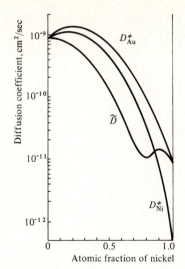

Fig. 13.4 Diffusion coefficients in gold–nickel alloys. $\tilde{D}$ is the interdiffusion coefficient and D^* the self-diffusion coefficient. (From J. E. Reynolds, B. L. Averbach, and M. Cohen, *Acta Met.* **5**, 29 (1957).)

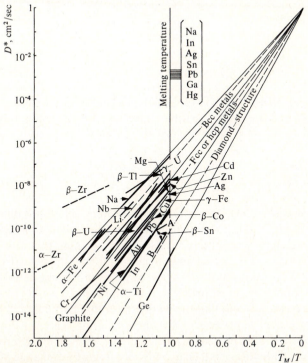

Fig. 13.5 Self-diffusion coefficients for pure metals, plotted as $\log D$ versus T_M/T, where T_M is the melting temperature. (From O. D. Sherby and M. T. Simnad, *Trans. ASM* **54**, 227 (1961).)

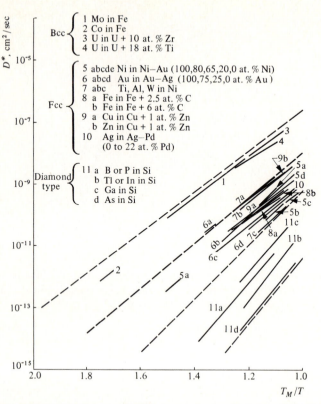

Fig. 13.6 Effect of dilute alloying on self-diffusion and rate of solute substitutional diffusion for a number of metallic systems. T_m is the temperature halfway between the liquidus and solidus of the alloy. (From O. D. Sherby and M. T. Simnad, *ibid.*)

within the lattice of a solid, and the literature provides plentiful data on it. Some data for pure solid metals are given in Fig. 13.5 and for dilute solid alloys, in Fig. 13.6.

A problem arises when we compare calculated or actual self-diffusion data to tracer data. In the development leading to Eq. (13.5), we assume that the atom jumps are completely random and uncorrelated, meaning that any of the atoms surrounding a vacancy may exchange places with the vacancy. However, if we follow the motion of atoms (not vacancies) by following the motion of tracer atoms moving via vacancies, we must realize that the tracer atom is only one of the many atoms around a vacancy and that the tendency for that particular atom (the radioactive atom) to exchange places with the vacancy is not as great as for any atom in general to exchange. This means that the vacancy is often likely to move away from the tagged atom, with diffusion then actually occurring by exchange of an untagged atom with the vacancy, but remaining undetected as long as we only follow the tagged atom displacement. The self-diffusion coefficient D_i^T, calculated from the results of tracer studies, is thus slightly lower than the true self-diffusion coefficient D_i^*. They are related by means of the *correlation coefficient, f*:

$$D_i^T = fD_i^*. \tag{13.6}$$

Table 13.1 gives f calculated for several different crystal structures. For interstitial diffusion, f is equal to 1.0, since the host lattice is not involved in the mechanism.

Table 13.1   Correlation coefficients for various structures*

Structure	f
Diamond	0.500
Simple cubic	0.655
Body-centered cubic	0.721
Face-centered cubic	0.781

* From K. Compaan and Y. Haven, *Trans. Faraday Soc.* **52**, 786 (1956).

13.2.2 Diffusion under the influence of a composition gradient

In commercial processes, diffusion occurs because a concentration gradient or driving force is provided. When this force is present, the diffusion coefficient used to calculate the flux of A or B atoms is *not* the self-diffusion coefficient, except under special circumstances.

Consider the diffusion of interstitial atoms (that is, atoms normally residing in interstitial sites) through a lattice via interstitial sites, for example, the diffusion of carbon through iron. The process is quite straightforward. In this instance, carbon is in dilute concentration, and we can imagine that it diffuses through the stationary iron lattice without displacing the iron atoms from their own sites. Thus, if we treat a problem involving the diffusion of carbon in iron, and obtain an analytical expression with a diffusion coefficient in it, we can safely use a value for carbon's diffusivity in iron from the literature or a handbook, even though it is not clearly specified what type of diffusion coefficient is exactly given.* We presume that the interstitial mechanism dominates if the solute atoms are considerably smaller in size than the solvent atoms, which results in little distortion of the lattice. If the solute radius begins to approach that of the solvent, this mechanism no longer operates, and movement of the solvent atoms as well as motion of the solute atoms takes place. In this case, then, diffusion involves interchange of solute and solvent atoms on sites in the alloy. In substitutional alloys, the influence of a chemical gradient is such that we must clearly recognize the right type of diffusion coefficient: either the *intrinsic diffusion coefficient* or *interdiffusion coefficient*. Both types are discussed below, but first it seems appropriate to discuss briefly the various

---

* However, the alloy composition and the temperature must still be specified.

mechanisms by which substitutional alloy elements are thought to diffuse. More detailed discussions of diffusion mechanisms in solids are available in most texts on materials science or physical metallurgy.

Vacancy mechanism. This is a mechanism by which an atom on a site adjacent to a vacancy jumps into the vacancy. While some distortion of the lattice is required for the atom to pass between neighboring atoms, the energy associated with this distortion is not prohibitive, and this mechanism is well established as the predominant one in many metals and ionic compounds.

Ring mechanism. In some bcc metals, it is thought there might exist a mechanism of diffusion in which a ring of three atoms may rotate, resulting in diffusion. This possibility is considered to be more plausible than an exchange of two atoms, since it involves less strain energy than a two-atom exchange. However, direct evidence of either of these mechanisms operating in metals is lacking.

Interstitialcy mechanism. This mechanism involves the addition of an extra solute atom to the lattice, by pushing an adjacent atom out of its normal site and into an interstitial site. The motion continues as the new, oversized interstitial atom pushes a further atom out of its normal site in a chain-reaction type of process. This mechanism is believed to operate in some compounds in which one atom is smaller than the other, for example, in AgBr, where silver diffuses via this mechanism. In most metals, however, this mechanism seems unlikely to operate. We find a possible exception in materials subjected to bombardment by radiation, in which case high energy particles may knock atoms out of their normal sites and into interstitial positions.

Thinking of diffusion in general, we can see now the basic difference between diffusion and heat flow quite clearly. Heat flow taking place in a medium does not cause the medium to move while diffusion, in itself, involves the movement of the

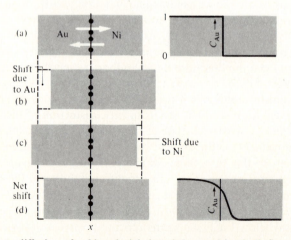

Fig. 13.7 The interdiffusion of gold and nickel, and the resulting bulk flow and composition profiles.

medium, and thus contributes to the velocity of the medium. Refer to Fig. 13.7, for example, and consider a diffusion couple made by joining a gold and a nickel bar together so that there is diffusion across the marked interface. The inert markers may be pieces of fine tungsten wire (insoluble in the alloy) which are located in the plane of joining. During a diffusion anneal of many hours at a sufficiently high temperature, such as 900°C, interdiffusion of the gold and nickel occurs and changes the concentration distribution as shown. But, because gold diffuses more rapidly than nickel, more gold atoms than nickel atoms diffuse past the inert markers. Figure 13.7(b) shows in a very exaggerated schematic form the transfer of gold atoms across the plane of markers without any nickel atoms crossing over. If the vacancy concentration remains uniform, that is, the volume is constant, this transfer of gold atoms requires that the bar of pure gold must shorten while the nickel bar (now containing the transferred gold atoms) lengthens by the same amount. In the same manner, the transfer of nickel atoms across the plane of markers without any gold atoms crossing over is shown, producing the shift due to Ni. The net result of these two processes, which occur simultaneously, is that the side that was originally pure gold is somewhat shorter, and the other side is correspondingly longer. An alternative description of the phenomenon could be that the inert markers have moved from their original position toward the gold end of the specimen. This is the viewpoint taken if one end of the specimen is the reference plane, and the movement is called the Kirkendall shift or the *Kirkendall effect*. The differences between the various "types" of diffusion coefficients mentioned above are related to the Kirkendall effect, and are pointed out below.

Due to bulk motion in the gold–nickel diffusion couple, we note that the solid bar moves with a velocity depending on the difference between the rates of diffusion of gold and nickel. An observer sitting on a lattice plane moving with the solid's velocity would notice a flux of gold atoms past that plane given by

$$j_{Au} = -D_{Au}\frac{\partial C_{Au}}{\partial x}.$$

However, an observer sitting on an unattached plane in space, would see the total flux of gold atoms passing by as

$$N_{Au} = -D_{Au}\frac{\partial C_{Au}}{\partial x} + v_x C_{Au}, \tag{13.7}$$

where the flux is expressed by N_{Au}, rather than by j_{Au}, to emphasize that it is relative to the stationary plane. The accumulation of gold as a function of time within a unit volume that straddles the stationary plane is given by the net difference between the gold entering and leaving. Thus, referring to Fig. 13.8, we may write for a unit area perpendicular to the page:

$$\Delta x\left(\frac{\partial C_{Au}}{\partial t}\right) = [N_{Au}|_x - N_{Au}|_{x+\Delta x}]. \tag{13.8}$$

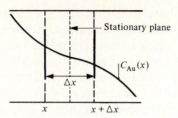

Fig. 13.8 Unit element for mass balance.

When the $\lim_{\Delta x \to 0}$ is taken, we obtain

$$\frac{\partial C_{Au}}{\partial t} = -\frac{\partial N_{Au}}{\partial x}, \tag{13.9}$$

which, on substituting Eq. (13.7), becomes

$$\frac{\partial C_{Au}}{\partial t} = \frac{\partial}{\partial x}\left(D_{Au}\frac{\partial C_{Au}}{\partial x} - v_x C_{Au}\right). \tag{13.10}$$

Similarly, we can obtain the accumulation of nickel in the same unit volume:

$$\frac{\partial C_{Ni}}{\partial t} = \frac{\partial}{\partial x}\left[D_{Ni}\frac{\partial C_{Ni}}{\partial x} - v_x C_{Ni}\right]. \tag{13.11}$$

If the vacancy concentration within the unit volume is constant (volume constant), then

$$\frac{\partial C}{\partial t} = \frac{\partial C_{Ni}}{\partial t} + \frac{\partial C_{Au}}{\partial t} = 0, \tag{13.12}$$

which requires, by Eqs. (13.10) and (13.11), that

$$v_x = \frac{1}{(C_{Au} + C_{Ni})}\left[D_{Au}\left(\frac{\partial C_{Au}}{\partial x}\right) + D_{Ni}\left(\frac{\partial C_{Ni}}{\partial x}\right)\right]. \tag{13.13}$$

Only in terms of the gold concentration gradient, this takes the form

$$v_x = \frac{1}{C}[D_{Au} - D_{Ni}]\left(\frac{\partial C_{Au}}{\partial x}\right). \tag{13.14}$$

Having obtained an expression for v_x, we can now write the rate of accumulation of gold solely in terms of diffusion coefficients and concentration gradients, by substituting Eq. (13.14) into (13.10):

$$\frac{\partial C_{Au}}{\partial t} = \frac{\partial}{\partial x}\left[(X_{Ni}D_{Au} + X_{Au}D_{Ni})\frac{\partial C_{Au}}{\partial x}\right]. \tag{13.15}$$

We developed Eq. (13.15) from the viewpoint of a stationary observer. However, this requires that the position of a reference plane fixed in space can be determined. In the experimental set-up of Fig. 13.7, the inert markers could provide such a reference plane. But in applying diffusion data to a commercial process, we cannot expect to be provided with inert markers. In addition, Eq. (13.15) is useful for determining D_{Au} and D_{Ni}, but is of no value in the sense that analytical solutions to practical diffusion problems cannot be conveniently developed from it. Instead, the basic complexity that arises from the bulk motion is avoided by merely using Fick's first law in its simplest form, that is, for gold

$$j_{Au} = -\tilde{D}\left(\frac{\partial C_{Au}}{\partial x}\right). \tag{13.16}$$

Note that the diffusion coefficient is written differently; $\tilde{D}$ includes the basic concept that the flux is proportional to the concentration gradient but side-steps the issue of bulk flow. With $\tilde{D}$ defined in this way, the accumulation of gold within the unit volume of Fig. 13.8 is

$$\Delta x\left(\frac{\partial C_{Au}}{\partial t}\right) = [j_{Au}|_x - j_{Au}|_{x+\Delta x}]. \tag{13.17}$$

When taking the $\lim_{\Delta x \to 0}$, we obtain the unsteady-state equation for unidirectional diffusion in solids:

$$\frac{\partial C_{Au}}{\partial t} = \frac{\partial}{\partial x}\left(\tilde{D}\frac{\partial C_{Au}}{\partial x}\right). \tag{13.18}$$

We use this equation, which is often called *Fick's second law*, to obtain analytical solutions for diffusion problems. By comparing Eqs. (13.15) and (13.18), the relation between the more fundamental quantities, D_{Au} and D_{Ni}—called *intrinsic* diffusion coefficients—and the more useful quantity, $\tilde{D}$—called the *interdiffusion*, *mutual diffusion*, or *chemical diffusion* coefficient*—is evident

$$\tilde{D} = X_{Ni}D_{Au} + X_{Au}D_{Ni}. \tag{13.19}$$

Figure 13.4 shows $\tilde{D}$ for the gold–nickel system.

13.2.3 Darken's equations

Fick's laws contain the implicit assumption that the driving force for diffusion is the concentration gradient. A more fundamental viewpoint assumes that the driving force is a chemical free energy gradient. There is usually a direct correspondence between the two gradients, but occasionally the relationship becomes inverted and so-called "up-hill" diffusion, that is, diffusion against a concentration gradient, occurs. Darken[2] provided the following analysis of this situation.

---

* $\tilde{D}$ as the quantity used in diffusion calculations is very often called simply the diffusion coefficient and denoted D.

[2] L. S. Darken, *Trans. AIME* **180**, 430 (1949).

The force F acting on an atom of species A may be expressed in terms of its partial molar free energy gradient, $\partial \bar{G}_A / \partial x$, and Avogadro's number, N_0, by*

$$F = -\frac{1}{N_0} \left(\frac{\partial \bar{G}_A}{\partial x} \right). \tag{13.20}$$

Under the influence of this force, the velocity of an A atom is

$$v_{Ax} = B_A F = \frac{-B_A}{N_0} \left(\frac{\partial \bar{G}_A}{\partial x} \right), \tag{13.21}$$

where B_A is the mobility of A atoms in the presence of an energy gradient. Then, the flux of A atoms passing through a unit area is $\dot{n}_{Ax}$, defined by

$$\dot{n}_{Ax} = v_{Ax} n_A, \tag{13.22}$$

or

$$\dot{n}_{Ax} = \frac{-n_A B_A}{N_0} \left(\frac{\partial \bar{G}_A}{\partial x} \right), \tag{13.23}$$

where n_A is the number of A atoms per unit volume. Recalling (from thermodynamics) that

$$\bar{G}_A = G_A^0 + RT \ln a_A,$$

then

$$\left(\frac{\partial \bar{G}_A}{\partial x} \right) = RT \left(\frac{\partial \ln a_A}{\partial x} \right). \tag{13.24}$$

Finally, by combining Eqs. (13.23) and (13.24), we can write the flux of A in terms of the mobility, B_A, and activity, a_A, as follows:

$$\dot{n}_{Ax} = \frac{-n_A B_A RT}{N_0} \left(\frac{\partial \ln a_A}{\partial x} \right). \tag{13.25}$$

Now comparing Eq. (13.25) to Fick's first law (Eq. 13.2), we find that

$$D_A \left(\frac{\partial \ln C_A}{\partial x} \right) = B_A \kappa_B T \left(\frac{\partial \ln a_A}{\partial x} \right), \tag{13.26}$$

with κ_B (Boltzmann's constant) $= R/N_0 = 1.38 \times 10^{-23}$ joules-deg^{-1}.

Since we can express the molar volume of A (cm^3 mol^{-1}) as X_A / C_A, then

$$\frac{\partial C_A}{\partial x} = \frac{C_A}{X_A} \left(\frac{\partial X_A}{\partial x} \right),$$

---

* A self-consistent set of units for use with this and following equations relating electrical and diffusive properties is $\bar{G}_i =$ joules-mol^{-1}; $B_i =$ cm^2-sec$^{-1} \cdot$ joules^{-1}; $\kappa_B =$ joules-deg^{-1}; $n_i =$ ions-cm^{-3}; $e =$ coulombs; $\mu_i =$ joules-atoms^{-1}.

and

$$\left(\frac{\partial \ln C_A}{\partial x}\right) = \left(\frac{\partial \ln X_A}{\partial x}\right). \tag{13.27}$$

Thus, referring to Eq. (13.26), we see that

$$D_A\left(\frac{\partial \ln C_A}{\partial x}\right) = B_A\kappa_B T\left(\frac{\partial \ln a_A}{\partial \ln X_A}\right)\left(\frac{\partial \ln X_A}{\partial x}\right), \tag{13.28}$$

and finally

$$D_A = B_A\kappa_B T\left(\frac{\partial \ln a_A}{\partial \ln X_A}\right). \tag{13.29}$$

Equation (13.29) is an expression for the intrinsic diffusivity of species A under the influence of its free energy gradient. Examine this expression as applied to a thermodynamically ideal solution:

$$\left(\frac{\partial \ln a_A}{\partial \ln X_A}\right) = 1,$$

and

$$D_A = B_A\kappa_B T, \tag{13.30}$$

known as the *Nernst–Einstein* equation. *If* the mobility B_A, in the presence of a driving force, is independent of composition, then D_A in an ideal solution alloy is independent of composition. However, it is reasonable to expect B_A to be a function of composition in alloys and non-stoichiometric compounds and so, even if the solutions are thermodynamically ideal, D_A may still be a function of composition. In addition, due to the fact that $D_A^* = B_A^*\kappa_B T$ for self-diffusion, it is often stated that Eq. (13.30) may be written as $D_A^* = B_A\kappa_B T$ for an ideal solution. However, this is *not* true, unless fortuitously $B_A^* = B_A$, for there is no reason to think that B_A under the influence of an energy gradient should be the same as B_A^* with no energy gradient present. Figure 13.4 shows sufficiently clearly that D_A^* is not a constant with composition; at best, one would have to know B_A^* as a function of composition, and since B values are usually computed *from* diffusion coefficient measurements, this becomes an exercise in rhetoric.

Only in one instance—in nonmetallic stoichiometric compounds—may B_A^* and B_A be the same, and D_A^* be calculated from an independent measurement (such as ionic electrical conductivity) of B_A. We shall learn more about this subject in Section 13.3.

Returning to Eq. (13.19), we can modify the expression. From the Gibbs-Duhem equation, we know that

$$\left(\frac{\partial \ln a_B}{\partial \ln X_B}\right) = \left(\frac{\partial \ln a_A}{\partial \ln X_A}\right). \tag{13.31}$$

Therefore

$$\tilde{D} = (X_A B_B \kappa_B T + X_B B_A \kappa_B T)\left(\frac{\partial \ln a_A}{\partial \ln X_A}\right). \tag{13.32}$$

Now *if, and only if*, $B_A = B_A^*$ *and* $B_B = B_B^*$, then

$$\tilde{D} = (X_A D_B^* + X_B D_A^*)\left(\frac{\partial \ln a_A}{\partial \ln X_A}\right), \tag{13.33}$$

so that in this case, if we know the self-diffusion coefficients, D_B^* and D_A^*, and the way in which the activity of A varies with composition, we can calculate the inter-diffusion coefficient, which is the useful value for engineering calculations. The thermodynamic term, which is sometimes called the *thermodynamic factor*, can also be expressed in the form

$$\frac{\partial \ln a_A}{\partial \ln X_A} = \left(1 + \frac{\partial \ln \gamma_A}{\partial \ln X_A}\right), \tag{13.34}$$

where γ_A is the activity coefficient. If the system is ideal, then $\ln \gamma_A$ is everywhere equal to 0, and

$$\tilde{D} = X_A D_B^* + X_B D_A^*. \tag{13.35}$$

But since the restriction that B_A equals B_A^*, and B_B equals B_B^*, still operates, Eqs. (13.35) and (13.33) are of limited application. However, Eq. (13.32) gives some insight into the conditions where diffusion may occur from the point of view of a concentration gradient. The best-known example is the experiment carried out by Darken.[3] He welded two pieces of carbon steel together and studied the diffusion of carbon, as schematically shown in Fig. 13.9. One of the bars contained 3.80% Si and 0.48% C, while the other contained only 0.44% C. Figure 13.9(a) shows the initial carbon distribution and we might expect that the carbon distribution would simply have evened out at 0.46% C. However, because silicon greatly increases the activity of carbon in iron, the carbon in the left-hand bar was at a much higher chemical potential than it would have been at if no silicon had been present. The result was that a sizeable amount of carbon diffused from left to right, and after a short time, the carbon gradient was as shown in Fig. 13.9(b). If the silicon did not diffuse, the final profile would look as in Fig. 13.9(c). If diffusion continues until both chemical potential curves are flat, not only for carbon but also for silicon, then the silicon will eventually diffuse to the right-hand side. The diffusion of silicon will be much slower than that of carbon, since it involves substitutional diffusion rather than the interstitial mechanism by which carbon diffuses. This means that the higher carbon concentration (created on the right-hand side before appreciable silicon diffusion occurred) will then decrease until ultimately both the silicon and carbon gradients become flat (Fig. 13.9e).

---

[3] L. S. Darken, *Trans. AIME*, Inst. Met. Div., Tech. Publ. 2443 (1948).

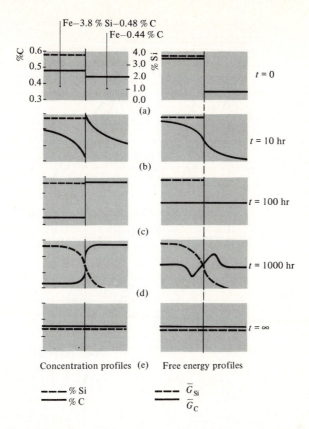

Fig. 13.9 Concentration and partial molar free-energy curves at various times for Fe–C and Fe–C–Si alloys welded together and annealed.

13.2.4 Temperature dependence of diffusion in solids

Temperature has a tremendous influence on diffusion in solids. It has empirically been found that the Arrhenius equation adequately describes the relationship between any type of diffusion coefficient and temperature:

$$D = D_0 e^{-Q/RT}, \tag{13.36}$$

in which Q is the activation energy, and D_0 is sometimes called the frequency factor, both essentially constant over a wide temperature range.

Much effort has gone into the development of theoretical expressions for D_0. Recall that the self-diffusion coefficient D^* was expressed earlier in terms of the jump frequency v according to

$$D^* = \tfrac{1}{6}\delta^2 v, \tag{13.5}$$

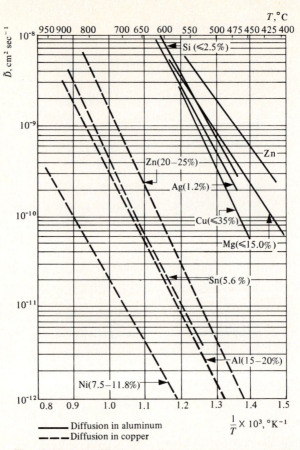

Fig. 13.10 Interdiffusion coefficients in nonferrous metals.

where δ is the interatomic distance. According to Zener,[4] if the jump process is an activated one, then v can be described by

$$v = v_0 Z e^{-\Delta G^{\ddagger}/RT} \tag{13.37}$$

where v_0 = vibrational frequency of the atom in the lattice, Z = coordination number, and $\Delta G^{\ddagger}$ = free energy of activation required for the atom to jump from one site into the next.

Since

$$\Delta G^{\ddagger} = \Delta H^{\ddagger} - T\Delta S^{\ddagger},$$

then

$$D^* = \frac{\delta^2 v_0 Z}{6} e^{+(\Delta S^{\ddagger}/R)} e^{-(\Delta H^{\ddagger}/RT)}, \tag{13.38}$$

---

[4] C. Zener, *J. Applied Physics* **22**, 372 (1951).

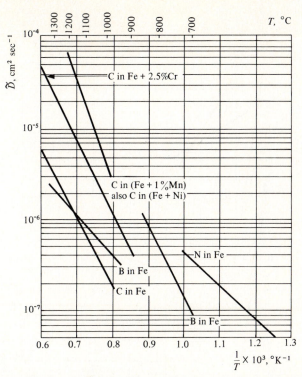

Fig. 13.11 Interdiffusion coefficients of interstitial elements through ferrous materials.

in which

$$D_0 = \frac{\delta^2 v_0 Z}{6} e^{(\Delta S^{\ddagger}/R)}. \tag{13.39}$$

$\Delta H^{\ddagger}$ is synonymous with Q, the activation energy for diffusion; δ, v_0, and $\Delta S^{\ddagger}$ do not vary significantly with temperature, so the result is that D_0 varies with temperature only slightly. This is somewhat different from saying that D_0 is a constant, but for practical purposes it is a close approximation. On the other hand, it should not be too surprising if a plot of log D versus T^{-1} is not a perfectly straight line.

There are many data available on metallic systems, and some of these are presented in Figs. 13.10, 13.11, and 13.12. In addition, there are several rules that may be used to estimate Q and D_0 if data are completely missing. Sherby and Simnad[5] developed a correlation equation for predicting self-diffusion data in pure metals:

$$D^* = D_0 e^{-(K_0 + V)(T_M/T)}, \tag{13.40}$$

[5] O. D. Sherby and M. T. Simnad, *Trans. ASM* **54**, 227 (1961).

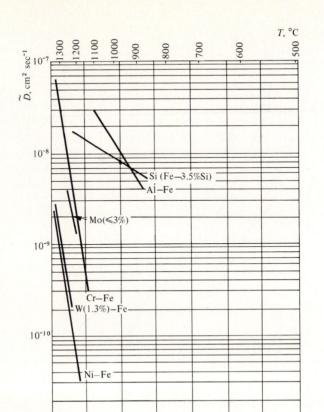

Fig. 13.12 Interdiffusion coefficients in ferrous materials.

where K_0 depends only on the crystal structure, V is the normal valence of the metal, and T_M is the absolute melting point. The values of K_0 are

Crystal structure	K_0
bcc	14
fcc	17
hcp	17
Diamond	21

For estimation purposes, D_0 is approximated as $1 \text{ cm}^2/\text{sec}$, as shown for most metals in Fig. 13.5. Furthermore, Q is predicted by

$$Q = RT_M(K_0 + V). \tag{13.41}$$

This correlation has been tested for self-diffusion in alloys, in which case T_M is taken as the temperature halfway between the liquidus and solidus for the alloy;

it also applies to intrinsic diffusivities of solutes in dilute concentrations in a variety of solvent hosts, with reasonably satisfactory results.

13.3 DIFFUSION IN SOLID NONMETALS

The distinction between diffusion in metals and nonmetals is mainly that non-metals show varying degrees of polarization between cations and anions, from oxides with only minor polarization and a narrow electron conduction band, to halides with strong polarization and no conduction bands. Diffusion in these materials takes place via the same atomic mechanisms as in metals, but because of the electrical structure of the crystals, there is an additional strong relation between electrical conductivity and diffusion mechanisms. Their diffusion mechanisms, like those of metals, are closely related to atomic defects such as vacancies in the crystals, because these point defects provide the paths by which diffusion usually occurs. To discuss defects in nonmetals and their relationships to conduction and diffusion in any detail would consume at least an entire chapter, and would not serve the purpose of this text; we shall give, however, a short introduction to the subject. For more details we refer the reader to the several books available on this subject.* Diffusivities of ions in some crystals are given in Table 13.2.

Stoichiometric crystals may contain several different types of point defects. The *Schottky defect* refers to equal numbers of cation and anion vacancies, that is, $[V_A] = [V_B]$.† The *Frenkel defect* may occur on either the cation or anion sublattice, when equal numbers of vacancies are balanced by interstitials of the same species, that is, $[V_A] = [A_i]$. An example of a material that exhibits Schottky defects is NaCl, in which $[V_{Na}] = [V_{Cl}]$. Frenkel defects predominate in AgBr, in which $[V_{Ag}] = [Ag_i]$. If diffusion is via vacancies, any chemical additions that increase the concentration of vacancies will tend to increase the diffusion coefficient. For example, if $CdCl_2$ is added to NaCl, then, since the cadmium cation has a $+2$ charge, electrical balance in the crystal can only be maintained if one cation site is left vacant for every cadmium atom added, thus increasing $[V_{Na}]$.

Many nonmetallic materials show deviations from stoichiometry, which may be due to unequal concentrations of cation and anion defects. This is particularly true of transition metal oxides, such as TiO_2, Nb_2O_5, NiO, CoO, and wustite, Fe_xO. In the latter case, the defects are predominantly cation vacancies, $[V_{Fe}]$,

---

* Jost, *Diffusion*, Academic Press, New York, 1960.

K. Hauffe, *Oxidation of Metals*, Plenum Press, New York, 1965.

O. Kubachewski and Hopkins, *Oxidation of Metals and Alloys*, Plenum Press, New York, 1962.

P. Kofstad, *High Temperature Oxidation of Metals*, Wiley, New York, 1966.

F. A. Kroger, *Imperfections in Crystals*, North–Holland, New Amsterdam, 1964.

† The notation $[V_A]$ refers to the concentration, cm^{-3}, of vacant A sites, $[A_i]$ refers to the concentration of A atoms on interstitial sites, and $[A_B]$ to the concentration of A atoms on sites normally occupied by B atoms.

with charge neutrality maintained by the creation of Fe^{3+} ions according to the equation

$$3Fe^{2+} \rightarrow V_{Fe} + 2Fe^{3+} + Fe^0 \text{ (removed)}.$$

Some of the trivalent iron ions may be ionized according to*

$$Fe^{3+} \rightarrow Fe^{2+} + \oplus,$$

which results in an increase in electronic conduction via electron holes as further deviations from stoichiometry occur.

Because of the electrical nature of the defects and of the diffusing species (ions), *ionic* electrical conduction and diffusion are inseparable processes. Consider an ionic material in which one ionic species makes the predominant contribution to electrical conduction (for example, Na^+ cations in NaCl). The general equation for the transport of the species in the x-direction under the influence of an electric field and a concentration gradient in the x-direction is:

$$\dot{n}_{ix} = -D_i\left(\frac{\partial n_i}{\partial x}\right) - B_i n_i z_i e\left(\frac{\partial \phi}{\partial x}\right), \qquad i \text{ ions cm}^{-2}\text{-sec}^{-1}, \quad (13.4)$$

where z_i = number of charges on the diffusing species,

$\quad\quad n_i$ = concentration of diffusing species, ions-cm^{-3},

$\quad\quad B_i$ = mobility of species i (steady-state velocity of the particle under the influence of a unit force), cm^2-sec^{-1}-volt^{-1} -coulomb^{-1},

$\quad\quad D_i$ = the diffusion coefficient of species i, cm^2-sec^{-1},

$\quad\quad e$ = 1.6×10^{-19} coulombs-charge^{-1},

$\quad\quad \phi$ = electrical potential, volts.

In the absence of a concentration gradient, the flux is

$$\dot{n}_{ix} = -B_i n_i z_i e\left(\frac{\partial \phi}{\partial x}\right), \qquad i \text{ ions cm}^{-2}\text{-sec}^{-1}. \tag{13.43}$$

Since the current density I is given by

$$I = \dot{n}_{ix} z_i e, \qquad \text{amp-cm}^{-2}, \tag{13.44}$$

then

$$I = -B_i n_i (z_i e)^2\left(\frac{\partial \phi}{\partial x}\right). \tag{13.45}$$

---

* The notation $\oplus$ is used to represent an electron hole in the valence band, contributing to *p*-type, or hole conduction; similarly, $\ominus$ is the symbol for electrons in the conduction band.

The electrical conductivity σ is defined by*

$$\sigma_i = \frac{I}{-\left(\frac{\partial \phi}{\partial x}\right)} = -B_i n_i (z_i e)^2, \qquad \text{ohm}^{-1}\text{-cm}^{-1}. \tag{13.46}$$

Now recalling Eq. (13.30)

$$D_i = B_i \kappa_B T,$$

we can relate the electrical conductivity and diffusivity by the simple expression:

$$\frac{\sigma_i}{D_i} = \frac{n_i (z_i e)^2}{\kappa_B T}, \tag{13.47}$$

where $\kappa_B = 1.38 \times 10^{-23}$ joules-deg^{-1}. This equation is known as the "extended" Nernst-Einstein equation.

In order to predict D_i from conductivity measurements, we must know the fraction of the total conductivity due to species i; this fraction is called the transference number of species i, denoted t_i. This number, obtained from electrolysis experiments, gives the fraction of the total current carried by a particular species, and the total of all transference numbers must equal one. Therefore

$$\sum_{\substack{\text{cation}\\\text{species}}} t_{\text{cation }i} + \sum_{\substack{\text{anion}\\\text{species}}} t_{\text{anion }i} + \sum t_{\text{electron}} = 1. \tag{13.48}$$

Thus

$$\sigma_i = t_i \sigma_{\text{total}}, \qquad \text{ohm}^{-1}\text{-cm}^{-1}. \tag{13.49}$$

Eq. (13.47) has been tested for several compounds. For example, in NaCl, $t_{\text{Na}^+} \cong 1$, so that $\sigma_{\text{measured}} \cong \sigma_{\text{Na}^+}$. Therefore, if we assume that all of the sodium ions in the crystal participate in the conduction process, then

$$D_{\text{Na}} = \frac{\sigma_{\text{total}} \kappa_B T}{n_{\text{Na}^+} e^2}, \tag{13.50}$$

where n_{Na^+} is the number of sodium ions per cubic centimeter. In order to test this, Mapother, Crooks, and Mauer[6] measured the conductivity of NaCl and then, using radioactive sodium, measured diffusion coefficients of sodium, both as functions of temperature. The values of D_{Na} calculated by means of Eq. (13.47) and the D_{Na}^T values are compared in Fig. 13.13. The agreement between the two

---

* Note that in the physics literature, $\sigma = ne\mu$, where μ is called the mobility. Since this is applied to electron or electron-hole conduction, $z = -1$ or $+1$, and therefore, $\mu = Be$. In this case, the units of σ are ohm^{-1} cm^{-1}, n are cm^{-3}, $e = 1.6 \times 10^{-19}$ coulombs, and $\mu = $ cm^2 sec^{-1} volt^{-1}.

[6] D. Mapother, H. N. Crooks, and R. Mauer, *J. Chem. Phys.* **18**, 1231 (1950).

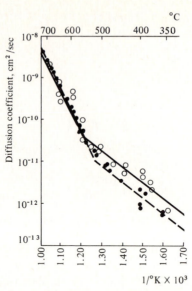

Fig. 13.13 Log D versus $1/T$ for sodium in NaCl as determined with radioactive sodium (o), and as calculated from the conductivity (•). (From D. Mapother, H. N. Crooks, and R. Mauer, *J. Chem. Phys.* **18**, 1231 (1950).)

values is excellent at the higher temperatures. The discrepancy at the lower temperatures points out the effect of $CdCl_2$ present as an impurity in NaCl in dilute concentration. By adding $CdCl_2$, $[V_{Na}]$ increases, thereby increasing D_{Na} over that in the pure material. However, the conductivity expression does not include this added effect, and the values differ when the concentration of vacancies due to $CdCl_2$ is significant, compared to the intrinsic concentration of vacancies.

Non-stoichiometry, resulting in vacancies on the cation sublattice, can also have an effect on D_i^* (and through Eqs. (13.33) or (13.35) on $\tilde{D}$) in the same manner as impurities, as illustrated in the following Example.

Example 13.1 Given the chemical diffusion coefficient $\tilde{D}$ as determined from electrical conductivity measurements and the thermodynamics of the Co_xO system, calculate the diffusion coefficient of Co^{++} ions and compare it with the measured diffusion coefficient D_{Co}. This is a problem studied by Price and Wagner.[7]

Solution. CoO exists as a non-stoichiometric oxide with defects on the cation sublattice in the form of vacancies. This amounts to the addition of an oxygen anion without addition of a cation, according to the reaction

$$\tfrac{1}{2}O_2(g) = O_O + V_{Co}^+ + \oplus,$$

where O_O indicates an excess oxygen ion on a regular oxygen site, V_{Co}^+ a singly-ionized cation vacancy, and $\oplus$ a free electron hole.

---

[7] J. B. Price and J. B. Wagner, Jr., *Zeit. f. physik. Chemie, Neue Folge* **49**, 3–4 (1966).

Presumably, diffusion of Co^{++} ions should be via a vacancy mechanism on the cation sublattice, with diffusivity increasing as the concentration of cation vacancies increases. The phase diagram (Fig. 13.14) shows the Co–CoO equilibrium. When P_{O_2} is increased above that existing at this lower boundary of the CoO phase field, the molar concentration of Co decreases as oxygen is added to the oxide, until finally an oxygen pressure is reached at which $X_{Co} = 0.492$ (corresponding to $Co_{0.969}O$) beyond which Co_3O_4 forms. This occurs at $P_{O_2} = 0.53$ atm at 1000°C. Price and Wagner measure chemical diffusion coefficients in CoO by equilibrating a single crystal of CoO with an atmosphere at low P_{O_2} (4.7×10^{-2} atm); then, at time zero, they change the oxygen potential to a higher level, near the CoO–Co₃O₄ boundary. The change in $X_O (= -X_{Co})$ is measured as a function of time by following the change in electrical conductivity, since the conductivity in this material is proportional to the concentration of electron holes [$\oplus$], which is dependent, in turn, on the oxygen content through the equilibrium constant for the reaction shown above;

$$K = \frac{[\oplus][V_O^+]}{P_{O_2}^{1/2}}.$$

Although $\tilde{D}$ is generally composition dependent in non-stoichiometric crystals, in this case the deviations from non-stoichiometry are small, and $\tilde{D}$ is found to be essentially constant with composition:

$$\tilde{D} = 4.33 \times 10^{-3} \exp(-24{,}000/RT).$$

Since the cobalt cation is much smaller than the oxygen anion, D_{Co} is very much

Table 13.2 Diffusivities in nonmetallic crystals

Diffusing ion	Crystal in which diffusion takes place	D_0, cm²/sec	Q, cal/mol
Ag^+	α-Cu_2S	38×10^{-5}	4,570
Cu^+	α-Ag_2S	12×10^{-5}	3,180
Ag^+	α-Cu_2Te	2.4	20,860
Cu^+	α-AgI	16×10^{-5}	2,250
Li^+	α-AgI	50×10^{-5}	4,570
Se^{--}	α-Ag_2S	17×10^{-5}	20,040
Pb^{++}	$PbCl_2$	7.8	35,800
Pb^{++}	PbI_2	10.6	30,000
O^{--}	Fe_2O_3	$1 \times 10^{+11}$	146,000
Fe^{+++}	Fe_2O_3	$4 \times 10^{+5}$	112,000
Co^{++}	CoO	2.15×10^{-3}	34,500
Ni^{++}	NiO	1.83×10^{-3}	45,900
O^{--}	NiO	1.0×10^{-5}	54,000
Cr^{+++}	Cr_2O_3	0.137	61,100

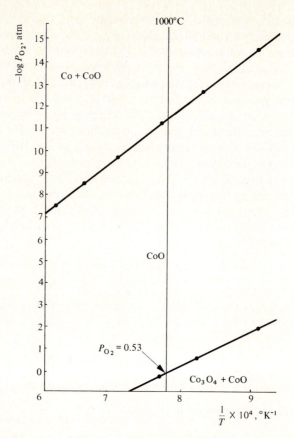

Fig. 13.14 The stable phase field of cobaltous oxide with respect to oxygen pressure and temperature.

greater than D_O in CoO. Then we may use Eq. (13.33), neglecting the term $D_O^* X_{Co}$, if we further assume that $B_{Co} = B_{Co}^*$ in order to calculate D_{Co}^* from the $\tilde{D}$ data:

$$\tilde{D} \cong D_{Co}^* X_O \left(\frac{\partial \ln a_{Co}}{\partial \ln X_{Co}} \right).$$

We may calculate the thermodynamic factor by noting that

$$\left(\frac{\partial \ln a_{Co}}{\partial \ln X_{Co}} \right) = - \left(\frac{\partial \ln a_O}{\partial \ln X_{Co}} \right),$$

and that

$$a_O = \left(\frac{P_{O_2}}{P_{O_2}^0} \right)^{1/2},$$

where $P_{O_2}^0$ is the standard oxygen pressure, 1 atm. We obtain the slope of the log a_O versus log X_O plot from the data of Eror and Wagner,[8] who measure $[V_{Co}]$ as a function of P_{O_2} across the CoO phase field:

$$-\left(\frac{\partial \ln a_O}{\partial \ln X_{Co}}\right) \cong -\left(\frac{\Delta \log a_O}{\Delta \log X_{Co}}\right) = -\left[\frac{\frac{1}{2}(\log P_{O_2,\text{initial}} - \log P_{O_2,\text{final}})}{\log X_{Co,\text{initial}} - \log X_{Co,\text{final}}}\right]$$

$$= -\left[\frac{\frac{1}{2}(\log(4.7 \times 10^{-2}) - \log(5.1 \times 10^{-1})}{\log(0.4963) - \log(0.4936)}\right]$$

$$= +1.45 \times 10^2.$$

At 1000°C, $\tilde{D} = 2.16 \times 10^{-7}$ cm^2/sec, and $X_O \cong 0.50$. Substituting into Eq. (13.33) as simplified above, we find D_{Co}^* to be 2.98×10^{-9} cm^2/sec. Applying the correlation coefficient correction, with $f = 0.78$ for the fcc cation sub-lattice, then

$$D_{Co}^T = 0.78 \, D_{Co}^* = 2.33 \times 10^{-9} \text{ cm}^2/\text{sec} \qquad \text{(calculated)},$$

which is in good agreement with Carter and Richardson's[9] measured value

$$D_{Co}^T = 2.60 \times 10^{-9} \text{ cm}^2/\text{sec} \qquad \text{(measured)}.$$

13.4 DIFFUSION IN LIQUIDS

As in the previous chapters, in which the transport properties of liquids were discussed, we find the problem of predicting, correlating, and extrapolating diffusion data in liquids very difficult because of our lack of understanding of the structure of liquids. The difficulties of making accurate experimental determinations of diffusion coefficients in liquids due to natural convection, and the problems of sampling, further complicate the diffusion problem. In the following sections, we shall discuss various theories of the liquid state as they relate to diffusion, and then present data for a variety of liquids.

13.4.1 Liquid state diffusion theories

Hydrodynamical theory. The earliest theoretical equation for prediction of self-diffusion in a liquid is that of Einstein.[10] In this theory, he assumes that the diffusing species are non-reacting spherical particles of radius R moving through a continuous medium of viscosity η with steady-state velocity V_∞. According to Eq. (2.123), the force on a sphere moving at steady-state in laminar flow is

$$F = 6\pi R \eta V_\infty, \qquad\qquad (2.123)$$

---

[8] N. Eror and J. B. Wagner, Jr., *J. Phys. Chem. Solids* **29**, 1597 (1968).

[9] R. E. Carter and F. D. Richardson, *J. Metals* **200**, 1244 (1954).

[10] A. Einstein, *Z. Electrochem.* **17**, 235 (1908).

and since the definition of the mobility B is V_∞/F, we have

$$B = \frac{1}{6\pi R\eta}.$$ (13.51)

Combining this result with Eq. (13.30), we get

$$D = \frac{\kappa_B T}{6\pi R\eta},$$ (13.52)

known as the Stokes-Einstein equation. This has been modified by Sutherland,[11] who adopted a slightly different form of the drag force, with the result that

$$D = \frac{\kappa_B T}{4\pi R\eta}.$$ (13.53)

Although the use of Stokes' law for this purpose may seem unjustified, since it neglects interatomic forces and considers only form drag, the results, in terms of predicted data, are remarkable. For example, in liquid metals, if we substitute the measured values of D and η into Eq. (13.53) and then calculate the radii of the diffusing species, we obtain very good agreement between these predicted radii and the radii of ions, as shown in Table 13.3.

Table 13.3   Comparison of calculated and measured radii for self-diffusion in liquid metals

Element	Temperature range, °C	Sutherland model	Measured radii*
Na	110–630	1.51–4.53 Å	1.57 Å
Hg	274–364	1.20–1.24	1.44
In	448–1013	1.36–3.06	1.50
Ag	977–1397	1.37–1.37	1.34
Zn	394–837	1.29–0.75	1.25
Sn	575–956	1.30–2.22	1.41

* The radii used for comparison are those reported by Pauling for single-bond metallic situation.[12]

The Sutherland equation is quite successful in predicting self-diffusion data for a wide variety of other substances, including molten semiconductors, polar liquids, associated liquids, and molten sulfur. If we assume that the atoms in the liquid are in a cubic array, then for materials generally the simple theory predicts that

$$2R = (\hat{V}/N_0)^{1/3},$$

and

$$D^* = \frac{\kappa_B T}{2\pi} \left(\frac{N_0}{\hat{V}}\right)^{1/3},$$

where $\hat{V}$ is the molar volume.

[11] W. Sutherland, *Philosophical Magazine* **9**, 781 (1905).
[12] L. Pauling, *Nature of the Chemical Bond*, Cornell Univ. Press, Ithaca, New York, 1960.

Hole theory. The oldest *structural* picture of a liquid is the hole theory which presumes the existence of holes or vacancies randomly distributed throughout the liquid and providing ready diffusion paths for atoms or ions. The concentration of these holes would have to be very great in order to account for the volume increase upon melting, thus resulting in much higher diffusion rates in liquids than in solids just below the melting point. This jump in D on melting is obvious in Fig. 13.5. The hole theory, however, does not result in a prediction of diffusion coefficients by itself, although it has been used to estimate the activation energy for self-diffusion in a liquid, by assuming that this energy is equal to that required to form a hollow sphere (hole) of a diameter on the order of a few angstroms.

Eyring theory. Eyring *et al.*[13] have applied their activated state theory of diffusion, which works reasonably well in solids, to liquid diffusion, assuming that the migrating atoms move from hole to hole in the liquid by a process of discrete jumps. If the liquid is considered quasi-crystalline, and the atoms are in a cubic configuration, then an expression relating D^* and η results:

$$D^* = \frac{\kappa_B T}{2R\eta}.$$ (13.54)

This expression obviously predicts D^* values six times larger than that given by the Sutherland equation. Subsequent revisions of this equation by Li and Chang,[14] and by Eyring *et al.*[15] have improved the agreement with the Sutherland equation, but the modifications suggested do not shed any more significant light on the structure of liquids, nor are they any more useful from a prediction standpoint.

Fluctuation theory. Because the ideas of a well-defined "activated state" as well as of discrete "holes" in the liquid have been difficult to accept, Cohen and Turnbull,[16] and later, Swalin[17] and Reynik[18] have developed a new theory known as the fluctuation theory. In this theory, the "extra" volume in the liquid (over that of the solid) is distributed evenly throughout the liquid, making the *average* nearest neighbor distance increase. One may imagine a diffusing atom contained in a cage whose dimensions are constantly fluctuating. Local fluctuations in density then occasionally open up holes or openings in the cage large enough to allow the atom to diffuse out of the cage. In their approach, Cohen and Turnbull consider that a critical fluctuation must occur before diffusion occurs, whereas Swalin and Reynik both think that a spectrum of fluctuations occurs, with cooperative motion of the

[13] H. Eyring, S. Glasstone, and K. Laidler, *Theory of Rate Processes*, McGraw-Hill, New York, 1941.

[14] J. C. M. Li and P. Chang, *J. Chem. Physics* **23**, 518 (1955).

[15] H. Eyring, T. Ree, D. Grant, and R. Hirst, *Z. Elektrochem.* **64**, 146–152 (1962).

[16] M. H. Cohen and D. Turnbull, *J. Chem. Physics* **31**, 1164 (1959).

[17] R. A. Swalin, *Acta Met.* **7**, 736 (1959).

[18] R. J. Reynik, *Trans. AIME* **245**, 75 (1969).

neighboring atoms resulting in diffusive motion of an atom. No activation energy is required, since there is virtually no difference between an atom moving forward as a result of cooperative movements and an atom not moving. The result of Swalin's theory* is that D exhibits a linear dependence on temperature.

Reynik has proposed a small fluctuation model which also predicts a linear temperature dependence of D on T (°K):

$$D = a + bT, \qquad cm^2 \ sec^{-1}, \tag{13.55}$$

where $a = 1.72 \times 10^{24} Z X_O^4 K$, $b = 2.08 \times 10^9 Z X_O$, Z = number of nearest neighbors, X_O = maximum diffusive displacement due to normal vibrations, Å, and K = a force constant.

Since
$$X_O = \bar{d} - 2R, \tag{13.56}$$

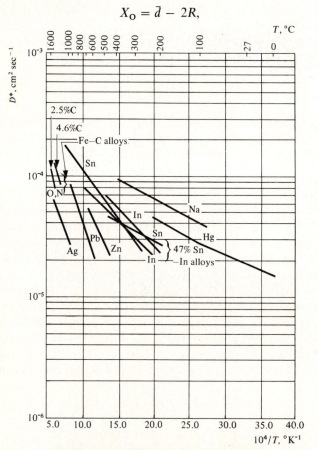

Fig. 13.15 Self-diffusion data in liquid metals.

---

* As corrected by R. J. Reynik, *Applied Physics Letters* **9**, 239 (1966).

[19] R. Swalin and V. G. Leak, *Acta Met.* **13**, 471 (1965).

where $\bar{d}$ = average interatomic distance in the liquid, Å and R = ion radius, Å, the evaluation of this theory relies on the use of x-ray data for $\bar{d}$ and Z, and b-values obtained from correlations of existing data using least squares data fitting techniques to fit an equation of the form of Eq. (13.55). From the calculated values of X_O and Eq. (13.56), R-values are obtained, which agree quite closely with Pauling's neutral atom radii, suggesting that in liquid metals the diffusing ion core carries its valence electrons with it.

The problem of all the theoretical approaches is that the tests of the theories rely on the test of the functional relationship between D and T. For example, while Reynik can reasonably fit liquid metal diffusion data to a linear relationship, Saxton and Sherby[20] have tested a large amount of liquid metal data and concluded that diffusion in liquid metals is a thermally activated process, obeying an Arrhenius relationship (Eq. 13.36), which is clearly at odds with a linear relationship with T. The activation energies Q are apparently not large, and so, when the RT term is on the same order of magnitude as Q, D is not a strong function of temperature. Thus no critical tests of these models will really be possible until more accurate data over much wider ranges of temperature are available. Until then, we shall have to rely on measured values and intelligent "guestimates."

13.4.2 Liquid diffusion data

The remarkable characteristic of diffusion in liquid metals, and in liquids in general, is the fact that the values of D are almost all approximately the same order of magnitude, 10^{-4}–10^{-5} cm²-sec^{-1}, even though their solid state properties differ widely. Furthermore, the activation energies are also approximately equal, usually in the range 1000–4000 cal/mol. Because the data are usually presented in the Arrhenius form and since there is still no clearcut evidence against this representation, we shall present the data in this section in the same way.

Diffusion in liquid metals. Figure 13.15 gives some data for self-diffusion in pure liquid metals, Fig. 13.16 for interdiffusion in common nonferrous binary alloys, and Fig. 13.17 for ferrous alloys. We refer the reader to Edwards et al.,[21] Yang and Derge,[22] or Elliott et al.[23] for more complete listings of the available data.

Diffusion in molten salts and slags. Again, we know so little of the detailed structure of molten halides, sulfides, oxides, and silicates, that no really satisfactory quantitative expression is available for predicting diffusivities in these materials. However, if we refer to Chapter 1 in which we discussed the relationship between bonding and viscosity, we realize that the same semiquantitative arguments can be

---

[20] H. Saxton and O. Sherby, *Trans. ASM* **55**, 826 (1962).

[21] J. B. Edwards, E. E. Hucke, and J. J. Martin, *Metallurgical Reviews* **13**, No. 120, 1 (1968).

[22] L. Yang and G. Derge, paper in *Physical Chemistry of Process Metallurgy, AIME* **7**, 195 (1961).

[23] J. F. Elliott, M. Gleiser, and V. Ramachrishna, *Thermochemistry for Steelmaking* **II**, Addison-Wesley, 1963.

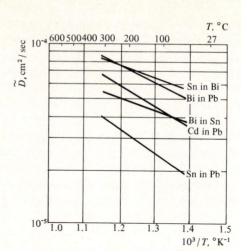

Fig. 13.16 Interdiffusion coefficients in liquid nonferrous alloys.

used with regard to diffusion in these materials, particularly in light of the Stokes-Einstein model of diffusion.

Consider first the molten salts. We may describe their structure as holes within a molecular assembly that has density fluctuations and short-range order, but with the added restriction of electroneutrality, so that the cation and anion charge densities must remain equal in any region of the melt. Diffusion coefficients measured in molten salts are remarkably close to values observed in liquid metals, even though there is an obvious difference in bonding. Molten salts have highly polarized ionic bonds, in contrast to the metallic bond. Table 13.4 gives some typical data for self-diffusion in molten salts. Note that the activation energies for the salts are, however, larger than the activation energies for self-diffusion in metals.

In molten slags, the picture is more complicated; from the relation between the degree of polymerization of the silicates and their viscosity, we can postulate that

Table 13.4 Self-diffusivities in molten salts†

Diffusate	Melt	Temp. range, °C	$D_0 \times 10^4$, cm^2-sec^{-1}	Q, cal-mole^{-1}	Typical D^* value, cm^2-sec^{-1}	°C
Na	NaCl	845–916	8	4,000	14.2×10^{-5}	906
Cl	NaCl	825–942	23	7,100	8.8×10^{-5}	933
Na	NaNO$_3$	315–375	12.88	4,970	2.00×10^{-5}	328
NO$_3$	NaNO$_3$	315–375	8.97	5,083	1.26×10^{-5}	328
Tl	TlCl	487–577	7.4	4,600	3.89×10^{-5}	502
Zn	ZnBr$_2$	394–650	790	16,060	0.22×10^{-5}	500

† Data from L. Yang and G. Derge, *ibid.*

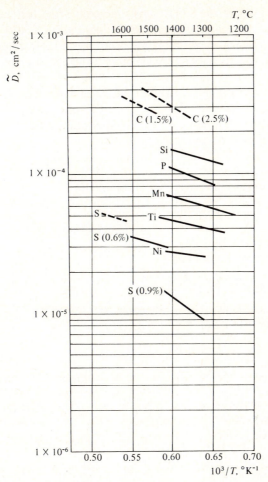

Fig. 13.17 Interdiffusion coefficients in ferrous liquid alloys (——— carbon-saturated; ------pure Fe).

the diffusivity of oxygen, for example, is lower in the more acid slags, and increases rapidly as the concentration of free oxygen ions increases near the orthosilicate composition. On the other hand, the diffusivity of basic cations should not depend on the slag composition to such an extent.

    There are relatively few diffusion data available for molten slags; most of them are presented in Table 13.5. The curious result, which has not yet been satisfactorily explained, is that the tracer self-diffusion coefficient of oxygen in a basic slag is greater than the self-diffusion coefficients of calcium in the same slag. It is also greater than that of silicon and aluminum, but we may explain this on the basis that silicon is tied up in SiO_4^{4-} ions, which clearly do not diffuse as easily as the smaller O^{2-} ions; aluminum is probably tied up in AlO_3^{3-} ions, which diffuse only a fraction more easily than the SiO_4^{4-} ions. Note also that the activation energies are now quite large, in correspondence to the increased difficulty of movement of these complex molecules, presumably requiring much more complicated relative movements of nearest neighbor ions in order to affect any net movement of a molecule.

Diffusion in common liquids. Table 13.6 gives some data for diffusivities in aqueous and organic solvents. Again, note that they are all on the order of 10^{-5} cm^2-sec^{-1}, even though there are obvious differences in bonding between these liquids and liquid metals or slags.

Table 13.5 Self-diffusivities in molten silicates and sulfides†

System	Temp. range, °C	$D_0 \times 10^4$, cm^2-sec^{-1}	Q, cal-mole^{-1}	Typical D^* value, cm^2-sec^{-1}	°C
Ca in CaO–Al$_2$O$_3$–SiO$_2$ (40/20/40)	1350–1450	—	70,000	0.062×10^{-5}	1400
Ca in CaO–Al$_2$O$_3$–SiO$_2$ (39/21/40)	1350–1540	—	70,000	0.067×10^{-5}	1400
Ca in CaO–Al$_2$O$_3$–SiO$_2$ (40/20/40)	1350–1450	—	70,000	0.067×10^{-5}	1400
Si in CaO–Al$_2$O$_3$–SiO$_2$ (40/20/40)	1365–1460	—	70,000	0.01×10^{-5}	1430
O in CaO–Al$_2$O$_3$–SiO$_2$ (40/20/40)	1370–1520	—	95,000	0.4×10^{-5}	1430
Fe in CaO–Al$_2$O$_3$–SiO$_2$ (30/15/55)	1500	—	—	0.24–0.31×10^{-5}	—
Fe in CaO–Al$_2$O$_3$–SiO$_2$ (43/22/35)	1500	—	—	0.21–0.50×10^{-5}	—
Fe in FeO–SiO$_2$ (61/39)	1250–1305	—	40,000	9.6×10^{-5}	1275
P in CaO–Al$_2$O$_3$–SiO$_2$ (40/21/39)	1300–1500	—	46,600	0.2×10^{-5}	1400
Fe in Fe–S (33.5%)	1150–1238	65.7	13,600	5.22×10^{-5}	1152
Fe in Fe–S (31.0%)	1164–1234	298	17,000	6.39×10^{-5}	1180
Fe in Fe–S (29.0%)	1158–1254	20200	27,700	11.91×10^{-5}	1158
Fe in Fe (48.1%)–Cu (20.5%)–S (31.9%)	1160–1250	1440	21,500	7.57×10^{-5}	1160
Fe in Fe (32.0%)–Cu (40.0%)–S (28.0%)	1168–1244	35.7	13,700	2.94×10^{-5}	1168
Cu in Cu–S (19.8%)	1160–1256	69.3	12,800	7.49×10^{-5}	1160
Cu in Fe (32.0%)–Cu (40.0%)–S (28.0%)	1160–1245	564	19,700	5.52×10^{-5}	1160

† Data from L. Yang and G. Derge, *ibid.*

Table 13.6 Diffusivities in common liquids at 25°C

Solute	Solvent	Concentration	D, cm^2-sec^{-1}
HCl	Water	0.1M	3.05×10^{-5}
NaCl	Water	0.1M	1.48×10^{-5}
CaCl$_2$	Water	0.1M	1.10×10^{-5}
H$_2$	Water	Dilute	5.0×10^{-5}
O$_2$	Water	Dilute	2.5×10^{-5}
SO$_2$	Water	Dilute	1.7×10^{-5}
NH$_3$	Water	Dilute	2.0×10^{-5}
Cl$_2$	Water	Dilute	1.44×10^{-5}
H$_2$SO$_4$	Water	Dilute	1.97×10^{-5}
Na$_2$SO$_4$	Water	0.01 M	1.12×10^{-5}
K$_4$Fe(CN)$_6$	Water	0.01 M	1.18×10^{-5}
Ethanol	Water	$X = 0.05$	1.13×10^{-5}
Glucose	Water	0.39%	0.67×10^{-5}
Benzene	CCl$_4$	Dilute	1.53×10^{-5}
CCl$_4$	Benzene	Dilute	2.04×10^{-5}
Br$_2$	Benzene	Dilute	2.7×10^{-5}
CCl$_4$	Kerosene	Dilute	0.96×10^{-5}
CO$_2$	Ethanol	Dilute	4.0×10^{-5}
CCl$_4$	Ethanol	Dilute	1.50×10^{-5}
Phenol	Ethanol	Dilute	0.89×10^{-5}

13.5 DIFFUSION IN GASES

Through the years, diffusion studies have been mostly concerned with measuring and predicting diffusion coefficients in gaseous mixtures. In gases, we are not faced with the problem of Kirkendall shift phenomena, since any tendency for more A molecules to go in one direction than B molecules in the opposite direction is immediately counteracted by the fact that a pressure gradient builds up; thus in gases, $D_A = D_B = D_{AB} = \tilde{D}$.† Of course, one must still superimpose the bulk flow term in Eq. (13.7) onto the Fick's first law expression to obtain the total flux if there are any convection effects present.

Chemists and chemical engineers have spent a great deal of time and effort in developing equations which predict diffusion coefficients in gases. For example, based on the kinetic theory of gases, the following equation applies to prediction of self-diffusivity of spherical A atoms diffusing in pure A:

$$D_{AA}^* = \frac{2}{3}\left(\frac{\kappa_B^3}{\pi^3 m_A}\right)^{1/2} \frac{T^{3/2}}{Pd^2}. \tag{13.57}$$

---

† Chemical engineers often use the notation D_{AB} to refer to the diffusion of A in an A–B mixture. This is the same as our $\tilde{D}$, which, if $D_A = D_B$, is also equal to D_A and to D_B.

For the interdiffusivity of two unequal size spherical atoms A and B, kinetic theory predicts

$$D_{AB} = \frac{2}{3}\left(\frac{\kappa_B^3}{\pi^3}\right)^{1/2}\left(\frac{1}{2m_A} + \frac{1}{2m_B}\right)^{1/2}\left(\frac{T^{3/2}}{P\left(\frac{d_A + d_B}{2}\right)^2}\right),$$ (13.58)

where κ_B = Boltzmann's constant, 1.38×10^{-16} ergs-molecule^{-1} °K^{-1},
   d = molecular diameter, cm,
   m = molecular mass, grams-molecule^{-1},
   P = pressure, dynes-cm^{-2},
   T = temperature, °K.
   Equations (13.57) and (13.58) give the proper pressure dependence up to about 10 atm; the predicted temperature dependence is only qualitatively correct in that D_{AB} actually varies more with temperature.[24]
   To predict D_{AB} more accurately, it is better to apply the Chapman-Enskog theory than the above equations. For monatomic gases, which act ideally, we have

$$D_{AB} = \frac{0.0018583 T^{3/2}}{P(\sigma_{AB})^2 \Omega_{D,AB}}\sqrt{\frac{1}{M_A} + \frac{1}{M_B}},$$ (13.59)

where $\sigma_{AB} = \frac{1}{2}(\sigma_A + \sigma_B)$ = collision diameter, Å,   (13.60)
   $\Omega_{D,AB}$ = collision integral for A–B mixture at dimensionless temperature, T^*_{AB}, for the Lennard–Jones potential.

$$T^*_{AB} = \left(\frac{\kappa_B}{\varepsilon}\right)_{AB} \cdot T,$$

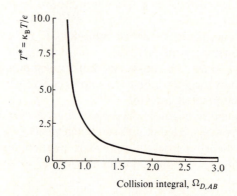

Fig. 13.18 Variation of collision integral with reduced temperature for the Lennard–Jones (6–12) potential.

[24] R. B. Bird, W. E. Stewart, and E. N. Lightfoot, *ibid.*

and

$$\left(\frac{\varepsilon}{\kappa_B}\right)_{AB} = \left[\left(\frac{\varepsilon}{\kappa_B}\right)_A \cdot \left(\frac{\varepsilon}{\kappa_B}\right)_B\right]^{1/2} = \text{average intermolecular force parameter, }^\circ K;$$

$$M_A, M_B = \text{molecular weights of species } A \text{ and } B;$$
$$T = \text{temperature, }^\circ K;$$
$$P = \text{pressure, atm.}$$

The values of the force parameters and the collision diameters for common gases are given in Table 1.1, and the collision integrals $\Omega_{D,AB}$ are given in Figure 13.18, from which we may calculate D_{AB} values for these gases. Note that the collision integral values for diffusion are only slightly different from those for gas viscosity calculations.

The force parameters for metal vapors are not well known, but Turkdogan[25] has estimated some of them. Remembering that Hirschfelder $et\ al.$ [26] showed that in the case of almost spherical molecules the reduced pressure P_r, volume V_r, and temperature T_r become almost constant at the critical points of the compound or element, he went back to earlier literature and found empirical equations which relate critical temperatures T_c, and volumes V_c, to characteristics such as boiling and melting temperatures, T_b and T_M, respectively. From these he developed several equations for estimating ε/κ_B and σ values for metal vapors. These have been used to develop Table 13.7 for σ_{AB} values, and we may use the following equations for ε/κ_B:

$$\varepsilon/\kappa_B = 1.15 T_b, ^\circ K, \tag{13.61}$$

$$\varepsilon/\kappa_B = 1.92 T_M, ^\circ K. \tag{13.62}$$

Although Eqs. (13.61) and (13.62) give somewhat differing values of ε/κ_B, the errors in D_{AB} values, particularly at temperatures greater than T_M, are not great, and the estimates for D_{AB} are valid within 20% of the true value.

Example 13.2 Calculate the diffusion coefficient for iron vapor diffusing through argon, at 1600°C, assuming that pure iron is the source of the vapor.

Solution. Using the Chapman-Enskog equation, we obtain

$$D_{\text{Fe–Ar}} = \frac{0.001858(1873)^{3/2}}{(1)(\sigma_{\text{Fe–Ar}})^2 \Omega_{D,\text{Fe–Ar}}} \sqrt{\frac{1}{55.85} + \frac{1}{39.54}}.$$

We evaluate $\sigma_{\text{Fe–Ar}}$ from the data in Tables 1.1 and 13.7:

$$\sigma_{\text{Fe–Ar}} = \frac{\sigma_{\text{Fe}} + \sigma_{\text{Ar}}}{2} = \frac{2.43 + 3.42}{2} = 2.92\,\text{Å}.$$

[25] E. T. Turkdogan, paper in $Steelmaking$, $The\ Chipman\ Conference$, J. F. Elliott (ed.), Addison-Wesley, Reading, Massachusetts, 1962.

[26] J. Hirschfelder, R. Curtiss, and R. B. Bird, $The\ Molecular\ Theory\ of\ Gases$, Wiley, New York, 1956.

Table 13.7 Estimated collision cross sections for metal vapors.

| Metal | Melting point, °C | Boiling point, °C | Molar volume, cm³/mol | | | σ, Å | |
			$\hat{V}$(sol.)	$\hat{V}$ (liq.)	$\hat{V}_b$(liq.)	From $\hat{V}$(sol.)	From $\hat{V}_b$(liq.)
Ag	960.8	2163	11.02	11.60	13.06	2.72	2.74
Al	660	2057	10.49	11.36	13.26	2.67	2.76
Bi	271	1477	—	20.76	24.07	—	3.37
Cd	321	765	13.34	14.01	14.89	2.90	2.87
Co	1493	2877	7.14	7.68	8.98	2.36	2.43
Cu	1083	2570	7.58	8.01	9.15	2.39	2.44
Fe	1535	2833	7.60	7.94	9.55	2.40	2.47
Ga	29.8	1983	—	11.43	13.87	—	2.80
Hg	−38.9	357	14.09	14.65	15.71	2.95	2.92
In	156.4	2087	15.94	16.33	19.98	3.07	3.16
K	63.7	760	46.0	47.10	59.36	4.37	4.54
Li	186	1317	13.27	13.40	16.76	2.89	2.98
Mg	651	1103	14.66	15.47	20.30	3.00	3.18
Na	97.9	883	24.04	24.79	31.07	3.52	3.66
Ni	1453	2816	7.11	7.57	9.02	2.34	2.43
Pb	327.4	1717	18.35	19.39	22.47	3.22	3.29
Pu	637	3300	14.20	14.49	18.78	2.95	3.11
Sb	630.5	1440	18.59	18.74	20.29	3.23	3.18
Sn	231.9	2770	16.51	17.04	21.63	3.11	3.25
Tl	303	1457	—	18.10	20.64	—	3.19
Zn	419.5	906	9.56	9.45	10.19	2.59	2.51

* From E. T. Turkdogan, *ibid.*

In order to evaluate the collision integral, we need to calculate $(\varepsilon/\kappa_B)_{\text{Fe–Ar}}$

$$\left(\frac{\varepsilon}{\kappa_B}\right)_{\text{Fe–Ar}} = \sqrt{\left(\frac{\varepsilon}{\kappa_B}\right)_{\text{Fe}}\left(\frac{\varepsilon}{\kappa_B}\right)_{\text{Ar}}} = \sqrt{(3521)(124)} = 655°\text{K}.$$

Then

$$T^* = \left(\frac{\kappa_B}{\varepsilon}\right)_{\text{Fe–Ar}} \cdot T = \frac{1873}{655} = 2.86,$$

and from Fig. 13.18, $\Omega_{D,\text{Fe–Ar}} = 0.92$, or

$$D_{\text{Fe–Ar}} = \frac{0.001858(1873)^{3/2}}{(1)(2.92)^2(0.92)} \cdot \sqrt{\frac{1}{55.85} + \frac{1}{39.54}} = 4.00 \text{ cm}^2/\text{sec}.$$

Many other correlations have been developed for nonmetallic gases. The most successful correlation, which makes the prediction of D_{AB} over a wide range of binary gas systems and temperatures possible, is that of Fuller et al.[27] They correlated 340 data points in a theoretically justified form, predicted D_{AB} to within 10% of the measured values 92.6% of the time, and had an average error of only 4.3% and a standard deviation of 6.17%. Their equation is

$$D_{AB} = \frac{(1 \times 10^{-3})T^{1.75}}{P(v_B^{1/3} + v_A^{1/3})^2} \sqrt{\frac{1}{M_A} + \frac{1}{M_B}}, \qquad cm^2/sec, \qquad (13.63)$$

where T = temperature, °K, P = pressure, atm, M_A, M_B = molecular weight of species A and B, gm/gm-mole, and v_A, v_B = diffusion volumes, given for simple molecules in Table 13.8. This approach is relatively simple and avoids the necessity to evaluate the collision integral.

Table 13.8 Diffusion volumes for simple molecules*

Molecule	v
H_2	7.07
D_2	6.70
He	2.88
N_2	17.9
O_2	16.6
Air	20.1
Ne	5.59
Ar	16.1
Kr	22.8
CO	18.9
CO_2	26.9
N_2O	35.9
NH_3	14.9
H_2O	12.7
Cl_2	37.7
Br_2	67.2
SO_2	41.1

* From E. N. Fuller, P. D. Schettler, and J. C. Giddings, *ibid.*

Values of D in gas phases are usually in the range 0.1–10.0 cm^2/sec. Figure 13.19 gives some representative data.

13.6 DIFFUSION THROUGH POROUS MEDIA

As we have pointed out in previous chapters, metallurgists often encounter situations where they must deal with the properties of solids containing some degree of porosity. Gas diffusion through porous media is important in such fields as ore

---

[27] E. N. Fuller, P. D. Schettler, and J. C. Giddings, *Indust. and Eng. Chem.* **58**, 19 (May 1966).

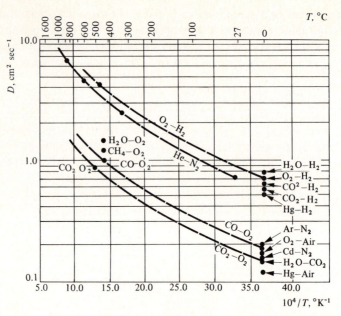

Fig. 13.19 Diffusion coefficients in gas mixtures.

reduction or roasting, vapor penetration into foundry sands, outgassing of powder metallurgy compacts, gas phase alloying of powders, and catalysis.

In general, gas phase diffusion through porous media occurs by one of two mechanisms: ordinary diffusion or Knudsen diffusion. (A third mechanism, surface diffusion, may be important at low temperatures, but for most elevated temperature processes, it is negligible.) The size of the pores through which diffusion is taking place determines whether ordinary or Knudsen diffusion predominates. With ordinary diffusion, the pores are large *relative to the mean free path* of the gas molecules. Knudsen diffusion is important when the pores are small in relation to the mean free path.

When ordinary diffusion prevails, the total diffusion path, from one plane in the porous media to a parallel plane, is longer than when no aggregate is present, because of the irregularity of the porosity. Even after correcting for the decrease in cross-sectional area available for diffusion, using the void fraction, ω, we must make a further decrease in the D_{AB} value, in order to yield the *effective interdiffusivity*. We accomplish this by introducing a new factor, the *tortuosity*, so that

$$D_{AB,\text{eff}} = \frac{D_{AB}\omega}{\tau}. \tag{13.64}$$

The tortuosity is a number greater than one, ranging from values of 1.5–2.0 for unconsolidated particles, to as high as 7 or 8 for compacted particles. It is not a function of the void fraction, but does depend on the particle size, particle size distribution, and particle shape. It is not possible to predict τ on the basis of material parameters, but reasonable estimates may be made by comparison with previous results using similar materials, such as presented in Fig. 13.20.

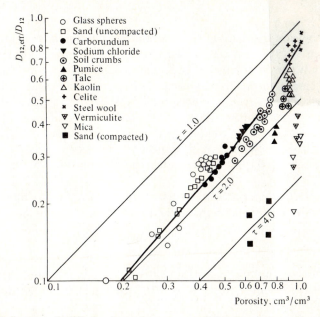

Fig. 13.20 Hydrogen–air diffusion in various unconsolidated porous media. (Adapted from L. N. Satterfield and T. K. Sherwood, *The Role of Diffusion in Catalysis*, Addison-Wesley, Reading, Massachusetts, 1963.)

When the gas density is low, or the pore size small, then the molecules have a higher probability of collision with the pore walls than with each other, similarly as in the flow of molecules in a vacuum system. An analysis of the flow of molecules through a cylindrical pore of radius r under a concentration gradient yields

$$j_A = \tfrac{2}{3} r \bar{V} \left(\frac{dC_A}{dx} \right).$$

By using Eq. (1.4) for $\bar{V}$, the average speed of the molecules, and comparing it with Fick's first law, we obtain the Knudsen diffusion coefficient:

$$D_K = 9700 r \sqrt{\frac{T}{M}}, \qquad cm^2/sec, \tag{13.65}$$

where r = pore radius, cm, T = temperature, °K, and M = molecular weight of diffusing species A. Since collisions between species are negligible at low pressures, this equation applies equally to any component present.

There are two tests that we may use to determine when ordinary diffusion no longer predominates and Knudsen diffusion becomes important. In the first test, we may estimate the pore radius and compare it to the mean free path of the gas λ, given by

$$\lambda = (\sqrt{2}\pi d^2 n)^{-1}, \qquad (1.5)$$

where d = collision diameter of the molecule, cm, and n = concentration of the molecules, cm^{-3}. If r and λ are of the same order of magnitude, or if the pore diameter is only one order larger, then Knudsen diffusion is presumably important. At elevated temperatures, this may occur when pore diameters are on the order of 1000 Å or less, that is, even when the molecules themselves are still quite small compared to the pore dimensions.

The other test involves comparing the calculated values of D_{AB} and D_K. If the ratio D_{AB}/D_K is small, then we take ordinary diffusion as the rate determining step. Conversely, if D_{AB}/D_K is large, then we presume that Knudsen diffusion controls.

Since even the micropores are not straight and smooth, we must also define tortuosity for those systems where Knudsen diffusion predominates:

$$D_{K,\text{eff}} = \frac{D_K \omega}{\tau}. \qquad (13.66)$$

The results of a large number of tests at relatively low temperatures where tortuosity has been calculated for systems in which Knudsen diffusion predominates are shown in Table 13.9.

Example 13.3 Consider the diffusion of metal vapors into molding sand when steel is cast. Calculate the diffusion coefficient for manganese vapor through silica sand at 1600°C, assuming that the only other gas present is argon.

Solution. The first step is to test for the importance of Knudsen diffusion. Calculating $D_{\text{Mn-Ar}}$, as in Example 13.2, we obtain 3.4 cm^2/sec for Mn at 1600°C. Then we calculate D_K, using Eq. (13.65) with r taken as 0.005 cm, and a reasonable inter-particle distance for sand grains on the order of 0.05 cm in diameter

$$D_K = (9700)(0.005)\sqrt{\frac{1873}{55}} = 283 \text{ cm}^2/\text{sec}.$$

Then

$$\frac{D_{\text{Mn-Ar}}}{D_K} = \frac{3.4}{2.83} = 0.0119,$$

which implies that in this situation ordinary diffusion controls.

As a further check, we calculate the mean free path of Mn vapor where $n = 4.48 \times 10^{-4}$ atoms/Å^3, $d = 2.4$ Å, and $\lambda = 1/\sqrt{2}\pi(4.48 \times 10^{-4})(5.76) = 87$Å. Based on the fact that the pore size is on the order of 10^5 Å and λ_{Mn} is much less

Table 13.9   Knudsen diffusion and flow in porous media*

Material	Technique	Gases	T, °K	r, Å	τ
Alumina pellets	Diffusion	N_2, He, CO_2	303	96	0.85
Fresh and regenerated silica–alumina cracking catalyst	Flow	H_2, N_2	298	31–50	2.1
Vycor glass	Flow	H_2, He, A, N_2	298	30.6	5.9†
Water-gas shift catalyst	Diffusion	O_2, N_2	298	177	2.7
Ammonia synthesis catalyst	Diffusion	O_2, N_2	298	203	3.8
Vycor glass	Flow	He, Ne, H_2, A, N_2, O_2, Kr, CH_4, C_2H_6	292, 294	30	5.9†
Vycor glass	Flow	H_2, He, A, N_2	298	50	5.9†
Silica–alumina cracking catalyst, various treatments	Reaction	Gas oil	755	28–71	3–10
Vycor glass	Flow	A, N_2	298	46	5.9†
Silica–alumina cracking catalyst	Flow	He, Ne, A, N_2	273–323	16	0.725
Silica–alumina cracking catalyst	Flow	He, Ne, N_2	273–323	24	0.285
Vycor glass	Flow	He, CO_2, N_2, O_2, A	298	44	5.9†
Silica–alumina cracking catalyst	Reaction	Cumene	420	(24)	5.6
Nickel–alumina**	Reaction	o-, p-H_2	77	(30)	1.8

* From L. N. Satterfield and T. K. Sherwood, *ibid.*
† Average value for the five sets of data on vycor.
** Data were obtained in the transition region. $D_{K,\text{eff}}$ was calculated from the observed D_{eff}.

than that, we can quite safely conclude that Knudsen diffusion is unimportant in this instance, and that we may analyze the problem in terms of ordinary diffusion.

It has been established[28] that τ for metal vapor diffusion at high temperatures through porous compacts has a range 3–6, with 4 considered to be a good average value. Therefore, for a sand mold with a void fraction of 0.45, we have

$$D_{\text{Mn–Ar,eff}} = \frac{D_{\text{Mn–Ar}}\omega}{\tau}$$

[28] J. M. Svoboda and G. H. Geiger, *Trans. AIME* **245**, 2363 (1969).

$$= \frac{(3.4)(0.45)}{(4.0)}$$

$$= 0.383 \text{ cm}^2/\text{sec},$$

or approximately one-tenth the value in the gas phase alone.

PROBLEMS

13.1 Show that the units in Eq. (13.1) are as indicated. Do the same for Eqs. (13.2) and (13.3).

13.2 Discuss the reasons why self-diffusion data must apply to homogeneous materials only.

13.3 Read one of the references of Footnote 1 and derive Eq. (13.5).

13.4 Look up the article by Compaan and Haven (Table 13.1) and summarize the method used to derive the correlation coefficients in Table 13.1.

13.5 Work out the units in Eq. (13.20).

13.6 Derive Eq. (13.37), after reading Zener's article.

13.7 Using the method of Fuller, Schettler, and Giddings, estimate the diffusion coefficient for a CO_2-O_2 gas mixture at 1-atm pressure and 700°K and compare the result to the data in Fig. 13.19.

13.8 Assuming a tortuosity of 2.0 and a void fraction of 0.25, calculate the effective diffusion coefficient for CO-CO_2 in a reduced iron oxide pellet at 800°K. Assume $r = 5 \times 10^{-3}$cm.

13.9 PbS has an NaCl-type structure. Would you expect the self-diffusion coefficient of Pb or S to be higher? How would you expect the addition of Ag_2S to affect the diffusion coefficient of Pb, knowing that the defects in PbS are predominantly Frenkel defects on the Pb sublattice, and that the undoped PbS is an n-type semiconductor? How would Bi_2S_3 additions affect it? [Ref.: G. Simovich and J. B. Wagner, Jr., *J. Chem. Phys.* **38**, 1368 (1963).]

13.10 Describe the conditions under which the following terms are applicable:
(1) self-diffusion,
(2) tracer diffusion,
(3) chemical diffusion,
(4) interstitial diffusion,
(5) substitutional diffusion,
(6) interdiffusion coefficient, and
(7) intrinsic diffusion.

14

DIFFUSION IN SOLIDS

This chapter, in which the primary aim is to obtain solutions to Fick's laws of diffusion, is almost identical to Chapter 9 where we presented solutions to heat conduction problems. Therefore, much of the groundwork laid in Chapter 9 will reappear in the following discussions. We shall begin by presenting some classical approaches used for determining diffusion coefficients in solids, and then consider some applied problems involving diffusion in solids as the rate-limiting step.

14.1 STEADY-STATE DIFFUSION EXPERIMENTS

As an example of the application of Fick's first law, consider an experiment in which an iron tube is held in the isothermal part of a furnace. A carburizing gas is passed through the inside of the tube, and a carburizing gas of a different composition is passed over the outside. Steady state is reached when the carbon concentration at each point in the tube wall no longer changes with time. By this time, the appropriate differential equation for steady-state diffusion through a cylinder can be derived from shell balances (Problem 14.7). If the diffusion coefficient of carbon in iron is a constant, independent of composition, then

$$\frac{1}{r}\frac{d}{dr}\left(r\frac{dC}{dr}\right) = 0. \tag{14.1}$$

The solution to this equation is given by its heat transfer analog from Eq. (7.65):

$$\frac{C - C_2}{C_1 - C_2} = \frac{\ln(r/r_2)}{\ln(r_1/r_2)}, \tag{14.2}$$

where r_1 and r_2 are the inside and outside radii of the tube, and C_1 and C_2 are the corresponding concentrations of carbon at these surfaces. Thus a plot of C versus $\ln r$ should be a straight line. However, for carbon diffusing in γ-iron, the slope of such a plot, as shown in Fig. 14.1, becomes smaller on passing from the low-carbon side to the high-carbon side. Therefore, the diffusion coefficient must be a function of composition, Eqs. (14.1) and (14.2) do not apply, and we must approach the problem somewhat differently.

In addition to the fact that $\partial C/\partial t = 0$, steady state also means that the quantity

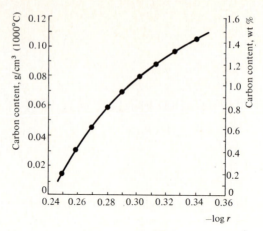

Fig. 14.1 Steady-state carbon concentration profile through a hollow cylinder of iron at 1000°C. (From R. P. Smith, *Acta Met.* **1**, 578 (1953).)

of carbon passing through the tube per unit time is constant and independent of r. Thus

$$J = 2\pi r l j_r,\tag{14.3}$$

where l = length of the cylinder, j_r = local flux, and J = quantity of carbon passing through the tube wall per unit time. Since

$$j_r = -D\frac{dC}{dr},\tag{14.4}$$

we may express Eq. (14.3) as

$$r\left[-D\frac{dC}{dr}\right] = \frac{J}{2\pi l},$$

or

$$\frac{dC}{d\ln r} = \frac{-J}{2\pi l D}.\tag{14.5}$$

For a given experiment, we can measure J and l, and if the carbon concentrations within the tube wall are determined by chemical analyses, then we can determine D from the slope of the plot of C versus $\ln r$.

Similar experiments have also been performed for determining the diffusion coefficient of gases through metals such as, for example, the diffusion of hydrogen through a metal foil in Fig. 14.2. According to Fick's first law, the flux of hydrogen through the metal is

$$j_x = -D\frac{dC}{dx}.\tag{14.6}$$

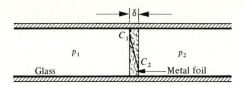

Fig. 14.2 Experiment for diffusion of hydrogen through metal foils.

The foils are very thin, and so it is extremely difficult to determine the concentration as a function of distance through the foil. The experimental results therefore consist of a measured steady-state flux, the hydrogen pressure drop across the foil, and the foil's thickness. To obtain D from the data, we take the value of C in the metal at each gas–metal interface as the solubility S that would exist in equilibrium with the gas.* From Sievert's law, we know that for equilibrium between gas and the metal

$$S_1 = K p_1^{1/2}, \tag{14.7a}$$

and

$$S_2 = K p_2^{1/2}, \tag{14.7b}$$

in which K includes the equilibrium constant for the reaction

$$H_2(g) = 2\underline{H}(\text{in solution}), \tag{14.8}$$

and p_1 and p_2 are the partial pressures of hydrogen on both sides of the foil with a thickness δ, as shown in Fig. 14.2.

The gradient dC/dx can then be expressed in terms of the pressures:

$$\frac{dC}{dx} = \frac{S_1 - S_2}{\delta} = \frac{K}{\delta}(\sqrt{p_1} - \sqrt{p_2}). \tag{14.9}$$

Combining Eqs. (14.6) and (14.9), we obtain the flux:

$$j_x = \frac{-DK}{\delta}(\sqrt{p_1} - \sqrt{p_2}). \tag{14.10}$$

In relation to the diffusion of gases through solids, the term *permeability, P,* is often used defined by

$$P = DS = DK\sqrt{p}, \tag{14.11}$$

so that

$$j_x = \frac{-(P_1 - P_2)}{\delta}, \tag{14.12}$$

---

* This is true under the conditions when the solution of gas in the surface of the metal occurs much more rapidly that the rate at which the diffusing species leaves the surface and enters the bulk metal. Experimentally, we check this assumption by determining the fluxes for two thicknesses of foil under the same pressure drop and temperature. If equilibrium does exist at the interface, then ΔC is the same for both cases, and the flux is inversely proportional to the thickness.

and permeability is then given by an equation of the form

$$P = Ap^{1/2} e^{-Q_p/RT}, \qquad (14.13)$$

which includes the temperature dependence of both S and D, with A and Q_p as constants. But there are also other ways to define permeability, which is rather confusing. For example, using a different definition, we obtain the flux:

$$j_x = -\frac{P^*}{\delta} (\sqrt{p_1} - \sqrt{p_2}), \qquad (14.14)$$

so that $P^* = P = DS$ only at 1 atm pressure, or

$$P^* = DK. \qquad (14.15)$$

In this case, we have

$$P^* = P_0^* e^{-Q_p/RT}, \qquad (14.16)$$

where $P_0^* = $ cm^3(STP)-sec^{-1} cm^{-2} measured for a cm thickness and at 1 atm pressure, and $Q_p = $ cal $\cdot$ mol^{-1} (activation energy for permeation).

There are other sets of units which apply to P_0^* as well, and therefore one should be extremely careful of the term *permeability* because so many different definitions and units are used. Table 14.1 gives some representative data for various systems, using Eqs. (14.15) and (14.16) to define permeability.

Table 14.1 Permeability data for gas–metal systems

Gas	Metal	P_0^*, cm^3(STP)/sec-cm-atm$^{1/2}$†	Q_p, cal/mol
H_2	Ni	1.2×10^{-3}	13,850
H_2	Cu	$1.5–2.3 \times 10^{-4}$	16,000–18,700
H_2	α-Fe	2.9×10^{-3}	8,400
H_2	Al	$3.3–4.2 \times 10^{-1}$	30,800
N_2	Fe	4.5×10^{-3}	23,800
O_2	Ag	2.9×10^{-3}	22,550

† The units in Eq. (14.14) are: $\delta = $ cm, $p = $ atm, and $j_x = $ cm^3 (STP)/ sec-cm^2.

Permeabilities have also been measured for gases diffusing through other materials. Table 14.2 contains some of these data.

Example 14.1 A pilot plant for hydrogenation of hydrocarbon vapor is to be constructed of a low-alloy steel. In designing, the question of the effect of wall thickness on the rate of hydrogen loss through the wall is raised. If the inside diameter of a vessel 100 cm long is 10 cm, calculate the rate of hydrogen loss as a function of wall thickness at 450°C and a pressure of 75 atm hydrogen, assuming that the gas diffusing through the wall is collected and removed at 1 atm.

Table 14.2 Permeability of gases through nonmetals at various temperatures

Gas	Material	Temperature, °C	P_0^*, $(\text{cm}^3\text{-cm}^{-2}\ \text{sec}^{-1}\ \text{cm thickness}^{-1} \cdot \text{cm Hg}^{-1})$
H_2	Neoprene®	17.5	8.5×10^{-10}
N_2	Neoprene®	27.1	1.37×10^{-10}
		65.4	1.06×10^{-9}
Ar	Neoprene®	36.1	6.8×10^{-10}
		52.2	1.44×10^{-9}
H_2	Cellophane	25.0	8.0×10^{-12}
H_2	Rubber	25.0	1.9×10^{-9}
H_2O	Cellophane	38.0	1.8×10^{-8}
H_2O	Rubber	25.0	5.2×10^{-8}
H_2	SiO_2	300	4.8×10^{-11}
		700	4.2×10^{-10}
He	SiO_2	300	9.9×10^{-12}
		700	2.5×10^{-10}
He	Pyrex®	100	2.6×10^{-12}
		500	1.6×10^{-10}

Solution. Combining Eqs. (14.2), (14.3), and (14.4), and assuming that D is constant, we have

$$J = -2\pi l D \frac{C_1 - C_2}{\ln (r_1/r_2)}.$$

In terms of permeability

$$J = \frac{-2\pi l D K(\sqrt{p_1} - \sqrt{p_2})}{\ln(r_1/r_2)}$$

$$= \frac{-2\pi l P^*(\sqrt{p_1} - \sqrt{p_2})}{\ln(r_1/r_2)}.$$

From Table 14.1 and Eq. (14.16), $P^* = 8.4 \times 10^{-6} \text{ cm}^3(\text{STP})/\text{sec-cm-atm}^{1/2}$. Therefore

$$J = \frac{-2(3.14)(100)(8.4 \times 10^{-6})(\sqrt{75} - \sqrt{1})}{2.303(\log 5 - \log r_2)}$$

$$= \frac{-1.755 \times 10^{-2}}{(0.699) - \log r_2}.$$

In more general terms, where r_1 is not specified, but l is still 100 cm,

$$J = \frac{-1.755 \times 10^{-2}}{\log(r_1/r_2)}.$$

Both results are shown below:

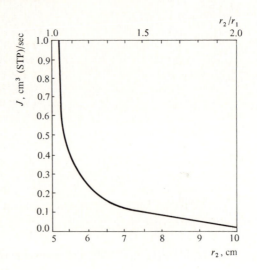

14.2 TRANSIENT DIFFUSION EXPERIMENTS

Under most circumstances it is not possible to carry out steady-state experiments in order to determine diffusion coefficients in solids. This means that we must use the results of transient experiments, solving Eq. (13.18) in its general form

$$\frac{\partial C}{\partial t} = \nabla(D\nabla C) \tag{14.17}$$

for various boundary conditions. In correspondence with heat conduction, a solution to Eq. (14.17) is usually either a series of error functions, which converge rapidly for short times, or a trigonometric series which converges rapidly for long times. In the following sections, we shall examine several solutions to Eq. (14.17) and their applications.*

14.2.1 Thin film source: infinite and semi-infinite sink

The solution and procedure that follows is often used for self-diffusion studies of substitutional atoms. Radioactive tracers are used as solutes since their concentration can be determined quite accurately, even at low concentrations. A small

---

* Extensive compilations of solutions are presented in J. Crank, *The Mathematics of Diffusion*, Oxford University Press, London, 1956, and W. Jost, *Diffusion in Solids, Liquids, and Gases*, Academic Press, New York, 1960.

quantity β of the tracer is plated as a thin film $\Delta x'$ thick on one end of a long rod of tracer-free material. The rod is then annealed at the diffusion temperature of interest. Since D^* is a self-diffusion coefficient and does not depend on position, for such an application, Fick's second law is

$$\frac{\partial C}{\partial t} = D^* \frac{\partial^2 C}{\partial x^2}. \qquad (14.18)$$

Suppose we take a second tracer-free rod and butt-weld it to the plated end (without any diffusion occurring), and then carry out the diffusion anneal. According to Eq. (9.89), we see that the solution is

$$C(x, t) = \frac{C_i}{2\sqrt{\pi D^* t}} \exp\left(\frac{-x^2}{4D^* t}\right) \Delta x'. \qquad (14.19)$$

Here C_i is the concentration of the tracer in the plated material whose thickness is $\Delta x'$.

Since $C_i \Delta x'$ is the quantity of tracer material plated as the thin film, we write the solution

$$C(x, t) = \frac{\beta}{2\sqrt{\pi D^* t}} \exp\left(\frac{-x^2}{4D^* t}\right), \qquad (14.20)$$

which describes the spreading by diffusion of a thin plate source into an infinite sink. This is illustrated in Fig. 14.3.

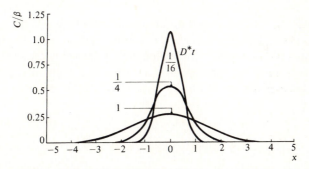

Fig. 14.3 Spreading of concentration from a plane source at $x = 0$.

Note that we may determine D^* without measurement or control of β, since a plot of log C versus x^2 at time t yields D^* directly.

Now suppose that the second bar has not been welded to the first. Then a *semi-infinite* bar would extend over the region $x > 0$ with an impermeable barrier at $x = 0$. In either case, infinite or semi-infinite,

$$\frac{\partial C}{\partial x} = 0 \qquad \text{at } x = 0. \qquad (14.21)$$

If diffusion is now allowed to take place, the material that would normally diffuse in the negative x-direction is *reflected* at the $x = 0$ plane, and moves in the positive x-direction. The concentration at any x in the $+x$ domain is then given by the superposition of the original solution for $x > 0$ and the reflected solution for $x < 0$, or

$$C(x, t) = \frac{\beta}{\sqrt{\pi D^* t}} \exp\left(\frac{-x^2}{4D^* t}\right). \tag{14.22}$$

14.2.2 Diffusion couple with constant $\tilde{D}$

The previous case has only a limited practical value because the initial distribution usually is not an infinitely thin source, but rather it occupies a finite region. This case is exemplified in Fig. 14.4a by butt-welding two bars of A–B alloy of concentrations C_1 and C_2. If $\tilde{D}$ is independent of composition, and the bars extend far enough in the positive and negative domains to be considered infinite, then Eq. (14.18) applies with the boundary conditions:

$$C(x, 0) = C_1 \quad \text{at} \quad x < 0, \tag{14.23a}$$

$$C(x, 0) = C_2 \quad \text{at} \quad x > 0, \tag{14.23b}$$

$$C(-\infty, t) = C_1, \tag{14.23c}$$

$$C(\infty, t) = C_2. \tag{14.23d}$$

Due to symmetry, the concentration at $x = 0$ immediately takes on the average value of C_1 and C_2. Let this average be C_s; then it is easy to see that in the positive x-domain, the solution is directly analogous to the temperature distribution of the semi-infinite solid, as discussed in Section 9.5.2 culminating in Eq. (9.110). Appropriately changing the symbols, we can write the solution:

$$\frac{C - C_s}{C_2 - C_s} = \text{erf} \frac{x}{2\sqrt{\tilde{D}t}}. \tag{14.24}$$

Figure 14.4b illustrates this result in a general form, from which one may obtain $\tilde{D}$-values from measurements of C, x, and t. Since $\text{erf}\, x = -\text{erf}(-x)$, the concentration profile given by Eq. (14.24) is symmetrical about $x = 0$. This is known as the *Grube solution*, and applies when $\tilde{D}$ is not a function of concentration. Since $\tilde{D}$ is usually a function of composition, the use of Eq. (14.24) is restricted to small differences between C_1 and C_2. For example, a good value for the diffusion coefficient of A in 50 A–50 B alloy could be determined from a couple of 45% A alloy welded to a 55% A alloy.

Apart from application to diffusion couples, we can use Eq. (14.24) to predict concentration–time curves for situations where the part in which diffusion occurs

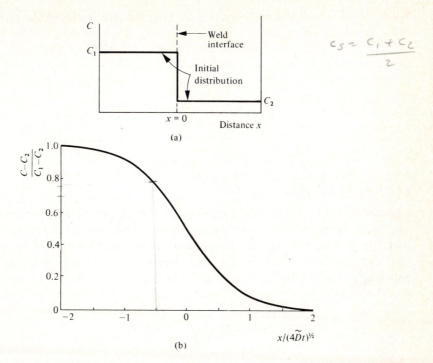

$$c_s = \frac{c_1 + c_2}{2}$$

Fig. 14.4 Interdiffusion between an alloy of composition C_1 and an alloy of composition C_2. (a) Initial distribution. (b) Diffusion profile when the initial distribution is as in (a).

is thick enough so that, within the time for diffusion, there is still a region of the part unchanged in composition.

Example 14.2 A piece of AISI 1020 steel is heated to 1800°F (in the austenite region) and subjected to a carburizing atmosphere such that the reaction

$$2CO = CO_2 + \underline{C}$$

is in equilibrium with 1.0% C in solution at the surface. Calculate the carbon profile after 1, 3, and 10 hours, assuming that diffusion within the solid is the rate-limiting step.

Solution. The initial condition is $C_2 = 0.2\%C$, and the boundary condition at the surface is $C_s = 1.0\%C$.

At 1800°F, $D_c = 2.0 \times 10^{-7}$ cm²/sec. Therefore, Eq. (14.24) for the distribution after 1 hour is

$$\frac{C - C_s}{C_2 - C_s} = \text{erf}\, \frac{x}{2\sqrt{Dt}}$$

$$C(x, 3600) = 1.0 + (0.2 - 1.0)\, \text{erf}\, \frac{x}{2\sqrt{2 \times 3.6 \times 10^{-4}}}.$$

Specifically at $x = 0.05$ cm, $C = 0.35\%$ C. The results for various locations as well as for the longer times are:

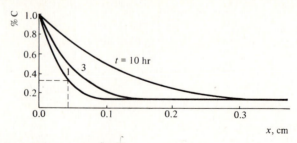

14.2.3 Diffusion couple—variable $\tilde{D}$

The analysis in Section 14.2.2 is valid only for $\tilde{D}$ independent of concentration. In general, however, the diffusion cofficient varies with composition, and since there is a concentration gradient, this means that $\tilde{D}$ changes with position. This variation in $\tilde{D}$ is particularly evident in diffusion couple experiments in which pure A is joined to pure B, and a continuous solid solution is formed. Fick's second law must be written over all compositions between A and B, and for such situations this is

$$\frac{\partial C}{\partial t} = \frac{\partial}{\partial x}\left(\tilde{D}\,\frac{\partial C}{\partial x}\right). \tag{14.25}$$

The solution to Eq. (14.25) that follows is useful for obtaining $\tilde{D}$ over a range of compositions, but not for the *a priori* task of predicting a concentration profile for a diffusion anneal. In other words, it does not give a solution $C(x, t)$, which is usually sought, but rather allows $\tilde{D}(C)$ to be calculated from an experimental plot of $C(x)$. This method of analyzing experimental data is called the *Boltzmann–Matano* technique.

We combine the position variable x and the time variable t into one variable $\lambda = x/\sqrt{t}$, so that we consider C to be a function of only the one variable, λ. Using this definition of λ, we transform Eq. (14.25) into an ordinary differential equation:

$$\frac{\partial C}{\partial t} = \frac{\partial \lambda}{\partial t}\left(\frac{dC}{d\lambda}\right) = -\frac{1}{2}\frac{x}{t^{3/2}}\left(\frac{dC}{d\lambda}\right) = -\frac{\lambda}{2t}\left(\frac{dC}{d\lambda}\right), \tag{14.26a}$$

and

$$\frac{\partial C}{\partial x} = \frac{\partial \lambda}{\partial x}\left(\frac{dC}{d\lambda}\right) = \frac{1}{t^{1/2}}\left(\frac{dC}{d\lambda}\right). \tag{14.26b}$$

Substituting into Eq. (14.25), we obtain

$$-\frac{\lambda}{2t}\left(\frac{dC}{d\lambda}\right) = \frac{\partial}{\partial x}\left[\frac{\tilde{D}}{t^{1/2}}\left(\frac{dC}{d\lambda}\right)\right] = \frac{1}{t}\frac{d}{d\lambda}\left(\tilde{D}\,\frac{dC}{d\lambda}\right),$$

or finally we can get

$$-\frac{\lambda}{2}\frac{dC}{d\lambda} = \frac{d}{d\lambda}\left(\tilde{D}\frac{dC}{d\lambda}\right). \tag{14.27}$$

Consider the diffusion couple depicted in Fig. 14.5a. Then for $C(\lambda)$, we recognize that

$$C = C_1 \qquad \text{for } \lambda = -\infty, \tag{14.28a}$$

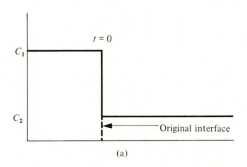

(a)

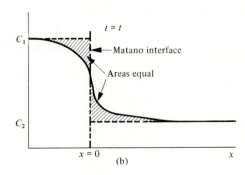

(b)

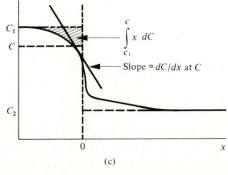

(c)

Fig. 14.5 (a) Initial conditions. (b) Definition of location of Matano interface after diffusion has taken place for a time t. (c) The integral and the slope obtained in order to calculate $\tilde{D}$ at composition C.

and

$$C = C_2 \qquad \text{for } \lambda = +\infty. \tag{14.28b}$$

We then solve Eq. (14.27) by integrating between $C = C_1$ and $C = C$:

$$-\tfrac{1}{2} \int_{C_1}^{C} \lambda \, dC = \tilde{D} \frac{dC}{d\lambda} \Big|_{C_1}^{C}. \tag{14.29}$$

Since the concentration gradient goes to zero as C approaches C_1, the right-hand side of Eq. (14.29) is simply $\tilde{D}(dC/d\lambda)$. Then

$$\tilde{D} = -\frac{1}{2\left(\dfrac{dC}{d\lambda}\right)} \int_{C_1}^{C} \lambda \, dC. \tag{14.30}$$

From Eq. (14.29), we obtain the definition of the *Matano interface*; since the concentration gradient also goes to zero as C approaches C_2, Eq. (14.29) gives us the additional condition that

$$\int_{C_1}^{C_2} \lambda \, dC = 0. \tag{14.31}$$

Since experimental data are available only at some constant time t, Eqs. (14.30) and (14.31) can be written in terms of x and t, and the relationships used for calculating $\tilde{D}$ from the measured concentration profile are

$$\tilde{D} = -\frac{1}{2t} \frac{1}{\left(\dfrac{dC}{dx}\right)} \int_{C_1}^{C} x \, dC, \tag{14.32}$$

and we choose the plane defining $x = 0$ such that

$$\int_{C_1}^{C_2} x \, dC = 0. \tag{14.33}$$

In Fig. 14.5(b), the plane $x = 0$ is given by the line that makes the two hatched areas equal. We calculate the value of $\tilde{D}$ at a given C by measuring the cross-hatched area $\int_{C_1}^{C} x \, dC$ and the reciprocal slope at that point, dx/dC. The diffusion coefficient $\tilde{D}$, found by applying Eq. (14.32) in this manner, is the interdiffusion coefficient discussed in Section 13.2.3.

In addition, if we place inert markers at the original plane of welding, we can also determine the intrinsic diffusion coefficients. This is left for the reader to do in

Problem 14.6. Furthermore, Jost[1] has pointed out that it is not necessary for a single phase to exist over the entire range of the diffusion region. A discontinuity in $C(x)$ and $\tilde{D}(C)$ exists where an intermediate phase is formed. The only condition required when applying the Boltzmann–Matano technique is that the concentrations in both phases on either side of the interface between the phases are independent of time.

Example 14.3 A diffusion couple made of pure Nb welded to pure U is held at 800°C for 49 days. The resulting concentration profile, shown below, is obtained with a microprobe analyzer. Note that an intermediate phase δ is formed between the two solid solutions, γ_1 and γ_2. Calculate $\tilde{D}$ at 90 at.% U.

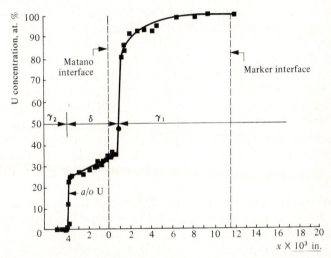

From N. L. Peterson and R. F. Ogilvie, *Trans. AIME* **218**, 439 (1960).

Solution. Let C represent the composition of U. First, the Matano interface is chosen, such that

$$\int_{-\infty}^{0} x\, dC = \int_{0}^{\infty} x\, dC.$$

This is found by trial and error, and placed on the figure as shown.

At 90% U, dC/dx is evaluated:

$$dC/dx = 1810 \text{ at.\%/cm}.$$

Then

$$\int_{C=100}^{C=90} x\, dC = -0.132 \text{ cm} - \text{at.\%, by graphical means};$$

---

[1] W. Jost, *Diffusion in Solids, Liquids, and Gases*, Academic Press, New York, 1960, page 76, and also W. Jost, *Z. Physik* **127**, 163 (1950).

$$t = \frac{49 \text{ days} \mid 3600 \text{ sec} \mid 24 \text{ hr}}{\mid \text{hr} \mid \text{day}} = 4.24 \times 10^6 \text{ sec.}$$

Thus

$$\tilde{D} \, (90\% \, \text{U}) = \frac{-(-0.132)}{2(4.24 \times 10^6)(1.810 \times 10^3)}$$

$$= 8.6 \times 10^{-12} \, \text{cm}^2/\text{sec.}$$

14.3 FINITE SYSTEM SOLUTIONS

The solutions to Fick's laws presented thus far represent useful cases in many situations, but there are also other problems of interest. This is the case when the system boundaries are close together, relatively speaking, and the sink cannot be considered infinite or semi-infinite, as the effect of the diffusion process on the composition is felt at the furthest point in the material prior to the end of the diffusion treatment. This condition can arise quite often when small parts are exposed to a gaseous environment, and there is diffusion of the gas species into the part. Conversely, metal parts must often be degassed. When dealing with these situations, it is usually safe to assume that the rate-limiting step of the overall mass transfer is the diffusion of the gas species into the solid. Hence, we seek solutions of Fick's second law with a surface concentration imposed at time zero and maintained constant.

To illustrate this case, consider the diffusion into or out of a slab of infinite length and thickness $2L$. Initially the slab has a uniform concentration C_i, and then its surfaces are raised or lowered to C_s and maintained constant. Hence, we are seeking a solution to Fick's second law with constant $\tilde{D}$ (or simply D):

$$\frac{\partial C}{\partial t} = D \frac{\partial^2 C}{\partial x^2}. \tag{14.34}$$

The initial and boundary conditions of interest are

$$C(x, 0) = C_i, \tag{14.35a}$$

$$\frac{\partial C}{\partial x}(0, t) = 0, \tag{14.35b}$$

and

$$C(L, t) = C_s. \tag{14.35c}$$

The solution to this problem can be determined in the same manner as the solution of the heat conduction equation in Section 9.4. By separation of variables it has the form:

$$\theta = X(x) \cdot G(t),$$

where θ is $(C - C_s)$, and all the boundary conditions can be written in a homogeneous form. According to Eqs. (9.78) and (9.79), we see that

$$X = c_1 \cos \lambda x + c_2 \sin \lambda x, \tag{14.36}$$

and

$$G = \exp(-\lambda^2 Dt). \tag{14.37}$$

Boundary condition (14.35b) requires that $c_2 = 0$, and then when we apply (14.35c) $c_1 \cos \lambda L = 0$ results. This is satisfied by $\lambda = (2n + 1)\pi/2L$ where n is any integer from 0 to ∞. Hence

$$\theta = \sum_{n=0}^{\infty} A_n \exp\left[\frac{-(2n+1)^2\pi^2}{4} \frac{Dt}{L^2}\right] \cos\left[\frac{(2n+1)\pi}{2} \frac{x}{L}\right], \tag{14.38}$$

where the A_n's are now the constants involved. The initial condition, $\theta(x, 0) = \theta_i = C_i - C_s$, remains to be satisfied and when substituted into Eq. (14.38), it yields:

$$\theta_i = \sum_{n=1,\text{odd}}^{\infty} A_n \cos\frac{(2n+1)\pi}{2} \frac{x}{L}. \tag{14.39}$$

If we apply Fourier's analysis to Eq. (14.39) as we demonstrated previously for Eqs. (9.48), (9.59), and (9.83), we obtain

$$A_n = \frac{(-1)^n}{(2n+1)} \frac{4}{\pi} \theta_i. \tag{14.40}$$

Thus the solution we seek is

$$\frac{\theta}{\theta_i} = \frac{C - C_s}{C_i - C_s} = \frac{4}{\pi} \sum_{n=0}^{\infty} \frac{(-1)^n}{2n+1} \exp\left[\frac{-(2n+1)^2\pi^2}{4} \frac{Dt}{L^2}\right] \cos\frac{(2n+1)\pi}{2} \frac{x}{L}. \tag{14.41}$$

Equation (14.41) is useful for describing concentration profiles as a function of time. However, the total amount of material that diffuses into or out of the slab is often of more interest, particularly in experimental work where this is the only measurable quantity. So, the average concentration $\bar{C}$ is required:

$$\bar{C} = \frac{1}{L}\int_0^L C \, dx. \tag{14.42}$$

Carrying out this operation, using Eq. (14.41) for C, we obtain the relative change in average composition for diffusion into a slab:

$$\frac{\bar{C} - C_s}{C_i - C_s} = \frac{8}{\pi^2} \sum_{n=0}^{\infty} \frac{1}{(2n+1)^2} \exp\left[\frac{-(2n+1)^2\pi^2}{4} \frac{Dt}{L^2}\right]. \tag{14.43}$$

This expression is good for diffusion into or out of a slab. If we take the first term in the series

$$\frac{\bar{C} - C_s}{C_i - C_s} = \frac{8}{\pi^2} \exp{(-t/\tau)}, \tag{14.44}$$

where τ is the time constant for the diffusion process ($\tau = 4L^2/\pi^2 D$), it is apparent that a graph of log $\bar{\theta}$ versus (t/τ) is a straight line and that we can obtain D from the slope. We plot Eq. (14.43) in Fig. 14.6, along with diffusion into or out of cylinders and spheres (see Table 14.3). For long times ($Dt/L^2 > 0.05$), the first term of the series is sufficient, and a straight line relationship is obeyed.

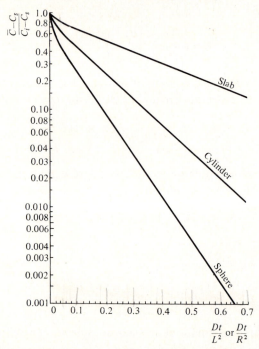

Fig. 14.6 The relative change in average composition for the basic shapes. R = radius, and L = semithickness.

For diffusion into or out of simple multidimensional shapes other than the infinite plate, infinite cylinder, or sphere, we handle the problem in the same manner as we treated heat transfer to or from these shapes in Section 9.5. We can combine product solutions to yield the solution for the shape of interest.

Example 14.4 Calculate the fraction of hydrogen remaining in
   a) a 4-in. thick slab of steel, 10 ft long × 4 ft wide,
   b) a 4-in^2 square billet of steel, 10 ft long, and
   c) a 4-in^2 square billet of steel, 8 in. long,

Table 14.3 The relative change in average composition for the basic shapes

Diffusion in a slab of semithickness, L

I.C.: $C(x, 0) = C_i$,

B.C.: $C(L, t) = C_s$,

$$\frac{\partial C}{\partial x}(0, t) = 0.$$

Solution

$$\frac{\bar{C} - C_s}{C_i - C_s} = \frac{8}{\pi^2} \sum_{n=0}^{\infty} \frac{1}{(2n + 1)^2} \exp\left[\frac{-(2n + 1)^2 \pi^2}{4} \frac{Dt}{L^2}\right]. \tag{14.43}$$

Diffusion in solid circular cylinder of radius, R

I.C.: $C(r, 0) = C_i$,

B.C.: $C(R, t) = C_s$,

$$\frac{\partial C}{\partial r}(0, t) = 0.$$

Solution

$$\frac{\bar{C} - C_s}{C_i - C_s} = \sum_{n=1}^{\infty} \frac{4}{\xi_n^2} \exp\left(\frac{-\xi_n^2 Dt}{R^2}\right), \tag{14.45}$$

where $\xi_n = 2.405, 5.520, 8.654, 11.792, 14.931$, when $n = 1, 2, 3, 4, 5$, etc.*

Diffusion in spheres of radius, R

The same set of initial and boundary conditions as for the cylinder above:

$$\frac{\bar{C} - C_s}{C_i - C_s} = \frac{6}{\pi^2} \sum_{n=1}^{\infty} \frac{1}{n^2} \exp\left(\frac{-n^2 \pi^2 Dt}{R^2}\right). \tag{14.46}$$

* ξ_n are roots of the equation $J_0(x) = 0$, where $J_0(x)$ is the Bessel function of zero order.

after 40 hours of vacuum outgassing treatment at a temperature where $D_H = 1.0 \times 10^{-5}$ cm^2/sec, assuming an initially uniform distribution.

Solution
a) Consider this to be an infinite plate. Then

$$\frac{Dt}{L^2} = \frac{(1 \times 10^{-5})(40 \times 3600)}{(2 \times 2.54)^2} = 0.056.$$

From Fig. 14.6, we have

$$\frac{\bar{C} - C_s}{C_i - C_s} = 0.74.$$

b) The desired solution can be obtained as the product of the infinite plate solutions for the two 4-in. dimensions:

$$\frac{\bar{C} - C_s}{C_i - C_s} = (0.74)(0.74) = 0.55.$$

c) In this case, first evaluate Dt/L^2 for the 8-in. dimension:

$$\frac{Dt}{L^2} = \frac{(1 \times 10^{-5})(40 \times 3600)}{(4 \times 2.54)^2} = 0.028.$$

From Fig. 14.6, we get

$$\left.\left(\frac{\bar{C} - C_s}{C_i - C_s}\right)\right|_{8'' \text{dim.}} = 0.80.$$

Therefore

$$\frac{\bar{C} - C_s}{C_i - C_s} = (0.80)(0.74)(0.74) = 0.437.$$

14.4 DIFFUSION-CONTROLLED PROCESSES WITH A MOVING INTERFACE

Many metallurgical reactions and processes involve diffusion steps in conjunction with reactions at phase boundaries. One result of these transient processes is the motion of the boundary between the phases. The analysis of the diffusion process is somewhat complicated due to the presence of more than one phase. First, we shall discuss some of the general aspects of the problem, including the boundary condition at the moving interface, and then present applications in carburizing and decarburizing.

14.4.1 General aspects

In the general situation, two phases are in contact as in Fig. 14.7. The moving phase

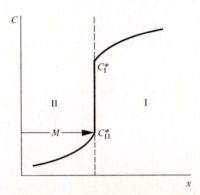

Fig. 14.7 The concentration profile of two phases coexisting during transient diffusion.

boundary is at $x = M$, and at this boundary C_{II}^* and C_I^* represent the equilibrium concentrations that coexist in phases II and I, respectively, at the temperature under consideration. In both phases, diffusion takes place, and there is exchange of mass from one phase to the other at the interface.

In both phases, Fick's second law applies. Hence

$$\frac{\partial C_I}{\partial t} = D_I \frac{\partial^2 C_I}{\partial x^2}, \qquad x > M, \tag{14.47a}$$

and

$$\frac{\partial C_{II}}{\partial t} = D_{II} \frac{\partial^2 C_{II}}{\partial x^2}, \qquad x < M. \tag{14.47b}$$

D_I and D_{II} are the diffusion coefficients of the diffusing substance; both are assumed to be independent of composition within each phase.

The next condition assumed to hold in this situation where diffusion is the rate-controlling step, is that the concentrations on either side of the interface are related by an equilibrium expression of the form

$$C_I^* = KC_{II}^*, \tag{14.48}$$

where K is the partition ratio between the phases. (Later, in Chapter 16, we shall consider the situation where equilibrium is not reached at the interface, that is, where diffusion does not completely control the overall rate.)

Finally, we shall always conserve mass as transfer across the interface from one phase to another occurs. The net difference in the flux of the diffusing solute entering and leaving the interface equals the solute added to phase II by virtue of the moving interface, that is,

$$D_I \left(\frac{\partial C_I}{\partial x} \right)_{x=M} - D_{II} \left(\frac{\partial C_{II}}{\partial x} \right)_{x=M} = (C_{II}^* - C_I^*) \frac{dM}{dt}. \tag{14.49}$$

14.4.2 Formation of a second phase on the surface of an initially homogeneous first phase

One application of the preceding basic assumption is the decarburization of γ-Fe with the formation of α-Fe on the surface in the temperature range 723–910°C, or the carburization of α-Fe to form γ-Fe in the same range, as originally presented by Wagner.[2]

We presume that Fick's second law holds in each phase, and assume equilibrium conditions at the α–γ interface. For reference, Fig. 14.8 gives a portion of the iron–carbon phase diagram. Consider the situation depicted by Fig. 14.9(a). Initially, the alloy contains C_i-carbon and is heated to the austenite region.

---

[2] C. Wagner, *unpublished notes from Course* 3.63, M.I.T., Cambridge, Massachusetts, 1955. See also W. Jost, *ibid.*

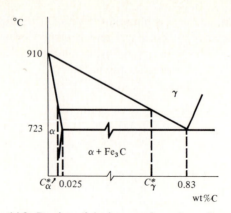

Fig. 14.8 Portion of the iron–carbon phase diagram.

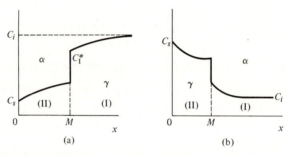

Fig. 14.9 Concentration profiles for cases where a second phase II forms on the surface of an initial homogeneous phase I.

Depending on the carbon potential of the atmosphere, a surface composition of C_s is established. Therefore the conditions to be satisfied, in addition to Eqs. (14.48) and (14.49), are

$$\text{I.C.:} \qquad C(x, 0) = C_i, \tag{14.50a}$$

and

$$\text{B.C.:} \qquad C(0, t) = C_s. \tag{14.50b}$$

Equations (14.47a, b) and (14.50a, b) are satisfied by

$$C_\mathrm{I} = C_i + B_1\left(1 - \operatorname{erf}\frac{x}{2\sqrt{D_\mathrm{I}t}}\right), \qquad x > M, \tag{14.51}$$

and

$$C_\mathrm{II} = C_s - B_2 \operatorname{erf}\frac{x}{2\sqrt{D_\mathrm{II}t}}, \qquad 0 < x < M. \tag{14.52}$$

B_1 and B_2 are constants of integration, and since at $x = M$, $C_{II} = C_{II}^*$ (constant), it follows that the argument of the error function must be constant at $x = M$. Hence

$$M = 2\beta \sqrt{D_{II} t}, \tag{14.53}$$

where β is a dimensionless parameter to be determined. Substituting Eqs. (14.51)–(14.53) into the mass balance equation (14.49), we get

$$C_{II}^* = C_s - B_2 \, \mathrm{erf}\,(\beta), \tag{14.54}$$

and

$$C_I^* = C_i + B_1 [1 - \mathrm{erf}\,(\beta\phi^{1/2})], \tag{14.55}$$

where

$$\phi = D_{II}/D_I,$$

and

$$C_{II}^* - C_I^* = \frac{B_2 e^{-\beta^2}}{\sqrt{\pi}\beta} - \frac{B_1 e^{-\beta^2\phi}}{\sqrt{\pi}\beta\sqrt{\phi}}. \tag{14.56}$$

Eliminating B_1 and B_2 between Eqs. (14.54)–(14.56), the parameter β is found by trial and error as that value which satisfies

$$(C_{II}^* - C_I^*) = \frac{(C_s - C_{II}^*)}{\sqrt{\pi}\beta e^{\beta^2} \, \mathrm{erf}\,\beta} - \frac{(C_I^* - C_i)}{\sqrt{\pi}\beta\sqrt{\phi} e^{\beta^2\phi} \, \mathrm{erfc}\,(\beta\phi^{1/2})}. \tag{14.57}$$

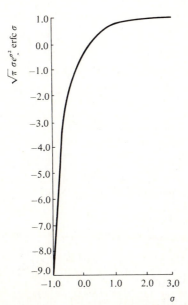

Fig. 14.10 Graph of σ values as a function of $\sqrt{\pi}\sigma e^{\sigma^2} \, \mathrm{erfc}\,\sigma$.

The denominator of the center term contains the function of β, βe^{β^2} erf β, which we have already encountered in Chapter 10, and we may evaluate it by means of Fig. 10.6. The right-hand term contains a function of the form $\sqrt{\pi}\sigma e^{\sigma^2}$ erfc σ (in this case $\beta\phi^{1/2} = \sigma$), and we can evaluate it from Fig. 14.10. Solution of the practical problem of predicting the rate of advance of the interface would require specifying the compositions and diffusion coefficients, and determining the value that satisfies Eq. (14.57). Then we simply relate M and t by Eq. (14.53). We can also determine B_1 and B_2 using Eqs. (14.54) and (14.55), respectively, and then the concentration profile, as given by Eqs. (14.51) and (14.52), is completely specified.

Example 14.5 A thin shell of 0.40% C alloy steel is austenitized at 800°C in an atmosphere of CO and CO_2 that is in equilibrium with 0.01% C. Under these conditions some α-Fe forms on the surface. After 30 min, how thick will the layer of α-Fe be?

Solution. In this case: $C_i = 0.40\%$, $C_s = 0.01\%$, $C_I^* = 0.24\%$, $C_{II}^* = 0.02\%$, $D_{II} = 2 \times 10^{-6}$ cm²/sec, $D_I = 3 \times 10^{-8}$ cm²/sec, and $\phi = 66.6$.

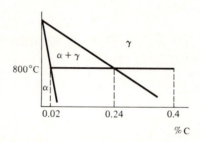

Eq. (14.53) gives the thickness of the α-Fe layer once β has been evaluated by means of Eq. (14.57):

$$(0.02 - 0.24) = \frac{(0.01 - 0.02)}{f(\beta)} - \frac{(0.24 - 0.40)}{f(\beta\phi^{1/2})},$$

$$0.22 = \frac{0.01}{f(\beta)} - \frac{0.16}{f(\sigma)}.$$

Trial and error is used, along with Figs. 14.10 and 10.6. Let $\beta = 0.2$; $f(\beta) = 0.045$ from Fig. 10.6. Then, $\sigma = \beta\phi^{1/2} = (0.2)(8.1) = 1.62$, and $f(\sigma) = 0.85$ from Fig. 14.10. Substituting these into Eq. (14.57), we have

$$\frac{0.010}{0.045} - \frac{0.16}{0.85} = 0.034 < 0.22.$$

Try $\beta = 0.1$; $f(\beta) = 0.011$, $\sigma = 1.215$, $f(\sigma) = 0.70$.

$$\frac{0.010}{0.011} - \frac{0.16}{0.70} = 0.68 > 0.22.$$

The correct value of β has been bracketed. Further trials eventually result in $\beta = 0.144$. Now we can calculate the layer thickness M:

$$M = 2\beta\sqrt{D_{II}t} = 2(0.144)(2 \times 10^{-6} \times 1.8 \times 10^3)^{1/2} = 0.0173 \text{ cm}.$$

14.4.3 Formation of a single-phase layer from an initial two-phase mixture

This situation is best exemplified by the carburization or decarburization of a steel that is initially a two-phase mixture; four possible cases are shown in Fig. 14.11. The interface M is now where the concentration of carbon in the surface phase reaches the concentration required for equilibrium with the second phase. In the two-phase region, the average composition C_0 is assumed to be uniform, which requires, in effect, that the grain size is small and that second-phase dispersion is uniform.

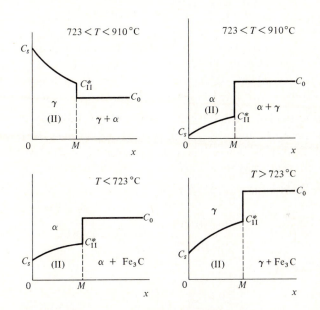

Fig. 14.11 Examples of the formation of a single-phase layer on a surface when the initial material is a two-phase mixture with average composition C_0.

Making the same assumptions as before, that is, D's are constant and interface reactions are so fast that only diffusion controls the rate of movement of M, we set down the conditions that describe the situation. Equation (14.47b) is obeyed in the region $0 < x < M$, with the boundary condition that $C_{II}(0, t) = C_s$.

Again we give the solution by Eq. (14.52). The equilibrium at M is simply described by

$$C_{II}(M, t) = C_{II}^*, \tag{14.58}$$

and the material balance at the interface takes the form

$$-D_{\text{II}}\left(\frac{\partial C_{\text{II}}}{\partial x}\right)_{x=M} = (C_{\text{II}}^* - C_0)\frac{dM}{dt}. \tag{14.59}$$

Equation (14.53) describes the locus of M with time. Then substituting Eqs. (14.53) and (14.52) into (14.58) and (14.59), we obtain

$$C_s - C_{\text{II}}^* = B_2 \operatorname{erf} \beta, \tag{14.60}$$

and

$$C_{\text{II}}^* - C_0 = \frac{B_2}{\sqrt{\pi\beta}} \exp(-\beta^2). \tag{14.61}$$

We can eliminate B_2 to give

$$\frac{1}{\sqrt{\pi}}\left(\frac{C_s - C_{\text{II}}^*}{C_{\text{II}}^* - C_0}\right) = \beta \exp \beta^2 \operatorname{erf} \beta. \tag{14.62}$$

We easily solve Eq. (14.62) for β using Fig. 10.6. Then we can calculate M if D and t are given, or we can calculate D if M and t are measured. In turn, we can determine B_2 from Eq. (14.60). These relationships have been tested against the data of Bramley and Beebe,[3] with satisfactory agreement.

Example 14.6 Calculate the depth of decarburization of an 0.20 % C, Cr–Mo steel, after 1 year of exposure to severe decarburizing conditions at 950°F. At 950°F the diffusion coefficient of carbon in this steel (ferrite) is 1.0×10^{-9} cm^2/sec,[4] and the steel is a two-phase mixture ($\alpha + Fe_3C$).

Solution. One year equals 3.02×10^7 sec. $C_s \cong 0.00\%$ C, $C_0 = 0.20\%$ C, and $C_{\text{II}}^* = 0.02\%$.

$$\frac{1}{\sqrt{\pi}}\frac{(0.00 - 0.02)}{(0.02 - 0.20)} = \beta e^{\beta^2} \operatorname{erf} \beta = 0.0626.$$

From Fig. 10.6, we get

$$\beta = \frac{M}{2\sqrt{Dt}} = 0.23,$$

$$M = (2 \times 0.23)(3.02 \times 10^{-2})^{1/2}$$

$$= 0.078 \text{ cm.}$$

14.5 HOMOGENIZATION OF ALLOYS

During solidification of alloys, *coring* occurs, because the rate of diffusion for most alloying elements in the solid state is too slow to maintain a solid of uniform

---

[3] A. Bramley and G. H. Beebe, *Carnegie Scholarship Memoirs* **15**, 71 (1926).
[4] R. R. Arnold, *M.S. Thesis*, Univ. of Wisconsin, 1968.

concentration in equilibrium with the liquid. A cast structure is exemplified in Fig. 14.12 showing the repetitive pattern of microsegregation. In the case at hand we presume the alloy to be a single phase.

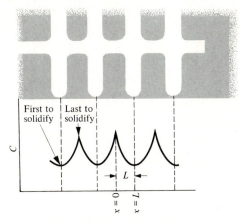

Fig. 14.12 A dendritic structure showing the *coring* or microsegregation of the alloying element. L is one-half of the dendrite arm spacing.

During homogenization, the alloy naturally tends toward a uniform concentration (Fig. 14.13). We need only examine what happens within one dendritic element since the profile is repetitive, and there is no net flow of solute from any dendritic region to the next.

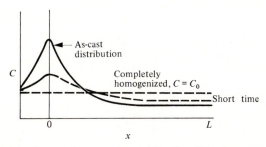

Fig. 14.13 Dendritic element, $0 < x < L$, used for describing homogenization kinetics.

To describe the homogenization kinetics, a solution to Fick's second law is needed that satisfies

$$\text{I.C.:} \qquad C(x, 0) = f(x), \tag{14.63a}$$

$$\text{B.C.:} \qquad \frac{\partial C}{\partial x}(0, t) = 0, \tag{14.63b}$$

$$\frac{\partial C}{\partial x}(L, t) = 0, \qquad t > 0. \tag{14.63c}$$

The solution can be obtained by applying the method of separation of variables. Here we simply present the solution as given by Crank:[5]

$$C(x, t) = C_0 + \sum_{n=1}^{\infty} A_n \exp\left(-n^2\pi^2 \frac{Dt}{L^2}\right) \cos \frac{n\pi x}{L}, \tag{14.64}$$

with

$$A_n = \frac{2}{L} \int_0^L f(x) \cos \frac{n\pi x}{L}\, dx. \tag{14.65}$$

In Eq. (14.64), C_0 is the overall or average alloy content, and we evaluate the A_n's by Eq. (14.65) using $f(x)$ as the initial solute distribution.

A useful parameter to describe the homogenization kinetics is the *residual segregation index* δ which is defined as

$$\delta \equiv \frac{C_M - C_m}{C_M^0 - C_m^0} \tag{14.66}$$

where C_M = maximum concentration, that is, $C_M = C(0, t)$; C_M^0 = initial maximum concentration, that is, $C_M^0 = C(0, 0)$; C_m = minimum concentration, that is, $C_m = C(L, t)$; and C_m^0 = initial minimum concentration, that is, $C_m^0 = C(L, 0)$. For no homogenization, $\delta = 1$, and after complete homogenization $\delta = 0$.

We can find $C_M - C_m$ by applying Eq. (14.64) to $x = 0$ and $x = L$, and performing the indicated subtraction. The residual segregation index can then be written as

$$\delta = \frac{2 \sum_{n=1,3\ldots\text{odd}}^{\infty} A_n \exp\left(-n^2\pi^2 \frac{Dt}{L^2}\right)}{C_M^0 - C_m^0}. \tag{14.67}$$

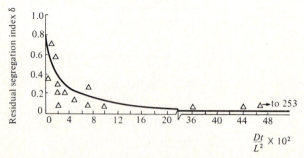

Fig. 14.14 The residual segregation index for chromium in cast 52100 steel. (From D. R. Poirier, R. V. Barone, H. D. Brody, and M. C. Flemings, *JISI*, 371, April 1970.)

[5] J. Crank, *The Mathematics of Diffusion*, Oxford University Press, London, 1957, page 58.

Equation (14.67) has been used to analyze the homogenization of chromium in cast 52100 steel (1 % C–1.5 % Cr). Figure 14.14 shows the results. The practical conclusions of such studies show that:

1. In commercial material, with relatively large dendrite arm spacings (200–400 μ), substitutional elements do not homogenize unless excessively high temperatures and long diffusion times are employed. For example, in laboratory cast 52100 ingots the dendrite arm spacing could typically be 300 μ which would have to be held at 2150°F for about 20 hours to reduce δ to 0.2 for chromium. In large commercial ingots, dendrite arm spacings are much larger and homogenization is even more difficult to achieve.

2. Significant homogenization of substitutional elements is possible at reasonable temperatures and times only if the material has fine dendrite arm spacings ($< 50\ \mu$) which result from rapid solidification.

3. Interstitial elements (such as carbon in steel) diffuse very rapidly at austenizing temperatures.

Homogenization studies have also been carried out for 4340 low-alloy steel[6] and 7075 aluminum alloy.[7] In Reference 7, the analysis and discussion of homogenization emphasizes the dissolution of a nonequilibrium second phase during solutionizing.

Example 14.7 An ingot of 52100 steel goes through the processing schedule indicated below. Estimate the residual segregation index of chromium for the material in the center and the outside surface of the finished bar. Neglect the small amount of diffusion that occurs during blooming, rolling, and final cooling. Near

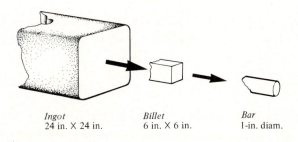

Ingot
24 in. × 24 in.

Billet
6 in. × 6 in.

Bar
1-in. diam.

[6] T. F. Kattamis and M. C. Flemings, *Trans. TMS–AIME* **233**, 992–999 (1965).
[7] S. N. Singh and M. C. Flemings, *Trans. TMS–AIME* **245**, 1803–1809 (1969).

the surface of the original cast ingot, the dendrite arm spacing is 40 μ, and in the center, it is 800 μ. The diffusion coefficient of chromium in this steel at these temperatures is given by

$$D = 2.35 \times 10^{-5} \exp\left[-17{,}300/T(°K)\right].$$

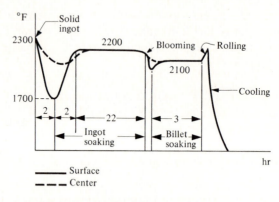

Reduction schedule for 52100 alloy steel bars.

Solution. Before proceeding directly to the solution of this problem, we should recognize that, although the basis for Eq. (14.67) assumes a constant diffusion coefficient (hence isothermal treatment), we can apply it to non-isothermal conditions.

For example, if a part is subjected to n heat treatment steps, each at a different temperature and for different times, we can compute the total magnitude of Dt/L^2 as

$$\frac{Dt}{L^2} = \frac{1}{L^2} \sum_{i=1}^{n} D_i t_i.$$

We can then use this value of Dt/L^2 in Eq. (14.67). For a continuous nonisothermal situation, we compute the total magnitude of Dt/L^2 to be used as

$$\frac{Dt}{L^2} = \frac{1}{L^2} \int_0^t D(t)\, dt.$$

For hot-working in which D and L are both time dependent, Dt/L^2 should be evaluated as

$$\frac{Dt}{L^2} = \int_0^t \frac{D(t)}{[L(t)]^2}\, dt.$$

First determine $D(t)$ given the thermal process schedule and $D(T)$.

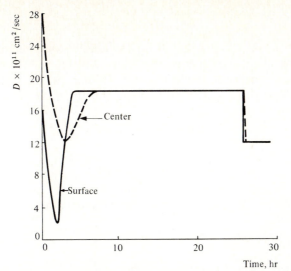

During processing, assume that the dendrite spacing decreases in proportion to the changes of linear dimensions. Based on this, $L(t)$ is given in the table below.

Time	Center of ingot L, cm	Surface of ingot L, cm
0–26 hr	0.040	0.0020
26–29 hr	0.010	0.0005

Now Dt/L^2 for the "surface" material can be evaluated by determining the area under the curve in the figure below.

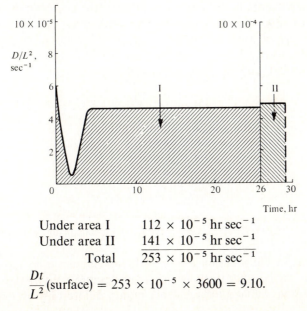

Under area I 112×10^{-5} hr sec^{-1}
Under area II 141×10^{-5} hr sec^{-1}
$\overline{\text{Total} \qquad 253 \times 10^{-5} \text{ hr sec}^{-1}}$

$$\frac{Dt}{L^2}(\text{surface}) = 253 \times 10^{-5} \times 3600 = 9.10.$$

According to Fig. 14.14, this value of Dt/L^2 indicates that the surface material would be homogeneous since $\delta \cong 0$.

In the center of the ingot, since the dendrite spacing is 20 times the spacing of the surface, then (neglecting small differences in the thermal history)

$$\frac{Dt}{L^2}(\text{center}) \cong \frac{Dt}{L^2}(\text{surface}) \times \frac{1}{20^2}$$

$$= \frac{9.10}{400} = 0.0228.$$

From Fig. 14.14, we see that $\delta \cong 0.27$, and a significant amount of microsegregation remains in the material.

14.6 FORMATION OF SURFACE TARNISH LAYERS

The rate of formation of oxide (sulfide) layers on metals and alloys exposed to oxidizing (sulfidizing) conditions is a matter of considerable technological importance, and has been the subject of a great deal of research. In general, it is not possible to say *a priori* that the rate of formation of a nonmetallic layer will be controlled by diffusion. For example, the initial rate of formation of the layer is often determined by the rate of an interface reaction between the gas and solid. As growth proceeds, if the specific volume of the oxide is much larger than that of the metal substrate, separation of the two phases may occur, causing an interruption in the growth of the oxide layer. However, in many cases this separation does not occur, and growth continues by diffusion of either the metal out through the oxide layer, or diffusion of oxygen into the metal–oxide interface, or a combination of both. In the case of adherent oxides, eventually the diffusion flux slows down to the point where it is considerably slower than the interface reaction, and diffusion controls the rate of growth of the layer from then on.

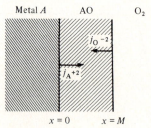

Fig. 14.15 The oxide layer on a metal showing the direction of flow of ions and electrons.

If we consider the situation in Fig. 14.15, we can derive an expression for the rate of increase in the oxide thickness M. In this example, the metal is divalent, and the oxygen anions, having a large ionic radius, diffuse only at a negligible rate through the oxide layer so that $j_{O^{2-}} = 0$. Thus the oxide thickens because the metal cations diffuse through the oxide so that the oxidation reaction can proceed. At any instant, the flux of cations through the oxide is given by Fick's law (Eq. 13.2):

$$j = (j_{A^{2+}}) = -D\left(\frac{\partial C}{\partial x}\right).$$

Here D is the diffusivity of the cation, and C is the cation concentration. If the oxide layer is thin, then we approximate the concentration profile of the cation as linear, and can integrate Fick's law as

$$j \int_0^M dx = -\int_{C_0}^{C_M} D\, dC.$$

Here, C_0 is the cation concentration at $x = 0$, and C_M is that at $x = M$. If the boundary conditions C_M and C_0 are unchanged with time, (that is, with M), then the instantaneous flux *at any thickness M* is

$$j = \frac{1}{M}\int_{C_M}^{C_0} D\, dC, \qquad (14.68)$$

or

$$j = \frac{k}{M}, \qquad (14.69)$$

where k is a constant, and has the units of mol cm^{-1}-sec^{-1}.

The flux is proportional to the rate of growth of the thickness of the oxide layer:

$$j \propto \frac{dM}{dt},$$

or

$$\frac{k}{M} \propto \frac{dM}{dt}, \qquad (14.70)$$

so that

$$\int_0^M M\, dM = \int_0^t k'\, dt,$$

or

$$M^2 = 2k't, \qquad (14.71)$$

where k' is a constant with units cm^2/sec, known as the *Tammann scaling constant*[8] or the *Pilling and Bedworth constant*.[9] The parabolic nature of the rate of change of the oxide thickness with time is apparent.

Experimentally, it is usually more convenient to measure weight gain rather than the oxide thickness. Then

$$\frac{\Delta m}{A} = M\rho_0, \qquad (14.72)$$

where $\Delta m/A = $ g (weight gained)/cm^2 (surface area), and $\rho_0 = $ density of oxygen in the oxide, g of oxygen/cm^3 of oxide.

If we substitute Eq. (14.72) into Eq. (14.71), we obtain

$$\left(\frac{\Delta m}{A}\right)^2 = \frac{2k'}{\rho_0^2}t,$$

or

$$\left(\frac{\Delta m}{A}\right)^2 = k_p t, \qquad (14.73)$$

where k_p is the *practical parabolic scaling constant*, $(g\,O_2)^2/cm^4$-sec. Usually, we obtain experimental data by weight gain measurements, and take straight-line behavior when we plot $(\Delta m/A)^2$ versus t as an indication of diffusion-controlled oxidation.

Such data alone, however, do not indicate what *species* is responsible for the major material flow, that is, whether the metal is diffusing out from the oxide–metal interface or the oxygen is diffusing in. Wagner[10] has extended this simple expression to express k_p in terms of diffusion coefficients of the migrating species.

Combining Eqs. (13.30), (13.47), and (13.49), we obtain the relation

$$n_i B_i = \frac{t_i \sigma}{(z_i e)^2}, \qquad (14.74)$$

where $n_i = $ concentration of the migrating species, $B_i = $ mobility of the species, $\sigma = $ total electrical conductivity of the compound, $t_i = $ transference number of species i, $z_i = $ valence, and $e = $ electronic charge. Now, using Eq. (13.42) and substituting Eqs. (13.47) and (14.74), we obtain the flux $\dot{n}_i$:

$$\dot{n}_i = -\frac{t_i \sigma}{(z_i e)^2}\left[\kappa_B T\left(\frac{\partial \ln n_i}{\partial x}\right) + z_i e\left(\frac{\partial \phi}{\partial x}\right)\right], \qquad (14.75)$$

---

[8] G. Tammann, *Z. anorg. u. allgem. Chem.* **111**, 78 (1920).

[9] N. B. Pilling and R. E. Bedworth, *J. Inst. Metals* **29**, 529 (1923).

[10] C. Wagner, *Atom Movements*, Amer. Soc. for Metals, Cleveland, Ohio, 1951, page 153.

or in terms of the free energy for an ideal solution

$$\dot{n}_i = -\frac{t_i \sigma}{(z_i e)^2}\left(\frac{\partial \mu_i}{\partial x} + z_i e \frac{\partial \phi}{\partial x}\right), \tag{14.76}$$

where μ_i is the chemical potential (per atom) of species i. If the compound formed has the stoichiometric composition $A_a B_b$ where A is the cation, then the scale consists of ions according to the dissociation reaction:

$$A_a B_b = aA^{z_{A^+}} + bB^{z_{B^-}},$$

where the z's are the respective valences. Since the scale cannot have a net charge, the fluxes of individual species are related by

$$z_A \dot{n}_{A^+} = z_B \dot{n}_{B^-} + \dot{n}_e. \tag{14.77}$$

Using Eq. (14.76) for each flux and recalling Eq. (13.48), we obtain an expression for $\partial \phi / \partial x$:

$$\frac{\partial \phi}{\partial x} = \frac{1}{e}\left(-\frac{t_A}{z_A}\frac{\partial \mu_{A^+}}{\partial x} + \frac{t_B}{z_B}\frac{\partial \mu_{B^-}}{\partial x} + t_e \frac{\partial \mu_e}{\partial x}\right). \tag{14.78}$$

In order to obtain an expression for the total flux in terms that we can measure, we must replace the expressions involving μ_{A^+}, μ_{B^-}, and μ_e. We first consider the equilibria

$$A = A^{z_{A^+}} + z_A e, \tag{14.79a}$$

and

$$B + z_B e = B^{z_{B^-}}. \tag{14.79b}$$

At equilibrium

$$\mu_A = \mu_{A^+} + z_A \mu_e, \tag{14.80a}$$

and

$$\mu_B = \mu_{B^-} - z_B \mu_e. \tag{14.80b}$$

For the compound in general, we get

$$d\mu_A = -\left|\frac{z_A}{z_B}\right| d\mu_B, \tag{14.81}$$

from the Gibbs-Duhem relationship. Then, eliminating μ_{A^+}, μ_{B^-}, and μ_e from Eqs. (14.78) and (14.76) and using Eqs. (14.80a, b) and (14.81), we obtain expressions for the particle fluxes:

$$\dot{n}_{A^+} = \frac{z_B t_A t_e \sigma}{z_A e^2 z_B^2}\left(\frac{\partial \mu_B}{\partial x}\right), \qquad A \text{ ions/cm}^2\text{-sec}, \tag{14.82}$$

and

$$\dot{n}_{B^-} = -\frac{t_B t_e \sigma}{e^2 z_B^2} \left(\frac{\partial \mu_B}{\partial x}\right), \qquad B \text{ ions/cm}^2\text{-sec.} \qquad (14.83)$$

Since the growth rate of the compound is the sum of the particle fluxes (although either may go to zero), we obtain

$$\dot{n}_{A_a B_b} = \frac{\dot{n}_{A^+}}{a} - \frac{\dot{n}_{B^-}}{b} = \frac{\sigma(t_A + t_B)t_e}{e^2 z_B^2 b} \left(\frac{\partial \mu_B}{\partial x}\right), \qquad \text{molecules } A_a B_b/\text{cm}^2\text{-sec.} \qquad (14.84)$$

If we use, for some reason, Eq. (14.84), we usually convert the units to give the flux in g-equivalents cm^{-2}-sec^{-1}. Denoting the number of equivalents per mole by the symbol r,* we obtain the growth rate:

$$\dot{n}_{A_a B_b} = \frac{r\sigma(t_A + t_B)t_e}{e^2 z_B^2 N_0} \left(\frac{\partial \mu_B}{\partial x}\right), \qquad \text{g-equivalents } A_a B_b/\text{cm}^2\text{-sec.} \qquad (14.85)$$

We define the *rational rate constant*, k_r, as $\int_0^x \dot{n}_{A_a B_b} \, dx$; then

$$k_r = \frac{r}{e^2 N_0 z_B^2} \int_{\mu_B^i}^{\mu_B^o} \sigma(t_A + t_B)t_e \, d\mu_B, \qquad (14.86)$$

where μ_B^i and μ_B^o are the chemical potentials of B at the inside and outside (metal–gas) interfaces of the oxide, respectively; k_r has the units of g-equivalents cm^{-1}-sec^{-1} and is related to the parabolic rate constant k_p by

$$k_p = \frac{2\rho_0(\text{at. wt } B)^2}{r(\text{mol. wt } A_a B_b)} \cdot k_r. \qquad (14.87)$$

The Tammann scaling constant is related to k_r by

$$k' = \frac{2(\text{at. wt } B)}{\rho_0 b z_B} \cdot k_r. \qquad (14.88)$$

If we assume that the anion exists as B_2 in the gaseous state, and there is ideal gas behavior, we can substitute

$$d\mu_B = \tfrac{1}{2}\kappa_B T \, d \ln p_{B_2}$$

into Eq. (14.86) and obtain

$$k_r = \frac{r\kappa_B T}{2e^2 N_0 z_B^2} \int_{p_{B_2}^i}^{p_{B_2}^o} t_e(t_A + t_B)\sigma \, d \ln p_{B_2}. \qquad (14.89)$$

We note that although t_e may approach 1.0, and $(t_A + t_B)$ may approach zero, the

---

* Note that *numerically* $r = z_B$, but that their *units* differ.

expression $(t_A + t_B)$ cannot equal zero even for an electronic oxide conductor because it would cause a charge imbalance. In an electronic semiconductor, the product $t_e(t_A + t_B)$ is a very small number, and k_r is small. In this case, where $t_e = 1.0$, *if* we assume that *mobilities* B_A and B_B are equal to B_A^* and B_B^*, respectively, we can substitute Eq. (13.47) into Eq. (14.89) and obtain

$$k_r = \frac{n_B r}{2N_0} \int_{p_{B_2}^i}^{p_{B_2}^o} \left(\left| \frac{z_A}{z_B} \right|^2 \cdot \frac{n_A}{n_B} D_A^* + D_B^* \right) d \ln p_{B_2}. \tag{14.90}$$

When D_A^* or D_B^* are very different in magnitude, we may further simplify this expression as shown in the following example.

Example 14.8 Given the diffusion data for self-diffusion of Ni^{2+} and O^{2-} ions in NiO at 1190°C, calculate the rational rate constant and the parabolic rate constant for the oxidation of Ni in pure oxygen. It is known that $D_O^* \ll D_{Ni}^*$.

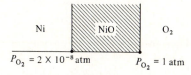

$$P_{O_2} = 2 \times 10^{-8} \text{ atm} \qquad P_{O_2} = 1 \text{ atm}$$

Solution. Since D_O^* is much less than D_{Ni}^*, we simplify Eq. (14.90) to

$$k_r = \frac{n_B r}{2N_0} \int_{p_{O_2}^i}^{p_{O_2}^o} D_{Ni}^* 2.3 \, d \log p_{O_2}.$$

The term $n_B r/N_0$ has the units of equivalents/cm³. In this case, we have

$$\frac{n_B r}{N_0} = \frac{2\rho_{NiO}}{\text{mol. wt}_{NiO}} = \frac{2(7.44)}{74.69} = 0.199 \text{ equivalents/cm}^3.$$

Alternatively, if we take z_B as electronic charges/B atom, and n_B is the concentration of B atoms/cm³, then we must divide by Faraday's constant, 96,487 couls/equivalent, and multiply by the charge on an electron, 1.602×10^{-19} couls/charge. In this case, we get

$$k_r = \frac{n_B z_B e}{\mathscr{F} 2} \int_{p_{O_2}^i}^{p_{O_2}^o} D_{Ni}^* 2.3 \, d \log p_{O_2}$$

$$= \frac{(6.00 \times 10^{22})(2)(1.602 \times 10^{-19})}{2 \times 96,487} \int_{p_{O_2}^i}^{p_{O_2}^o} D_{Ni}^* 2.3 \, d \log p_{O_2}.$$

Thus either way

$$k_r = \frac{0.199}{2} \int_{p_{O_2}^i}^{p_{O_2}^0} (D_{Ni}^*) 2.3 \, d \log p_{O_2}.$$

We evaluate the integral graphically. The figure below gives D_{Ni}^* as a function of p_{O_2} in NiO. The area under the curve is equal to the integral/2.3. The area is

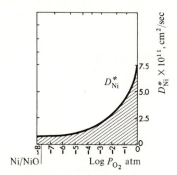

equal to 16×10^{-11}. Therefore

$$k_r = \left(\frac{0.199}{2}\right)(2.3)(16 \times 10^{-11}) = 3.66 \times 10^{-11} \text{ equivalents/cm-sec.}$$

From Eq. (14.87), the parabolic rate constant is

$$k_p = \frac{2\rho_0(\text{mol. wt}_O)^2}{r(\text{mol. wt}_{NiO})} k_r = \frac{(2)(7.44)(16)^2}{(2)(74.69)} (3.66 \times 10^{-11})$$

$$= 9.45 \times 10^{-11} \, (\text{g } O_2)^2/\text{cm}^4\text{-sec.}$$

Many metals exhibit parabolic oxidation kinetics, among them iron, nickel, cobalt, manganese, copper, and aluminum. Table 14.4 lists some typical values of k_p. Some, however, do not form compact, adherent oxides, and the kinetics of their oxidation are governed by either gas–solid reaction rates or gas–phase transport. For example, both molybdenum and tungsten form oxides that volatilize immediately upon reaction, and continually expose fresh metal to further oxidation, with no limit by solid-state diffusion on the rate.

Certain metals absorb a significant amount of oxygen in solid solution during the process of oxidation; zirconium, for example, can dissolve up to 29 at. % oxygen in solid solution prior to the formation of ZrO_2. Once the oxide layer has formed, its thickness increases in proportion to $\sqrt{k't}$, according to Eq. (14.83), and the thickness of the oxide as measured from the original surface is

$$M = \frac{2\hat{V}_{Zr}}{\hat{V}_{ZrO_2}} \sqrt{k't}, \tag{14.91}$$

Table 14.4 Parabolic oxidation constants for various metals

Metal	Oxide	$k_p, (\text{g O}_2)^2/\text{cm}^4\text{-sec}$	Conditions Temp., °C	p_{O_2}, 1 atm
Co	CoO	2.43×10^{-8}	1000	1.0
Cu	Cu_2O	6.3×10^{-9}	1000	0.083
Ni	NiO	3.8×10^{-10}	1000	1.0
Fe	FeO	1.6×10^{-7}	1000	3×10^{-14}
Fe	$FeO/Fe_3O_4/Fe_2O_3$	1.4×10^{-6}	1000	1.0
Ni–10% Cr	Complex	5.0×10^{-10}	1000	1.0
Cr	Cr_2O_3	1.3×10^{-11}	900	0.1
Fe–1% Ti	Complex	1.6×10^{-7}	1000	1.0
Al	Al_2O_3	$*8.5 \times 10^{-16}$	600	1.0

* There is a considerable variation in this number, since the surface condition appears to have a strong effect.

where $\hat{V}_{Zr}$ is the molar volume of Zr and $\hat{V}_{ZrO_2}$ is the molar volume of ZrO_2. The diffusing species in the oxide is the oxygen anion moving in from the gas phase to the ZrO_2–Zr interface. At the interface, some of the oxygen dissolves in the metal, and some reacts to form more ZrO_2. Beyond the interface and into the metallic phase, for $x > M$, we have

$$\frac{\partial C}{\partial t} = D_0 \frac{\partial^2 C}{\partial x^2}, \tag{14.92}$$

where C is the oxygen concentration, mol/cm^3. If equilibrium is established at all times at the interface, and C_e is the oxygen content of the metal in equilibrium with ZrO_2, then

$$C(M, t) = C_e, \tag{14.93a}$$

$$C(\infty, t) = 0, \tag{14.93b}$$

$$C(x, 0) = 0. \tag{14.93c}$$

The solution to Eq. (14.92) subject to Eqs. (14.91), (14.93a), (14.93b), and (14.93c) is

$$C = C_e \frac{\text{erfc}\left(\dfrac{x}{2\sqrt{D_0 t}}\right)}{\text{erfc}\left(\dfrac{\hat{V}_{Zr}}{\hat{V}_{ZrO_2}}\sqrt{\dfrac{k'}{D_0}}\right)}, \tag{14.94}$$

where x is the distance from the original interface, and $x' = x - M$, where x' is the distance from the oxide–metal interface. Thus

$$C = C_e \frac{\text{erfc}\left[\dfrac{x'}{2\sqrt{D_O t}} + \dfrac{\hat{V}_{Zr}}{\hat{V}_{ZrO_2}}\sqrt{\dfrac{k'}{D_O}}\right]}{\text{erfc}\left[\dfrac{\hat{V}_{Zr}}{\hat{V}_{ZrO_2}}\sqrt{\dfrac{k'}{D_O}}\right]}. \tag{14.95}$$

14.7 SURFACE COATINGS

The processes of galvanizing, tin-plating, chromizing, electroless nickel plating, and other surface treatments utilize elevated reaction temperatures or annealing temperatures, to obtain a specified degree of reaction of the coating material with the base material or a specific coating thickness. In most instances, this involves formation of an intermediate phase and growth of this phase to some minimum thickness. Prediction of growth rates in such systems is very difficult because about half of the cases which have been studied thus far have shown diffusion-controlled kinetics, but the others have not exhibited the same behavior. In these cases, kinetics appear to be controlled by the interface reaction rates.

When diffusion controls the rate of intermediate phase formation, then the thickness of the phase increases in proportion to $t^{1/2}$. This has been observed when Nb is clad on U,[11] Sn is plated on Fe[12] (except for early stages of the growth of FeSn$_2$), and Al is clad on Zn.[13] Theoretical analyses of formation and growth of multiphase coatings have been made,[14,15] but all end up with a parabolic relationship between thickness and time and a constant that is often useless in terms of prediction or analysis, because of its complexity and/or unavailability of numerical values for required parameters.

PROBLEMS

14.1 A diffusion couple of a bar of copper consisting of 5 at. % Zn welded to a bar containing 25 at. % Zn interdiffuses for 50 hr at 1000°C. The interdiffusion coefficient is assumed to be

[11] N. L. Peterson and R. E. Ogilvie, *Trans. AIME* **218**, 439 (1960).
[12] D. R. Gabe, *JISI* **204**, 95 (1966).
[13] A. U. Seybolt, *Trans. ASM* **29**, 937 (1941).
[14] J. S. Kirkaldy, *Can. J. Phys.* **36**, 899 (1958).
[15] J. S. Kirkaldy, *Can. J. Phys.* **37**, 30 (1959).

constant (independent of concentration) and equals 2×10^{-10} cm²/sec at 1000°C. Markers are inserted at the original interface and move along during the diffusion process at a composition of 20.205 atom fraction Zn. Determine the intrinsic diffusion coefficients of copper and zinc at 20.205 atom fraction Zn.

14.2 The term "banding" is used to describe chemical heterogeneity in rolled steels that shows up as closely spaced light and dark bands in the microstructure of steel. These bands represent areas of segregation of alloying elements during freezing of the ingot. During rolling the segregated areas are elongated and compressed into narrow bands. Assume that the alloy concentration varies sinusoidally with distance after rolling according to the sketch below.

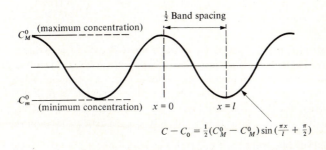

$$C - C_0 = \tfrac{1}{2}(C_M^0 - C_m^0)\sin\left(\tfrac{\pi x}{l} + \tfrac{\pi}{2}\right)$$

If the steel is now heated to the austenite range and held at some constant temperature, then

a) schematically sketch the concentration profile as time passes;

b) write a differential equation for concentration (state assumptions) and

c) write the boundary conditions (for time and space) that apply;

d) solve for the concentration as a function of time and space.

14.3 The solubility of hydrogen in solid copper at 1000°C is 1.4 ppm (by weight) under a pressure of hydrogen of 1 atm. At 1000°C, $D_H = 10^{-6}$ cm²/sec.

a) Determine the time for hydrogen to reach a concentration of 1.0 ppm at a depth of 0.1 mm in a large chunk of copper initially with null hydrogen if the copper is subjected to 2 atm pressure of H_2 at 1000°C.

b) Copper foil, 0.2 mm thick, is equilibrated with hydrogen at a pressure of 4 atm of hydrogen at 1000°C. The same foil is then placed in a perfect vacuum at 1000°C and held for 10 hr. Calculate the concentration of hydrogen at the center of the foil after the 10-hr period.

14.4 One side of an iron sheet, 0.01 cm thick, is subjected to a carburizing atmosphere at 1700°F such that a surface concentration of 1.2% carbon is maintained. The opposite face is maintained at 0.1% carbon. At steady state, determine the flux (g-moles/cm² sec) of carbon through the sheet

a) if the diffusion coefficient is assumed to be independent of concentration ($D = 2 \times 10^{-7}$ cm²/sec);

b) if the diffusion coefficient varies as shown in the graph on page 512.

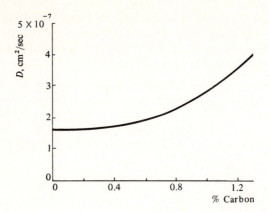

14.5 A composite foil made of metal A bonded to metal B, each 0.01 cm thick, is subjected to $\frac{1}{2}$ atm of pure hydrogen on metal A's face; the other side, metal B's face, is subjected to a perfect vacuum. At the temperature of interest and 1 atm of hydrogen, the solubility of hydrogen in metal A is 4×10^{-4} g per cm^3 of A and in B it is 1×10^{-4} g per cm^3 of B. It is also known that hydrogen diffuses four times faster in A than B and that A and B do not diffuse in each other. At steady state draw the concentration profile of hydrogen across the composite foil.

14.6 A gold–nickel diffusion couple of limiting compositions $X_{Ni} = 0.0974$ and $X_{Ni} = 0.4978$ is heated at 925°C for 2.07×10^6 sec. Layers 0.003 in. thick and parallel to the original interface are machined off and analyzed.

a) Using the data tabulated below, calculate the diffusion coefficient at 20, 30, and 40 at.% nickel.

b) Suppose that markers are inserted at the original interface and move along during the diffusion process at a composition of 30 atom fraction nickel. From this, determine the intrinsic coefficients of gold and nickel at 30 atom fraction nickel.

Slice No.	a/o Ni	Slice No.	a/o Ni	Slice No.	a/o Ni	Slice No.	a/o Ni
11	49.78	22	33.17	29	21.38	39	12.55
12	49.59	23	31.40	30	20.51	41	11.41
14	47.45	24	29.74	31	19.12	43	10.48
16	44.49	26	25.87	32	17.92	45	9.99
18	40.58	27	24.11	33	16.86	47	9.74
19	38.01	28	22.49	35	15.49		
20	37.01			37	13.90		
21	35.10			38	13.26		

14.7 By making use of the shell balance technique, derive an expression for diffusion through a hollow cylinder.

14.8 A thin sheet of iron at 800°C is subjected to different gaseous atmospheres on both of its surfaces such that the composition of one face is at 4 atom percent carbon and the other is at zero atom fraction carbon. At steady state, make a plot of the composition profile in the sample indicating *clearly* compositions and respective distances.

The original thickness is 1 mm and density changes during the experiment may be neglected. At 800°C, it is known that the diffusion coefficient of carbon in iron is given by:

$$D = 10^{-6} \text{ cm}^2/\text{sec in ferrite } (\alpha),$$

$$D = 10^{-8} \text{ cm}^2/\text{sec in austenite } (\gamma).$$

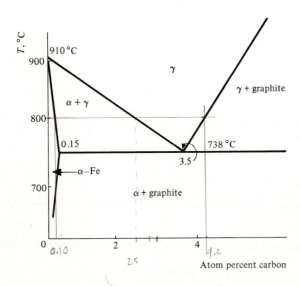

14.9 In order to make a transformer steel with the proper hysteresis loop, a low silicon steel sheet is to be exposed on both sides to an atmosphere of $SiCl_4$ which dissociates to Si(g) and Cl_2(g). The Si(g) dissolves in the steel up to 3% at equilibrium.

a) Indicate what equation and what boundary and initial conditions would apply in order to calculate how long you would have to hold the sample before the composition is essentially uniform at 3% across the sample.

b) Using the data in Fig. 13.12, calculate the time required to reach 95% of the 3% Si at 1800°F.

15

MASS TRANSFER IN FLUID SYSTEMS

In Chapters 2 and 7, we demonstrated the development of differential equations pertinent to momentum and energy transport in simple fluid systems. In this chapter, we shall consider how to formulate and describe elementary diffusion and mass transfer problems in fluid systems. We use practically the same procedure in this situation as we did previously; we develop a differential equation, and a solution containing arbitrary constants evolves. These constants are evaluated by applying boundary conditions that specify the concentration or the mass flux at the bounding surfaces. Again, we demonstrate the principles involved by considering specific examples, but first let us reconsider the general situation, as outlined in Section 13.2.2. $P6, 437$

Species A in a gas stream moving in the x-direction is under the influence of a concentration gradient, also in the x-direction. The molar flux of A relative to stationary coordinates is then made up of two parts: $C_A v_x^*$ which is the molar flux of A resulting from the bulk motion, and $j_{Ax} = -CD_A(\partial X_A/\partial x)$ which is the diffusive contribution. Thus

$$N_{Ax} = -CD_A \frac{\partial X_A}{\partial x} + C_A v_x^* = C_A v_{Ax}. \tag{15.1}$$

Here v_x^* is the local *molar average velocity* in the x-direction, and v_{Ax} is the velocity of A in the x-direction with respect to stationary coordinates, and C is the local total molar concentration in the solution. Thus, we define v_x^* so that the *total molar flux* of all components in the x-direction is made up of the sum of the n component fluxes in the same direction:

$$C v_x^* = \sum_{i=1}^{n} C_i v_{ix}. \tag{15.2}$$

For a binary $A-B$ system, we write

$$v_x^* = \frac{1}{C}\left(C_A v_{Ax} + C_B v_{Bx}\right)$$

$$= \frac{1}{C}\left(N_{Ax} + N_{Bx}\right). \tag{15.3}$$

When we combine Eq. (15.3) with Eq. (15.1), we obtain a form of Fick's first law for a binary solution:

$$N_{Ax} = X_A(N_{Ax} + N_{Bx}) - CD_A \frac{\partial X_A}{\partial x}. \qquad (15.4)$$

15.1 DIFFUSION THROUGH A STAGNANT GAS FILM

Consider the system shown in Fig. 15.1 where liquid A is evaporating into gas B, and a constant liquid level at $x = 0$ is maintained. At the liquid–gas phase interface $(x = 0)$, the gas phase concentration of A is that corresponding to the vapor pressure of A at that temperature.* For simplicity, also assume that the solubility of B in liquid A is negligible and that the entire system is maintained at a constant temperature and pressure. At the top of the tube, a stream of A–B gas flows past slowly, thereby maintaining a constant concentration of A at $x = l$ which is less than the liquid–gas interface concentration. Therefore, a concentration difference of A exists between $x = 0$ and $x = l$, which causes diffusion. When the system attains a steady state, there is a net motion of A away from the evaporating surface and the vapor B is stationary.

Under these conditions, despite the fact that gas B is stationary, there is bulk motion of fluid since A itself is moving, and its motion contributes to the average velocity. Thus we refer to Eq. (15.4) for the flux of A with $N_{Bx} = 0$. Solving for N_{Ax}, we obtain

$$N_{Ax} = -\frac{CD_A}{1 - X_A}\frac{dX_A}{dx}. \qquad (15.5)$$

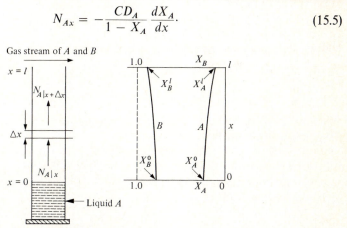

Fig. 15.1 Diffusion of A through B at steady state. B is not in motion, but note that the graph shows how its concentration profile is not linear because of the motion of A.

---

* This, of course, implies that equilibrium is maintained at the interface, i.e., from a very simplified mechanistic viewpoint, atoms (or molecules) can readily leave the liquid state and enter the gas phase. This assumption is valid except at very high diffusion rates where the rate of transfer of atoms across the liquid–gas interface is not able to keep pace with the exhaustion of the atoms away from the interface.

A mass balance on a unit volume Δx of column height (see Fig. 15.1) for steady state is

$$SN_{Ax}|_x - SN_{Ax}|_{x+\Delta x} = 0, \tag{15.6}$$

in which S is the cross-sectional area. In the expected manner, we divide by Δx and take the limit as $\Delta x \to 0$:

$$\frac{dN_{Ax}}{dx} = 0. \tag{15.7}$$

Substitution of Eq. (15.6) for N_{Ax} yields

$$\frac{d}{dx}\left(\frac{CD_A}{1 - X_A} \cdot \frac{dX_A}{dx}\right) = 0. \tag{15.8}$$

As pointed out in Chapter 13, diffusion coefficients for isothermal gas solutions are very nearly independent of concentration; also, C is constant for an ideal gas mixture at constant temperature and pressure. Hence we simplify the derivative further:

$$\frac{d}{dx}\left(\frac{1}{1 - X_A} \cdot \frac{dX_A}{dx}\right) = 0. \tag{15.9}$$

Two successive integrations can be made directly, resulting in

$$-\ln(1 - X_A) = c_1 x + c_2, \tag{15.10}$$

and we determine the constants by use of the boundary conditions:

$$\text{B.C.1:} \qquad \text{at } x = 0, \qquad X_A = X_A^0; \tag{15.11a}$$

$$\text{B.C.2:} \qquad \text{at } x = l, \qquad X_A = X_A^l. \tag{15.11b}$$

When we evaluate the constants and substitute them into Eq. (15.10), we obtain the concentration profile:

$$\ln\left(\frac{1 - X_A}{1 - X_A^0}\right) = \frac{x}{l}\ln\left(\frac{1 - X_A^l}{1 - X_A^0}\right), \tag{15.12}$$

or

$$\ln\left(\frac{X_B}{X_B^0}\right) = \frac{x}{l}\ln\left(\frac{X_B^l}{X_B^0}\right). \tag{15.13}$$

Figure 15.1 shows these solutions, where the slope dX_A/dx is not uniform with x although the flux N_{Ax} is. A gradient of A in the gas must be accompanied by a gradient of B. Consequently, B has a tendency to diffuse down the column, but this diffusive tendency is exactly compensated by the opposing bulk motion of the gas in the direction of diffusion of gas A.

The information that is most often sought after is the rate of mass transfer at the liquid–gas interface. We obtain this by using Eq. (15.5):

$$N_{Ax}|_{x=0} = -\frac{CD_A}{1 - X_A^0}\left(\frac{dX_A}{dx}\right)_{x=0} = \frac{CD_A}{l}\ln\left(\frac{1 - X_A^l}{1 - X_A^0}\right), \tag{15.14}$$

or

$$N_{Ax}|_{x=0} = \frac{CD_A}{(X_B)_{\ln}}\left(\frac{X_A^0 - X_A^l}{l}\right), \tag{15.15}$$

where $(X_B)_{\ln}$ is the *log mean* of the terminal values of X_B.

$$(X_B)_{\ln} = \frac{X_B^l - X_B^0}{\ln(X_B^l/X_B^0)}. \tag{15.16}$$

Equation (15.15) is somewhat more appealing because a characteristic concentration difference $X_A^0 - X_A^l$ over a distance l is evident. For a gas in which species A is dilute, Eq. (15.15) reduces to

$$N_{Ax}|_{x=0} = D_A\left(\frac{C_A^0 - C_A^l}{l}\right), \tag{15.17}$$

which could have resulted from originally ignoring bulk motion and expressing the flux of A simply as

$$j_{Ax} = -D_A\frac{dC_A}{dx}. \tag{15.18}$$

Typical applications of Eqs. (15.14) and (15.15) are evaporation and sublimation processes which involve diffusion of the vapor being created (gas A) through a stationary gas (gas B). Also, a method for measuring diffusion coefficients is to measure the rate of fall of liquid A in a small glass tube as gas B passes over the top. Furthermore, these results find use in the "film theories" of mass transfer.

Example 15.1   In order to determine the diffusivity of Mn in the gas phase, a melt of pure Mn is held in a chamber at 1600°C through which pure Ar flows. The level of the Mn is 2.0 cm below the edge of the crucible. The weight of the crucible is monitored continuously, and when the rate of weight loss is steady with time, that rate is found to be 2.65×10^{-7} mol cm^{-2}-sec^{-1}. Calculate $D_{\text{Mn–Ar}}$.

Solution. At 1600°C, $P_{\text{Mn}}^0 = 0.03$ atm, which may be taken as the pressure just above the liquid surface.

The pressure of manganese may be taken as zero at the crucible edge, as the argon flowing across the opening removes it immediately.

Since the concentration of manganese is clearly dilute, $D_{\text{Mn-Ar}}$ is obtained directly from Eq. (15.17). The concentration is expressed in mol cm^{-3}.

$$C_A^0 = \frac{P_A^0}{RT} = \frac{0.03 \text{ atm}}{}\left|\frac{\text{mol-}^\circ\text{K}}{0.08205 \text{ l-atm}}\right|\frac{1}{10^3 \text{ cm}^3}\left|\frac{}{1873^\circ\text{K}}\right.$$

$$= 1.61 \times 10^{-7} \text{ mol cm}^{-3};$$

$$D_{\text{Mn-Ar}} = \frac{2.65 \times 10^{-7} \times 2.0}{1.61 \times 10^{-7}} = 3.3 \text{ cm}^2/\text{sec}.$$

15.2 DIFFUSION IN A MOVING GAS STREAM

Figure 15.2 illustrates one technique which we use to determine the vapor pressure of a metal (liquid or solid). Argon, as a carrier gas, passes over the sample which is at the temperature corresponding to the vapor pressure being determined. This gas, containing the saturation concentration of the metal vapor, enters the exit tube at $z = 0$, and at the cool end of the exit tube the metal condenses out and deposits where we can collect it for subsequent mass determination.

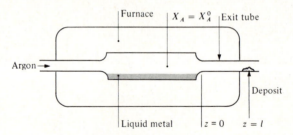

Fig. 15.2 Diffusion in a moving gas stream.

A mass balance applied to a section Δz long for steady state yields

$$\frac{dN_{Az}}{dz} = 0. \tag{15.19}$$

We may choose either Eq. (15.1) or Eq. (15.4) to represent N_{Az}; here we select Eq. (15.1) from which we write

$$v_z^* \frac{dC_A}{dz} - \frac{d}{dz}\left(CD_A \frac{dX_A}{dz}\right) = 0. \tag{15.20}$$

We can certainly consider that the argon–metal gas solution is ideal, and if the temperature variation between $z = 0$ and $z = l$ is small, then C and D_A are constants, and Eq. (15.20) takes the form

$$\frac{d^2 X_A}{dz^2} - \frac{v_z^*}{D_A}\frac{dX_A}{dz} = 0. \tag{15.21}$$

The boundary conditions can be represented as:

$$\text{B.C.1:} \qquad \text{at } z = 0, \qquad X_A = X_A^0 \text{ (saturation value)}, \qquad (15.21\text{a})$$

$$\text{B.C.2:} \qquad \text{at } z = l, \qquad X_A = 0. \qquad (15.21\text{b})$$

The second boundary condition implies that the temperature at l is low enough so that the vapor pressure of A is negligible. We can obtain the solution to Eq. (15.21) directly by integrating twice, or by treating it as a linear homogeneous differential equation with constant coefficients. Applying the latter method, the solution is

$$X_A = c_1 e^{r_1 z} + c_2 e^{r_2 z}, \qquad (15.22)$$

where r_1 and r_2 are the roots of

$$r^2 - \frac{v_z^*}{D_A} r = 0. \qquad (15.23)$$

Thus, $r_1 = 0$ and $r_2 = v_z^*/D_A$, and the solution is

$$X_A = c_1 + c_2 \exp\left(\frac{v_z^*}{D_A} z\right). \qquad (15.24)$$

We evaluate the arbitrary constants by using Eqs. (15.24) and (15.21a and b):

$$c_2 = -\frac{X_A^0}{\left[\exp\left(\dfrac{v_z^* l}{D_A}\right) - 1\right]}, \qquad (15.25)$$

and

$$c_1 = X_A^0 \left[1 + \frac{1}{\exp\left(\dfrac{v_z^* l}{D_A}\right) - 1}\right]. \qquad (15.26)$$

The concentration profile can then be written:

$$\frac{X_A}{X_A^0} = \frac{\exp\left(\dfrac{v_z^* l}{D_A}\right) - \exp\left(\dfrac{v_z^* z}{D_A}\right)}{\exp\left(\dfrac{v_z^* l}{D_A}\right) - 1}. \qquad (15.27)$$

We evaluate the flux at which the metal vapor enters the exiting gas stream at $z = 0$.

$$N_{Az}|_{z=0} = C\left[-D_A\left(\frac{\partial X_A}{dz}\right)_{z=0} + X_A^0 v_z^*\right],$$

or

$$N_{Az}|_{z=0} = Cv_z^* X_A^0 \left[1 - \frac{1}{1 - \exp\left(\dfrac{v_z^* l}{D_A}\right)} \right]. \tag{15.28}$$

If S is the cross-sectional area of the tube, then $SN_{Az}|_{z=0}$ represents the amount of A passing through the tube; experimentally, we determine this quantity by weighing the condensate formed at $z = l$ over a measured period of time. The product SCv_z^* is the total molar flow down the tube, and simply represents the number of moles of argon passed per unit time plus the moles of condensate collected per unit time. The vapor pressure of A is related to these experimental quantities by

$$\frac{P_A^0}{P} = X_A^0 = \frac{SN_{Az}|_{z=0}}{SCv_z^*} \left[\frac{\exp\left(\dfrac{v_z^* l}{D_A}\right) - 1}{\exp\left(\dfrac{v_z^* l}{D_A}\right)} \right], \tag{15.29}$$

where P_A^0 is the vapor pressure of A, and P is the total pressure. Preferably, the effect of diffusion should be negligible for best experimental results due to the uncertainty of the diffusion coefficient and because we would have to assume an experimental set up that corresponds to the mathematical formulation.

The real value of the analysis lies in the group $v_z^* l/D_A$ which indicates how to set up the experiment, so that the effects of diffusion may be ignored. If $v_z^* l/D_A \geq 5$, the effect of diffusion may be ignored because the last term in Eq. (15.29) is ≥ 0.993 and < 1.0. To insure sufficiently high values of $v_z^* l/D_A$, the experimentalist should provide a small diameter tube between $z = 0$ and $z = l$, and use argon or nitrogen as the carrier gas, rather than hydrogen or helium, since D in the lighter gases is larger than in the heavier gases.

Example 15.2 An experimental apparatus is being constructed to study the thermodynamics of Mn–Cu alloys by measuring the Mn vapor pressure over the molten alloys at 1400°K. In order to use the transport technique, what exit tube dimensions and argon gas flow rates should be used?

Solution. The criterion that provides the most direct experimental measurement of P_{Mn}^0 (the equilibrium pressure over the alloy) is

$$P_{Mn}^0 = P \frac{N_{Mn}}{N},$$

where N_{Mn} is the number of moles of Mn condensed out and $N = N_{Mn} + N_{Ar}$ = total moles of gas passing through the exit tube over some period of time. This is essentially true when $v^* l/D_{Mn-Ar} \geq 5.0$.

We evaluate $D_{\text{Mn–Ar}}$ at $1400°\text{K}$ as $2.6\ \text{cm}^2\text{-sec}^{-1}$. Therefore, v^*l must be greater than $13.0\ \text{cm}^2\text{-sec}^{-1}$. Since $v^* \cong 4\dot{n}RT/\pi d^2 P$, and $\dot{n} \cong \dot{n}_{\text{Ar}}$ (mol Ar sec^{-1}), then

$$\frac{4\dot{n}_{\text{Ar}}RTl}{\pi P} \geq 13.0 d^2.$$

The figure below shows the results, with the preferred design being to the right-hand side of each curve.

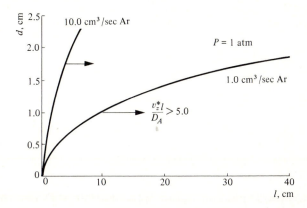

At temperatures where P^0_{Mn} is greater than negligible values, and $\dot{n} > \dot{n}_{\text{Ar}}$, the curves should be shifted even further to the right for better results.

15.3 DIFFUSION INTO A FALLING LIQUID FILM

In this section, we shall consider a fluid system moving in such a way that the velocity distribution is unaffected by diffusion into the fluid. Figure 15.3 shows a film of liquid B falling in laminar flow down a wall. Gas A is absorbed at the liquid–gas interface; we shall restrict the situation to that where the penetration distance of A into B is small relatively to the film thickness. We wish to calculate the amount of gas absorbed after the film travels a distance L.

First, we develop a mass balance on component A. The gas–liquid interface concentration of A at all points along the film, is the saturation value C^0_A; thus A diffuses into the liquid which initially contains less than the saturation amount of A. As the film drops, the liquid is exposed to C^0_A for a longer time and more penetration of A into the film results. We see, therefore, that C_A changes both with x and z, and we select the unit volume: Δx by Δz by unity in the y-direction. Then the mass balance for A is simply

$$N_{Az}|_z \cdot \Delta x - N_{Az}|_{z+\Delta z} \cdot \Delta x + N_{Ax}|_x \cdot \Delta z - N_{Ax}|_{x+\Delta x} \cdot \Delta z = 0. \quad (15.30)$$

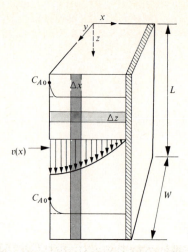

Fig. 15.3 Absorption into a falling film.

Dividing by $\Delta x \Delta z$, and performing the usual limiting process, we get

$$\frac{\partial N_{Az}}{\partial z} + \frac{\partial N_{Ax}}{\partial x} = 0. \tag{15.31}$$

We now have to insert into this equation the expressions for N_{Az} and N_{Ax}. For the molar flux in the z-direction, we write

$$N_{Az} = -CD_A \frac{\partial X_A}{\partial z} + C_A v_z^*, \tag{15.32}$$

and neglect the diffusive contribution, realizing that A moves in the z-direction primarily due to bulk flow. In addition, for small increases in the concentration of A, $v_z^* \cong v_z$. Finally, we give N_{Az} by the simplified expression

$$N_{Az} = C_A v_z. \tag{15.33}$$

For the molar flux in the x-direction, we write

$$N_{Ax} = -D_A \frac{\partial C_A}{\partial x} + C_A v_x^*. \tag{15.34}$$

That is, in the x-direction, A is transported primarily by diffusion, there being almost no bulk flow in the x-direction due to the small solubility of A in B. Substitution of Eqs. (15.32) and (15.34) into Eq. (15.31) yields the differential equation for $C_A(x, z)$:

$$v_z \frac{\partial C_A}{\partial z} = D_A \frac{\partial^2 C_A}{\partial x^2}. \tag{15.35}$$

When dealing with energy transport in a convective system, it is necessary to obtain the velocity profile $v_z(x)$; similarly, we need to describe the velocity for the analogous situation of mass transfer in a convective system. For a falling film, we have already worked this out in Chapter 2 in the absence of mass transfer at the fluid surface, and we know that the results for fully developed flow are

$$v_z = v_{max}\left[1 - \left(\frac{x}{\delta}\right)^2\right],$$

where δ is the film thickness, and x is the distance into the film from the gas–liquid surface.

If, as indicated in Fig. 15.3, A has penetrated only a slight distance into the film, then for the most part species A sees only v_{max}. Further, since it does not penetrate very far, we can consider that the liquid is semi-infinite. These conditions would hold, for example, for short contact times. With these approximations we write the differential equation and the boundary conditions:

$$v_{max}\frac{\partial C_A}{\partial z} \cong D_A\frac{\partial^2 C_A}{\partial x^2}. \tag{15.36}$$

At $z = 0$, $C_A = C_A^i$, $x \geq 0$,   (15.37a)

at $x = 0$, $C_A = C_A^0$, $L \geq z \geq 0$,   (15.37b)

at $x = \infty$, $C_A = C_A^i$, $L \geq z \geq 0$.   (15.37c)

Note that we may alternatively view z/v_{max} as the time t, over which a moving slice of liquid has been subjected to the surface concentration C_A^0. Thus, we recognize the solution to Eq. (15.36), subject to Eqs. (15.37a), (15.37b), and (15.37c), as the solution for the temperature distribution in a semi-finite solid, initially at a uniform temperature, which is suddenly subjected to a new constant surface temperature. Then, referring to Eq. (9.110), we have

$$\frac{C_A - C_A^0}{C_A^i - C_A^0} = \text{erf}\frac{x}{2\sqrt{D_A z/v_{max}}}. \tag{15.38}$$

Now knowing the concentration profile, we proceed to determine the local diffusion mass flux at the surface, $x = 0$:

$$N_{Ax}|_{x=0} = -D_A\left(\frac{\partial C_A}{\partial x}\right)_{x=0} = (C_A^0 - C_A^i)\sqrt{\frac{D_A v_{max}}{\pi z}}, \tag{15.39}$$

the average rate of A transferred per unit across the entire surface between $z = 0$ and $z = L$ being

$$\bar{N}_{Ax}|_{x=0} = \frac{1}{L} \int_0^L N_{Ax}|_{x=0}\, dz$$

$$= 2(C_A^0 - C_A^i)\sqrt{\frac{D_A v_{max}}{\pi L}}. \tag{15.40}$$

Pigford[1] solved the complete equation

$$v_{max}\left[1 - \left(\frac{x}{\delta}\right)^2\right]\frac{\partial C_A}{\partial z} = D_A\frac{\partial^2 C_A}{\partial x^2}, \tag{15.41}$$

and obtained the result:

$$\frac{\bar{C}_A^L - C_A^0}{C_A^i - C_A^0} = 0.7857 e^{-5.1213\eta} + 0.1001 e^{-39.318\eta} + \cdots, \tag{15.42}$$

where $\eta = D_A L/\delta^2 v_{max}$, and $\bar{C}_A^L$ = bulk average composition of the liquid at L.

For small values of η, corresponding to short contact times or very thick films, we obtain

$$\frac{\bar{C}_A^L - C_A^0}{C_A^i - C_A^0} = \sqrt{\frac{6}{\pi}}\sqrt{\frac{D_A L}{\delta^2 \bar{V}}}, \tag{15.43}$$

and for long times, we have

$$\frac{\bar{C}_A^L - C_A^0}{C_A^i - C_A^0} = 0.7857 e^{-5.1213\eta}. \tag{15.44}$$

15.4 THE MASS-TRANSFER COEFFICIENT

In Chapter 7, we analyzed simple problems of heat transfer with laminar convection, and formulated the temperature distribution from which we calculated heat-transfer rates by evaluating the heat fluxes at the fluid–solid boundary. With the fluxes, we wrote expressions for the heat-transfer coefficients, and we saw that $\mathrm{Nu} = f(\mathrm{Re}, \mathrm{Pr})$, and gained insight into what was to follow in Chapter 8 where we presented the correlations for heat transfer in turbulent convective systems.

Having considered diffusion in the presence of forced convection in Section 15.3, it is convenient to introduce the *mass-transfer coefficient*. As we have mentioned, we may treat the movement of a species as the sum of a diffusional contribution and a bulk flow contribution (see Eq. 15.1). To be analogous with heat

---

[1] R. L. Pigford, *Ph.D. Thesis*, University of Illinois, 1941.

transfer, a mass-transfer coefficient for transfer of A into or out of a phase is defined in terms of the *diffusive* contribution normal to the interface:

$$k_M = \frac{j_A^0}{C_A^0 - C_{A\infty}} = -\frac{D_A(\partial C_A/\partial x)_{x=0}}{C_A^0 - C_{A\infty}}. \qquad (15.45)$$

Here, the superscript 0 refers to quantities evaluated at the interface, and $C_{A\infty}$ to some concentration of A within the fluid, usually the bulk concentration. Note that, while k_M in Eq. (15.45) is defined in terms of the diffusion flux at the surface, in general, at interfaces involving a fluid phase, there is the additional contribution to mass transfer caused by bulk flow. We define the mass-transfer coefficient here only in terms of the diffusive contribution, rather than of the total flux N_A^0. This is because the coefficient so defined is somewhat more fundamental, since we might expect the diffusion flux to be approximately proportional to a characteristic concentration difference as indicated by Eq. (15.45), whereas the bulk flow contribution can be relatively independent of any concentration difference. Similarly, when both heat and mass transfer occur, it is advantageous to retain the definition of the heat-transfer coefficient given by Eq. (8.1), which considers only the conduction flux.

In the limit of low mass-transfer rates, as is often the case, we may neglect the distortion of the velocity and concentration profiles by mass transfer, and the bulk flow term is negligible. Then

$$k_M = \frac{N_A^0}{C_A^0 - C_{A\infty}}. \qquad (15.46)$$

This equation is definitional only, and we must evaluate it by means of various analytical expressions for the flux. As the first example of applying Eq. (15.46), consider the results of diffusion into a falling film in Section 15.3. We evaluate the *local* mass-transfer coefficient relating the rate of mass transfer of A into the liquid when the time of contact is short, by substituting Eq. (15.39) into Eq. (15.46).

$$k_{M,z} = \frac{N_{Ax}|_{x=0}}{C_A^0 - C_{A\infty}} = \sqrt{\frac{D_A v_{max}}{\pi z}}, \qquad (15.47)$$

or in terms of dimensionless groups, and recalling that $v_{max} = \frac{3}{2}\bar{V}$,

$$\frac{k_{M,z}z}{D_A} = \sqrt{\frac{3}{2\pi}}\sqrt{\frac{\bar{V}z}{v}}\sqrt{\frac{v}{D_A}}. \qquad (15.48)$$

The group $k_{M,z}z/D_A$ is called the *Sherwood number*, Sh, or alternatively the *mass transfer Nusselt number*, Nu_M. The Reynolds number should be easily recognized. The Schmidt number, Sc, which is defined by

$$Sc = \frac{v}{D}, \qquad (15.49)$$

is the analog of the Prandtl number encountered in heat transfer. Most available forced-convection mass-transfer correlations are in the form

$$\text{Sh} = f(\text{Re, Sc, geometry}),* \tag{15.50}$$

as, for example, in the situation above. Specifically, we could write Eq. (15.48) as

$$\text{Sh}_z = \sqrt{\frac{3}{2\pi}}\,\text{Re}_z^{1/2}\text{Sc}^{1/2}. \tag{15.51}$$

In this case, by subscripting with z, we emphasize that local values are being considered. If we used Eq. (15.40) instead of Eq. (15.39), we would define an *average* mass-transfer coefficient over the film length L.

$$
\begin{aligned}
k_M &= \frac{(C_A^0 - C_A^i)\left(\dfrac{4D_A v_{\max}}{\pi L}\right)^{1/2}}{(C_A^0 - C_A^i)} \\[2mm]
&= \sqrt{\frac{6D_A \bar{V}}{\pi L}}.
\end{aligned} \tag{15.52}
$$

We can then write this as

$$\frac{k_M L}{D_A} = \sqrt{\frac{6}{\pi}}\sqrt{\frac{\bar{V}L}{v}}\sqrt{\frac{v}{D_A}},$$

or in dimensionless form

$$\text{Sh}_L = \sqrt{\frac{6}{\pi}}\,\text{Re}_L^{1/2}\text{Sc}^{1/2}, \tag{15.53}$$

where the subscript L indicates that the quantities are averaged over the entire film length.

In the case of long contact times, where Eq. (15.44) applies, the rate at which A is absorbed in the distance dz is

$$\bar{V}\delta\,d\bar{C}_A = k_M(C_A^0 - \bar{C}_A)\,dz.$$

Over the entire length of the film, the absorption rate of A is

$$\bar{V}\delta \int_{C_A^i}^{\bar{C}_A^L} \frac{d\bar{C}_A}{C_A^0 - \bar{C}_A} = k_M \int_0^L dz.$$

The integration yields

$$k_M = \frac{\bar{V}\delta}{L}\ln\frac{C_A^0 - C_A^i}{C_A^0 - \bar{C}_A^L}. \tag{15.54}$$

---

* The product ReSc which we often encounter in the literature on mass transfer, is sometimes called the Peclet number, Pe.

Substituting Eq. (15.44), we obtain

$$k_M = \frac{\bar{V}\delta}{L} \left[\ln\left(e^{5.1213\eta} - \ln 0.7857\right)\right]$$

$$= \frac{\bar{V}\delta}{L} \left[5.1213\eta + 0.241\right]$$

$$\cong 3.42 \frac{D_A}{\delta}. \tag{15.55}$$

By rearranging this expression, we get

$$\frac{k_M \delta}{D_A} = \text{Sh} \cong 3.42, \tag{15.56}$$

which is similar to the results given in Table 7.1 where the Nusselt number was found to be a constant for fully developed laminar flow. We consider that this equation is applicable at $\text{Re}\,(= \Gamma/\eta) \le 25$, where Γ is the mass flow rate per unit width of film.

Having been tested for absorption of gases into liquids flowing down wetted-wall columns, Eq. (15.56) has been found to somewhat underestimate the actual mass-transfer coefficient. At low Reynolds numbers, this is now understood to be partly due to the so-called *Marangoni effect*, in which upward-directed surface tension forces counteract the downward-directed gravitational forces, causing rippling and turbulence on the surface and an increase in transfer which is not anticipated in the simplified development described above.

Example 15.3 A method for degassing molten metals involves exposing a thin film of metal to vacuum by allowing it to flow continuously over an inclined plate. Calculate the average hydrogen concentration of a ferrous alloy with an initial concentration of $1\ cm^3\ H_2$ (STP) per cm^3 of alloy flowing down a plate 100 cm long and 15 cm wide, which is inclined at an angle of 1 deg from the horizontal. The concentration of hydrogen at the surface exposed to the vacuum may be taken as zero. The desired film thickness is 1 mm. Data are as follows: $\rho = 8.32\ g/cm^3$, $\eta = 6\ cP$, $D_H = 1.3 \times 10^{-4}\ cm^2/sec$.

Solution. Using Eq. (2.14), we find the average velocity:

$$\bar{V} = \frac{(8.32\ g/cm^3)(980\ cm/sec^2)(0.1\ cm)^2(\cos 89°)}{3(6 \times 10^{-2}\ g/cm\text{-}sec)} = 7.90\ cm/sec.$$

The contact time is

$$\frac{L}{\bar{V}} = \frac{100\ cm}{7.90\ cm/sec} = 12.7\ sec.$$

Since this is very short, we make use of Eq. (15.52) to calculate an average k_M :

$$k_M = \sqrt{\frac{6D_H \bar{V}}{\pi L}} = \sqrt{\frac{(6)(1.3 \times 15^4)(7.9)}{(3.14)(100)}}$$

$$= 1.71 \times 10^{-3} \text{ cm/sec.}$$

Then

$$j_{H_2} = (1.71 \times 10^{-3} \text{ cm/sec})(1 \text{ cm}^3 \text{ H}_2/\text{cm}^3 \text{ alloy}) = 1.71 \times 10^{-3} \text{ cm}^3 \text{ H}_2/\text{cm}^2\text{-sec.}$$

Total content removed per cm^2 of exposed surface is $= j_{H_2}$ (contact time)

$$= 2.16 \times 10^{-2} \text{ cm}^3 \text{ H}_2/\text{cm}^2 \text{ film.}$$

Initial total content $= (1 \text{ cm}^3 \text{ H}_2/\text{cm}^3 \text{ alloy}) \cdot (0.1 \text{ cm}^3 \text{ alloy/cm}^2 \text{ film})$

$$= 0.1 \text{ cm}^3 \text{ H}_2/\text{cm}^2 \text{ film.}$$

Final total content $= 0.1000 - 0.0216 = 0.0784 \text{ cm}^3 \text{ H}_2/\text{cm}^2$ film, or the average content of the metal is reduced to $0.784 \text{ cm}^3 \text{ H}_2/\text{cm}^3$ alloy.

Under many circumstances encountered in interphase mass transfer, the bulk flow term is not important, and the diffusive contribution in the mass flux equation is all that we need to consider. On the other hand, there may be occasions, particularly where transfer to and from gas phases is involved, in which this contribution is not negligible. In this case, we write

$$N_A = \theta k_M (C_A^0 - C_A^i), \tag{15.57}$$

where N_A is the total interphase flux, and θ is a correction factor that depends on N_A, N_B, and k_M according to

$$\theta = \frac{N_A + N_B}{k_M \left[\exp \left(\dfrac{N_A + N_B}{k_M} \right) + 1 \right]}. \tag{15.58}$$

Figure 15.4 gives a graph of θ as a function of $(N_A + N_B)/k_M = \phi$. A limiting case is equimolar counter-diffusion, in which $N_A = -N_B$ and $\phi = 0$, so that $\theta = 1.0$ and no correction is involved.

In case we do not know N_A, and $N_B = 0$, the expression

$$1 + \frac{C_A^0 - C_A^i}{\dfrac{N_A}{N_A + N_B} - C_A^0} = \exp \left(\frac{N_A + N_B}{k_M} \right) \tag{15.59}$$

may be used to evaluate N_A at high mass-transfer rates.

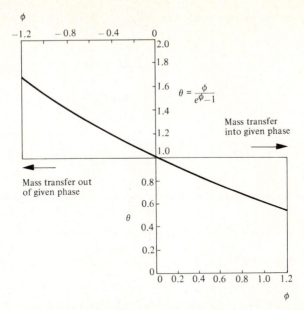

Fig. 15.4 The variation of coefficients with mass transfer rate. (From R. B. Bird, W. E. Stewart, and E. N. Lightfoot, *Transport Phenomena*, Wiley, New York, 1960, page 664.)

Thus, forced convection mass transfer at high mass-transfer rates (large N_A and/or N_B) is generally correlated by

$$\text{Sh} = f(\text{Re}, \text{Sc}, N_A, \text{ and geometry}). \tag{15.60}$$

For the most part, we shall not make use of this latter form.

15.5 FORCED CONVECTION OVER A FLAT PLATE—APPROXIMATE INTEGRAL TECHNIQUE

In Chapters 2 and 7, we developed expressions for the thickness of a momentum boundary layer and a thermal boundary layer of a fluid flowing past a plate. In this section, we shall again apply the approximate integral technique to obtain a solution for the thickness of a concentration boundary layer developing as a component diffuses from the solid plate into the fluid.

As in Section 7.2.2, consider the process of transfer of A from the solid (pure A) into the fluid (B). We disregard diffusion in the x-direction, it being negligible with respect to the velocity component of mass-transfer in the x-direction. Figure 15.5 depicts the unit volume to which we apply the integral mass balance.

The amount of A flowing into the element is

$$W_{A,x} = \int_0^l C_A v_x \, dy, \tag{15.61}$$

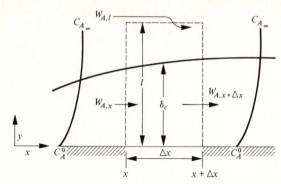

Fig. 15.5 Integrated convective contributions to the concentration boundary layer over a flat plate.

and the amount leaving is

$$W_{A,x+\Delta x} = W_{A,x} + \frac{d}{dx}\left(\int_0^l C_A v_x \, dy\right)\Delta x. \qquad (15.62)$$

To satisfy continuity, there is also fluid entering at $y = l$; this amount is

$$W_{A,l} = \frac{d}{dx}\left(C_{A\infty} \int_0^l v_x \, dy\right)\Delta x. \qquad (15.63)$$

The mass transfer into the unit element across the phase boundary at $y = 0$ is

$$j_{Ay}|_{y=0}\,\Delta x = -D_A\left(\frac{\partial C_A}{\partial y}\right)_{y=0}\Delta x. \qquad (15.64)$$

We now formulate the mass balance for component A:

$$j_{Ay}|_{y=0}\,\Delta x + W_{A,x} + W_{A,l} = W_{A,x+\Delta x}. \qquad (15.65)$$

Substituting Eqs. (15.61)–(15.64) into Eq. (15.65) yields

$$-D_A\left(\frac{\partial C_A}{\partial y}\right)_{y=0} = \frac{d}{dx}\left(\int_0^l C_A v_x \, dy - C_{A\infty}\int_0^l v_x \, dy\right). \qquad (15.66)$$

The integrals in Eq. (15.66) are split, that is,

$$\int_0^l = \int_0^{\delta_c} + \int_{\delta_c}^l$$

and then by simplifying, we arrive at

$$D_A\left(\frac{\partial C_A}{\partial y}\right)_{y=0} = \frac{d}{dx}\left[\int_0^{\delta_c}(C_{A\infty} - C_A)v_x\,dy\right]. \tag{15.67}$$

Equation (15.67) is an exact analog of Eq. (7.36). If we assume a concentration profile analogous to Eq. (7.37), that is,

$$\frac{C_A - C_A^0}{C_{A\infty} - C_A^0} = \frac{3}{2}\left(\frac{y}{\delta_c}\right) - \frac{1}{2}\left(\frac{y}{\delta_c}\right)^3, \tag{15.68}$$

which satisfies the conditions

$$\text{B.C.1:} \qquad \text{at } y = 0, \qquad C_A = C_A^0, \tag{15.69a}$$

$$\text{B.C.2:} \qquad \text{at } y = \delta_c, \qquad C_A = C_{A\infty}, \tag{15.69b}$$

then by substituting the assumed velocity distribution, Eq. (2.105), and the concentration distribution, Eq. (15.68), into Eq. (15.67), and following the procedure outlined previously for developing Eq. (7.42), we obtain the expression for the ratio of boundary layers:

$$\frac{\delta_c}{\delta} = \frac{1}{1.026\sqrt[3]{Sc}}. \tag{15.70}$$

Using Eq. (15.68), the local mass-transfer coefficient is

$$k_{M,x} = \frac{-D_A\left(\dfrac{\partial C_A}{\partial y}\right)_{y=0}}{C_A^0 - C_{A\infty}} = \frac{3}{2}\frac{D_A}{\delta_c}. \tag{15.71}$$

Thus, we combine Eqs. (15.70), (2.107), and (15.71) to obtain

$$Sh_x = 0.323\sqrt[3]{Sc}\sqrt{Re_x}. \tag{15.72}$$

These solutions are valid for most fluids, including liquid metals, because $(\delta_c/\delta) \ll 1$ for metals. Table 15.1 gives typical magnitudes of Schmidt numbers for fluids.

Table 15.1 Typical magnitudes of Prandtl numbers and Schmidt numbers

	Pr	Sc
Gases	0.6–1.0	0.1–2.0
Liquids	1–10	10^2–10^3
Liquid metals	10^{-2}	10^3

The average mass-transfer coefficient for transfer from a flat plate to a fluid is given by

$$k_M = \frac{1}{L} \int_0^L k_{M,x}\, dx = 0.646 \frac{D_A}{L} \mathrm{Sc}^{1/3} \mathrm{Re}_L^{1/2}, \tag{15.73}$$

or

$$\mathrm{Sh}_L = 0.646\, \mathrm{Sc}^{1/3} \mathrm{Re}_L^{1/2}. \tag{15.74}$$

According to Eq. (15.70), the concentration boundary layer and velocity layers for gases are about the same as in heat transfer, where δ_T and δ were similar because Pr was about unity. For liquid metals, however, while we found that in the case of heat transfer $\delta_T \gg \delta$, now we see that $\delta_c \ll \delta$. This means that we can use temperature profiles to predict mass-transfer profiles and rates, or vice versa, for gas phase transfer, but we cannot do likewise for liquid metals, because their concentration and temperature boundary layer profiles differ widely.

The *exact solution* of the problem of describing mass transfer in the above system requires simultaneous solution of the equations of momentum and continuity for both the total material flux and each individual component. We shall discuss the results of such a study in Section 15.7.

15.6 GENERAL EQUATION OF DIFFUSION WITH CONVECTION

In this section, we summarize the general approach to the law of mass conservation in the volume element $\Delta x\, \Delta y\, \Delta z$, depicted in Fig. 2.4, through which a fluid containing A in solution is flowing. In the following expressions, ρ_A is the mass *concentration* (for example, g of A/cm^3 of total solution) as defined by Eq. (13.1); W_{Ax} is the total *mass* flux of A in the x-direction and is composed of a diffusive term and a convective term. Specifically

$$W_{Ax} = -\rho D_A \left(\frac{\partial \rho_A^*}{\partial x} \right) + \rho_A v_x = \rho_A v_{Ax}, \tag{15.75}$$

where ρ is the density of the entire solution, ρ_A^* is the mass *fraction* of A, and v_x is the local *mass average velocity* in the x-direction; thus, we define v_x such that the total mass flux of all components in the x-direction is made up of the sum of the n component fluxes in the same direction:

$$\rho v_x = \sum_{i=1}^n \rho_i v_{ix}. \tag{15.76}$$

For a binary, in order to illustrate, we write

$$v_x = \frac{1}{\rho} (\rho_A v_{Ax} + \rho_B v_{Bx})$$

$$= \frac{1}{\rho}(W_{Ax} + W_{Bx}).$$

Now we can proceed to develop a mass balance for the volume element. The various contributions to the mass balance of component A are

Accumulation of mass of A in the volume element $\Delta x\, \Delta y\, \Delta z\, \dfrac{\partial \rho_A}{\partial t}$,

Input of A across face at x $\Delta y\, \Delta z\, W_{Ax}|_x$,

Output of A across face at $x + \Delta x$ $\Delta y\, \Delta z\, W_{Ax}|_{x+\Delta x}$.

There are also input and output terms in the y- and z-directions. When we write the entire mass balance for species A, divide through by $\Delta x\, \Delta y\, \Delta z$, and take the limits in the usual manner, we obtain the *general equation of continuity for component A*:

$$\frac{\partial \rho_A}{\partial t} + \frac{\partial W_{Ax}}{\partial x} + \frac{\partial W_{Ay}}{\partial y} + \frac{\partial W_{Az}}{\partial z} = 0. \tag{15.77}$$

The quantities W_{Ax}, W_{Ay}, W_{Az} are the rectangular components of the *mass flux vector*, $W_A = \rho_A v_A$, which includes motion of A due to diffusion and bulk flow:

$$W_A = -\rho D_A \nabla \rho_A^* + \rho_A v = \rho_A v_A. \tag{15.78}$$

Finally, by combining Eqs. (15.77) and (15.78), we develop the *diffusion equation* for component A:

$$\frac{\partial \rho_A}{\partial t} + \nabla \cdot \rho_A v = \nabla \cdot \rho D_A \nabla \rho_A^*. \tag{15.79}$$

As is usually the case, simplifications are utilized more frequently than general equations. Often, one can assume constant mass density and D_A, and make some simplification. For *constant ρ and D_A*, Eq. (15.79) becomes

$$\frac{\partial \rho_A}{\partial t} + v \nabla \rho_A = D_A \nabla^2 \rho_A, \tag{15.80}$$

or if divided by M_A (molecular weight of A), we get

$$\frac{\partial C_A}{\partial t} + v \nabla C_A = D_A \nabla^2 C_A. \tag{15.81}$$

The left-hand side of this equation is DC_A/Dt, showing direct similarity with Eq. (7.90) which is the basis for the numerous analogies between heat and mass transport in fluids with constant ρ.

    The above analysis could have been made equally well in terms of molar fluxes such as we have used previously.

Equation of continuity for component A:

$$\frac{\partial C_A}{\partial t} + (\nabla N_A) = 0. \tag{15.82}$$

Diffusion equation for A in solution:

$$\frac{\partial C_A}{\partial t} + \nabla \cdot C_A v^* = \nabla \cdot C D_A \nabla X_A. \tag{15.83}$$

Table 15.2 The equation of continuity of A in various coordinate systems

Rectangular coordinates:

$$\frac{\partial C_A}{\partial t} + \left(\frac{\partial N_{Ax}}{\partial x} + \frac{\partial N_{Ay}}{\partial y} + \frac{\partial N_{Az}}{\partial z} \right) = 0 \tag{A}$$

Cylindrical coordinates:

$$\frac{\partial C_A}{\partial t} + \left(\frac{1}{r} \frac{\partial}{\partial r} (r N_{Ar}) + \frac{1}{r} \frac{\partial N_{A\theta}}{\partial \theta} + \frac{\partial N_{Az}}{\partial z} \right) = 0 \tag{B}$$

Spherical coordinates:

$$\frac{\partial C_A}{\partial t} + \left(\frac{1}{r^2} \frac{\partial}{\partial r} (r^2 N_{Ar}) + \frac{1}{r \sin \theta} \frac{\partial}{\partial \theta} (N_{A\theta} \sin \theta) + \frac{1}{r \sin \theta} \frac{\partial N_{A\phi}}{\partial \phi} \right) = 0 \tag{C}$$

Table 15.3 The equation of diffusion of A for constant ρ and D_A

Rectangular coordinates:

$$\frac{\partial C_A}{\partial t} + \left(v_x \frac{\partial C_A}{\partial x} + v_y \frac{\partial C_A}{\partial y} + v_z \frac{\partial C_A}{\partial z} \right) = D_A \left(\frac{\partial^2 C_A}{\partial x^2} + \frac{\partial^2 C_A}{\partial y^2} + \frac{\partial^2 C_A}{\partial z^2} \right) \tag{A}$$

Cylindrical coordinates:

$$\frac{\partial C_A}{\partial t} + \left(v_r \frac{\partial C_A}{\partial r} + v_0 \frac{1}{r} \frac{\partial C_A}{\partial \theta} + v_z \frac{\partial C_A}{\partial z} \right)$$

$$= D_A \left(\frac{1}{r} \frac{\partial}{\partial r} \left(r \frac{\partial C_A}{\partial r} \right) + \frac{1}{r^2} \frac{\partial^2 C_A}{\partial \theta^2} + \frac{\partial^2 C_A}{\partial z^2} \right) \tag{B}$$

Spherical coordinates:

$$\frac{\partial C_A}{\partial t} + \left(v_r \frac{\partial C_A}{\partial r} + v_\theta \frac{1}{r} \frac{\partial C_A}{\partial \theta} + v_\phi \frac{1}{r \sin \theta} \frac{\partial C_A}{\partial \phi} \right)$$

$$= D_A \left(\frac{1}{r^2} \frac{\partial}{\partial r} \left(r^2 \frac{\partial C_A}{\partial r} \right) + \frac{1}{r^2 \sin \theta} \frac{\partial}{\partial \theta} \left(\sin \theta \frac{\partial C_A}{\partial \theta} \right) + \frac{1}{r^2 \sin^2 \theta} \frac{\partial^2 C_A}{\partial \phi^2} \right) \tag{C}$$

For constant C and D_A, Eq. (15.83) takes the form

$$\frac{\partial C_A}{\partial t} + \boldsymbol{v}^* \nabla C_A = D_A \nabla^2 C_A. \tag{15.84}$$

This equation is usually applied to low-density gases at constant temperature and pressure. The left-hand side of this equation cannot be written as DC_A/Dt because of the appearance of $\boldsymbol{v}^*$ rather than of $\boldsymbol{v}$. A more simplified form of the above equations, which is used for diffusion in solids or stationary liquids ($\boldsymbol{v} = 0$ in Eq. 15.81), or for equimolar counterdiffusion in gases ($\boldsymbol{v}^* = 0$ in Eq. 15.84), is *Fick's second law of diffusion*:

$$\frac{\partial C_A}{\partial t} = D_A \nabla^2 C_A. \tag{15.85}$$

In Tables 15.2 and 15.3, we summarize the most important equations of this discussion in rectangular, cylindrical, and spherical coordinates. Fick's second law of diffusion can be obtained by setting the velocity components in Table 15.3 equal to zero.

15.7 FORCED CONVECTION OVER A FLAT PLATE—EXACT SOLUTION

As an application of the above equations, consider the flow system discussed in Section 15.5. A thin, semi-infinite plate of solid A dissolves very slowly under steady-state conditions, into an unbounded fluid stream of A and B. The flow is initially at uniform velocity, concentration, and temperature. For constant properties of the fluid, we may write the boundary layer equations of momentum, energy, mass, and continuity:

$$\text{Continuity,} \qquad \frac{\partial v_x}{\partial x} + \frac{\partial v_y}{\partial y} = 0, \tag{15.86}$$

$$\text{Momentum,} \qquad v_x \frac{\partial v_x}{\partial x} + v_y \frac{\partial v_x}{\partial y} = v \frac{\partial^2 v_x}{\partial y^2}, \tag{15.87}$$

$$\text{Energy,} \qquad v_x \frac{\partial T}{\partial x} + v_y \frac{\partial T}{\partial y} = \alpha \frac{\partial^2 T}{\partial y^2}, \tag{15.88}$$

$$\text{Mass,} \qquad v_x \frac{\partial C_A}{\partial x} + v_y \frac{\partial C_A}{\partial y} = D_A \frac{\partial^2 C_A}{\partial y^2}. \tag{15.89}$$

Equations (15.86)–(15.88) were obtained in Chapter 7 for zero mass transfer. The assumption that the same equations are valid in the presence of mass transfer means that any additional momentum and energy fluxes associated with mass transfer are negligible. Equation (15.89) is derived from Eq. (A) in Table 15.3 with $\partial C_A/\partial t = 0$, $\partial^2 C_A/\partial z^2 = 0$, $v_z = 0$, and by neglecting the negligible amount of diffusion in the x-direction.

A typical set of boundary conditions that may be specified is:

B.C. 1: $x \le 0$, $v_x = V_\infty$, $v_y = 0$, $T = T_\infty$,

$$C_A = C_{A\infty} \qquad \text{for all } y, \tag{15.90a}$$

B.C. 2: $y = \infty$, $v_x = V_\infty$, $v_y = 0$, $T = T_\infty$,

$$C_A = C_{A\infty} \qquad \text{for } x > 0, \tag{15.90b}$$

B.C. 3: $y = 0$, $v_x = 0$, $v_y = v_0$, $T = T_0$,

$$C_A = C_A^0 \qquad \text{for } x > 0. \tag{15.90c}$$

The fact that $v_y = v_0$ at the wall accounts for the bulk motion accompanying diffusion from the wall. The method of solving Eqs. (15.86)–(15.89) subject to the conditions (Eqs. 15.90 a, b, and c) is not given here, but Fig. 15.6 presents the results for certain values of Pr and Sc. Note that the differential equations and boundary conditions for temperature and concentration are analogous; therefore, when Pr = Sc = 1, the velocity, temperature, and concentration profiles within the boundary layer must coincide. Figure 15.6 shows these results, along with the results for Pr = Sc = 0.7; velocity profiles remain unchanged.

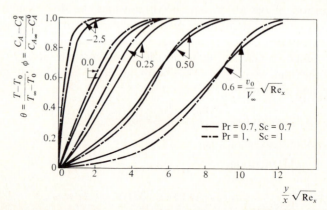

Fig. 15.6 Temperature and concentration profiles in a laminar boundary layer on a flat plate for Pr = Sc = 0.7, and Pr = Sc = 1. Curves for Pr = Sc = 1 also represent velocity profiles. (From J. P. Hartnett and E. R. G. Eckert, *Trans. ASME* **79**, 247 (1957).)

These profiles show a dependence on the mass flux $(v_0\sqrt{\text{Re}_x}/V_\infty)$. Mass transfer away from the plate (positive v_0) gives flatter profiles as would be true if the solid surface were porous and a gas diffused upward through the plate, or if a liquid passed through the porous plate and evaporated. On the other hand, mass transfer towards the plate (negative v_0) gives steeper profiles; this situation can be obtained if condensation occurs at the surface, or suction is applied to a porous

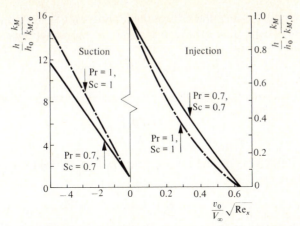

Fig. 15.7 Heat- and mass-transfer coefficients for laminar flow over a flat plate. Subscript zero indicates the respective coefficients for zero bulk flow normal to wall. (From J. P. Hartnett and E. R. G. Eckert, *ibid.*)

plate. Figure 15.7 shows the local heat- and mass-transfer coefficients plotted against the parameter $v_0\sqrt{Re_x}/V_\infty$. For $v_0 = 0$ (no mass transfer, or, more realistically, at low mass-transfer rates), the local mass-transfer coefficient is given by Eq. (7.26) with a simple change of notation, namely:

$$h_x \rightarrow k_{M,x},$$

$$k \rightarrow D_A,$$

$$Pr \rightarrow Sc.$$

Then, the local Sherwood number is

$$Sh_x = 0.332 \, Sc^{0.343} \, Re_x^{1/2}, \tag{15.91}$$

and the average Sherwood number

$$Sh_L = 0.664 \, Sc^{0.343} \, Re_L^{1/2}. \tag{15.92}$$

15.8 CORRELATIONS OF MASS-TRANSFER COEFFICIENTS FOR TURBULENT FLOW

We have seen in the previous sections that many forced-convection mass-transfer situations are completely analogous to heat-transfer situations, and the appropriate heat-transfer solutions apply with simple changes of notation, namely:

$$\alpha \rightarrow D_A,$$

$$T \rightarrow C_A,$$

$$Pr \rightarrow Sc,$$

$$Nu \rightarrow Sh.$$

In addition, we may assume that the results for natural convection resulting from density differences caused by mass transfer may be correlated by a mass-transfer Grashof number,

$$\mathrm{Gr}_M = g\xi(X_A - X_{A\infty})L^3/v^3,$$

where ξ is the concentration coefficient of volumetric expansion defined as:

$$\xi = \frac{1}{\rho}\left(\frac{\partial\rho}{\partial X_A}\right)_T.$$

In this case, we may use correlations for heat transfer to yield mass-transfer data if the substitution

$$\mathrm{Gr} \to \mathrm{Gr}_M$$

is made. The flow parameters such as Re and position parameters such as L/D remain the same.

We noted in Chapter 8 that in turbulent flow there is a parallel between the friction factor f for turbulent flow in tubes and heat transfer, in terms of a quantity known as the Chilton–Colburn "j-factor", j_H:

$$j_H = \frac{\mathrm{Nu}}{\mathrm{RePr}}(\mathrm{Pr})^{2/3} = \frac{f}{2}. \tag{8.6}$$

Continuing the analogy, we define a mass-transfer j-factor, j_M, for fully developed flow in round tubes:

$$j_M = \frac{\mathrm{Sh}}{\mathrm{ReSc}}(\mathrm{Sc})^{2/3} = \frac{f}{2}. \tag{15.93}$$

If flow is not fully developed, we use Fig. 8.2 where we take L/D into account, and substitute j_M for j_H. Epstein[2] used Eq. (15.93) to compute the corrosion rate of an iron tube by molten mercury; apparently, the mass transfer of iron into the mercury stream determines the rate of this process.

In the case of flow past a flat plate, from the results of Chapter 8, we write for average values in *laminar* flow:

$$j_H = j_M = \frac{f}{2} = \frac{0.664}{\sqrt{\mathrm{Re}_L}}, \tag{15.94}$$

and in *turbulent* flow

$$j_H = j_M = \frac{f}{2} = \frac{0.037}{(\mathrm{Re}_L)^{0.2}}. \tag{15.95}$$

---

[2] L. F. Epstein, *Chem. Engr. Progress Symposium Series* **53**, No. 20, 67.

This relationship has been found to adequately describe the rate of deposition of metallic solutes from liquid to solid,[3] and the rate of dissolution of carbon into iron melts.[4]

In flow around curved surfaces, such as spheres and cylinders, $f/2$ greatly exceeds j_H and j_M. However, the analogy still holds between heat and mass transfer, so that j_H and j_M should be equivalent. To illustrate this, consider the heat-transfer correlation given in Chapter 8 for forced convection around a sphere of radius R:

$$2hR/k_f = 2.0 + 0.60\,\mathrm{Re}_f^{1/2}\,\mathrm{Pr}_f^{1/3}. \tag{8.10}$$

Translated into the j_H form it becomes

$$j_H = \frac{2.0}{\mathrm{Re}\,\mathrm{Pr}^{1/3}} + \frac{0.60}{\mathrm{Re}^{1/2}}.$$

By analogy we get

$$j_M = \frac{2.0}{\mathrm{Re}\,\mathrm{Sc}^{1/3}} + \frac{0.6}{\mathrm{Re}^{1/2}}. \tag{15.96}$$

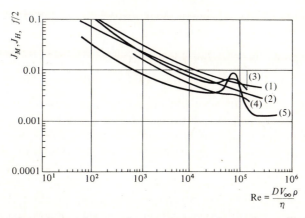

Fig. 15.8 Comparison of mass-, heat-, and momentum-transfer to spheres. (1) $f/2$; (2) Chilton–Colburn factor j_H; (3) j_M for cinnamic acid–water system; (4) j_M for 2-naphthol–water system; (5) j_M for uranium dissolving in cadmium. (Data from E. D. Taylor, L. Burris, and C. J. Geankoplis, *I. & E. C. Fundamentals* **4**, 119 (1965); and T. R. Johnson, R. D. Pierce, and W. J. Walsh, *ANL Report on Contract W-31-109-eng-38*, 1966.)

Figure 15.8 illustrates the results of experiments where j_M is plotted as a function of Re for dissolution of uranium spheres in flowing cadmium[5], cinnamic

[3] W. N. Gill, R. P. Vanek, R. V. Jelinek, and C. S. Grove, *AIChEJ* **6**, 139 (1960).
[4] M. Kosaka and S. Minowa, *Trans. Japan Iron and Steel Inst.* **8**, 393 (1968).
[5] F. D. Taylor, L. Burris, and C. J. Geankoplis, *I. & E.C. Fundamentals* **4**, 119 (1965).

acid spheres in flowing water, 2-naphthol spheres in flowing water[6], and for comparison purposes, j_H[7], and $f/2$[8]. Note that the data from the liquid metal experiments show lower values of j_M at low Re values than those observed in the organic system experiments. This may be partially corrected for by the presence of the Schmidt number in the expression given in Eq. (15.96), which is not plotted in Fig. 15.8. No equation accounts for the peak in j_M near Reynolds numbers of 10^5, so usually the graph is preferred rather than an equation in this range.

For natural convection, the mass transfer analog of

$$\text{Nu}_L = 0.13\,(\text{Gr}_L\text{Pr})^{1/3} \tag{8.18}$$

has been applied to both the rate of dissolution of carbon in molten iron[9] and the rate of dissolution of steel in molten iron–carbon alloys[10], and found to be quite reasonable as an approximation, with the best fit of the data yielding the equation

$$\text{Sh}_L = 0.11\,(\text{Gr}_{M,L}\text{Sc})^{1/3}. \tag{15.97}$$

Many metallurgical processes depend on gas–liquid contact and mass transfer between the phases. Although this is a very complex area,[11] several relationships have been found that describe the process of mass transfer *on the liquid side* of a gas bubble–liquid interface. One of the most useful is that of Hughmark.[12] His expression is

$$\frac{k_M d}{D} = 2.0 + a\left[\left(\frac{V_t d}{v}\right)^{0.48}\left(\frac{v}{D}\right)^{0.339}\left(\frac{g^{1/3}d}{D^{2/3}}\right)^{0.072}\right]^b, \tag{15.98}$$

where d is the bubble diameter, V_t is the terminal velocity of the rising bubble, and a and b are constants that depend on whether the bubbles act singly or in swarms. For single bubbles, $a = 0.061$ and $b = 1.61$. This relationship has been tested in molten copper–carbon monoxide systems, and was found satisfactory for describing the rate of transfer of oxygen in the copper to the interface.[13]

Note that the Chilton-Colburn analogy is applicable only at relatively low mass-transfer rates, and that the best analogous results are obtained when similar materials are utilized in both the heat- and analogous mass-transfer situations. Turbulent flow mass-transfer correlations based on studies using common liquids appear to be directly translatable into liquid metal systems, but heat transfer

---

[6] L. R. Steele and C. J. Geankoplis, *AIChEJ* **5**, 178 (1959).

[7] T. K. Sherwood, *Ind. Engr. Chem.* **42**, 2077 (1950).

[8] F. H. Garner and R. D. Suckling, *AIChEJ* **4**, 114 (1958).

[9] M. Kosaka and S. Minowa, *ibid.*

[10] M. Kosaka and S. Minowa, *Tetsu-to-Hagane* **53**, 983 (1967).

[11] For reviews, see P. H. Calderbank, *Trans. Inst. Chem. Engrs.* **212**, 209 (1967); F. H. H. Valentin, *Absorption in Gas–Liquid Dispersions*, Spon Ltd, London, 1967.

[12] G. A. Hughmark, *I. & E. C. Process Design and Development* **6**, 218 (1967).

[13] C. R. Nanda and G. H. Geiger, *Metallurgical Transactions* **2**, 1101 (1971).

data are not quite as accurate in predicting mass transfer, although usually satisfactory to within an order of magnitude.

Also, many of the tests of applicability of heat transfer correlations to mass transfer involving the gas phase have not been made at conditions likely to be of interest to metallurgists, particularly at elevated temperatures and high rates of transfer, so they should be used with caution.

Example 15.4 Graphite particles are often added to molten cast iron in order to increase the carbon content when scrap steel is used as starting material. The time required to dissolve the particles is of interest. Determine the time to dissolve particles of graphite as a function of the bath's carbon content.

The particles have a shape factor λ of 1.5, and a characteristic dimension $\bar{D}_p$ of 0.14 cm. The particles float, and are swept to the side of the surface of the melt by the magnetically induced stirring action resulting in a relative metal velocity of approximately 25 cm/sec. Due to the displacement of metal by graphite, the surface area of an individual particle in contact with the metal is calculated as $\frac{3}{8}$ of the total particle surface area. In addition, a portion of the particle exposed to the atmosphere is burned to CO due to the air circulating over the bath surface. Thus the recovery of carbon in the melt is less than 100%, and experience shows that, in fact, the recovery is only 50%. Therefore, the mass of an individual particle (of density ρ_s) that dissolves in the melt is

$$m = \frac{1}{2}\left(\frac{\pi}{6}\bar{D}_p{}^3\rho_s\right),$$

and its surface area exposed to the melt is

$$A = \tfrac{3}{8}\,\pi\bar{D}_p{}^2\lambda.$$

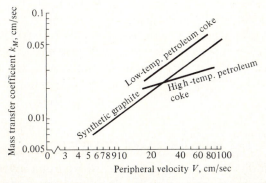

Fig. 15.9 Mass-transfer coefficients for carbon dissolution in Fe–C melts. (From R. G. Olsson, V. Koump, and T. F. Perzak, *Trans. AIME* **236**, 426 (1965); O. Angeles, G. H. Geiger, and C. R. Loper, *Trans. AFS* **76**, 629 (1968); and M. Kosaka and S. Minowa, *Trans. Japan Iron Steel Inst.* **8**, 393 (1968).)

Solution. The mass flow, in terms of g C/sec, is

$$\frac{dm}{dt} = -k_M A \rho_L (C_0 - C_\infty),$$

where ρ_L is liquid density, g/cm³, C_0 is weight fraction of carbon at the particle–melt interface, and C_∞ = weight fraction of carbon in the bulk melt.

We evaluate the mass-transfer coefficient from Fig. 15.9 which has been developed by several investigators for the dissolution of rotating carbon rods in Fe–C melts. For a velocity of 25 cm/sec, k_M for graphite is 0.02 cm/sec. Now, since

$$\frac{dm}{dt} = -\tfrac{1}{4}\pi \bar{D}_p{}^2 \rho_s \frac{d\bar{D}_p}{dt},$$

and the area is as given above, we can determine the time to dissolve a particle:

$$\int_{\bar{D}_p}^{0} d\bar{D}_p = \tfrac{3}{2} k_M \lambda \frac{\rho_L}{\rho_s} (C_0 - C_\infty) \int_0^t dt,$$

or

$$t = \frac{2\rho_s \bar{D}_p}{3\rho_L \lambda k_M (C_0 - C_\infty)}, \text{ sec.}$$

The results are plotted below.

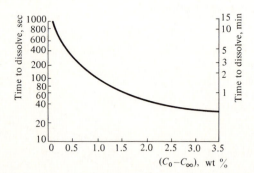

15.9 MODELS OF THE MASS-TRANSFER COEFFICIENT

Since most conditions in which mass transfer is important involve fluids undergoing turbulent motion, we must usually rely on the preceding correlations, with experimental studies giving the necessary empirical coefficients. However, there are many situations to which these test data do not directly apply, such as beyond the experimental scope of variables, and it is desirable to know whether and how the experimental data can be extended to the new situation. Several theories of the process of mass-transfer have been developed, and they attempt to present models of what actually happens at the interface between two fluids or between a fluid and solid from a fundamental viewpoint, in order to aid in intelligent extrapolation of data.

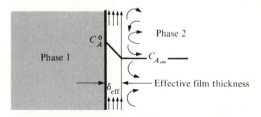

Fig. 15.10 The "effective film thickness" model.

The oldest theory is the film theory of Lewis and Whitman[14] who suggested that there is an unmixed layer or film in the fluid next to the actual interface, continuously exposed to the completely mixed bulk fluid on one side and to the other phase on the other. This layer, devoid of any fluid motion, is supposed to offer all the resistance to the transfer of component A from the interface into the bulk solution, as depicted in Fig. 15.10. The transfer takes place purely by atomic or molecular diffusion through the film. Figure 15.10 indicates the concentration profile assumed in the model. Since the entire concentration change from $C_{A\infty}$ to C_A^0 is assumed to take place within the film in a steady-state manner, and since the mass-transfer coefficient is defined by

$$j_A = k_M(C_A^0 - C_{A\infty}),$$

we can compare this to

$$j_A = -D\frac{dC_A}{dx} = +D\frac{(C_A^0 - C_{A\infty})}{\delta_{eff}},$$

with the result that $k_M = D/\delta_{eff}$, where δ_{eff} is the effective film thickness. This result often appears in the metallurgical literature in cases where the flux is measured and the overall concentration change $(C_A^0 - C_{A\infty})$ is known, and then either the diffusivity is known (or more often assumed) and δ_{eff} is calculated, or *vice versa*. As noted below, the significance of δ_{eff} is dubious, at best.

From a fluid mechanics standpoint, it was recognized at an early stage that interfaces between fluids are bound to be unstable with time, and that any given element of fluid at the interface does not remain there for long. Thus, the film

---

[14] W. K. Lewis and W. Whitman, *Ind. Engr. Chem.* **16**, 1215 (1924).

theory is much too crude to be really meaningful. Higbie[15] proposed a model to describe the contact between two fluids, in which he assumed that one fluid exposes a "particle" of fluid to the other phase for an average time θ, which is taken to be extremely short, such that the particle is subject only to unsteady-state diffusion or "penetration" by the transferred species during its contact time with the other phase. The particle is assumed to be stagnant internally during this time, and well mixed before and after. Figure 15.11 gives a schematic picture of the situation. This theory results in a prediction that

$$k_M = 2\sqrt{\frac{D}{\pi\theta}}. \tag{15.99}$$

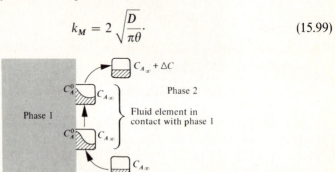

Fig. 15.11 Schematic diagram of fluid motion in penetration theory.

The logical extension of this theory was performed by Danckwerts[16] who suggested that the idea of a constant time of exposure θ ought to be replaced by an average time of exposure calculated from an assumed distribution of residence times of the "particles" at the surface. The result is again a relationship of the form

$$k_M \propto \sqrt{D}. \tag{15.100}$$

The constants in his equation, like the constant θ in Higbie's equation, are not readily obtainable, with the exception of bubbles rising through a liquid in which case we may estimate θ to be the time required for a bubble to rise a distance equal to its diameter.

When considering the two theories, it is apparent that the dependence of k_M on D is different. Experimentally, it has been found that k_M is proportional to D^n where n varies from 0.5 to 1.0, depending on the fluids and the circumstances. In order to resolve this discrepancy, Dobbins[17] and Toor and Marchello[18] proposed a combined *film penetration theory* in which the residence time of the surface

[15] R. Higbie, *Trans. AIChEJ* **31**, 365 (1935).

[16] P. U. Danckwerts, *Ind. Engr. Chem.* **43**, 1460 (1951).

[17] W. E. Dobbins, *Int. Conference on Water Pollution Research*, London, 1962, Pergamon Press, New York, 1964, page 61.

[18] H. L. Toor and J. M. Marchello, *AIChEJ* **4**, 97 (1958).

elements (Higbie model) is long enough to allow the concentration gradient to approach steady-state across the finite thickness of the element (film model). This theory approaches each of the other theories as limiting cases. When D is large or the rate of surface renewal is small (θ is large), then n approaches 1.0 and so the film theory is applicable. When D is small or θ is small (rapid surface renewal rate), n approaches 0.5, and the penetration theory results. In any case, there are still parameters that must be specified in order to use the theory for predictive purposes, and they are not readily obtainable.

PROBLEMS

15.1 At 1000°F metal A is soluble in liquid B but B is not soluble in solid A as shown below in the pertinent part of the phase diagram.

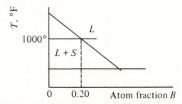

A 2-in. diameter cylinder of A is rotated at 1000 rpm in a large melt of 0.5 atom fraction B at 1000°F, and it is noted that after 15 min the bar diameter is 1.90 in. For the same temperature, estimate the bar diameter after 15 min if another 2-in. diameter cylinder of A is rotated in a large melt of 0.25 atom fraction B. We can assume that the molar volume of liquid $A–B$ alloys is constant.

15.2 Use dimensional analysis to show

a) $\text{Sh} = f(\text{Re}, \text{Sc})$ for forced convection;

b) $\text{Sh} = f(\text{Gr}, \text{Sc})$ for natural convection.

15.3 Levitation melting is a means of supporting a metallic melt by an electromagnetic field. No impurities are added in melting and operation under an inert atmosphere removes dissolved gases. At 3000°F and 1 atm hydrogen pressure, the solubility of hydrogen in iron is 31 cm³ per 100 g of iron. Estimate the rate at which hydrogen can be removed from a levitated drop of iron that initially contains 10 ppm in the set-up shown below. Assume that no convection occurs within the iron drop, and that the gas temperature is 3000°F so that Eq. (8.10), in mass-transfer form, applies.

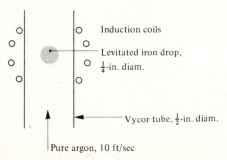

15.4 Derive expressions for diffusion through a spherical shell that are analogous to Eq. 15.12 (concentration profile) and Eq. 15.14 (molar flux).

15.5 Hydrogen gas is being absorbed from a gas in an experimental set-up shown in the figure below. The absorbing liquid is aluminum at 1400°F which is falling in laminar flow with an average velocity of 6 in. per min.

What is the hydrogen content of the aluminum leaving the tube if it enters with no hydrogen? Assume that at $T = 1400°F$, and 1 atm hydrogen pressure, the solubility of hydrogen is 1 cm^3 per 100 g of aluminum, the density of Al $= 2.5$ g/cm^3, and $D_H = 1 \times 10^{-5}$ cm^2 sec.$^{-1}$

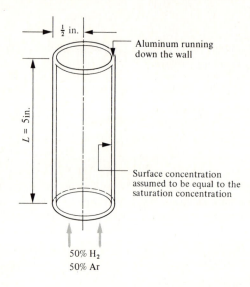

$\frac{1}{2}$ in.

Aluminum running down the wall

$L = 5$in.

Surface concentration assumed to be equal to the saturation concentration

50% H$_2$
50% Ar

16

INTERPHASE MASS TRANSFER

In Chapter 15 we undertook much of the analytical presentation to familiarize ourselves with mass-transfer coefficients as they arise from considerations of diffusion with convection in a single phase only. There are many situations, however, in which two fluids or a solid and fluid are in contact, and interphase transfer by diffusion with convection takes place in one or both phases. In some cases, convective mass transfer may be important in controlling the overall rate; in others it is not important. If there is a reaction at the interface, it may very well be the controlling step.

Experimentally, it is difficult to study interphase transfer, and at the same time separate the individual phase resistances; so an *overall transfer coefficient* is usually measured. Having determined this coefficient, we then attempt, via a mathematical model, to *deduce* the individual phase coefficients, such as those presented in Chapter 15. In some cases the results are clear cut, and we are then able to adjust process conditions to optimize the process. However, in other cases, we cannot differentiate between models even though they may be based on important differences in basic assumptions, because their predictions can be numerically similar within experimental error. Then, we deal with the overall coefficients and utilize them, but only if the fluid conditions are similar in both the prototype and model cases.

The objective of this chapter is to indicate the relationships between the individual phase transfer coefficients and the overall coefficients, and to examine several cases in detail, showing how different fluid conditions can influence the overall coefficient through their effect on the individual coefficients.

16.1 TWO-RESISTANCE MASS-TRANSFER THEORY

Let us investigate the situation as it might exist in a gas–liquid physical reaction. Figure 16.1 depicts the phases in contact and the concentration profiles in each phase. The *mole fraction* of A in the bulk gas phase is $Y_{A\infty}$, and it decreases to Y_A^* at the interface. In the liquid, the mole fraction drops from X_A^* at the interface to $X_{A\infty}$ in the bulk liquid. The bulk concentrations $X_{A\infty}$ and $Y_{A\infty}$ are obviously not in equilibrium, otherwise diffusion of the solute would not occur. To determine the rate of mass transfer between the phases, it is necessary to consider the sequential "steps" in the process.

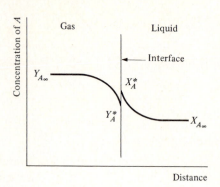

Fig. 16.1 The two-resistance mass transfer concept.

For example, assume that the only resistances to the overall process are diffusional resistances within the fluids themselves, and that there is no resistance to the transfer of A across the interface. Consequently, the interface concentrations, X_A^* and Y_A^*, are equilibrium values. In some instances, of course, the atoms cannot be accommodated into the new phase as rapidly as they might jump from their original phase across the interface. If the accommodation process is slow relative to the diffusional processes, then X_A^* and Y_A^* would not represent equilibrium concentrations. In most situations, X_A^* and Y_A^* are the equilibrium values, and we proceed with the "two-resistance" theory.

For steady-state transfer, the rate at which A reaches the interface from the gas must equal the rate at which it diffuses into the bulk liquid. Thus, if $k_{M,g}$ and $k_{M,l}$ are the local coefficients, then the flux of A is

$$N_A = k_{M,g}(Y_{A\infty} - Y_A^*) = k_{M,l}(X_A^* - X_{A\infty}). \qquad (16.1)$$

In this instance, since N_A is given in moles/cm^2 sec, and the concentrations are expressed as mole fractions, the mass-transfer coefficients ($k_{M,g}$ and $k_{M,l}$) have the units of moles/cm^2 sec. The units of the mass-transfer coefficients, as presented in Chapter 15 and found within the Sherwood number, are cm/sec. Of course, k's are derivable from k; specifically, we have $k_A = k_A C$, recalling that C is the total molar concentration of the solution in question.

In experimental determinations of the rate of mass transfer, it is usually possible to determine solute concentrations in the bulk fluids by sampling. But, because the concentration boundary layers are usually extremely thin, it is physically impossible to approach the interface sufficiently closely to measure X_A^* and Y_A^*. Under these circumstances, only the overall effect in terms of either $X_{A\infty}$ or $Y_{A\infty}$ can be determined.

Consider the equilibrium of solute A as partitioned between the gas phase and a solvent, depicted in Fig. 16.2. The curve for the system is unique at fixed

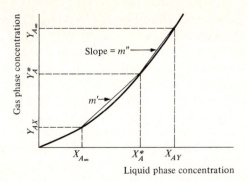

Fig. 16.2 The equilibrium curve for solute A partitioned between gas and liquid phases.

temperature and partial pressure of A in the gas phase. On the curve, we identify the four concentrations referred to in Eq. (16.1), specifically X_A^*, Y_A^*, $X_{A\infty}$, and $Y_{A\infty}$. In addition, we show X_{AY} and Y_{AX} which represent equilibrium concentrations corresponding to $Y_{A\infty}$ and $X_{A\infty}$, respectively. Y_{AX}, in equilibrium with $X_{A\infty}$, is as good a measure of X_A as X_A itself, and moreover it is on the same basis as $Y_{A\infty}$. We may then measure the entire two-phase mass-transfer effect in terms of an overall mass-transfer coefficient, K_g:

$$N_A = K_g(Y_{A\infty} - Y_{AX}). \tag{16.2}$$

From the geometry of the figure, we can show that the relationship among the individual phase coefficients and the overall coefficient is

$$N_A/K_g = N_A/k_{M,g} + m'N_A/k_{M,l},$$

or

$$1/K_g = 1/k_{M,g} + m'/k_{M,l}. \tag{16.3}$$

In a similar manner, X_{AY} is a measure of Y_A, and we may use it to define another overall coefficient K_l:

$$N_A = K_l(X_{AY} - X_{A\infty}). \tag{16.4}$$

It readily follows that

$$1/K_l = 1/m'' k_{M,g} + 1/k_{M,l}. \tag{16.5}$$

Equations (16.3) and (16.5) lead to the following relationships among the mass-transfer *resistances*:

$$\frac{\text{gas phase resistance}}{\text{total resistance}} = \frac{1/k_{M,g}}{1/K_g}, \tag{16.6}$$

and

$$\frac{\text{liquid phase resistance}}{\text{total resistance}} = \frac{1/k_{M,l}}{1/K_l}. \tag{16.7}$$

If the individual phase coefficients, $k_{M,g}$ and $k_{M,l}$ have roughly the same values, then we can demonstrate the importance of the shape of the equilibrium–partition curve. If m' is small (equilibrium–partition curve is fairly flat and solute A is very soluble in the liquid), Eq. (16.3) shows that the major resistance is really $1/k_{M,g}$, and the rate of mass transfer is gas-phase controlled.

Under such circumstances, even large increases in $k_{M,l}$ do not significantly change K_g, and efforts to increase the mass-transfer rate are best directed towards decreasing the gas-phase resistance. Conversely, when m'' is large (solute A is relatively insoluble in the liquid) and $k_{M,g}$ and $k_{M,l}$ are nearly equal, the major resistance is in the liquid. In these instances, efforts to bring about substantial increases in the rate of mass transfer are best directed toward increasing the liquid coefficient $k_{M,l}$.

When overall coefficients are used in practice, they are frequently synthesized through the correlations developed (such as those in Chapter 15) for the individual coefficients of the phases in contact. It is important to recognize the limitations in this procedure. For example, the hydrodynamic circumstances must be the same as those for which the correlations were developed. This is especially true in the case of two fluids, where motion in one may influence motion in the other. In other situations, absorption of surface-active substances at the interface may slow down reactions (chemical or physical) to such an extent that the estimated overall coefficient in no way reflects the magnitudes of the individual phase coefficients.

On the other hand, most experimental studies of reaction rates measure overall coefficients, from which attempts are often made to deduce individual phase resistances. Again, if experimentally measured *overall* coefficients are to be used in practice, the conditions with respect to fluid motion and reactant supply must also be similar, unless it is clear that the controlling resistance of the overall transfer lies in a nonfluid phase.

Example 16.1 In order to remove hydrogen from molten copper at 1150°C, the copper is brought in contact with pure argon at 1 atm pressure. Hydrogen diffuses to the argon, and undergoes the reaction

$$\underline{H} = \tfrac{1}{2}H_2(g),$$

in which $\underline{H}$ represents hydrogen dissolved in liquid copper. At 1150°C and 1 atm hydrogen pressure, the solubility of hydrogen in molten copper is 7.0 cm³H₂ (STP)/100 g of copper. Assuming that the mass-transfer coefficients (k's) within the individual phases are roughly equal in any process involving contact of argon and the copper, determine whether the rate of mass transfer would be gas-phase or liquid-phase controlled.

Solution. We proceed by examining the equilibrium of hydrogen between the gas phase (Ar–H₂) and the liquid (Cu–$\underline{H}$). We ignore the presence of Cu in the gas phase since its vapor pressure is very low at this temperature. To examine the relative values of the mass-transfer rates, we need to express the concentrations in mole fractions. In the liquid phase, we convert the solubility as follows

$$X_H = \frac{7.0\ cm^3 H_2(STP)}{100\ g\ Cu}\left|\frac{1\ liter\ H_2}{1000\ cm^3 H_2}\right|\frac{1\ g\text{-}mol\ H_2}{22.4\ liters\ (STP)H_2}\left|\frac{63.5\ g\ Cu}{1\ g\text{-}mol\ Cu}\right|\frac{1\ g\text{-}mol\ \underline{H}}{0.5.g\text{-}mol\ H_2}$$

$$= 3.96 \times 10^{-4}.$$

(Note: X_H is the mole fraction of $\underline{H}$ in the liquid.)
We can express the equilibrium constant for the reaction as

$$K_{eq} = \frac{X_H}{\sqrt{Y_{H_2}}},$$

where Y_{H_2} is the mole fraction of hydrogen in the gas; it follows that $K_{eq} = 3.96 \times 10^{-4}$ at 1150°C. Therefore, we can express the equilibrium partition of hydrogen between the phases as

$$X_H = 3.96 \times 10^{-4}\sqrt{Y_{H_2}}.$$

By referring to Fig. 16.2, we can estimate m' and m'' with $Y_{H\infty} = 0$ and $X_{H\infty} = 3.96 \times 10^{-4}$. In addition, Eq. (16.1) indicates that

$$Y_{H\infty} - Y_H^* \cong X_H^* - X_{H\infty},$$

since we are assuming $k_{M,g} = k_{M,l}$. Therefore

$$Y_H^* \cong X_{H\infty} - X_H^* \cong 10^{-4},$$

$$m' = \frac{Y_{H_2}^* - Y_{HX}}{X_H^* - X_{H\infty}} \cong \frac{(10^{-4}) - 1}{0 - (4 \times 10^{-4})} \cong 10^4,$$

$$m'' = \frac{Y_{H\infty} - Y_H^*}{X_{HY} - X_H^*} = \frac{0 - Y_H^*}{0 - X_H^*} = \frac{\sqrt{Y_H^*}}{K_{eq}} = \frac{10^{-2}}{4 \times 10^{-4}} \cong 10^2.$$

If we examine Eq. (16.3), since $m' \cong 10^4$, it is easy to see that major resistance to mass transfer is in the liquid phase and the overall rate of mass transfer is liquid-phase controlled.

Alternatively, if we examine Eq. (16.5), since $m'' \cong 10^2$, we again conclude that the major resistance to mass transfer is in the liquid phase.

16.2 MIXED CONTROL IN GAS–SOLID REACTIONS

In Chapter 14 we examined several solutions for diffusion within solids; we found that some of these solutions were appropriate to situations involving gas–solid reactions at surfaces. Specifically, we examined those cases in which the solute surface concentration in the solid could be considered to be in equilibrium with the gas-phase environment. In most metallurgical situations, this is indeed the case, because most reactions do proceed readily at the high temperatures involved in processing. There are situations, however, in which exceptions to this generalization should be made, as the two following examples demonstrate.

16.2.1 Carburization of iron with surface reaction and diffusion as controlling factors

If an iron plate is exposed to a CH_4–H_2 atmosphere, carburization occurs at the surface according to

$$CH_4(g) = \underline{C} + 2H_2(g), \tag{16.8}$$

and consequently carbon diffuses into the plate. The objective here is to describe the kinetics of carbon diffusion.

It has been proposed[1] that the surface reaction proceeds at a rate given by

$$\frac{1}{A}\frac{dn_c}{dt} = r_1 \frac{P_{CH_4}}{P_{H_2}^v} - r_2 P_{H_2}^{2-v} C_s, \tag{16.9}$$

where dn_c/dt is the amount of carbon in g-atoms taken up by the surface area A per unit time, r_1 and r_2 are reaction rate constants, P_{CH_4} and P_{H_2} are the partial pressures in the gas phase, presumed to be constant, C_s is the surface concentration of carbon in g-atoms/cm^3, and v is a number between zero and two, depending on the details of the reaction mechanism.

At equilibrium, $dn_c/dt = 0$; therefore, from Eq. (16.9) we can obtain the equilibrium concentration of carbon

$$C_e = \frac{r_1}{r_2} \cdot \frac{P_{CH_4}}{P_{H_2}^2}. \tag{16.10}$$

Substitution of Eq. (16.10) into Eq. (16.9) gives

$$\frac{1}{A}\frac{dn_c}{dt} = r(C_e - C_s), \tag{16.11}$$

where

$$r \equiv r_2 P_{H_2}^{2-v}.$$

We can incorporate Eq. (16.11) into one of the boundary conditions for the diffusion of carbon in the plate. Consider the region $0 \lesssim x \lesssim L$ in which $x = 0$ is the center of the plate, and $x = L$ is the semithickness of the plate. The appropriate differential equation is

$$\frac{\partial C}{\partial t} = D \frac{\partial^2 C}{\partial x^2}, \tag{16.12}$$

where C is the carbon concentration and D is the diffusion coefficient of carbon

---

[1] C. Wagner, *Notes from Course 3.63*, MIT, Cambridge, Massachusetts, 1955.

in iron. We shall assume that only a single phase of iron exists. Therefore, the initial and boundary conditions are

$$C(x, 0) = C_i \text{ (uniform)}, \tag{16.13a}$$

$$\frac{\partial C}{\partial x}(0, t) = 0, \tag{16.13b}$$

and

$$\frac{\partial C}{\partial x}(L, t) + \frac{r}{D}(C_s - C_e) = 0. \tag{16.13c}$$

Condition (16.13c) merely states that the amount of carbon furnished by the surface reaction must equal the amount diffusing into the interior. Note that Eq. (16.13c) is exactly the same condition existing at the surface of a solid losing heat to a surrounding environment at T_f, at a rate dependent on the heat-transfer coefficient. In fact, Eq. (16.12) and its boundary conditions, Eqs. (16.13a), (16.13b), and (16.13c), are exactly the same as Eq. (9.66) and its boundary conditions, Eqs. (9.67)–(9.69). Hence, Eq. (9.85) and Fig. 9.11 are proper solutions to the problem at hand if we merely replace

$$\frac{T - T_f}{T_i - T_f} \quad \text{by} \quad \frac{C - C_e}{C_i - C_e},$$

$$\alpha t / L^2 \quad \text{by} \quad Dt / L^2,$$

and

$$hL/k \quad \text{by} \quad rL/D.$$

16.2.2   Transport in the gas phase and diffusion as controlling factors

In this section, we examine a situation in which the gas phase composition does not remain constant, because the reaction rate at the surface of the solid is so rapid that the gas cannot be replenished instantaneously by the incoming fresh gas. As an example, suppose we wish to carburize the surface of low-carbon sheet steel with a CH_4–H_2 atmosphere. The steel is in the form of an open coil so that the gas can flow between parallel layers as depicted in Fig. 16.3. We assume that the thickness of the sheet in the y-direction is sufficiently large so that diffusion into a semi-infinite solid holds. However, there is a depletion of methane with increasing x because it is consumed by the reaction, Eq. (16.8). Hence, the concentration of carbon in the steel varies with x as well as with t (time).

    To simplify the problem considerably, we may disregard the variation of methane concentration normal to the flow direction; in the solid, the diffusion of carbon in the x-direction is ignored since the diffusive component in the y-direction is so much greater.

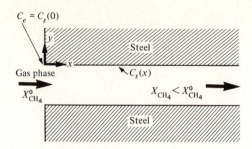

Fig. 16.3 Carburization of steel by gas flowing between parallel sheets.

Therefore, Fick's second law for diffusion in the steel applies, and is written

$$\frac{\partial C}{\partial t} = D \frac{\partial^2 C}{\partial y^2}, \tag{16.14}$$

where C is the carbon concentration, and D is its diffusion coefficient in the steel. Initially, the steel contains C_i of carbon; hence

$$C(x, y, 0) = C_i. \tag{16.15}$$

Along the plane $y = 0$, we presume that equilibrium exists between the surface carbon and the atmosphere according to

$$K_{eq} = \frac{C_s P_{H_2}^2}{P_{CH_4}},$$

where K_{eq} is the thermodynamic equilibrium constant for the reaction, Eq. (16.8). Thus, we can represent the relationship between the local methane mole fraction, X_{CH_4}, and the local surface concentration as

$$C(x, 0, t) = X_{CH_4}(x, t) \frac{K_{eq}}{P} = C_s, \tag{16.16}$$

where X_{CH_4} is the mole fraction of methane, P is the total pressure, and $X_{CH_4} \ll 1$. At the entrance, we have the original mole fraction, $X_{CH_4}^0$; thus

$$C(0, 0, t) = X_{CH_4}^0 \frac{K_{eq}}{P} = C_e, \tag{16.17}$$

where C_e equals the equilibrium concentration for the original gas composition.

As the gas flows between the parallel plates from x to $x + dx$, the loss of methane per unit time equals

$$- \dot{n} \left(\frac{\partial X_{CH_4}}{\partial x} \right) dx,$$

where $\partial X_{CH_4}/\partial x$ is negative, and $\dot{n}$ is the flow rate of the gas in mol/sec. The amount of carbon diffusing from the surface, bdx, per unit time into the two surfaces equals

$$- 2(bdx)\left[D\left(\frac{\partial C}{\partial y}\right)_{y=0} \right],$$

where b is the sheet width normal to the plane of Fig. 16.3, and C is given in g-atom/cm^3. Equating this value to the loss of methane per unit time, and substituting Eq. (16.16), we get

$$\left(\frac{\partial C}{\partial y}\right)_{y=0} = \frac{\dot{n}P}{2bDK_{eq}}\left(\frac{\partial C}{\partial x}\right)_{y=0} \tag{16.18}$$

for $x > 0$. To summarize, we wish to solve Eq. (16.14), subject to Eqs. (16.15), (16.17), and (16.18). A method of accomplishing this has been presented by Wagner,[2] and is as follows.

Assuming that C depends only on $x/\sqrt{t}$ and $y/\sqrt{t}$, we define the dimensionless variables:

$$\xi = \frac{2bDK_{eq}}{\dot{n}P}\frac{x}{2\sqrt{Dt}}; \qquad \eta = \frac{y}{2\sqrt{Dt}}.$$

Substitution of ξ and η into Eq. (16.14) gives

$$\frac{\partial^2 C}{\partial \eta^2} + 2\left(\xi \frac{\partial C}{\partial \xi} + \eta \frac{\partial C}{\partial \eta}\right) = 0. \tag{16.19}$$

Substitution of ξ and η into Eqs. (16.17), (16.15), and (16.18), respectively, yields

$$C(\xi = 0, \eta = 0) = C_e, \tag{16.20a}$$

$$C(\xi = \infty, \eta = \infty) = C_i, \tag{16.20b}$$

$$\left(\frac{\partial C}{\partial \xi}\right)_{\eta=0} = \left(\frac{\partial C}{\partial \eta}\right)_{\eta=0}. \tag{16.20c}$$

We can transform Eq. (16.19) into an ordinary differential equation if we consider C a function of the sum $\sigma = \xi + \eta$.

$$\frac{d^2 C}{d\sigma^2} + 2\sigma \frac{d\sigma}{dC} = 0. \tag{16.21}$$

The solution to Eq. (16.21) subject to Eqs. (16.20a, b, and c) is

$$\frac{C - C_i}{C_e - C_i} = 1 - \text{erf}\left[\frac{\dfrac{2bDK_{eq}}{\dot{n}P}x + y}{2\sqrt{Dt}}\right]. \tag{16.22}$$

Example 16.2 Calculate the distance x over which the surface concentration C_s does not vary excessively after carburizing. Specifically, it is required that the ratio

[2] C. Wagner, *ibid.*, and *Zeitsch. f. physik. Chem.* **192**, 157 (1943).

$(C_s - C_i)/(C_e - C_i) \geq 0.75$.   Surface carburization is carried out with dilute methane/argon–hydrogen at 1000°C and 1 atm, and the appropriate equilibrium constant is 0.7 atm g-atom/cm³. The gas flows at a linear velocity of 10 cm/sec through a separation of 2 cm between steel sheets for 5 hr. At 1000°C, the diffusion coefficient of carbon in γ-iron is 3×10^{-7} cm²/sec.

Solution. Seeking the surface concentration C_s, we obtain from Eq. (16.22)

$$\frac{C_s - C_i}{C_e - C_i} = 1 - \mathrm{erf}\left[\frac{\dfrac{2bDK_{eq}}{\dot{n}P}x}{2\sqrt{Dt}}\right].$$

We require the left-hand side of the equation to be 0.75. Hence

$$0.25 = \mathrm{erf}\left[\frac{\dfrac{2bDK_{eq}}{\dot{n}P}x}{2\sqrt{Dt}}\right].$$

We find the argument of the error function to be 0.225, so that

$$x = 0.450\sqrt{\frac{t}{D}\frac{\dot{n}}{2b}\frac{P}{K_{eq}}}.$$

We can substitute all values directly, except for $\dot{n}/2b$ which is evaluated simply as

$$\dot{n} = \frac{10\ \mathrm{cm}}{\mathrm{sec}}\left|\frac{b \cdot 2\ \mathrm{cm}^2}{}\right|\frac{273°\mathrm{K}}{1273°\mathrm{K}}\left|\frac{1\ \mathrm{g\text{-}mol}}{22{,}400\ \mathrm{cm}^3(\mathrm{STP})}\right..$$

Therefore

$$\frac{\dot{n}}{2b} = \frac{(10)(273)}{(22{,}400)(1273)}\ \mathrm{g\text{-}mol/sec\text{-}cm}.$$

Then, by substituting $t = 18{,}000$ sec, $D = 3 \times 10^{-7}$ cm²/sec, $P = 1$ atm, and $K_{eq} = 0.7$ atm g-atom/cm³, we calculate x to be 15 cm.

   Of practical significance, this section shows that excessive nonuniformity of the surface concentration can only be prevented by a sufficient concentration of the reactant species in the exit gas phase. For example, we saw in the above problem that there was a limit to the distance that could be properly carburized for the specifications put forth. If the specifications had been more severe, for example, $[C_s - C_i/C_e - C_i] \geq 0.90$, then a higher velocity of carburizing gas would have had to be supplied to achieve the same value of x, thus raising the concentration of the reactant species in the exit gas. If the active species is valuable, recovery of this component or recirculation of the exit gas must be used. This is an important economic consideration in chromizing steel with $CrCl_2(g)$.

16.2.3 Mixed control in oxide reduction reactions

The rates of reduction reactions are of practical interest to metallurgists, and many studies on this subject have been reported in the literature. In most cases, some attempt has been made either to ascertain the rate-controlling step or mass-transfer coefficient for a specific step in the overall process. However, what is usually measured is the overall coefficient, and in many instances, unless conditions are clearly such that one step does not control and only one other step is involved, it is very difficult to be sure of the rate-limiting step.

As a classic example, consider the reduction of iron oxides by gases, which has been the object of scores of papers. The measurements made are usually those of weight loss of a specimen as a function of time. If we assume that the overall rate is controlled by the rate of the chemical reaction

$$FeO + CO(g) = Fe + CO_2(g)$$

at the oxide–metal interface, then, for a spherical particle of initial radius r_0 and density ρ_0,

$$r_0\rho_0[1 - (1 - f)^{1/3}] = k_c t \qquad (16.23)$$

should relate the fractional reduction f to time t. From the plots of test data, we obtain the values of the rate constant k_c, and plot them as $\ln k_c$ versus T^{-1} in order to obtain activation energies for the process. On the basis of agreement of test data with Eq. (16.23), such as in Fig. 16.4, one might be tempted to conclude that the rate of reduction of iron oxide is controlled by the rate of the chemical reaction step. However, it happens that in another study the data fit another model that presumes that the rate of reduction is entirely controlled by equimolar countercurrent diffusion of the reactant gas and product gas through the porous metal product layer that forms. These data are shown in Fig. 16.5. Another study of data based on many investigations again concludes that the rate is controlled by the chemical

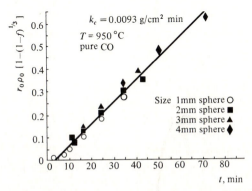

Fig. 16.4 Experimental data for hematite reduction by CO gas plotted according to Eq. (16.23), which is based on the assumption of chemical reaction control at the oxide-metal *interface*. (From W. M. McKewan, *Trans. AIME* **212**, 791 (1958).)

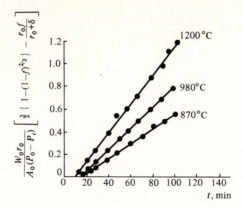

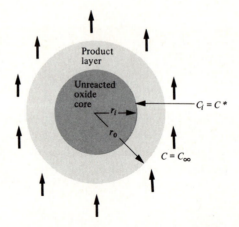

Fig. 16.5 Experimental data for hematite reduction by CO gas plotted according to an equation based on the assumption of control by diffusion through the product layer. Particle diameter was 2.8 cm. (From E. Kawasaki, J. Sanscrainte, and T. J. Walsh, *AIChEJ* **8**, 48 (1962).)

Fig. 16.6 Schematic model for oxide reduction.

reaction at the metal–oxide interface,[3] but in this case an entirely different activation energy than in the previous work is obtained.

More recently, Olsson and McKewan[4] have shown that in fact there exists mixed control. Their model, as defined in Fig. 16.6, assumes that there are three steps in the overall process: gas-phase mass transfer to and from the bulk gas stream and the solid, diffusion of the reactant and product gases across the solid, but porous, reaction product layer, and, finally, the chemical reaction at the

[3] N. J. Themelis and W. H. Gauvin, *Trans. AIME* **227**, 290 (1963).

[4] R. G. Olsson and W. M. McKewan, *Trans. AIME* **236**, 1518 (1966).

oxide–metal interface itself. If the mass transfer and diffusion steps are considered together, the mass flow to the interface from the bulk gas stream is

$$J_D = \frac{4\pi(C_\infty - C^*)}{\dfrac{1}{r_0^2 k_m} + \dfrac{r_0 - r_i}{r_0 r_i D_{\text{eff}}}}, \text{ mol/sec,} \tag{16.24}$$

where C_∞ = bulk gas concentration of reducing gas, mol/cm^3,
C^* = gas concentration at metal–oxide interface,
r_0 = initial particle radius, cm,
r_i = unreacted core radius, cm,
k_m = gas phase mass-transfer coefficient, cm/sec,
D_{eff} = effective diffusivity of reducing gas through the product layer, as given by Eq. (13.64), cm^2/sec.

The rate of weight loss is

$$R_D = 16J_D, \text{ g/sec,} \tag{16.25}$$

since 16 g of oxygen are removed by each mole of reducing gas that reaches the reaction interface. When we compare the experimental rate of weight loss, R_e, to the rate of weight loss predicted by Eqs. (16.24) and (16.25), we see from Fig. 16.7 that the ratio R_e/R_D increases rapidly toward 1.0 as the thickness of the porous iron layer increases. Thus, it is apparent that there is an interaction between the mass transfer steps on the one hand, and the chemical reaction step on the other, with the latter contributing to the control of the overall rate only at early stages of the reduction process at high temperatures, or at lower temperatures generally.

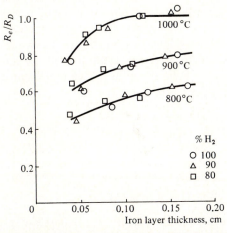

Fig. 16.7 The ratio of the experimental to the calculated diffusion limited rate of weight loss. (From R. G. Olsson and W. M. McKewan, *Trans. AIME* **236**, 1518 (1966).)

Wen[5] has discussed in detail the mathematical solutions to the kinetic equations for many different models of gas–solid reactions; he has shown how similar some of the results are in terms of experimental weight loss versus time data for models differing widely in terms of assumptions made. From his paper and the example above, it is clear that misleading conclusions concerning rate-controlling mechanisms can easily be made, unless much care is taken to ensure that experimental conditions eliminate control by steps being ignored in the analysis. However, it should be kept in mind that overall coefficients can be useful information from the design engineering standpoint.

16.3 MASS TRANSFER WITH VAPORIZATION

As pointed out in Chapter 5, there are several situations in metallurgy that involve heating materials in a vacuum. In vacuum melting or heat treating of ferrous alloys, valuable elements, such as manganese or chromium, may be lost, or conversely undesirable impurities, such as zinc or lead, may be removed. Experimentally, in Knudsen effusion cells, use is made of the measured rate of vaporization of an element to determine thermodynamic properties of the material.

In all of these cases, there are several steps involved in the overall mass transfer process: 1. transport of the volatile species to the surface of the condensed phase; 2. formation of volatile compounds at the surface; 3. evaporation from the surface into the gas or vacuum; and 4. transport away from the surface into the gas or vacuum. Depending on conditions, any one of these steps may be rate limiting, or several may comprise a mixed-control process.

16.3.1 Knudsen effusion cells

The Knudsen cell is used to study thermodynamic properties of alloys by taking advantage of the high volatility of one component relative to the others, and determining the vapor pressure of this species as a function of alloy content. The vapor pressure is determined by measuring the force exerted by a molecular stream emanating into a vacuum from a very small hole in the end of a hollow cell containing the alloy. Under these conditions, free molecular diffusion occurs; then, by referring to Section 5.4.1, we have

$$Z_{\text{net}} = \sqrt{\frac{\kappa_B T}{2\pi m}} (n_1 - n_2), \text{molecules/cm}^2\text{-sec,}$$

or

$$J = \frac{A_e(P_1 - P_2)}{\sqrt{2\pi MRT}}, \text{moles/sec,}$$

where P_1 is in g/cm sec^2, and R has units of g cm^2/sec^2-deg · mol.

---

[5] C. Y. Wen, *I. & E. C.* **60**, No. 9, 34 (1968).

Table 16.1 Clausing factors for effusion cells*

l/r	W_e
0	1.00
0.1	0.9524
0.2	0.9092
0.3	0.8699
0.4	0.8341

* From P. Clausing, *Annalen der Physik* **12**, 976 (1932).

These expressions give the net rate at which atoms or molecules leave the cell through an infinitely thin orifice opening. Because of the difficulty of constructing an infinitely thin opening, that is, with no walls, there is a correction to the theoretical flux accounting for the fact that there is a small probability of molecules rebounding and not passing through the opening. The Clausing factor, W_e, gives the probability of passing and is a function of the ratio of the orifice thickness to radius, l/r, as given in Table 16.1. When this factor is included, then

$$J = \frac{A_e W_e (P_1 - P_2)}{\sqrt{2\pi MRT}}. \tag{16.26}$$

At the same time as vapor leaves through the orifice, the solid within the cell loses alloy from its surface initially having an alloy concentration of C_A^0. If the area of the orifice, A_e, is small relative to the exposed surface area of the alloy, A_s, then we may take the partial pressure within the cell, P_1, as uniform. Furthermore, P_1 is the pressure of the volatile component in equilibrium with its concentration on the surface of the solid, C_A^s. Therefore, Fig. 16.8 gives the situation, and the diffusive flux within the solid to the surface equals the effusive flux

$$J = +DA_s \left(\frac{\partial C_A}{\partial x}\right)_{x=0} = \frac{A_e W_e (P_1 - P_2)}{\sqrt{2\pi MRT}}, \tag{16.27}$$

where C_A is in mol/cm^3.

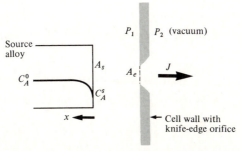

Fig. 16.8 The conditions in a Knudsen effusion cell.

In order for free molecular diffusion to occur, P_2 must be very low ($< 10^{-3}$ atm). In fact, the flux J is obtained more quickly if the cell is placed in a vacuum, especially since P_1 is then on the order of only $10^{-3} - 10^{-6}$ atm; P_1 is related to the surface composition by

$$P_1 = \gamma_A X_A P_A^0,$$

where γ_A is the activity coefficient relative to the pure component, and P_A^0 is the vapor pressure of the pure component at the temperature in question. Thus, with $P_2 = 0$ in a vacuum,

$$DA_s \left(\frac{\partial C_A}{\partial x} \right)_{x=0} = \frac{A_e W_e \gamma_A P_A^0 C_A{}^s}{\rho \sqrt{2\pi MRT}}, \tag{16.28}$$

where $\rho = $ mol alloy/cm^3, the molar density of the alloy. Fick's second law describes diffusion within the solid; the boundary condition at the surface of the solid is Eq. (16.28).

If we consider the alloy as a semi-infinite solid, then we seek a solution to

$$\frac{\partial C_A}{\partial t} = D \frac{\partial^2 C_A}{\partial x^2}, \tag{16.29}$$

with

$$C_A(x, 0) = C_A^0, \tag{16.30a}$$

$$\frac{\partial C_A(0, t)}{\partial x} - Y C_A(0, t) = 0, \tag{16.30b}$$

and

$$C(\infty, t) = C_A^0, \tag{16.30c}$$

where

$$Y \equiv A_e W_e \gamma_A P_A^0 / A_s D \rho \sqrt{2\pi MRT}, \text{cm}^{-1}.$$

Carslaw and Jaeger[6] give the solution to Eq. (16.29) satisfying Eq. (16.30a, b, and c):

$$\frac{C_A}{C_A^0} = \text{erf} \frac{x}{2\sqrt{Dt}} + \exp(Yx + Y^2 Dt) \cdot \text{erfc} \left(\frac{x}{2\sqrt{Dt}} + Y\sqrt{Dt} \right). \tag{16.31}$$

The surface concentration of the solid C_A^s, obtained by letting $x = 0$ in Eq. (16.31), is

$$C_A^s / C_A^0 = \exp(Y^2 Dt) \cdot \text{erfc}(Y\sqrt{Dt}). \tag{16.32}$$

Using Eq. (16.32), Schroeder[7] has calculated the maximum experimental times

---

[6] H. S. Carslaw and J. C. Jaeger, *Conduction of Heat in Solids*, second edition, Oxford, 1959, page 71.

[7] D. L. Schroeder, *Sc.D. Thesis*, MIT, 1966; also *Trans. AIME* **236**, 1091 (1967).

allowable in order to limit the surface depletion to 5% for Fe–Mn alloys. Figure 16.9 gives these results for one particular alloy.

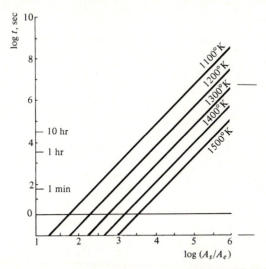

Fig. 16.9 Calculated times for 5% depletion of the surface of 5 wt % Mn–Fe alloys. (Adapted from D. L. Schroeder and J. F. Elliott, *Trans. AIME* **236**, 1091 (1967).)

16.3.2 Alloy vaporization during melting

In general, vaporization from liquid metals is not very different than that from solid metals. If the melting is carried out under a vacuum, there are again two steps: mass transfer to the free surface, and vaporization from the surface into the vacuum. If there is an inert gas pressure above the surface, there is an additional resistance to mass transfer, and as the gas density increases (or vacuum decreases), this may become significant relative to the resistance of the other transport steps.

Consider the first step. Mass transfer within the liquid to the surface could be calculated from a correlation of the form $Sh = f(Re, Sc)$ if this were available for the particular physical arrangement and stirring conditions involved. If there is no fluid motion, the problem reduces to the same conditions as in Section 16.3.1. In the case of induction melting, there is no general correlation available, so we shall make use of an hypothesis by Machlin,[8] which states that flow across the surface of an inductively stirred melt may be considered to be a free streamline slug flow without any shear gradients taking place within the surface layer. The volatile solute is supplied to the surface solely by diffusion from within this surface layer. This hypothesis is exactly the same as Higbie's "penetration theory" discussed in Section 15.9. The only new assumption is that the "lifetime" θ of a surface element is equal to the average distance from the center of the melt to the

---

[8] E. S. Machlin, *Trans. AIME* **218**, 314 (1960).

edge of the crucible, divided by the average velocity of the melt at the surface. In the induction stirring case, θ is of the order of 1 sec or less.

For this step in the process then, the average mass-transfer coefficient is

$$k_{M,l} = 2\sqrt{\frac{D}{\pi\theta}}, \text{ cm/sec,} \qquad (15.99)$$

and the flux to the surface from the bulk liquid is

$$j_A = k_{M,l}(C_{A\infty} - C_A^s), \text{ mol } A/\text{cm}^2\text{-sec,} \qquad (16.33)$$

where $C_{A\infty}$ is the bulk concentration and C_A^s is the surface concentration.

The second step is the evaporation itself. Returning to Section 16.3.1 and assuming that no chemical reactions are involved,* we have

$$j_A = \frac{\gamma_A P_A^0 C_A^s}{\rho\sqrt{2\pi MRT}}, \qquad (16.34)$$

or

$$k_{M,e} = \frac{\gamma_A P_A^0}{\rho\sqrt{2\pi MRT}}, \qquad (16.35)$$

where $k_{M,e}$ is the evaporation mass-transfer coefficient.

The final step in the sequence involves mass transfer within the vapor phase, as influenced either by convection (when the gas pressure is $> 10^{-2}$ atm), or by molecular flow mechanics, as in vacuums. In the former case, since $k_{M,g} \propto D$, and since Eqs. (13.57) and (13.58) indicate that D increases directly with reciprocal pressure, it is apparent that with a good vacuum $k_{M,g}$ becomes very large.

Now consider the overall process in a good vacuum where the gas phase resistance is negligible. The fluxes due to liquid phase mass transfer (Eq. 16.33), and evaporation (Eqs. 16.34 and 16.35) are equal:

$$j_A = k_{M,l}(C_{A\infty} - C_A^s) = k_{M,e}C_A^s.$$

Solving for C_A^s, we get

$$C_A^s = \frac{k_{M,l}}{k_{M,l} + k_{M,e}} C_{A\infty}.$$

Substituting, we obtain

$$j_A = \left(\frac{k_{M,e}k_{M,l}}{k_{M,l} + k_{M,e}}\right)C_{A\infty} = KC_{A\infty}, \qquad (16.36)$$

---

* We often see an additional multiplier on the right-hand side of Eq. (16.34). This is the *condensation coefficient* α, which is a measure of the proportion of atoms leaving a surface that do not return to it. If α is unity, then all of the atoms leaving that surface do not return to the surface, but are condensed instead on cool surfaces away from the melt. Typically, α is taken as unity for most metallurgical vaporization problems.

where K is the overall rate constant,* defined as:

$$1/K = 1/k_{M,l} + 1/k_{M,e}.$$

By evaluating $k_{M,l}^{-1}$ and $k_{M,e}^{-1}$, we can either determine the greater resistance, or calculate the rate at which the bulk concentration changes by using $k_{M,l}$ and $k_{M,e}$ to calculate K. In the latter case,

Rate of decrease of solute A in melt = Mass flow of A from surface,

or

$$-V\frac{dC_{A\infty}}{dt} = A_s K C_{A\infty},$$

where A_s and V are the surface area of melt exposed to vacuum and volume of the melt, respectively. Integrating, we have

$$\int_{C_{A\infty}=C_{A\infty}}^{C_{A\infty}=C_A^0} \frac{dC_{A\infty}}{dt} = \frac{A_s}{V} K \int_0^t dt,$$

or

$$\ln\frac{C_A^0}{C_{A\infty}} = \frac{A_s}{V} Kt. \tag{16.37}$$

These effects are illustrated in the following example.

Example 16.3 We wish to test for the rate-controlling step in the loss of manganese from iron, induction melted under vacuum, given that $D_{Mn} = 9 \times 10^{-5}$ cm^2/sec, $T = 1600°C$, $\gamma_{Mn} = 1.0$, $P_{Mn}^0 = 46.65 \times 10^3$ g/cm-sec^2, and $M = 54.9$ g/mol.

Solution. If the lifetime θ is taken as 1 sec, then

$$k_{M,l} = 2\sqrt{\frac{D}{\pi\theta}} = 1.08 \times 10^{-2} \text{ cm/sec.}$$

This term is unaffected by the gas phase pressure. Next

$$k_{M,e} = \frac{\gamma P_{Mn}^0}{\rho\sqrt{2\pi MRT}} = \frac{(1.0)(46.65 \times 10^3)}{(0.128)\sqrt{(2\pi)(54.9)(8.31 \times 10^7)(1873)}}$$

$$= 4.98 \times 10^{-2} \text{ cm/sec.}$$

Thus

$$\frac{1}{K} = \frac{1}{0.0108} + \frac{1}{0.0496} = 92.5 + 21.4 = 113.9 \text{ cm}^{-1} \text{ sec,}$$

---

* The constant K is sometimes called the *specific evaporation constant*.

or

$$K = 8.8 \times 10^{-3} \text{ cm/sec.}$$

It is apparent that when there is a good vacuum, liquid phase mass transfer offers 81 % of the total resistance; thus we may say that it essentially controls the rate.

This changes, however, if the stirring is increased to decrease θ. For example, if θ decreases to 0.1 sec, then

$$k_{M,l} = 3.39 \times 10^{-2} \text{ cm/sec,}$$

and

$$\frac{1}{k_{M,l}} = 29.5 \text{ cm}^{-1} \text{ sec.}$$

Now the liquid phase mass transfer and the evaporation process are almost equivalent in their contribution to K, and there is mixed control. Furthermore, K has increased to 1.97×10^{-2} cm/sec.

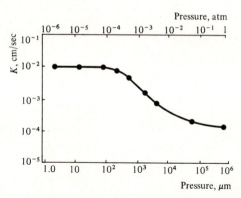

Fig. 16.10 Overall mass-transfer coefficients for Mn vaporization from 0.25 % C steel at 1580°C as a function of the pressure of Ar over the melt. (From R. G. Ward, *JISI* **201**, 11 (1963).)

Since neither $k_{M,l}$ nor $k_{M,e}$ depend on the external pressure, K is independent of vacuum pressure as long as $k_{M,g}$ remains very large. When $k_{M,g}$ becomes about an order of magnitude smaller than $k_{M,l}$ or $k_{M,e}$, it is rate controlling. Since $k_{M,g}$ is proportional to P^{-1}, then K decreases with P, as long as δ_{eff} in the gas phase remains constant. When the gas phase is dense, then $k_{M,g}$ is controlled by natural or forced convection. Figure 16.10 illustrates the effect of system pressure on K for vaporization of manganese from steel. Figure 16.11 gives the results of the application of Eq. (16.37) to calculate the rate of loss of manganese at 1600°C for various (A_s/V) values, using $K = 0.01$ and 0.001. One apparent result is that the use of an argon atmosphere to increase pressure decreases K to the point where the loss of manganese is negligible.

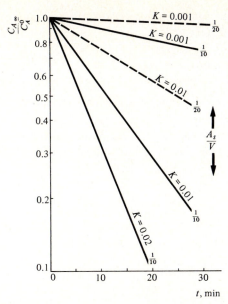

Fig. 16.11 Decrease in concentration of alloying element as a function of mass-transfer coefficient and A_s/V ratio.

Some other experimental values of K for the loss of various elements from iron–base melts into good vacuums are given in Table 16.2.

16.4 THE EFFECT OF TEMPERATURE AND THE CONCEPT OF THERMAL STABILITY

For any reaction between a gas and a solid, there are several temperature regions where control of the reaction may be by different mechanisms. In the low temperature region, the chemical reaction is often the slow step and controls the overall rate. The rate of change of the chemical reaction rate with temperature is exponential, since in general

$$k_c = Z \exp(-\Delta H/RT), \tag{16.38}$$

Table 16.2 Overall evaporation constants for iron–base melts at 1600°C*

Element	K, cm/sec
Mn	8.4×10^{-3}
Cu	4.8×10^{-3}
Sn	2.3×10^{-3}
Cr	2.1×10^{-4}
S	7.0×10^{-4}

* From R. Ohno and T. Ishida, *J. Iron and Steel Inst.*, London **206**, 904 (1968).

where k_c = chemical reaction rate constant, cm/sec,
       Z = frequency factor,
    ΔH = activation enthalpy,
      R = gas constant,
      T = absolute temperature.

Thus, as temperature increases, the chemical reaction rate usually increases rapidly to the extent that it no longer is the slow step, and diffusion of either the reactants or products to or away from the reaction surface becomes the slow step. This is depicted in Fig. 16.12, where, as the temperature increases, the proportion of control by diffusion increases.

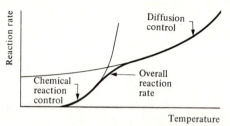

Fig. 16.12 Schematic reaction rate as a function of temperature for a gas–solid reaction.

Consider a simple model in which the rate of the reaction is proportional to the concentration C_s of the gaseous reactant at the reaction surface:

$$j_c = k_c C_s. \tag{16.39}$$

The rate of diffusion of the reactant from the turbulent bulk gas stream of concentration C_∞ to the reaction site is given simply by

$$j_D = \frac{D}{\delta_{\text{eff}}}(C_\infty - C_s), \tag{16.40}$$

where δ_{eff} is the effective thickness of the fluid through which reactants diffuse to the reaction site.

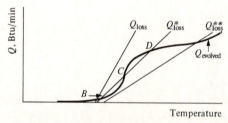

Fig. 16.13 Interaction of heat loss and heat evolution curves for various heat loss conditions.

Since these two rates are always equal, we can eliminate C_s and express the rate as

$$j = \frac{C_\infty}{1/k_c - \delta_{\text{eff}}/D}. \tag{16.41}$$

Remember that k_c increases exponentially with temperature, whereas D varies approximately as T^2 for gaseous diffusion. Therefore, at low temperatures, $k_c \ll D/\delta_{\text{eff}}$, and

$$j \cong k_c C_\infty.$$

On the other hand, at high temperatures, $k_c \gg D/\delta_{\text{eff}}$, and

$$j \cong \frac{D}{\delta_{\text{eff}}} C_\infty.$$

These are the limiting cases, and in between, the control is mixed. The absolute value of the temperature where this transition occurs depends, of course, on the parameters involved in each particular case. As an example, take Fig. 16.7 for the case of hematite reduction. The upper limit of mixed control in this case appears to be 1000°C, and the lower limit is at some temperature less than 800°C.

If we compare the reaction rate dependence on temperature with the rate of heat loss from a given system, also as a function of temperature, we can obtain a semiquantitative picture of what is involved in the thermal stability of a process. The shape of the reaction rate curve is important because the rate of heat release (in an exothermic reaction) is directly proportional to the reaction rate. Consider the situation in a packed bed reactor. If the gas entering the bed is at T_g, and the solids are at a uniform temperature T_s, then the rate of heat loss per unit volume from the solids (which are presumed to be overheated by the evolution of heat of reaction on or within them) is

$$Q_{\text{loss}} = hS(T_s - T_g). \tag{16.42}$$

Therefore, the rate of heat loss from the solids is a linear function of T_s, for a given T_g, as shown in Fig. 16.13 for several values of hS. The heat evolution curve, on the other hand, follows the same shape as the reaction rate curve of Fig. 16.12.

Stable steady-state conditions occur when the two curves cross. Note, however, that in the cases of multiple intersections, such as where the rates of heat addition and loss are equal at point C, any slight increase in temperature results in more heat being generated than removed, causing the temperature to rise to point D. Any further temperature increase results in more heat being lost than generated, thus returning the system to D. A similar line of reasoning shows that a slight decrease from point C reduces the reaction temperature down to point B. Therefore, in the case with a heat loss rate of Q^*, there are two stable states, one corresponding to the slow, chemical-reaction-controlled kinetics, and one to the fast, diffusion-controlled kinetics.

Considering this case further, there is a range of possible situations for a given hS value, depending only on the gas temperature, as indicated in Fig. 16.14. In the case where $T_g = T_g^*$, the stable steady-state conditions are those described above. However, if the entering gas temperature is decreased, the heat loss curve shifts to the left until a situation is encountered such as that for $T_g = T_g$. If the process operates at point D, it decreases in rate and temperature until point X has been reached. At that point, any further slight decrease in process temperature greatly decreases the rate of reaction, because the system rapidly approaches point Y.

On the other hand, an increase in entering gas temperature from T_g^* to T_g^{**} results in slight increases in rate and process temperature if the process is initially at either point B or point D, and in the possibility of rapid and uncontrollable temperature rise from point Z, once it has been reached. T_g^{**} is known as the *minimum gas temperature* for ignition of cold solids, and the *minimum ignition temperature* of the solids is T_Z. By the same token, T_X is the *minimum combustion temperature* or *critical extinction point*. These temperatures depend on the shapes of the heat loss and heat evolution curves, which in turn depend on the gas flow rates, surface areas, heat-transfer coefficients, diffusivities, gas compositions, etc. Thus, there are certain regions of operation of gas–solid systems in which stable operation is impossible.

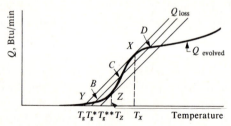

Fig. 16.14   The effect of variation in T_g on thermal stability for a constant value of hS.

PROBLEMS

16.1 Derive Eq. (16.24) leading to the diffusive and mass transfer flux to the reaction interface.

16.2 Estimate the rate of loss of Zn from 70–30 brass by evaporation alone as a function of temperature.

16.3 Some ferritic stainless steels are vacuum heat-treated in order to maintain surface finish. If AISI Type 410 parts are heat-treated in a furnace with an exhaust duct area of 200 cm^2 and a vacuum of 10 μ of Hg at 1600°F for 2 hr and their surface area is 2000 cm^2, what will the surface concentration of chromium of the parts be? Assume that the initial concentration is 12% Cr and $D_{Cr} = 10^{-12}$ cm^2 sec.

16.4 The basic oxygen process of making steel involves melting some steel scrap in the furnace. Occasionally, large pieces of scrap remain unmelted at the end of a heat.

  a) Which is more important in determining the rate of scrap melting, the weight of the piece of scrap or its dimensions?

b) Describe the conditions that characterize the rate of melting when the surrounding liquid is carbon-saturated pig iron at 2600°F.

c) Repeat (b) when the bath temperature is 2900°F and the carbon content is 0.20% C.

APPENDIX I

Table I.1 Physical constants*

Quantity	Symbol	Value
Gas constant	R	1.98725 ($\pm$ 0.00008) cal/°K · mol
		8.31467 ($\pm$ 0.00034) joule (absolute)/°K · mol
		82.0594 ($\pm$ 0.0034) cm³ · atm/°K · mol
		0.0820571 ($\pm$ 0.0000034) liter · atm/°K · mol
		1.985857 ($\pm$ 0.00008) Btu/°F lb · mol
Boltzmann constant	κ_B	1.38044 ($\pm$ 0.00007) × 10^{-16} erg/°K · mol
Planck constant	h	6.62517 ($\pm$ 0.00023) × 10^{-27} erg · sec
Avogadro constant	N_0	6.02320 ($\pm$ 0.00016) × 10^{23} atoms/g · atom
Faraday constant	$\mathscr{F}$	96,495.4 ($\pm$ 1.1) coul/equiv.
		23,063.0 ($\pm$ 0.3) cal/v equiv.

* *Note:* Throughout this book, the term *calorie*, unless otherwise qualified, means *thermochemical calorie*. It is arbitrarily defined by the relationship: *1 thermochemical calorie = 4.184 absolute joules*, or *4.193 international joules*.

The absolute joule is defined by the equation: *1 absolute joule = 10⁷ ergs*. One erg is the work done when a force of one dyne acts through a distance of one centimeter. One absolute or mechanical watt is the rate of working of one absolute joule per second.

The British thermal unit and the International calorie are in common industrial use. They are defined by the relationship: *1 international calorie/gm = 1 Btu/lb when 1 international calorie = 4.186 international joules*.

Table II.1 Densities and thermal properties of various substances

Name — Description	Normal state	Other classification	Density in lb/cu ft	Specific heat in Btu/lb°F (or gm cal/gm°C) — Solid state: Specific heat	Solid state: Temp range °F	Liquid state: Specific heat	Liquid state: Temp range °F	Gaseous state: Specific heat	Gaseous state: Temp range °F	Melting (fusion) point °F	Boiling point, °F (at std barometric pressure)	Latent heat in Btu/lb of fusion	Latent heat in Btu/lb of vaporization
Acetic acid (CH_3COOH)	Liquid	O	65.8	0.487	32	0.51	32-212	0.3468	79-230	62.6	244.4	80.5	174
Acetone (CH_3COCH_3)	Liquid	O	111.8			0.514	32-212	0.64	59	-138.2	128-134	42	239
Acetylene (C_2H_2)	Gas	O, F	0.0691							-113.8	-118.8		
Air	Gas		0.0763					0.2394 0.2469 0.2562	-22+50 68-824 68-1472		-311.0		
Alcohol, ethyl (C_2H_5OH)	Liquid	O, F	49.3			0.648	104	0.4534	226-428	-173.2	172.4	46	369
Alcohol, ethyl (90%) and water	Liquid	S	51.4			0.718							
Alcohol, ethyl (50%) and water	Liquid	S	57.3			0.923							
Alcohol, ethyl (10%) and water	Liquid	S	61.4			0.99							
Alcohol, methyl (CH_3OH)	Liquid	O, F	49.6			0.601	59-63	0.4580	214-333	-142.6	150.8	29.5	480.6
Alcohol, methyl (90%) and water	Liquid	S	51.4			0.643							
Alcohol, methyl (50%) and water	Liquid	S	57.3			0.846							
Alcohol, methyl (10%) and water	Liquid	S	61.4			0.986							
Alumina—see Aluminum Oxide, Alumina (fused) refractory, high-alumina refractory													
Alumina (fused) refractory (see also high-alumina refractory)	Solid	R	153-181	0.20	60-1200					3390+			
Aluminum	Solid	M, E	166.7	0.225	61-579					1220	3272	167.5	360.0
Aluminum foil	Solid	I		0.24	290								
Aluminum oxide (alumina)	Solid		243.5	0.183	32-212					3668			
Ammonia (NH_3)	Gas		0.046			1.08 0.778	-8.4+5	0.525	70-220	-103	-28.3	150	589
Ammonium chloride (10%) and water	Liquid	S	64.4										
Ammonium sulfate [$(NH_4)_2SO_4$]	Solid		110	0.284						956b			
Andalusite (Al_2SiO_5)	Solid		199.8	0.228						3290			
Aniline (C_6H_7N)	Liquid	O, F	63.9			0.514	59			17.6	363	38	198.0
Antimony (Sb)	Solid	M, E	422	0.052	392					1166	2624	68.9	703
Argon (A)	Gas	E	0.105					0.1233	68-194	-306.4	-303	12	68
Arsenic, gray	Solid	E	358	0.0822						c	c		
Asbestos	Solid	I	124-174	0.20	32-212								
Astes	Solid			0.20	32-212								
Asphalt, Bermudez	Solid		67.4	0.55						180			
Gilsonite	Solid		64.9	0.55						300			
Oil	Solid		61.7-64.9	0.55						140-180			
Trinidad	Solid		87.4	0.55						190			

a—A=alloy, E=element, F=fuel or fuel component, I=insulation, M=metal, O=organic compound, R=refractory, S=solution.
b—decomposes.
c—sublimes at 1038°F, melts at 1562°F under 36 atmospheres of pressure.

* Adapted from *North American Combustion Handbook*, published by North American Manufacturing Co., Cleveland, Ohio, 1952.

Table II.1 Densities and thermal properties of various substances (*continued*)

Name — Description	Normal state	Other classification	Density in lb/cu ft	Specific heat, Solid state	Temp range °F (solid)	Specific heat, Liquid state	Temp range °F (liquid)	Specific heat, Gaseous state	Temp range °F (gaseous)	Melting (fusion) point °F	Boiling point °F (at std barometric pressure)	Latent heat of fusion in Btu/lb	Latent heat of vaporization in Btu/lb
Babbitt, lead base	Solid	A		0.039	60-462	0.038				462		26.2	
tin base	Solid	A	465	0.071	60-464	0.063				464		34.1	
Bakelite (see Resin, phenol)	Solid												
Barium (Ba)	Solid	E	218	0.068						1562	2080		1120
Basalt (see lava)	Solid												
Beeswax	Solid	O, F	59.9	0.82	-121-+68					143.6		76.2	
Benzene (C_6H_6)	Liquid		54.9			0.423	60-144	0.3325	95-356	41.8	176.3	55.6	169.4
Benzoic acid ($C_7H_6O_2$)	Solid	O	81.2	0.287	60					249.8	480.2	61.0	
Benzol (C_6H_6) (water white)	Liquid	O, F	55			0.423	104			41.7	176.3	55.08	169.4
Beryllium (Be)	Solid	E, M	113.6	0.52		0.425		0.277		2732	5020	570	
Bismuth (Bi)	Solid	E, M	612	0.0302	68-212	0.036	535-725			519.8	2606	22.4	395
Bismuth (63.8)—Tin (36.2) alloy	Solid	A		0.040	68-210								
Blast furnace gas	Gas	F	0.0778										
Boron (B)	Solid	E	152	0.307	32-212					3992-4532	4600		
Borax ($Na_2B_4O_7 \cdot 10H_2O$)	Solid		107	0.385	95					1366			
Brass, Muntz metal (60 Cu, 40 Zn)	Solid	A	524	0.105	60-1630					1630		69.0	
Red (85 Cu, 15 Zn)	Solid	A	546	0.104	60-1952					1952		86.5	
Yellow (67 Cu, 33 Zn)	Solid	A	528	0.105	60-1688					1688		71.0	
Brick, red	Solid		118	0.22	32-212								
Britania metal (90 Sn, 10 Pb)	Solid	A	193	0.0843	57-208							50.4	
Bromine (Br)	Liquid	E		0.0862	-108-+4	0.107	34-89	0.0555	181-442	18.86	141.8	29.16	82.1
Bronze, 80 Cu, 20 Sn	Solid	A	510	0.126									
Aluminum	Solid	A	556	0.095	60-1922					1922		98.6	
Bearing	Solid	A	540	0.100	60-1832					1832		79.9	
Bell metal	Solid	A	550	0.107	60-1634					1634		76.3	
Gun metal	Solid	A	525	0.107	60-1850					1850		84.2	
Tobin	Solid	A			60-1625					1625		73.5	
Butane (C_4H_{10})	Gas	O, F	0.149			0.55 / 0.538	60 / 12	0.458	60	-210	30.9		165.5
Cadmium (Cd)	Solid	E, M	540	0.057	212					609.6	1412	23.8	409
Calcium (Ca)	Solid	E	96.6	0.170	32-358					1560	2625		
Calcium carbonate ($CaCO_3$)	Solid		168-184	0.210	32-212					1517b			
Calcium chloride ($CaCl_2$)	Solid		134	0.292	60		104			1425.2	>2910		
Calcium chloride (30%) and water	Liquid	S	78.7			0.676							
Camphor ($C_{10}H_{16}O$)	Solid	O	62.4	0.44	68-353	0.61	353-410			353	408.2	19.4	

a—A=alloy, E=element, F=fuel or fuel component, I=insulation, M=metal, O=organic compound, R=refractory, S=solution.
b—decomposes.

Table II.1 Densities and thermal properties of various substances (continued)

Name — Description	Normal state	Other classification [a]	Density in lb/cu ft	Specific heat in Btu/lb°F (or gm cal/gm°C)						Melting (fusion) point °F	Boiling point, °F (at std barometric pressure)	Latent heat in Btu/lb	
				Solid state		Liquid state		Gaseous state				of fusion	of vaporization
				Specific heat	Temp range °F	Specific heat	Temp range °F	Specific heat	Temp range °F				
Carbon (C) (graphite)	Solid	E, F	138	0.160 / 0.467	52 / 1789					6332c	8730		
Carbon bisulfide (see carbon disulfide)													
Carbon dioxide (CO₂)	Gas		0.117			0.0615	−82.6 / −73.8	0.2169	52-417		−109c		
Carbon disulfide (CS₂)	Liquid	F	79.3			0.232	60°	0.1596	187-374	−166	115.0		150.8
Carbon monoxide (CO)	Gas	F	0.0741				122	0.2426	79-388	−340	−314		83.5
Carbon tetrachloride (CCl₄)	Liquid		98.8			0.215				−9	170		
Castor oil	Liquid		60.1			0.434							
Cellulose	Solid		94.97	0.32	32-212								
Cerium (Ce)	Solid	E	430	0.0448	32-212					1184	2540		
Cesium (Cs)	Solid	E	118.6	0.0482	32-79					83	1238	6.77	
Chalk	Solid			0.215	32-212								
Charcoal	Solid	F	18-38	0.165-0.25	75								
Chlorine (Cl)	Gas	E	0.190			0.229	−82	0.1125	61-649	−150.7	−28.5	41.3	121
Chloroform (CHCl₃)	Liquid	O	95.5			0.23	32-212	0.1489	72-172	−85	142.1		105.3
Chrome refractory, burned	Solid	R	188	0.20	60-1200					3580+			
Chrome refractory, unburned	Solid	R	193	0.21	60-1200					3580+			
Chromite (chrome ore) (FeCr₂O₄)	Solid	R	281	0.22						3956			
Chromium (Cr)	Solid	E, M	449	0.1039 / 0.1121 / 0.1872	32 / 212 / 1112					2929	4500		
Cinders	Solid			0.18	32-212								
Clay	Solid		112-162	0.224	68-208					3160			
Coal	Solid	F		0.3	32-212								
Coal tar oil	Liquid	F				0.34					390-910		
Cobalt (Co)	Solid	E	556	0.1542 / 0.204	932 / 1832					2723	5250	115.2	
Coke	Solid	F		0.203 / 0.376	32-212 / 100-2200								
Columbium (Cb)	Solid	E	535	0.065	32-212					4380			
Concrete	Solid		137	0.156-0.27	32-212								
Constantan	Solid	A		0.098	59-450								
Copper (Cu)	Solid	M, E	559	0.0951 / 0.1259	1652					1981.4 ±5.4	4700	75.6	
Copper Sulfate (CuSO₄) (16.7%) and water	Liquid	S	73.7	0.848	53.6-59	…							
Cork, natural	Solid	I	15	0.419	77								
Cork, granulated	Solid	I	5.4-7.3	0.43	77								

a—A=alloy, E=element, F=fuel or fuel component, I=insulation, M=metal, O=organic compound, R=refractory, S=solution.

c—sublimes.

Table II.1 Densities and thermal properties of various substances (*continued*)

Specific heat columns are in Btu/lb°F (or gm cal/gm°C).

Name — Description	Normal state	Other classification[a]	Density in lb/cu ft	Solid: Specific heat	Solid: Temp range °F	Liquid: Specific heat	Liquid: Temp range °F	Gaseous: Specific heat	Gaseous: Temp range °F	Melting (fusion) point °F	Boiling point °F (at std barometric pressure)	Latent heat of fusion Btu/lb	Latent heat of vaporization Btu/lb
Corkboard	Solid	I	6.9-20.7	0.204	-115								
Corundum (Al_2O_3)	Solid		250	0.417	+150					3722			
Cottonseed oil	Liquid		58	0.1976	42-208	0.474				32			
Cream	Liquid			0.780	68-212								
Cupric oxide (CuO)	Solid		374-405	0.227	32-212					1944			
Cuprous oxide (Cu_2O)	Solid		375	0.111	32-212					2254			
D'Arcet's metal (50 Bi, 25 Pb, 25 Sn)	Solid	A		0.050									
Decane ($C_{10}H_{22}$)	Liquid	O, F	45.3			0.42	60°			-21.5	346	10.4	110
Diatomaceous earth	Solid	R	12.5-25	0.21	77								
Die casting metal, Aluminum base	Solid	A	176	0.236	60-1150	0.241				1150		163.1	
Lead base	Solid	A		0.038	60-600	0.037				600		17.4	
Tin base	Solid	A		0.070	60-450	0.062				450		30.3	
Zinc base	Solid	A		0.103	60-780	0.138				780		48.3	
Diphenyl ($C_6H_5C_6H_5$)	Liquid	O	62	0.693	104	0.481	159-492			159	492	47	
Diphenylamine ($C_6H_5NHC_6H_5$)	Liquid	O	72.3	0.443	133					129	576	45.4	136.5
Dolomite	Solid		181	0.222	68-208								
Dowtherm A	Liquid		58.8	0.53		0.41	122			180	500		
Earth (see also humus)	Solid			0.44	32-212								
Ebonite	Solid			0.33	32-212								
Ether, ethyl ($C_4H_{10}O$)	Liquid	O	45.9			0.529	32			-180.4	94.3		159.1
Ethyl acetate ($CH_3CO_2CH_2CH_3$)	Liquid	O	55.8			0.478	32-212	0.48	156-435	-118.3			
Ethyl bromide (CH_3CH_2Br)	Liquid	O	90.5			0.21	60			-182	100.7		108.7
Ethyl iodide (CH_3CH_2I)	Liquid	O	120			0.25	122			-163	159.8		84.6
Ethylene glycol ($C_2H_6O_2$)	Liquid	O	68.6			0.602	32-212				387		
Fiberglas board	Solid	I	2-6	0.129 / 0.192 / 0.236	-117 / -22 / 111								
Firebrick, fireclay	Solid	R	137-150	0.243	60-2195					2900-3200			
insulating (2600°F)	Solid	R	38.4	0.22	60-1200					2980-3000			
silica	Solid	R	144-162	0.258	60-2195					2100+			
Fluorine (F_2)	Gas	E	0.10					0.32	80	-369.4	-304.6		
Fosterite refractory	Solid	R	153	0.25	60-1200					3430			
Fuel oil	Liquid	F	52-61			0.50							145-150

[a]—A=alloy, E=element, F=fuel or fuel component, I=insulation, M=metal, O=organic compound, R=refractory, S=solution.

Table II.1 Densities and thermal properties of various substances (*continued*)

Name — Description	Normal state	Other[a] classification	Density in lb/cu ft	Specific heat in Btu/lb°F (or gm cal/gm°C)						Melting (fusion) point °F	Boiling point °F (at std barometric pressure)	Latent heat in Btu/lb	
				Solid state: Specific heat	Solid state: Temp range °F	Liquid state: Specific heat	Liquid state: Temp range °F	Gaseous state: Specific heat	Gaseous state: Temp range °F			of fusion	of vaporization
Fusel oil	Liquid					0.56	32-212						
Galena (PbS)	Solid		467	0.0466	32-212					2050			
Gallium (Ga)	Solid	E	367	0.080	54-235					86.18	3090		
Gasoline (commercial)	Liquid	F	41-43			0.5135					158-194		128-146
Germanium (Ge)	Solid	E	335	0.0737	32-212					1756.4			
German silver	Solid	A		0.0946	32-212	0.123				1850		86.2	
Glass	Solid		144-187	0.109 0.15-0.23 0.132 0.157 0.179 0.210 0.279	60-1850 32-212 -117 -20 112 150 624					2190±			
Glass block, expanded, foamglas	Solid	I	10.6										
Glass wool	Solid		1.4-4.8										
Glycerine ($C_3H_5O_3$) (glycerol)	Liquid	O	78.7			0.576	59-212			-68	554	76.5	29
Gneiss	Solid			0.196	63-210								
Gold (Au)	Solid	E, M	1205	0.0316	32-212	0.034				1945.4	5380	28.7	
Granite	Solid		162-175	0.192	54-212								
Graphite	Solid		138.3	0.201 0.38	32-212 70-2200					6300	8720		
Gypsum	Solid		145	0.259	50-212					2480			
Hairfelt	Solid	I		0.334	148								
Helium (He)	Gas	E	0.01043					1.25	456	← -456	-448.6		
Hematite (Fe_2O_3)	Solid			0.1645	59-210								
Heptane (C_7H_{16})	Liquid	O, F	42.4			0.55	122°			-130	194		140.0
Hexane (C_6H_{14})	Liquid	O, F	40.8			0.600	68-212			-138	158		142.6
High-alumina refractory	Solid	R	128	0.23	60-1200					3290			
Hornblende	Solid			0.195	32-212								
Hydrochloric acid (HCl) (45.2%)+H_2O	Liquid		92			1.75-2.33	(-111.3)			4.5	248		
Hydrofluoric acid (HF) (35-35%)+H_2O	Liquid		72				(-103.0)			-28			
Hydrogen (H_2)	Gas	E, F	0.0053					3.410	70-212	-434.2	-422.6	27	194
Hydrogen chloride (HCl)	Gas		0.0967					0.1940	55-212°	-168.3	-117		
Hydrogen fluoride (HF)	Gas		0.0754							-134.14	-33.8		
Hydrogen sulfide (H_2S)	Gas		0.0907					0.2451	68-403	-125	-81		
Humus (soil) (see also earth)	Solid			0.44	32-212								

a—A=alloy, E=element, F=fuel or fuel component, I=insulation, M=metal, O=organic compound, R=refractory, S=solution.

Table II.1 Densities and thermal properties of various substances (*continued*)

Name — Description	Normal state	Other[a] classification	Density in lb/cu ft	Specific heat in Btu/lb°F (or gm cal/gm°C) Solid state — Specific heat	Solid state — Temp range °F	Liquid state — Specific heat	Liquid state — Temp range °F	Gaseous state — Specific heat	Gaseous state — Temp range °F	Melting (fusion) point °F	Boiling point, °F (at std barometric pressure)	Latent heat in Btu/lb — of fusion	Latent heat in Btu/lb — of vaporization
Ice (H$_2$O)	Solid	....	55.8–57.4	0.463	–103–0					32		144	
India rubber (Para) (see also rubber)	Solid	....		0.27–0.48	32–212								
Indium (In)	Solid	E	456	0.0570	32–212					313.5	2630		
Insulation, high temp block type	Solid	I	14–24	0.203 / 0.269	149 / 710								
Iodine (I)	Solid	E	308	0.0541	46–208					236.3	364	28.4	42.3
Iridium (Ir)	Solid	E	1400	0.0323	54–212					4449	9600		
Iron	Solid	E, M	491	0.11	68–212					2802	5430	117	
gray cast	Solid	M	443	0.119	68–212					2330		41.4	
white cast	Solid	M	480	0.119	68–212					2000		59.5	
wrought	Solid	M	487–493	0.115	59–212								
Kaolin	Solid	R	131	0.224 / 0.22	68–208 / 60–1200					3200			
Kapok fiber	Solid	I	0.9	0.320	65								
Kerosene	Liquid	F	50			0.50	32–212						105–110
Krypton (Kr)	Gas	E	0.215	0.1069						–272.2	–241		
Lanthanum (La)	Solid	E	384	0.0448	32–212					1490	3270		
Lava	Solid			0.197	77–212								
Lead (Pb)	Solid	M, E	708	0.0019	61–493	0.041	590–680			620.6±4.5	2950	9.9	
Lead-antimony alloy (62.9Pb,37.1 Sb)	Solid	A		0.0388	50–208								
Lead-bismuth alloy (40Pb, 60Bi)	Solid	A		0.0317	32–212								
Lead oxide (PbO)	Solid		574–593	0.049	60–212					1749			
Lead slag wool	Solid	I		0.178 / 0.235	150 / 718								
Light oil	Liquid	F	50			0.50							145–150
Limestone	Solid		168–175	0.216	59–212								
Linotype	Solid	A		0.036	60–486					486		21.5	
Linseed oil	Liquid		58			0.441	60–140			–5	600		
Lipowitz's metal (Pb26, Sn13, Cd10, Bi51)	Solid	A		0.041	212					140		17.2	
Lithium (Li)	Solid	E	33	1.0407	212					366.8	2552	286	
Lodestone (magnetite)	Solid		322	0.156	32–212								
Machine oil	Liquid	I		0.40	32–212								
Magnesia (85%)	Solid		11–13	0.276 / 0.283	150 / 279								

a—A=alloy, E=element, F=fuel or fuel component, I=insulation, M=metal, O=organic compound, R=refractory, S=solution.

Table II.1 Densities and thermal properties of various substances (*continued*)

Name — Description	Normal state	Other classification	Density in lb/cu ft	Specific heat in Btu/lb°F (or gm cal/gm°C) — Solid state: Specific heat	Solid: Temp range °F	Liquid state: Specific heat	Liquid: Temp range °F	Gaseous state: Specific heat	Gaseous: Temp range °F	Melting (fusion) point °F	Boiling point, °F (at std barometric pressure)	Latent heat of fusion, Btu/lb	Latent heat of vaporization, Btu/lb
Magnesite refractory (unburned)	Solid	R	171	0.27	60-1200					3580+			
Magnesite refractory (burned)	Solid	R	183	0.26	60-1200					3580+			
Magnesite refractory (fused)	Solid	R	179	0.27	60-1200					3580+			
Magnesium (Mg)	Solid	M, E	108.6	0.2492	68-212	0.266				1203.8	2048	160	
Magnesium oxide	Solid		228	0.234 / 0.295	86-100 / 86-1800					5070	6580		
Magnetite (Lodestone)	Solid	E, M	322	0.156	32-212					2246.0	3452	115	
Manganese (Mn)	Solid	E, M	464	0.1211	68-212								
Marble	Solid		162-175	0.210	32-212								
Mercuric chloride (HgCl₂)	Solid		339	0.05	68					539.6	581		
Mercurous chloride (Hg₂Cl₂)	Solid		446	0.032						576	722		
Mercury (Hg)	Liquid	E, M	847		−121 to +68	0.033	32-212			−37.97	674.6	5.07	117.0
Methane (CH₄)	Gas	O, F	0.0243			0.992	−172	0.5929	64-406		−260		
Methyl chloride (CH₃Cl)	Gas	O	.135			0.385	60			−144	−11		
Mica	Solid	I		0.10	68								
Mica, expanded (vermiculite)	Solid	I		0.205 / 0.236	149 / 290								
Milk	Liquid	I	64.2	0.095	−120	0.847	150						
Mineral wool board with binder	Solid	I	14.3	0.247									
Molasses	Liquid	O	87.3			0.60							
Molybdenum (Mo)	Solid	E, M	637	0.0647	68-212					4595	6690	126	
Monel metal	Solid	A	550	0.129	60-2415					2415		117.4	
Naphtha	Liquid		41.2			0.493					306		184
Naphthalene (C₆H₄C₄H₄)	Solid	O	71.8	0.325	68-140	0.427	262			175.8	424.2	64.1	135.7
Neat's-foot oil	Liquid			0.457	68-86					32			
Neodymium (Nd)	Solid	E	431							1544			
Neon (Ne)	Gas	E	0.0514			0.111		0.443	any	−423	−398.2		
Nichrome	Solid	A	517										
Nickel (Ni)	Solid	E, M	556	0.109 / 0.1608	64-212 / 1832					2645	4950	133	2670
Nickel steel	Solid	A		0.109	32-212								
Nitric acid (HNO₃)	Liquid		96.1			0.445				−43.6	186.8		207.2
Nitric acid (10%) and water	Liquid	S				0.768							
Nitric acid (2%) and water	Liquid	S				0.930							
Nitric acid (1%) and water	Liquid	S				0.963							

¹ —A = alloy, E = element, F = fuel or fuel component, I = insulation, M = metal, O = organic compound, R = refractory, S = solution.

Table II.1 Densities and thermal properties of various substances (*continued*)

Name — Description	Normal state	Other[a] classification	Density in lb/cu ft	Solid state Specific heat	Solid state Temp range °F	Liquid state Specific heat	Liquid state Temp range °F	Gaseous state Specific heat	Gaseous state Temp range °F	Melting (fusion) point °F	Boiling point, °F (at std barometric pressure)	Latent heat of fusion Btu/lb	Latent heat of vaporization Btu/lb
Nitric oxide (NO)	Gas		0.079			0.580	{−249, −252}	0.2317	55–342		−240		
Nitrobenzene ($C_6H_5O_2N$)	Liquid	O	74.7			0.38	122			41	411.8		
Nitrobenzole	Liquid					0.35	57.2						
Nitrogen (N_2)	Gas		0.0741			0.475	{−322, −344}	0.2419	68–824	−347.8	−319	11.1	85.6
Nitrous oxide (N_2O)	Gas		0.117					0.2262	61–405		−129		
Octane (C_8H_{18})	Liquid	O, F	43.6			0.52	60				266.0		126.0
Oil (see castor oil, coal tar oil, cottonseed oil, fuel oil, fuel oil, light oil, linseed oil, machine oil, oil of citron, oil of juniper, oil of orange, oil of turpentine, olive oil, paraffin oil, petroleum, and Chapter II.)													
of citron	Liquid		53.2			0.438	42						
of juniper	Liquid					0.477							
of orange	Liquid					0.489							
of turpentine	Liquid		53.7			0.411	32			14	320		184
Olive oil	Liquid		57.4			0.471	44			68±	572±		
Osmium (Os)	Solid	E	1402	0.0311	68–208					4892±	9900		
Oxalic acid ($C_2H_2O_4 \cdot 2H_2O$)	Solid	O	103.8	0.338; 0.416	0; 100					372	302c		
Oxygen (O_2)	Gas	E	0.0847			0.398	{−88.0, −79.2}	0.2175	55–405	−360.4	−296.9	5.98	91.6
Palladium (Pd)	Solid	E	749	0.0714	32–2309					2820±9	3992	64.6	
Paper, expanding blanket ("Kimsul")	Solid	I		0.349	148								
Paraffin	Solid		54–57	0.622	95–104	0.712	140–145			100–133	662–806	63–70	
Paraffin oil	Liquid	O, F				0.52	32–212						
Pentane n-C_5H_{12}	Liquid	F	38.8								97		154.4
Petroleum	Liquid		47–55			0.511	69.8–136.4						
Phenol (C_6H_6O) (see also resin)	Solid	E	66.8	0.1829	32–124					105.6	360	45.0	
Phosphorus (P)	Solid	F	113.8	0.45	60–212					111.6	550.4	9.05	
Pitch (coal tar)	Solid		62–81	0.20		0.35–0.45				86–302	325		
Plaster	Solid		90										
Platinum (Pt)	Solid	E, M	1335	0.0359	68–2372					3191±9	7750	48.96	
Porcelain	Solid	R	143–156	0.26	59–1742								
Porcelain, refractory	Solid			0.23	60–1200					2140–3000			
Potassium (K)	Solid	E	53.6	0.170	{−301, +68}					144.1	1400	26.2	

a—A=alloy, E=element, F=fuel or fuel component, I=insulation, M=metal, O=organic compound, R=refractory, S=solution.
c—sublimes.

Table II.1   Densities and thermal properties of various substances (*continued*)

Name — Description	Normal state	Other[a] classification	Density in lb/cu ft	Specific heat in Btu/lb°F (or gm cal/gm°C) — Solid state: Specific heat	Solid state: Temp range °F	Liquid state: Specific heat	Liquid state: Temp range °F	Gaseous state: Specific heat	Gaseous state: Temp range °F	Melting (fusion) point °F	Boiling point °F (at std barometric pressure)	Latent heat of fusion (Btu/lb)	Latent heat of vaporization (Btu/lb)
Potassium chlorate ($KClO_3$)	Solid		145	.205	122					701.6			
Potassium hydroxide ($KOH + 30H_2O$)	Liquid		129.2			0.876	64.4			646		88	
Potassium nitrate (KNO_3)	Solid			0.19	59-212								
Praseodymium (Pr)	Solid	E								1724			
Propane (C_3H_8)	Gas	O, F	0.1196					0.576	32	−310	−44.2		
Pyrex	Solid			0.196	68-212								
Pyrites, copper	Solid			0.1291	59-210								
Quartz	Solid		165	0.17-0.28	32-212					1742	2080		
Quicklime	Solid			0.217	32-212								
Radium (Ra)	Solid	E	312										
Redwood bark, shredded ("Palco Bark")	Solid	I	4.0	0.172, 0.246	−127, 109								
Resin, phenol, pure	Solid		75-81	0.33-0.37						167-212f			
Resin, phenol, wood flour filled	Solid		81-87	0.30-0.36						257-266f			
Resin, phenol, asbestos filled	Solid		112-125	0.38-0.40						266-302f			
Resin, copalite	Solid		65-71	0.38-0.40						300-680			
Rhodium (Rh)	Solid	E	777	0.058, 0.201, 0.250	50-207, 149, 652					3542	>4580		
Rock wool	Solid		7-12										
Rose's metal (28Pb, 25Sn, 50Bi)	Solid	A		0.043	60-230	0.041				230		18.3	
Rosin	Solid		68	0.525	68-450					170-212			
Rubber	Solid		62-125	0.481	60-212					248			
Rubber board, expanded ("Rubatex")	Solid		4.9	0.152, 0.273	−125, 111								
Rubidium (Rb)	Solid	E	95.5	0.0802	32					100.4	1284.8		
Ruthenium (Ru), black	Solid	E	537	0.0611	32-212					3530			
Ruthenium (Ru), gray	Solid		760							4442			
Salt, rock	Solid		135	0.219	55-113					1495			
Samarium (Sm)	Solid	E	481							2372-2552			
Sand	Solid		162	0.195	59-212								
Sea water	Liquid	S	64			0.938	63.5						

a—A=alloy, E=element, F=fuel or fuel component, I=insulation, M=metal, O=organic compound, R=refractory, S=solution.

d—Manufactured under trade names: Bakelite, Redmanol, Condensite, etc.

e—Resins used in varnish making: Kauric, Congo, Zanzibar, and Manila copals.

f—Softening point under load.

Table II.1 Densities and thermal properties of various substances (*continued*)

Name — Description	Normal state	Other[a] classification	Density in lb/cu ft	Specific heat in Btu/lb°F (or gm cal/gm°C)						Melting (fusion) point °F	Boiling point, °F (at std barometric pressure)	Latent heat in Btu/lb	
				Solid state		Liquid state		Gaseous state				of fusion	of vaporization
				Specific heat	Temp range °F	Specific heat	Temp range °F	Specific heat	Temp range °F				
Selenium (Se)	Solid	E	300	0.068	−306–+64					422.6–428.0	1274		
Serpentine	Solid		75–76	0.25	32–212								
Shellac (Lac)	Solid		180	0.40	60–212					170–180			
Silica (SiO₂)	Solid			0.1910	32–212					3182			
Silica aerogel ("Santocel")	Solid	I	5.3	0.205 0.274	147 630								
Silica refractory	Solid	R	111	0.23	60–1200					3060–3090			
Silicon (Si)	Solid	E	145	0.1833 0.2029	135 450					2588	4149	607	
Silicon carbide (SiC)	Solid		199	0.23	60–950					4092			
Silicon carbide (clay-bonded) refractory	Solid	R	136–159	0.20	60–1200					3390			
Silk, raw	Solid		81–87	0.33	32–212								
Sillimanite (mullite) refractory	Solid	R	145–202	0.23	60–1200					3310–3340			
Silver (Ag)	Solid	E, M	655	0.05987	63–945					1760.9	3551	45.1	
Slag, blast furnace (powdered)	Solid	I	22.5	0.17	77								
Slag wool	Solid		9.4–18.7	0.231	32–212								
Soda, baking	Solid		137	0.231								90.0	
Sodium (Na)	Solid	E	60.6	0.253	−301–+68					207.5	1614	49.3	
Sodium carbonate (Na₂CO₃)	Solid		151.5	0.306						1565.6			
Sodium carbonate (2%) and water	Liquid	S				0.896							
Sodium chloride (NaCl) (see also rock salt)	Liquid	S	135							1472			
Sodium chloride (10%) and water (see also sea water)	Liquid	S	67			0.791	64.4						
Sodium hydroxide (2% NaOH) and water	Liquid	S	63.8			0.942	64.4						
Sodium nitrate (NaNO₃)	Solid		140.5	0.231						597	716	116.8	
Sodium sulfate (Na₂SO₄)	Solid									1623.2			
Solder (Pb and Sn)	Solid	A	580	0.040–0.051						361–594		11.6–50.6	
Spermaceti (whale oil)	Solid									113±		66.56	
Steel	Solid	A	490	0.165	60–2900							26.2	
Stereotype	Solid	A	670	0.036		0.036				500			
Stones, all kinds (see also Marble, Granite, Limestone, Sandstone)	Solid		168	0.18–0.23	54–212								
Sugar, cane, amorphous	Solid			0.342	68					320			
Sugar, cane, crystalline	Solid		102	0.301	68								
Sugar, cane, (4%) and water	Liquid	S		0.7558									

a—A=alloy, E=element, F=fuel or fuel component, I=insulation, M=metal, O=organic compound, R=refractory, S=solution.

Table II.1 Densities and thermal properties of various substances (*continued*)

Name — Description	Normal state	Other[a] classification	Density in lb/cu ft	Specific heat in Btu/lb°F (or gm cal/gm°C)						Melting (fusion) point °F	Boiling point, °F (at std barometric pressure)	Latent heat in Btu/lb	
				Solid state		Liquid state		Gaseous state				of fusion	of vaporization
				Specific heat	Temp range °F	Specific heat	Temp range °F	Specific heat	Temp range °F				
Sulfur (S)	Solid	E	119-130	0.190	59-130	0.2337	235-840	……	……	239	832.5	16.87h	651.5
Sulfur dioxide (SO₂)	Gas	……	0.1733	……	……	0.36	122	0.1544	61-396	-104.8	14	……	……
Sulfuric acid (H₂SO₄)	Liquid	……	115.8	……	……	0.336	32-212	……	……	50.9	640.4b	……	……
Talc	Solid	……	56.8-60.5	……	……	……	……	……	……	80-100	……	……	……
Tallow, beef	Solid	……	……	……	……	0.2092 / 0.79 / 0.54	68-208 / 79-108 / 151-216	……	……	……	……	……	……
Tantalum (Ta)	Solid	E	1035	0.043	2552	……	……	……	……	5252	……	……	……
Tartaric acid (C₄H₆O₆)	Solid	E	104	0.287	97	……	……	……	……	338	……	……	……
Tellurium (Te)	Solid	E	389	0.0483	59-212	……	……	……	……	845.6	2534	13.1	……
Thallium (Tl)	Solid	E	740	0.0326	68-212	……	……	……	……	575.6	3000	……	……
Thorium (Th)	Solid	E	699	0.0276	32-212	……	……	……	……	3350	>5432	……	……
Tile, hollow	Solid	……	75	0.15	……	……	……	……	……	……	……	……	……
Tin (Sn)	Solid	E, M	455	0.0551	70-228	0.05799 / 0.0758	482 / 2012	……	……	449.4±.4	4118	……	……
Titanium (Ti)	Solid	E	283	0.1125	32-212	……	……	……	……	3263	……	……	……
Toluene (C₇H₈)	Liquid	O	53.6	……	……	0.40	32-212	……	……	-133.6	230.5	……	150.3
Toluol (C₆H₈)	Liquid	O	……	……	……	0.490	149	……	……	……	230.5	……	154.8
Tufa	Solid	……	……	0.33	32-212	……	……	……	……	……	……	……	……
Tungsten (W)	Solid	E, M	1202	0.0336 / 0.0337	32-212 / 1832	……	……	……	……	6152	10526	79	……
Turpentine	Liquid	A	53.6	0.42	32-212	……	……	……	……	……	318.8	……	133.3
Type metal	Solid	……	……	0.0388	32-212	……	……	……	……	……	……	……	……
Uranium (U)	Solid	E	1167	0.028	32-208	……	……	……	……	<3344	6330	……	……
Vanadium (V)	Solid	E	375	0.1153	32-212	……	……	……	……	3128	5430	……	……
Varnish (see resins)	……	E	……	……	……	……	……	……	……	……	……	……	……
Vegetable fiberboard ("Celotex")	Solid	I	14.4	0.171 / 0.279	-116 / 109	……	……	……	……	……	……	……	……
Vermiculite (see mica)	……	I	……	……	……	……	……	……	……	……	……	……	……
Vulcanite	Solid	……	……	0.3312	68-212	……	……	……	……	……	……	……	……
Water (H₂O) (see also sea water)	Liquid	……	62.37	0.480	<32	1.00	60	……	……	32	212	144	970.2
Wood (see also redwood bark)	Solid	……	19-56	0.33-0.67	……	……	……	……	……	……	……	……	……

a—A=alloy, E=element, F=fuel or fuel component, I=insulation, M=metal, O=organic compound, R=refractory, S=solution.
b—decomposes.
h—transformation from rhombic to monoclinic absorbs 5.06 Btu/lb.

Table II.1 Densities and thermal properties of various substances (*continued*)

Name — Description	Normal state	Other[a] classification	Density in lb/cu ft	Specific heat in Btu/lb°F (or gm cal/gm°C)						Melting (fusion) point °F	Boiling point, °F (at std barometric pressure)	Latent heat in Btu/lb	
				Solid state		Liquid state		Gaseous state				of fusion	of vaporization
				Specific heat	Temp range °F	Specific heat	Temp range °F	Specific heat	Temp range °F				
Wood fiber blanket ("Balsam Wool")	Solid	I	2.6	0.330	150	.....	.....	.....	.....	.....	.....	.....	.....
Wood fiberboard	Solid	I	12-19	0.341	148	.....	.....	.....	.....	.....	.....	.....	.....
Wood, oak	Solid	.....	48	0.57	32-212	.....	.....	.....	.....	.....	.....	.....	.....
Wood, pine	Solid	.....	30	0.67	32-212	.....	.....	.....	.....	.....	.....	.....	.....
Wood's metal (26Pb, 13Sn, 12Cd, 49Bi)	Solid	A	.....	0.041	60-158	0.042	.....	.....	.....	158	.....	17.2	.....
Wool (see also glass wool, mineral wool, rock wool, lead slag wool, slag wool, etc.)	Solid	.....	80-83	0.325	.....	.....	.....	.....	.....	.....	.....	.....	.....
Xenon (Xe)	Gas	E	0.346	.....	.....	.....	.....	.....	.....	-220	-164.4	.....	.....
Xylene	Liquid	.....	54.3	0.42	122	.....	.....	.....	.....	-18	288	.....	147
Yttrium (Y)	Solid	E	343	.....	.....	.....	.....	.....	.....	2714	.....	.....	.....
Zinc (Zn)	Solid	E, M	445	0.0931 / 0.1040	68-212 / 572	.....	.....	.....	.....	786.9	1663	46.8	7.58
Zinc chloride (ZnCl$_2$)	Solid	.....	181.5	0.125	32-212	.....	.....	.....	.....	689	1350	.....	.....
Zinc oxide (ZnO)	Solid	.....	350	.....	.....	.....	.....	.....	.....	>3240	.....	.....	.....
Zinc sulfate (ZnSO$_4$)	Liquid	S	234	.....	.....	0.174	106°F	.....	.....	1330b	.....	.....	.....
Zircon	Solid	.....	293	0.132	32-212	.....	.....	.....	.....	4622	.....	.....	.....
Zirconium (Zr)	Solid	E	405	0.0660	32-212	.....	.....	.....	.....	3092	9122	.....	.....

a—A=alloy, E=element, F=fuel or fuel component, I=insulation, M=metal, O=organic compound, R=refractory, S=solution.
b—decomposes.

APPENDIX III*

Table III.1 Thermal properties of metals

Metal	Composition	Density, lb/cu ft	Mean specific heat, 60°– melting point, Btu/lb°F	Latent heat of fusion, Btu/lb	Mean specific heat of liquid, Btu/lb°F	Melting point, °F
Aluminum	Al	166.7	0.248	169.0	0.26	1215
Babbit, Lead Base	75 Pb, 15 Sb, 10 Sn	—	0.039	26.2	0.038	462
Babbit, Tin Base	83.3 Sn, 8.4 Sb, 8.3 Cu	462	0.071	34.1	0.063	464
Bismuth	Bi	612	0.033	18.5	0.035	518
Brass, Muntz Metal	60 Cu, 40 Zn	524	0.105	69.0	0.125	1630
Brass, Red	90 Cu, 10 Zn	546	0.104	86.5	0.115	1952
Brass, Yellow	67 Cu, 33 Zn	528	0.105	71.0	0.123	1688
Bronze, Aluminum	90 Cu, 10 Al	510	0.126	98.5	0.125	1922
Bronze, Bearing	80 Cu, 10 Pb, 10 Sn	556	0.095	79.9	0.109	1832
Bronze, Bell Metal	78 Cu, 22 Sn	540	0.100	76.3	0.119	1634
Bronze, Gun Metal	90 Cu, 10 Sn	550	0.107	84.2	0.106	1850
Bronze, Tobin	60 Cu, 39.2 Zn, 0.8 Sn	525	0.107	73.5	0.124	1625
Copper	Cu	559	0.104	91.0	0.111	1982
Die Casting Metal	92 Al, 8 Cu	176	0.236	163.0	0.241	1150
Die Casting Metal	80 Pb, 10 Sn, 10 Sb	—	0.038	17.5	0.037	600
Die Casting Metal	90 Sn, 4.5 Cu, 5.5 Sb	—	0.070	30.2	0.062	450
Die Casting Metal	87.3 Zn, 8.1 Sn, 4.1 Cu, 0.5 A	—	0.103	48.0	0.138	780

* Adapted from *North American Combustion Handbook*, published by North American Manufacturing Co., Cleveland, Ohio, 1952.

Table III.1 Thermal properties of metals (*continued*)

Metal	Composition	Density, lb/cu ft	Mean specific heat 60°–melting point, Btu/lb°F	Latent heat of fusion, Btu/lb	Mean specific heat of liquid, Btu/lb°F	Melting point, °F
German Silver	60 Cu, 25 Zn, 15 Ni	—	0.109	86.2	0.123	1850
Gold	Au	1205	0.033	28.5	0.034	1945
Iron 60°–2786°F	Fe	491	0.165.	89.0	0.150	2786
Lead	Pb	708	0.032	10.0	0.034	621
Linotype	86 Pb, 11 Sb, 3 Sn	—	0.036	21.5	0.036	486
Magnesium	Mg	108.6	0.272	83.7	0.266	1204
Manganese	Mn	464	0.171	66.0	0.192	2246
Monel Metal	67 Ni, 28 Cu ; Fe, Mn, Si	550	0.129	117.4	0.139	2415
Nickel 60°–2644°F	Ni	556	0.134	131.5	0.133	2644
Silver	Ag	665	0.063	46.8	0.070	1762
Solder, Bismuth	40 Pb, 20 Sn, 40 Bi	—	0.040	16.4	0.039	232
Solder, Plumbers'	50 Pb, 50 Sn	580	0.051	23.0	0.049	414
Tin	Sn	455	0.069	25.0	0.0637	450
Zinc	Zn	445	0.107	48.0	0.146	786

APPENDIX IV*

CONVERSION FACTORS

Table IV.1 Linear measure equivalents

Multiply by value in table to obtain these units → Given in these units ↓	angstrom (Å)	centimeter (cm)	foot (ft)	inch (in)	kilometer (km)
angstrom (Å)	1	10^{-8}	3.2808×10^{-10}	3.937×10^{-9}	10^{-13}
centimeter (cm)	10^8	1	3.2808×10^{-2}	3.937×10^{-1}	10^{-5}
foot (ft)	3.48×10^9	30.48	1	12	3.048×10^{-4}
inch (in)	2.54×10^8	2.54	8.333×10^{-2}	1	2.54×10^{-5}
kilometer (km)	10^{13}	10^5	3.2808×10^3	3.937×10^4	1
meter (m)	10^{10}	10^2	3.2808	3.937×10^1	10^{-3}
micron (μ)	10^4	10^{-4}	3.2808×10^{-6}	3.937×10^{-5}	10^{-9}
mile (mi)	1.61×10^{13}	1.61×10^5	5.28×10^3	6.336×10^4	1.61
millimeter (mm)	10^7	10^{-1}	3.2808×10^{-3}	3.937×10^{-2}	10^{-6}
yard (yd)	1.044×10^{10}	9.144×10^1	3	36	9.144×10^{-4}

* From J. F. Elliott *et al., Thermochemistry for Steelmaking,* **II,** Addison-Wesley Publishing Co., Reading, Massachusetts, 1963.

Table IV.1 Linear measure equivalents (*continued*)

meter (m)	micron (μ)	mile (mi)	millimeter [mm]	yard (yd)
10^{-10}	10^{-4}	6.2×10^{-14}	10^{-7}	1.0936×10^{-10}
10^{-2}	10^4	6.2×10^{-6}	10	1.0936×10^{-2}
3.048×10^{-1}	3.048×10^5	1.8939×10^{-4}	3.048×10^2	3.333×10^{-1}
2.54×10^{-2}	2.54×10^4	1.58×10^{-5}	25.4	2.778×10^{-2}
10^3	10^9	6.2137×10^{-1}	10^6	$1.0936 \times 10^{+3}$
1	10^6	6.2137×10^{-4}	10^3	1.0936
10^{-6}	1	6.2137×10^{-10}	10^{-3}	1.0936×10^{-6}
1.61×10^3	1.61×10^9	1	1.61×10^6	1.760×10^3
10^{-3}	10^3	6.2×10^{-7}	1	1.0936×10^{-3}
9.144×10^{-1}	9.144×10^5	5.682×10^{-4}	$9.144 \times 10^{+2}$	1

Table IV.2 Volume equivalents

Multiply by value in table to obtain these units → Given in these units ↓	cubic centimeter* (cc or cm³)	cubic feet (ft³)	cubic inch (in³)	cubic meter (m³)	gallons (U.S.) (gal)	liters* (l)	ounces (U.S. fluid oz)
cubic centimeter (cc or cm³)	1	3.531×10^{-5}	6.103×10^{-2}	10^{-6}	2.642×10^{-4}	10^{-3}	3.381×10^{-2}
cubic feet (ft³)	2.832×10^{4}	1	1.728×10^{3}	2.832×10^{-2}	7.481	28.32	9.575×10^{2}
cubic inch (in³)	16.39	5.787×10^{-4}	1	1.639×10^{-5}	4.329×10^{-3}	1.639×10^{-2}	5.541×10^{-1}
cubic meter (m³)	10^{6}	35.31	6.103×10^{4}	1	2.642×10^{2}	10^{3}	3.381×10^{4}
gallons (U.S.) (gal)	3.785×10^{3}	1.337×10^{-1}	2.31×10^{2}	3.785×10^{-3}	1	3.785	1.28×10^{2}
liters (l)	10^{3}	3.531×10^{-2}	6.103×10^{-1}	10^{-3}	2.642×10^{-1}	1	33.81
ounces (U.S. fluid oz)	29.57	1.044×10^{-3}	1.805	2.957×10^{-5}	7.812×10^{-3}	2.957×10^{-2}	1

* Note: 1 ml = 1.000027 cc

Table IV.3 Mass equivalents

Given in these units ↓ / Multiply by value in table to obtain these units →	grains	gram (gm)	kilogram (kg)	pound (lb)	ton, long	ton, short	ounces (oz)
grain	1	6.48×10^{-2}	6.48×10^{-5}	1.429×10^{-4}	6.378×10^{-8}	7.143×10^{-8}	2.286×10^{-3}
gram (gm)	15.43	1	10^{-3}	2.20×10^{-3}	9.84×10^{-7}	1.602×10^{-6}	3.527×10^{-2}
kilogram (kg)	1.543×10^{4}	10^{3}	1	2.205	9.842×10^{-4}	1.102×10^{-3}	35.27
pound (lb)	7000	4.536×10^{2}	4.536×10^{-1}	1	4.464×10^{-4}	5.0×10^{-4}	16
ton, long	1.568×10^{7}	1.016×10^{6}	1.016×10^{3}	2.24×10^{3}	1	1.12	3.584×10^{4}
ton, short	1.40×10^{7}	9.0718×10^{5}	9.072×10^{2}	2.00×10^{3}	8.929×10^{-1}	1	3.20×10^{4}
ounces (oz)	4.375×10^{2}	28.35	2.835×10^{-2}	6.25×10^{-2}	2.79×10^{-5}	3.125×10^{-5}	1

Table IV.4 Density equivalents

Given in these units ↓ / Multiply by value in table to obtain these units →	$gm \cdot cm^{-3}$	$gm \cdot liter^{-1}$	$kg \cdot m^{-3}$	$lb \cdot ft^{-3}$	$lb \cdot in^{-3}$	$lb \cdot US\ gal^{-1}$
$gm \cdot cm^{-3}$	1	10^3	10^3	62.43	3.613×10^{-2}	8.345
$gm \cdot liter^{-1}$	10^{-3}	1	1	6.243×10^{-2}	3.613×10^{-5}	8.345×10^{-3}
$kg \cdot m^{-3}$	10^{-3}	1	1	6.243×10^{-2}	3.613×10^{-5}	8.345×10^{-3}
$lb \cdot ft^{-3}$	1.602×10^{-2}	16.02	16.02	1	5.787×10^{-4}	1.337×10^{-1}
$lb \cdot in^{-3}$	27.68	2.768×10^4	2.768×10^4	1.728×10^3	1	2.31×10^2
$lb \cdot US\ gal^{-1}$	1.198×10^{-1}	1.198×10^2	1.198×10^2	7.481	4.329×10^{-3}	1

Table IV.5 Force equivalents

Multiply by value in table to obtain these units → Given in these units ↓	dyne $(\text{gm} \cdot \text{cm} \cdot \text{sec}^{-2})$	newton $(\text{kg} \cdot \text{m} \cdot \text{sec}^{-2})$	poundal $(\text{lb} \cdot \text{ft} \cdot \text{sec}^{-2})$	pound force (lb_f)
dyne $(\text{gm} \cdot \text{cm} \cdot \text{sec}^{-2})$	1	10^{-5}	7.233×10^{-5}	2.248×10^{-6}
newton $(\text{kg} \cdot \text{m} \cdot \text{sec}^{-2})$	10^5	1	7.233	2.248×10^{-1}
poundal $(\text{lb} \cdot \text{ft} \cdot \text{sec}^{-2})$	1.3826×10^4	1.3826×10^{-1}	1	3.108×10^{-2}
pound force (lb_f)	4.448×10^5	4.448	32.17	1

Table IV.6 Energy equivalents

Given in these units ↓ \ Multiply by value in table to obtain these units →	Btu	cal	ergs	ft-lb	hp-hr	joule	kcal	kg-m	kw-hr	liter-atm
Btu	1	$2.52 \times 10^{+2}$	1.055×10^{10}	7.7816×10^{2}	3.93×10^{-4}	1.055×10^{3}	2.520×10^{-1}	1.0758×10^{2}	2.93×10^{-4}	10.41
cal	3.97×10^{-3}	1	4.184×10^{7}	3.086	1.558×10^{-6}	4.184	10^{-3}	4.267×10^{-1}	1.162×10^{-6}	4.129×10^{-2}
erg	9.478×10^{-7}	2.39×10^{-8}	1	4.376×10^{-8}	3.725×10^{-14}	10^{-7}	2.39×10^{-11}	1.0197×10^{-8}	2.773×10^{-14}	9.869×10^{-10}
ft-lb	1.285×10^{-3}	3.241×10^{-1}	1.356×10^{7}	1	5.0505×10^{-7}	1.356	3.241×10^{-4}	1.383×10^{-1}	3.766×10^{-7}	1.338×10^{-2}
hp-hr	2.545×10^{3}	6.4162×10^{5}	2.6845×10^{13}	1.98×10^{6}	1	2.6845×10^{6}	6.4162×10^{2}	2.7375×10^{5}	7.455×10^{-1}	2.6494×10^{4}
joule	9.478×10^{-4}	2.39×10^{-1}	10^{7}	7.376×10^{-1}	3.725×10^{-7}	1	2.39×10^{-4}	1.0197×10^{-1}	2.773×10^{-7}	9.869×10^{-3}
kcal	3.9657	10^{3}	4.184×10^{10}	3.086×10^{3}	1.558×10^{-3}	4.184×10^{3}	1	4.267×10^{2}	1.162×10^{-3}	41.29
kg-m	9.296×10^{-3}	2.3438	9.8067×10^{7}	7.233	3.653×10^{-6}	9.8067	2.344×10^{-3}	1	2.724×10^{-6}	9.678×10^{-2}
kw-hr	3.4128×10^{3}	8.6057×10^{5}	3.6×10^{13}	2.655×10^{6}	1.341	3.6×10^{6}	8.6057×10^{2}	3.671×10^{5}	1	3.5534×10^{4}
liter-atm	9.604×10^{-2}	24.218	1.0133×10^{9}	74.73	3.774×10^{-5}	1.0133×10^{2}	2.422×10^{-2}	10.333	2.815×10^{-5}	1

1 therm = 100,000 Btu; also 1 therm = 10^{6} cal

Table IV.7 Pressure equivalents

Multiply by value in table to obtain these units → / Given in these units ↓	atmosphere (atm)	kg · cm^{-2}	kg · m^{-2}	lb · ft^{-2}	lb · in^{-2}	Column of Hg at 0°C		Column of H$_2$O at 15°C		
						in	mm	ft	in	mm
atmosphere (atm)	1	1.0332	1.033×10^4	2.1162×10^3	14.969	29.92	7.60×10^2	33.93	4.0714×10^2	1.034×10^4
kg · cm^{-2}	9.678×10^{-1}	1	10^4	2.048×10^3	14.22	28.96	7.335×10^2	32.84	3.9405×10^2	1.001×10^4
kg · m^{-2}	9.678×10^{-5}	10^{-4}	1	2.048×10^{-1}	1.422×10^{-3}	2.896×10^{-3}	7.355×10^{-2}	3.284×10^{-3}	3.9405×10^{-2}	1.001
lb · ft^{-2}	4.725×10^{-4}	4.883×10^{-4}	4.883	1	6.944×10^{-3}	1.414×10^{-2}	3.591×10^{-1}	1.603×10^{-2}	1.924×10^{-1}	4.887
lb · in^{-2}	6.804×10^{-2}	7.031×10^{-2}	7.031×10^2	144	1	2.036	51.71	2.309	27.70	7.037×10^2
Hg column, in	3.342×10^{-2}	3.453×10^{-2}	3.453×10^2	70.727	4.912×10^{-1}	1	25.4	1.134	13.61	3.456×10^2
Hg column, mm	1.316×10^{-3}	1.3596×10^{-3}	13.596	2.7845	1.934×10^{-2}	3.937×10^{-2}	1	4.464×10^{-2}	5.357×10^{-1}	13.61
H$_2$O column, ft	2.947×10^{-2}	3.045×10^{-2}	3.045×10^2	62.372	4.332×10^{-1}	8.819×10^{-1}	22.4	1	12	3.048×10^2
H$_2$O column, in	2.456×10^{-3}	2.538×10^{-3}	25.38	5.1977	3.61×10^{-2}	7.349×10^{-2}	1.867	8.333×10^{-2}	1	25.4
H$_2$O column, mm	9.67×10^{-5}	9.991×10^{-5}	9.991×10^{-1}	2.046×10^{-1}	1.421×10^{-3}	2.893×10^{-3}	7.349×10^{-2}	3.281×10^{-3}	3.937×10^{-2}	1

1·gm·cm^{-2} = 9.8066×10^2 dynes·cm^{-2} = 4.5762×10^{-1} poundals·in^{-2}

1 dyne·cm^{-2} = 1.0197×10^{-3} gm·cm^{-2} = 4.6664×10^{-4} poundals·in^{-2}

1 poundal·in^{-2} = 3.1081×10^{-2} lb·in^{-2} = 2.1854 gm·cm^{-2} = 2.143×10^3 dynes·cm^{-2}

1 Torr = 1 mm Hg

Table IV.8 Viscosity equivalents

Multiply by value in table to obtain these units → Given in these units ↓	centipoise	$kg \cdot m^{-1} \cdot sec^{-1}$	$gm \cdot cm^{-1} \cdot sec^{-1}$ (poise)	$lb_m \cdot ft^{-1} \cdot sec^{-1}$	$lb_m \cdot ft^{-1} \cdot hr^{-1}$	$lb_f \cdot sec \cdot ft^{-2}$
centipoise	1	10^{-3}	10^{-2}	6.7197×10^{-4}	2.4191	2.0886×10^{-5}
$kg \cdot m^{-1} \cdot sec^{-1}$	10^3	1	10	6.7197×10^{-1}	2.4191×10^3	2.0886×10^{-2}
$gm \cdot cm^{-1} \cdot sec^{-1}$ (poise)	10^2	10^{-1}	1	6.7197×10^{-2}	2.4191×10^2	2.0886×10^{-3}
$lb_m \cdot ft^{-1} \cdot sec^{-1}$	1.4882×10^3	1.4882	14.882	1	3.6×10^3	3.1081×10^{-2}
$lb_m \cdot ft^{-1} \cdot hr^{-1}$	4.1338×10^{-1}	4.1338×10^{-4}	4.1338×10^{-3}	2.7778×10^{-4}	1	8.6336×10^{-6}
$lb_f \cdot sec \cdot ft^{-2}$	4.788×10^4	47.88	4.788×10^2	32.174	1.1583×10^5	1

Table IV.9 Thermal conductivity equivalents

Multiply by value in table to obtain these units → Given in these units ↓	$\dfrac{\text{Btu}}{\text{hr}\cdot\text{ft}\cdot{}^\circ\text{F}}$	$\dfrac{\text{cal}}{\text{sec}\cdot\text{cm}\cdot{}^\circ\text{K}}$	$\dfrac{\text{ergs}}{\text{sec}\cdot\text{cm}\cdot{}^\circ\text{K}}$	$\dfrac{\text{poundals}}{\text{sec}\cdot\text{ft}\cdot{}^\circ\text{F}}$	$\dfrac{\text{lb}_f}{\text{sec}\cdot{}^\circ\text{F}}$	$\dfrac{\text{watt}}{\text{m}\cdot{}^\circ\text{K}}$
$\dfrac{\text{Btu}}{\text{hr}\cdot\text{ft}\cdot{}^\circ\text{F}}$	1	4.1365×10^{-3}	1.7307×10^{5}	6.9546	2.1616×10^{-1}	1.7307
$\dfrac{\text{cal}}{\text{sec}\cdot\text{cm}\cdot{}^\circ\text{K}}$	2.4175×10^{2}	1	4.1840×10^{7}	1.6813×10^{3}	52.256	4.1840×10^{2}
$\dfrac{\text{ergs}}{\text{sec}\cdot\text{cm}\cdot{}^\circ\text{K}}$	5.7780×10^{-6}	2.3901×10^{-8}	1	4.0183×10^{-5}	1.2489×10^{-6}	10^{-5}
$\dfrac{\text{poundals}}{\text{sec}\cdot\text{ft}\cdot{}^\circ\text{F}}$	1.4379×10^{-1}	5.9479×10^{-4}	2.4886×10^{4}	1	3.1081×10^{-2}	2.4886×10^{-1}
$\dfrac{\text{lb}_f}{\text{sec}\cdot{}^\circ\text{F}}$	4.6263	1.9137×10^{-2}	8.0068×10^{5}	32.174	1	8.0068
$\dfrac{\text{watt}}{\text{m}\cdot{}^\circ\text{K}}$	5.7780×10^{-1}	2.3901×10^{-3}	10^{5}	4.0183	1.2489×10^{-1}	1

Table IV.10 Diffusivity equivalents

Multiply by value in table to obtain these units → Given in these units ↓	$cm^2 \cdot sec^{-1}$	$ft^2 \cdot hr^{-1}$	$m^2 \cdot sec^{-1}$	centistokes
$cm^2 \cdot sec^{-1}$	1	3.8750	10^{-4}	10^2
$ft^2 \cdot hr^{-1}$	2.5807×10^{-1}	1	2.5807×10^{-5}	25.807
$m^2 \cdot sec^{-1}$	10^4	3.8750×10^4	1	10^6
centistokes	10^{-2}	3.8750×10^{-2}	10^{-6}	1

LIST OF PRINCIPAL SYMBOLS

A	area
A	constant, defined in Eq. (1.18)
B	particle mobility
Bi	Biot number
C	molar concentration of entire solution
C	Chvorinov's constant for solidification time
C	aperture conductance in vacuum
C_i	molar concentration of species i in solution
C_M	maximum concentration
C_m	minimum concentration
C_p	heat capacity at constant pressure
$\hat{C}_p$	molar heat capacity at constant pressure
C_t	conductance of vapor trap in vacuum
C_v	heat capacity at constant volume
$\hat{C}_v$	molar heat capacity at constant volume
c	heat capacity per molecule
c	speed of light $= 2.997902 \times 10^{10}$ cm/sec
D, d	diameter
$D, \tilde{D}$	interdiffusion coefficient, or simply diffusion coefficient
D^*	self-diffusion coefficient
D_b	bubble diameter
D_e	equivalent diameter, defined by Eq. (3.21)
D_i	intrinsic diffusion coefficient of species i
D_i^*	self-diffusion coefficient of species i
D_0	frequency factor for diffusion coefficient
D_P	particle diameter
$\bar{D}_{P_i}$	average particle diameter of size fraction i
D_T	tracer determined diffusion coefficient
$\bar{D}_{vs}$	volume–surface mean diameter, defined by Eq. (3.54)
d	crystal lattice dimension
d	molecular collision diameter
E_f	energy loss due to friction

E_g	energy gap
E_K	kinetic energy
E_P	potential energy
E_{tot}	total energy
e	charge on electron
e	emissive power, or radiant energy flux
e_b	emissive power of black body
e_f	friction loss factor
e_λ	monochromatic emissive power
F	force
F_{ij}	view factor between black bodies i and j, defined in Eq. (11.31)
$\bar{F}_{ij}$	combination of view factors; e.g., see Eq. (11.43)
F_K	drag force
Fo	Fourier number
Fr	Froude number
F_S	buoyant force
$\mathscr{F}_{ij}$	combination of view factors and surface radiation properties; e.g., see Eq. (11.44)
f	friction factor
f	correlation coefficient in diffusion, see Eq. (13.6)
f	fractional reduction of oxides, see Eq. (16.23)
G	total irradiation
G_b	mass rate of vapor bubbles per unit area
Gr	Grashof number, defined by Eq. (7.57)
ΔG^*	free energy of activation
g	gravitational acceleration
(g)	gaseous state
H	enthalpy per unit mass
H	"severity of quench", h/k
H_f	latent heat of fusion per unit mass
H'_f	latent heat of fusion plus superheat
H_v	latent heat of evaporation per unit mass
H'_v	latent heat of evaporation plus sensible heat
h	height
h	Planck's constant $= 6.62517 \times 10^{-27}$ erg-sec
h	heat-transfer coefficient
h_r	radiation heat-transfer coefficient
hS	volumetric heat-transfer coefficient; e.g., see Eq. (12.14)
h_t	total heat-transfer coefficient; i.e., sum of convection and radiation heat-transfer coefficients
h_v	volumetric heat-transfer coefficient; e.g., see Eq. (12.14)
I	current density
J	total radiosity
j	diffusion flux
j_H	"j-factor" used in Colburn's analogy
j_M	mass-transfer "j-factor"
K	equilibrium partition ratio between phases
K	characteristic kinetic energy
K	flow coefficient in flow meters

K_{eq}	thermodynamic equilibrium constant
k	thermal conductivity
k'	Tammann scaling constant for oxidation of metals
k_b	effective thermal conductivity of a packed bed or porous solid
k_D	permeability coefficient, defined by Eq. (3.30)
k_{eff}	effective thermal conductivity of packed bed
k_{el}	electronic contribution to thermal conductivity
k_{mix}	thermal conductivity of a mixture
k_p	parabolic scaling constant for oxidation of metals
k_r	radiant energy conductivity
k_r	rational rate constant for oxidation of metals
L, l	length
L	semithickness
L	Lorentz number
L	effective beam length for radiation
M, m	mass
M	thickness solidified
M	mechanical work
M	molecular or atomic weight
M	Mach number
M_e	jet momentum at nozzle exit
m_e	electron mass
N	total mass flux (diffusive plus convective contributions)
N_0	Avogadro's number $= 6.023 \times 10^{23}$ molecules (or atoms)/g-mol
Nu	Nusselt number
n	number in general
n	concentration of molecules, atoms, or electrons per unit volume
n	rotational speed of fans
P	permeability
P	pressure
p	partial pressure
P_B	brake horsepower
Pr	Prandtl number
P_v	vapor pressure
P_r	reduced pressure
$\mathscr{P}$	specific permeability
Q	heat or thermal energy
Q	volume flow rate
Q	activation energy for diffusion coefficient
Q	throughput in vacuum system
Q_a, Q_l	leakage in vacuum system
Q_R	heat of reaction
q	heat flux, i.e., rate of heat flow per unit area
q_x, q_y, q_z	heat flux components
R, r	radius
R	gas constant
Re	Reynolds number
R_h	hydraulic radius, defined by Eq. (3.36)

S	surface area
S	cross-sectional area
S	solubility
S	effective speed of vacuum pump
Sc	Schmidt number
S_0	intrinsic speed of vacuum pump
S_p	rated speed of vacuum pump
S_R	energy generation
St	Stanton number
T	temperature
T^*	reduced temperature, defined in Eq. (1.23)
T_b	boiling temperature
T_c	critical temperature
T_M	freezing point
T_m	mixed mean temperature
T_{sat}	saturation temperature, i.e., boiling temperature
t	time
t_i	transference number of species i
U, u	velocity
U	internal energy per unit mass
V, v	velocity
V	volume
$\hat{V}$	molar volume
$\overline{V}$	average velocity
V_c	critical volume
$\overline{V}_F$	electron velocity at the Fermi surface
V_{max}	maximum velocity; center line velocity
V_0	superficial velocity, defined by Eq. (3.33)
V_r	reduced volume
V_s	speed of sound
V_t	terminal velocity
V_∞	bulk stream velocity, steady-state velocity
v_x, v_y, v_z	velocity components
$\overline{v}_x, \overline{v}_y, \overline{v}_z$	components of the temporal mean velocity
v_x^*, v_y^*, v_z^*	local *molar average* velocity components
W	width
W	mass flow rate
W_e	Clausing factor for effusion cells
X_i	mole fraction of component i
x	distance
Y	thickness, distance between parallel plates
Y	expansion factor in flow meters
y	distance
Z	frequency
Z	flux of molecules
Z	coordination number
z	valence number
α	angle

α	linear thermal expansion coefficient
α	thermal diffusivity, $k/\rho C_p$
α	absorptivity for radiation
β	angle
β	compressibility
β	thermal coefficient of volume expansion
Γ	efficiency
γ	Gruneisen's constant, defined by Eq. (6.13)
γ	shear strain
$\dot{\gamma}$	shear strain rate
γ_i	activity coefficient of species i
δ	distance or thickness
δ	interatomic spacing in crystals
δ	momentum boundary layer thickness
δ_c	concentration boundary layer thickness
δ_{eff}	effective concentration boundary layer thickness
δ_T	thermal boundary layer thickness
ε	emissivity
ε	characteristic energy parameter in the Lennard-Jones potential function
ε	height of protuberances inside a rough tube
ε_F	Fermi energy
η	viscosity
η^*	reduced viscosity, defined by Eq. (1.22)
η_P	plastic viscosity (or coefficient of rigidity), Eq. (1.26)
θ	mean residence time
θ	time as a fundamental dimension
θ	dimensionless temperature ratio
κ_B	Boltzmann's constant $= 1.38044 \times 10^{-16}$
λ	mean free path of atoms or molecules
λ	shape factor, defined by Eq. (3.56)
λ	wavelength of radiation
μ_i	chemical potential of species i
ν	kinematic viscosity, η/ρ
ν	index of refraction
ν	atomic jump frequency in solids
ρ	mass density
ρ	reflectivity for radiation
ρ_A	mass concentration of A
ρ_A^*	mass fraction of A
σ	characteristic diameter of molecules in the Lennard-Jones potential function
σ	Stefan-Boltzmann constant $= 0.1713 \times 10^{-8} \text{Btu/ft}^2\text{hr}^\circ\text{R}^4$
σ	electrical conductivity
σ	surface tension
τ	transmissivity for radiation
τ	tortuosity
τ, τ_{yx}, etc.	shear stress components
Φ	inertial permeability in porous media
Φ	dissipation function, defined by Eq. (7.81)

ϕ	electrical potential
ϕ	viscous permeability in porous media
ψ	stream function in fluid flow, defined by Eq. (2.83)
Ω_η	collision integral for viscosity
$\Omega_{D, AB}$	collision integral for diffusion
ω	void fraction in porous media
ω_{mf}	void fraction at minimum fluidization

INDEX